Electromagnetic Coupling in the Polar Clefts and Caps

NATO ASI Series

Advanced Science Institutes Series

A Series presenting the results of activities sponsored by the NATO Science Committee, which aims at the dissemination of advanced scientific and technological knowledge, with a view to strengthening links between scientific communities.

The Series is published by an international board of publishers in conjunction with the NATO Scientific Affairs Division

A	**Life Sciences**	Plenum Publishing Corporation
B	**Physics**	London and New York
C	**Mathematical**	Kluwer Academic Publishers
	and Physical Sciences	Dordrecht, Boston and London
D	**Behavioural and Social Sciences**	
E	**Applied Sciences**	
F	**Computer and Systems Sciences**	Springer-Verlag
G	**Ecological Sciences**	Berlin, Heidelberg, New York, London,
H	**Cell Biology**	Paris and Tokyo

Series C: Mathematical and Physical Sciences - Vol. 278

Electromagnetic Coupling in the Polar Clefts and Caps

edited by

P. E. Sandholt

and

A. Egeland
Department of Physics,
University of Oslo,
Oslo, Norway

Kluwer Academic Publishers

Dordrecht / Boston / London

Published in cooperation with NATO Scientific Affairs Division

Proceedings of the NATO Advanced Research Workshop on
Electromagnetic Coupling in the Polar Clefts and Caps
Lillehammer, Norway
September 20–24, 1988

Library of Congress Cataloging in Publication Data

```
Electromagnetic coupling in the polar clefts and caps : proceedings of
  the NATO advanced research workshop held at Lillehammer, Norway,
  20-24 September 1988 / edited by P.E. Sandholt and A. Egeland.
      p.   cm. -- (NATO ASI series. Series C, Mathematical and
  physical sciences ; vol. 278)
    Papers from the NATO Advanced Research Workshop on Electromagnetic
  Coupling in the Polar Clefts and Caps.
    Includes index.
    ISBN-13: 978-94-010-6929-8
    1. Polar ionosphere--Congresses.  2. Solar wind--Congresses.
  3. Magnetosphere--Congresses.  4. Electrodynamics--Congresses.
  I. Sandholt, P. E. (Per Even), 1951-   . II. Egeland, Alv, 1932-
  . III. NATO Advanced Research Workshop on Electromagnetic Coupling
  in the Polar Clefts and Caps (1988 : Lillehammer, Norway)  IV. North
  Atlantic Treaty Organization.  V. Title: Polar clefts and caps.
  VI. Series: NATO ASI series.  Series C, Mathematical and physical
  sciences ; no. 278.
  QC994.75.E34   1989
  538'.767'0911--dc20                                      89-11060
```

ISBN-13: 978-94-010-6929-8   e-ISBN-13: 978-94-009-0979-3
DOI: 10/1007-978-94-009-0979-3

Published by Kluwer Academic Publishers,
P.O. Box 17, 3300 AA Dordrecht, The Netherlands.

Kluwer Academic Publishers incorporates the publishing programmes of
D. Reidel, Martinus Nijhoff, Dr W. Junk and MTP Press.

Sold and distributed in the U.S.A. and Canada
by Kluwer Academic Publishers,
101 Philip Drive, Norwell, MA 02061, U.S.A.

In all other countries, sold and distributed
by Kluwer Academic Publishers Group,
P.O. Box 322, 3300 AH Dordrecht, The Netherlands.

Printed on acid free paper

CONTENTS

PREFACE

These proceedings are based upon introductory talks, research reports and discussions at the NATO Advanced Workshop on ELECTROMAGNETIC COUPLING IN THE POLAR CLEFTS AND
CAPS, held at Lillehammer, Norway, 20-24th September 1988.
By this book we will make the information which was provided to the participants of the workshop, accessible to a
wider audience.
      Electromagnetic processes governing particle, momen-
tum, and energy transfer from the solar wind via the magnetosphere and into the earth's upper atmosphere are the main
topics of solar-terrestrial research. Due to the peculiar
magnetic field configuration in the magnetosphere, resulting from the interaction with the shocked solar wind, the
sunward-side boundary is mapped along magnetic field lines
into a thin, arc-like band of the dayside polar ionosphere
at the boundary of the polar cap; i.e. the ionospheric
cleft region. The polar cusp is a separate, more limited
region near magnetic noon, as defined by electron and
proton precipitation detected from polar orbiting satellites.
      The basic physics of the different coupling modes at
the dayside magnetopause is a matter of great controversy.
This is an important problem to solve, also because similar
boundaries exist in stellar objects throughout the Uni-
verse.
      It is expected that ground-based remote sensing tech-
niques, with their ability to continuously monitoring the
temporal and spatial variations of the ionospheric signatures, will have a great impact on this problem, in particular when combined with in situ measurements.
      Partly because of the inaccessibility of ground sites
which satisfy essential observation criteria; i.e. correct
distances to the geomagnetic and geographic poles, the
dayside part of the near earth space was neglected for a
long time. The first direct evidence of plasma entry into
the dayside cleft goes only back to the early 70'ies. Since
then, and particularly during the 80'ies, it has been
realized that the energy transfer mechnisms at the dayside
magnetospheric boundary layers and their ionospheric signatures are as important to solar-terrestrial research as the
nightside processes.
      Four years ago a NATO Workshop with a similar purpose
was also held at Lillehammer. By comparing the proceedings
from that meeting [The Polar Cusp (Eds. J.A. Holtet and A.

Egeland), 1985] and the present one, one will fortunately discover that the geophysical research related to these interesting regions are progressing at a breathtaking pace. It is also interesting to note that continuous, coordinated ground-based observations – which at least temporarily were put aside by several research groups – again receive high priority. The increasing number of new methods and data available from ground level observations, as well as from coordinated satellite- and rocketborne measurements, now provide new bases for more detailed and differentiated studies of physical processes in the polar clefts and caps.

Workshops arranged as the present one – at a fairly isolated place – provide ample opportunities for formal and informal discussions and for personal contact which are of significant value for planning further research.

We will take this opportunity to thank Mrs. A.-S. Andresen who was in charge of the Workshop Secretariat. The assistance received from Norwegian Space Center in organizing this workshop is acknowledged. The local organizing committee appreciated Professor R. Smith's enthusiastic and experienced help in planning the different sessions.

We gratefully acknowledge the funds provided for the ARW by the NATO Scientific Affairs Division as well as many research institutes and universities for covering travel and living expenses for the majority of the participants.

Finally, we thank all the participants and the "camera-ready" manuscript contributors who made these Proceedings possible.

Oslo, February 1989.

P.E. Sandholt   A. Egeland.

PARTICIPANTS

Dr. S. Basu   Emmanuel College
                             4000 The Fenway
                             Boston, MA 02115
                             USA

Dr. E.A. Bering III   University of Houston
                             Department of Physics
                             Houston, Texas 77204-5504
                             USA

Dr. A. Berthelier   CRPE
                             CNET-CNRS
                             4. Avenue de Neptune
                             94107 Saint-Maure-des-Fossés
                             France

Dr. B. Bostrøm   Swedish Inst. of Space Phys.
                             Uppsala Division
                             S-755 90 Uppsala
                             Sweden

Dr. P.F. Bythrow   The Johns Hopkins University
                             Applied Physics Laboratory
                             Johns Hopkins Road
                             Laurel, Maryland 20707
                             USA

Dr. H.C. Carlson, Jr.   AFGL/LIS
                             Hanscom AFB
                             Bedford, Mass 01731
                             USA

Dr. L. Eliasson   Swedish Inst. of Space Phys.
                             Box 812
                             S-981 28 Kiruna
                             Sweden

Dr. M.J. Engebretson   Augsburg College
                             731 21st Avenue South
                             Minneapolis, MN 55454
                             USA

Dr. J.C. Foster

M.I.T. Haystack Observatory
Westford, Massachusetss 01886
USA

Dr. E. Friis-Christensen

Danish Meteorological Inst.
Lyngbyvej 100
DK-2100 København
Denmark

Dr. C.-G. Fälthammar

The Royal Inst. of Tech.
Department of Plasma Physics
S-100 44 Stockholm
Sweden

Dr. C.S. Gardner

University of Illinois
College of Engineering
103 Engineering Hall
1308 West Green Hall
Urbana, IL 61801
USA

Dr. K.H. Glassmeier

Universität zu Köln
Inst. für Geophys. und Met.
Zülpicher Strasse 49
D-5000 Köln 41
West Germany

Dr. R.A. Greenwald

Johns Hopkins University
Johns Hopkins Road
Laurel, Maryland 20707
USA

Dr. S. Gussenhoven

AFGL/PHP
Hanscom Air Force Base
MA 03731
USA

Dr. G. Gustafsson

Swedish Inst. of Space Phys.
Uppsala Division
S-755 90 Uppsala
Sweden

Dr. C. Haldoupis

University of Crete
Physics Department
714 09 Irakleon, Crete
Greece

Dr. K. Hayashi

The University of Tokyo
Faculty of Science
Geophysics Research Lab.
Bunkyo-Ku, Tokyo 113
Japan

Dr. R.A. Heelis

The Univ. of Texas at Dallas
Box 830688 MS F022
Richardson, Texas 75083-0688
USA

Dr. C. Huang

The University of Iowa
Department of Phys. and Astr.
Iowa City, Iowa 52242
USA

B. Jacobsen

Department of Physics
University of Oslo
P.O.Box 1048 Blindern
0316 Oslo 3
Norway

Dr. T.B. Jones

Department of Physics
University of Leicester
University Road
Leichester LE1 7RH
United Kingdom

Dr. T.S. Jörgensen

Meteorological Institute,
Lyngbyvej 100
2100 København Ø
Denmark

Dr. J. Lemaire

I.A.S
3 Avenue Circulaire
B-1180 Bruxelles
Belgium

Dr. M. Lockwood

Rutherfor Appleton Lab.
Chilton. Didcot, Oxon
OX11 OQX
United Kingdom

Dr. B. Lybekk

Department of Physics
University of Oslo
Box 1048 Blindern
0316 Oslo
Norway

Dr. D.L. Matthews

University of Maryland
College Park Campus
Maryland 20742
USA

N. Mattin

British Antarctic Survey
High Cross
Maddingly Road
Cambridge CB8 OET
United Kingdom

Dr. N.C. Maynard

Space Plasma and Field Branch
Space Physics Division
Department of the Air Force
AFSC, Hanscom Air Force Base
Mass 01731-5000
USA

Dr. S.B. Mende

Lockheed Palo Alto Res. Labs
3251 Hanover Street,
B/255, 091-20
Palo Alto, CA 94304
USA

Dr. C.-I. Meng

The Johns Hopkins University
Applied Physics Laboratory
Johns Hopkins Road
Laurel, MD 20707
USA

K. Måseide

Department of Physics
University of Oslo
P.O.Box 1048 Blindern
0316 Oslo 3
Norway

Dr. P.T. Newell

The Johns Hopkins University
Applied Physics Lab.
Johns Hopkins Road
Laurel, Maryland 20707-6099
USA

Dr. T. Oguti

Geophysical Research Lab.
University of Tokyo
Tokyo 113
Japan

A. Ohta

Department of Physics
University of Oslo
P.O.Box 1048 Blindern
0316 Oslo 3

Dr. J. Olson

Geophysical Institute
University of Alaska
Fairbanks, Alaska 99775-0800
USA

Dr. W.K. Peterson

Lockheed Palo Alto Res. Lab.
3251 Hanover Street
Palo Alto, Calif. 94304-1191
USA

Dr. T.A. Potemra

The Johns Hopkins University
Applied Physics Laboratory
Johns Hopkins Road
Laurel, Maryland 20707
USA

Dr. F. Primdahl

Danish Space Research Inst.
Lundtoftevej 7,
DK-2800 Lyngby
Denmark

Dr. D. Rees

University College London
Dep. of Physics and Astronomy
Gower Street
London WC1E 6BT
United Kingdom

Dr. P. Reiff

Rice University
Space Physics
Box 1892
Houston Texas 77251
USA

Dr. H. Rishbeth

Department of Physics
The Univ. of Southampton
Southampton SO9 5NH
United Kingdom

Dr. P.E. Sandholt

Department of Physics
University of Oslo
Box 1048 Blindern
0316 Oslo 3
Norway

Dr. N. Sato

National Inst. of Polar Res.
9-10 Kaga 1-Chome, Itabashi-ku
Tokyo 173
Japan

Dr. R. Smith

Geophysical Institute
University of Alaska
Fairbanks, Alaska 99701
USA

Dr. I. Steen Mikkelsen

Meteorological Institute
Lyngbyvej 100
2100 København Ø
Denmark

Dr. M.J. Taylor

Department of Physics
The Univ. of Southampton
Southamton SO9 5NH
United Kingdom

Dr. M.A. Temerin

Univ. of Calif., Berkeley
Space Sciences Lab.
Berkeley, California 94720
USA

J.P. Thayer

The University of Michigan
Space Physics Research Lab.
Space Research Building
Ann Arbor,
Michigan 48109-2143
USA

Dr. Y. Tulunay

I.T.U, Ucak ve Usay Bil. Fak.
Maslak 80626 Istanbul
Turkey

Dr. V.M. Vasyliunas

Max-Planck Inst. für Aeronom.
Postfach 20
D-3411 Kathlenburg-Lindau 3,
West Germany

Dr. R.R. Vondrak

Lockheed Palo Alto Res. Lab.
3251 Hanover Street
Palo Alto, Calif. 94304-1191
USA

Dr. R.J. Walker

Inst. of Geophys. and
Planet. Sci.
UCLA
Los Angeles, Calif. 90024
USA

Dr. J.D. Winningham

Southwest Research Institute
P.O. Drawer 29510
San Antonio, Texas 28294
USA

Dr. A. Wolfe

AR&T Bell Laboratories
600 Mountain Avenue
Murray Hill,
New Jersay 07974-2070
USA

ELECTRODYNAMICS OF THE IONOSPHERE/MAGNETOSPHERE/ SOLAR WIND SYSTEM AT HIGH LATITUDES

V. M. VASYLIUNAS
Max-Planck-Institut für Aeronomie
D-3411 Katlenburg-Lindau
Federal Republic of Germany

ABSTRACT. Within the region threaded by magnetic field lines from the polar cap, the interaction between the solar wind and the magnetosphere manifests itself primarily in the structure of magnetic and electric fields, which then provide the electromagnetic boundary conditions for the ionosphere as well as for the lower-latitude particle-dominated regions, the auroral oval and the plasma sheet. The basic physical processes are thought to be understood, at least qualitatively, but many unknown and/or controversial aspects remain, particularly concerning the configuration in the case of northward interplanetary magnetic field.

1. Introduction

A large part of the high-latitude magnetosphere, the region threaded by magnetic field lines extending out from the polar cap to form the magnetotail is populated by plasmas of a very low density. With local plasma stresses thus playing for the most part only a minor role, this region is dominated by electrodynamical processes associated with the basic interaction between the solar wind and the magnetosphere.

There is a two-fold connection between the electrodynamics of the polar cap and magnetotail and the general theory of magnetosphere-ionosphere coupling. When the theory is applied to the lower-latitude auroral oval and ring current regions, boundary conditions at high latitudes (either electric fields or currents) must be specified, and these are fixed by the high-latitude electrodynamical processes. The theory can also be applied to the polar cap and magnetotail region itself. In view of the low plasma densities, the equations describing the magnetospheric side of the coupling can then be reduced to two statements: mapping of the electric field $\mathbf{E}$ along magnetic field lines, and neglect of the current density $\mathbf{J}_\perp$ perpendicular to the magnetic field $\mathbf{B}$, with consequent mapping of Birkeland currents ($J_\parallel/B = $ const.) along magnetic field lines.

Given the commonly accepted assumption that the magnetosphere is magnetically open and the field lines of the magnetotail extend ultimately into the solar wind, mapping of $\mathbf{E}$ implies that the electric field in the polar cap is an extension of $\mathbf{E} = -\mathbf{V} \times \mathbf{B}$ in the solar wind and that its spatial pattern is fixed by the pattern of magnetic field line connection between the two regions. Such mapping of $\mathbf{E}$ along $\mathbf{B}$ has been widely studied as the basic process of electric coupling between the solar wind and the magnetosphere/ionosphere

1

P. E. Sandholt and A. Egeland (eds.), Electromagnetic Coupling in the Polar Clefts and Caps, 1–9.
© *1989 by Kluwer Academic Publishers.*

2

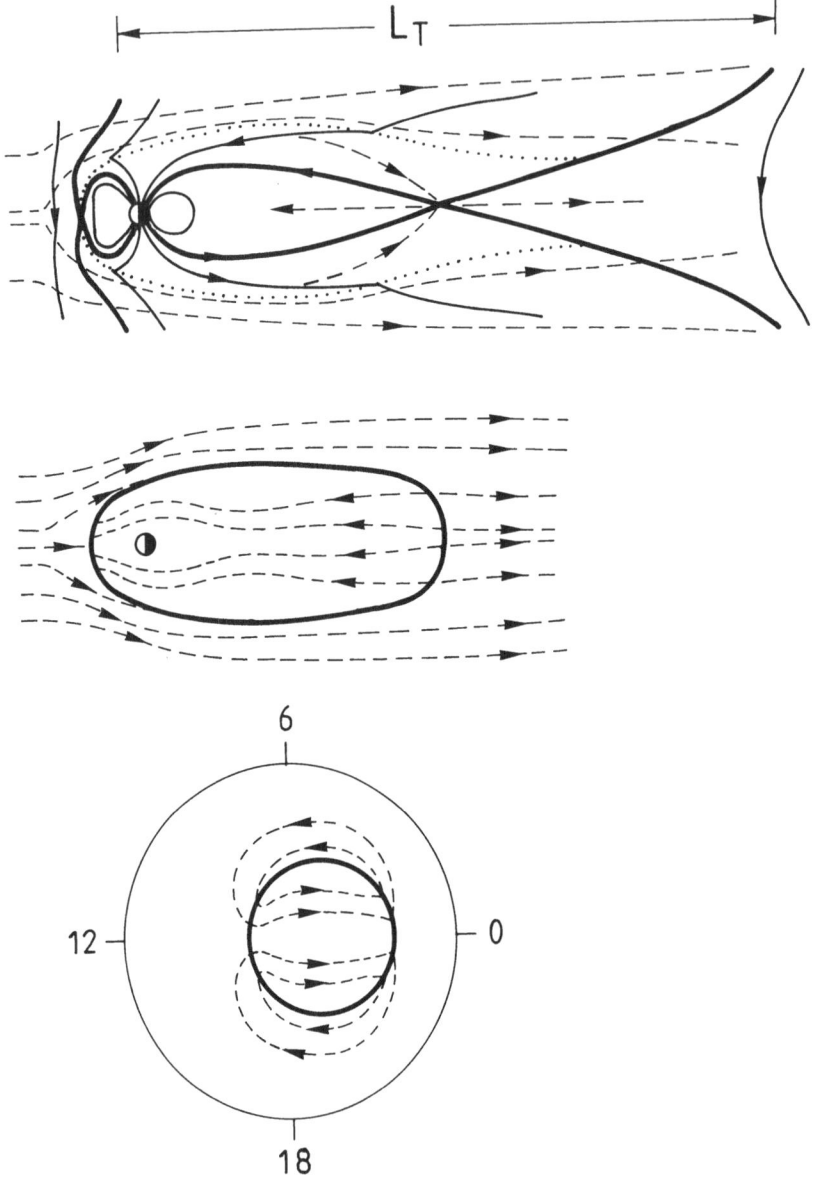

Figure 1. Schematic model of an open magnetosphere (qualitative and not to scale).
(a) Top: view in noon-midnight meridian plane; (b) middle: view in equatorial plane;
(c) bottom: projection on polar cap. Dashed lines: flow streamlines and electric equipotentials. Thin solid lines: magnetic field lines. Thick lines: boundaries of open field line region
(in (b), the magnetic X-line; in (c) the boundary of the polar cap). Only the geometrically
simplest case is shown; for a discussion of possible complexities and alternatives, see e.g.
Vasyliunas (1984).

system, the driving agency of magnetospheric convection; this paper is largely devoted to concepts and questions of this process. The mapping of Birkeland currents between the polar cap and the magnetopause has received much less attention, partly because the most intense Birkeland currents occur at or near the interface to the auroral oval, Birkeland currents within the polar cap itself being generally much weaker although by no means absent.

2. Implications of a Long Magnetotail

The region of open magnetic field lines, identified with the polar cap, maps into a bundle of interplanetary magnetic field lines in the solar wind with a well-defined length L_T (Figure 1a) which, following Dungey (1965), can be loosely thought of as the length of the magnetospheric tail and estimated by equating the plasma flow times through the solar-wind and the polar-cap ends of the bundle:

$$L_T/V = 2R_E\theta_{pc}/V_{pc} \qquad (1)$$

where V is the solar wind speed, $R_E \equiv$ Earth radius, θ_{pc} is the colatitudinal extent of the polar cap, and V_{pc} is the plasma flow speed across the polar cap, which can be related to the cross-polar-cap electric potential,

$$V_{pc} \approx c\Phi_{pc}/2R_E\theta_{pc}B_p \qquad (2)$$

with B_p the vertical magnetic field strength at the polar cap surface. Inserting (2) into (1) gives

$$
\begin{aligned}
L_T/V &= (2R_E\theta_{pc})^2 B_p/c\Phi_{pc} \\
&= (4/\pi)F_T/c\Phi_{pc}
\end{aligned}
\qquad (3)
$$

where

$$F_T = \pi\left(R_E\theta_{pc}\right)^2 B_p \qquad (4)$$

is the open magnetic flux.

The value of L_T can be estimated by scaling the polar cap potential to the solar wind potential across the radius R_M of the dayside magnetosphere,

$$c\Phi_{pc} = \epsilon V B R_M, \qquad (5)$$

and scaling the open flux to the dipole flux beyond a distance R_M,

$$F_T = f\pi B_E R_E^3/R_M. \qquad (6)$$

Here ϵ and f are dimensionless scaling factors, B is the appropriate component of the interplanetary magnetic field, and B_E is the Earth's equatorial surface field; with $B_p = 2B_E$, (4) and (6) imply

$$\theta_{pc}{}^2 = (f/2)R_E/R_M. \qquad (7)$$

Inserting (5) and (6) into (3) yields

$$L_T/R_M = (4f/\epsilon)(R_E/R_M)^3 B_E/B \tag{8}$$

so that the length of the magnetotail exceeds the radius of the dayside magnetosphere by the ratio of the dipole field at the magnetopause to the interplanetary magnetic field (because of pressure balance this ratio is of course closely related to the Alfvén Mach number of the solar wind), times a combination of dimensionless scaling factors.

The usually quoted empirical value for ϵ is $\lesssim 0.1$, while f is generally not much smaller than 1. With B $\lesssim$ 3 nT, compared to a dipole value of 30 nT, it is clear that (8) predicts $L_T \gg R_M$, i.e. a long magnetotail, in agreement with observations. This implies that the time scale τ_1 over which the magnetosphere/ionosphere system comes to electrodynamic equilibrium after changes in the solar wind (i.e. the time scale to reestablish complete mapping of the electric field from the solar wind through the magnetotail to the polar cap) is much longer than the time scale τ_2 over which the magnetosphere and the ionosphere respond to such changes (i.e. begin to vary in correlation with the changes in the solar wind). The time scale τ_1 is determined by the flow time of the solar wind past the open field line region; thus

$$\tau_1 \approx L_T/V \tag{9}$$

and its empirical value (which can, of course, be estimated from polar cap size and ionospheric plasma flow by means of equation (1), with no explicit use of L_T) is typically of order many hours. On the other hand, τ_2 is governed by the propagation time of Alfvén waves across the magnetosphere, which at least for field lines through the polar cusp is very roughly of the order of R_M/V; thus τ_2 is typically some ten minutes or less.

As a consequence of a long magnetotail, therefore, mapping of the electric potential from the solar wind to the polar cap can be assumed only when the interplanetary magnetic field is steady on a time scale of hours (or alternatively when dealing with observations averaged over such a time scale).

3. The Standard Open Model

Many electrodynamic phenomena at high latitudes can be interpreted on the basis of the widely accepted open model of the magnetosphere, developed as a synthesis of ideas on reconnection between interplanetary and geomagnetic fields (Dungey, 1961, 1963), magnetospheric convection (Axford and Hines, 1961), and magnetosphere-ionosphere coupling (Fejer, 1964; Swift, 1967; systematized by Vasyliunas, 1970), schematically illustrated in Figure 1. The geometry of the magnetic field line connection between the solar wind and the polar cap is assumed to yield a direct mapping of the electric field, with plasma flow across the polar cap that corresponds more or less directly to the solar wind flow and is thus basically antisunward (aside from secondary features such as a possible enhancement on the dawn or the dusk side or a narrowing to form a "throat" on the dayside). At lower latitudes there is then a sunward return flow, and we have the familiar two-cell pattern of magnetospheric convection flow. At the boundary between antisunward and sunward flow, the electric field reversal implies a divergence of horizontal electric currents in the ionosphere which must continue as Birkeland (field-aligned) currents, with the pattern and direction of the observed Region 1 currents. At lower latitudes, azimuthal pressure gradients induced by the magnetospheric convection flow (ring current polarization) give rise to a second set of Birkeland currents, with the pattern and direction of the observed Region 2 currents.

While the open model is thus quite successful in accounting for many of the observed large-scale features of high-latitude electrodynamics, several limitations should be kept in mind:

(1) Straightforward predictions of the model appear to be confirmed by observations predominantly during periods when the interplanetary magnetic field has a southward component. The more complex state of the magnetosphere during periods with a northward interplanetary field is discussed in Section 4.

(2) To date, the predictions of the model have been mostly qualitative. Although considerable quantitative work has been done on particular aspects (e.g. magnetosphere-ionosphere coupling at lower latitudes, inference of Birkeland currents from global magnetic and ionospheric data, empirical correlations of polar cap potential with solar wind parameters), the model as a whole is as yet nowhere near being able to make quantitative predictions from first principles.

(3) The basic magnetic geometry of the model at the interface between the magnetosphere and the solar wind plasma, specifically the distribution of normal magnetic field components at the magnetopause and of the electric fields tangential to the X-line (cf. Figure 1), remains essentially unknown.

(4) Time variations have been treated for the most part as a succession of quasi-equilibrium states; in view of the long time scale required to reach true equilibrium, as discussed in Section 2, this may not be adequate.

4. Configuration with Northward Interplanetary Field

When the north-south component of the interplanetary magnetic field remains northward for a prolonged period (up to a few hours), the electric field and plasma flow in the polar cap can assume a configuration considerably more complex than that associated with a simple two-cell convection pattern. Observations of ion flow above the high-latitude ionosphere have been interpreted as indicating convection patterns with three or even four cells as well as with regions of sunward flow within what is ostensibly the polar cap, although alternative interpretations based on a highly distorted two-cell pattern have also been suggested (for various views see, e.g., Burke et al., 1979; Reiff and Burch, 1985; Heelis et al., 1986; Rasmussen and Schunk, 1988).

Sunward flow within what might be part of the open field line region raises the question whether the magnetic connection from the polar cap to the solar wind might be such as to produce reverse mapping of electric fields, i.e. antisunward flow at the solar-wind end corresponding to sunward flow at the polar-cap end. Field geometries with this property were proposed by Russell (1972), who combined Dungey's (1963) sketch of magnetospheric topology for a northward interplanetary field (Figure 2a) with the assumption of a magnetosphere open at all times (Figure 2b), and were further developed particularly by Reiff (1982, 1984; also Reiff and Burch, 1985), who named the flow patterns associated with the geometries of Figure 2a and Figure 2b "reclosure" and "stirring" types of reverse merging, respectively (Figure 3).

However, the model sketched in Figure 2a is magnetically closed (as recognized by Dungey and by Russell) and is in fact topologically identical with the simple closed teardrop magnetosphere of Johnson (1960), the points N corresponding to the Chapman-Ferraro neutral points and the portion of the field line to the right of them corresponding to the downstream boundary of the teardrop. With no open field lines there is of course no mapping of electric fields from the solar wind, hence no obvious reason (other than viscous drag) for <u>any</u> flow

6

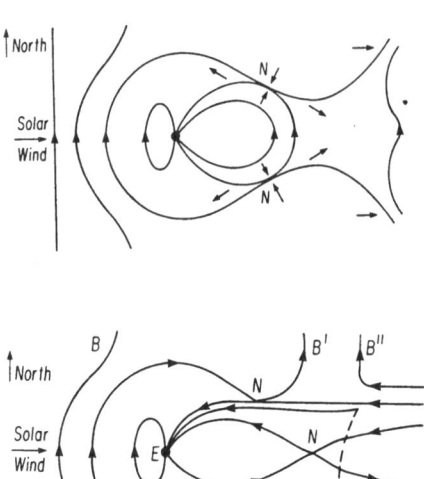

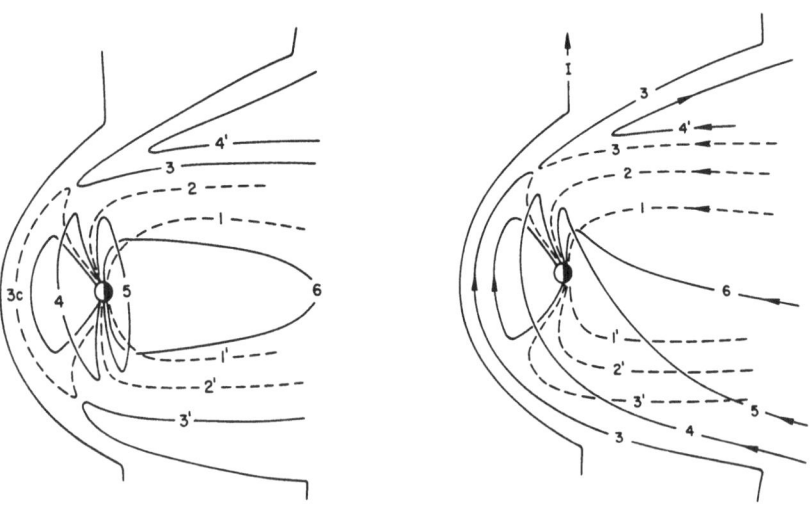

Figure 2. Magnetic topology of magnetosphere for northward interplanetary magnetic field.
(a) Top: after Dungey (1963). (b) Bottom: after Russell (1972).

Figure 3. Magnetic topology of magnetosphere for northward interplanetary magnetic field,
after Reiff (1982). (a) Left: reclosure type of reverse magnetic merging (corresponds to Figure 2a). (b) Right: stirring type of reverse magnetic merging (corresponds to Figure 2b).

in the magnetosphere. To my mind it is questionable whether the proposed "reclosure" merging exists at all.

The "stirring" type of merging involves reconnection of interplanetary magnetic field lines with open field lines of the magnetotail (rather than with closed field lines as in the standard open model). The topological distinctions implicit in this concept (magnetic reconnection is often defined as plasma flow across a separatrix surface between topologically different magnetic fields; cf. Vasyliunas, 1975) have not yet been fully elucidated.

When the plasma flow pattern is represented by electric equipotential contours over the ionosphere (or some equivalent surface), every convection cell is associated with a maximum or minimum of potential. Also, Maxwell's theorem "on hills and dales" states that

(number of maxima) + (number of minima) − 2 = (number of saddle points)

Now maxima, minima, and saddle points in an equipotential contour plot are all points where **E** and therefore the perpendicular plasma flow vanish. It is difficult to imagine such points existing, in a steady state, on open field lines, where the solar wind flows continually across the more distant portions of the field line. (In the standard open model, the two maxima and two minima – one of each in each hemisphere – all lie on the boundaries between open and closed field lines, and the two saddle points, in the simplest models, lie on the magnetic equator.)

Figure 4, taken from Heelis et al. (1986), illustrates both the need for topological boundaries and the difficulty of having the center of a circulation cell in the polar cap. It is evident that, because field line 1 splits into two branches, it must lie on some discontinuous topological boundary. As we approach the center of the circulation shown, the footprints of the field lines 1, 2, 3, 4 coalesce and we have a single field line with one end fixed in the ionosphere and the other end moving with the solar wind – an object not consistent with magnetohydrodynamic constraints in the steady state.

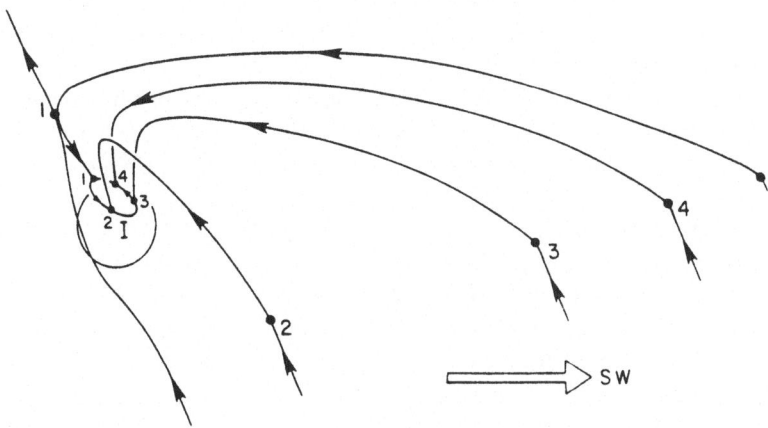

Figure 4. Sketch by Heelis et al. (1986) of how a high-latitude convection cell (supposedly entirely on open field lines) could be produced by solar wind flow.

5. Conclusion

The observed electrodynamic phenomena in the high-latitude regions of the magneto-sphere/ionosphere system show significant differences between periods of southward and northward interplanetary magnetic fields. A major unanswered question is whether the magnetosphere has two essentially different modes of interaction with the solar wind, or whether instead the interaction is the same in both cases but the detailed geometry in the case of northward fields is much more complex so that the simple intuitive qualitative models that sufficed for the case of southward fields are no longer adequate.

6. References

Axford, W.I. and Hines, C.O. (1961) 'A unifying theory of high-latitude geophysical phe-nomena and geomagnetic storms', Canadian J. Phys. 39, 1433-1464.

Burke, W.J., Kelley, M.C., Sagalyn, R.C., Smiddy, M., and Lai, S.T. (1979) 'Polar cap electric field structures with a northward interplanetary magnetic field', Geophys. Res. Lett. 6, 21-24.

Dungey, J.W. (1961) 'Interplanetary magnetic field and the auroral zones', Phys. Rev. Lett. 6, 47-48.

Dungey, J.W. (1963) 'The structure of the exosphere or adventures in velocity space', in C. DeWitt, J. Hieblot, and A. Lebeau (eds.), Geophysics/The Earth's Environment, Gordon and Breach Science Publishers, New York, pp. 503-550.

Dungey, J.W. (1965) 'The length of the magnetospheric tail', J. Geophys. Res. 70, 1753.

Fejer, J.A. (1964) 'Theory of the geomagnetic daily disturbance variation', J. Geophys. Res. 69, 123-137.

Heelis, R.A., Reiff, P.H., Winningham, J.D., and Hanson, W.B. (1986) 'Ionospheric con-vection signatures observed by DE 2 during northward interplanetary magnetic field', J. Geophys. Res. 91, 5817-5830.

Johnson, F.S. (1960) 'The gross character of the geomagnetic field in the solar wind', J. Geophys. Res. 65, 3049-3052.

Rasmussen, C.E. and Schunk, R.W. (1988) 'Ionospheric convection inferred from interplan-etary magnetic field-dependent Birkeland currents', J. Geophys. Res. 93, 1909-1921.

Reiff, P.H. (1982) 'Sunward convection in both polar caps', J. Geophys. Res. 87, 5976-5980.

Reiff, P.H. (1984) 'Evidence of magnetic merging from low-altitude spacecraft and ground-based experiments', in E.W. Hones, Jr. (ed.), Magnetic Reconnection in Space and Laboratory Plasmas, AGU Geophysical Monograph 30, pp. 104-113.

Reiff, P.H. and Burch, J.L. (1985) 'IMF B_y-dependent plasma flow and Birkeland currents in the dayside magnetosphere 2. A global model for northward and southward IMF', J. Geophys. Res. 90, 1595-1609.

Russell, C.T. (1972) 'The configuration of the magnetosphere', in E.R. Dyer (ed.), Critical Problems of Magnetospheric Physics, National Academy of Sciences, Washington, D.C., pp. 1-16.

Swift, D.W. (1967) 'Possible consequences of the asymmetric development of the ring current belt', Planetary Space Sci. 15, 835-862.

Vasyliunas, V.M. (1970) 'Mathematical models of magnetospheric convection and its coupling to the ionosphere', in B.M. McCormac (ed.), Particles and Fields in the Magnetosphere, D. Reidel Publishing Co., Dordrecht, pp. 60-71.

Vasyliunas, V.M. (1975) 'Theoretical models of magnetic field line merging, 1', Rev. Geophys. Space Phys. 13, 303-336.

Vasyliunas, V.M. (1984) 'Steady state aspects of magnetic field line merging', in E.W. Hones, Jr. (ed.), Magnetic Reconnection in Space and Laboratory Plasmas, AGU Geophysical Monograph 30, pp. 25-31.

THE FORMATION OF ISOLATED MAGNETIC FLUX TUBES ON THE DAYSIDE MAGNETOPAUSE

RAYMOND J. WALKER
Institute of Geophysics and Planetary Physics
University of California, Los Angeles
Los Angeles, CA. 90024-1567
USA

TATSUKI OGINO
Research Institute of Atmospherics
Nagoya University
Toyokawa, 442, Japan

ABSTRACT. We have studied dayside magnetic reconnection by using a three-dimensional global magnetohydrodynamic simulation of the interaction between the solar wind and the magnetosphere. We found that two types of magnetic flux tubes were formed which depend on the orientation of the interplanetary magnetic field (IMF). The dayside magnetic flux tubes occur only when the IMF has a southward component. When the IMF has a large B_y component as well, a strongly twisted and localized magnetic flux tube similar to a magnetic flux rope forms at the magnetopause. When the IMF B_y component is small, twin flux tubes appear at the dayside magnetopause. Both types of flux tubes are consistent with several observational features of flux transfer events and are generated by antiparallel magnetic reconnection.

1. Introduction

Observations indicate that when the interplanetary magnetic field (IMF) is southward, reconnection on the dayside magnetopause is often temporally and spatially limited rather than steady [Haerendel et al., 1978; Russell and Elphic, 1978; Elphic and Russell, 1979]. The localized reconnection events are called flux transfer events (FTE's) and are a common feature of the dayside magnetopause [Rijnbeek et al., 1984; Berchem and Russell, 1984].

FTE's were first identified by a bipolar signature in the component of the magnetic field normal to the magnetopause and frequently are associated with an increase in the magnetic field magnitude [Russell and Elphic, 1979; Berchem and Russell, 1984; and Rijnbeek et al., 1984]. North of the equator the change in the normal component is first away from the Earth and then towards the Earth. The opposite signature is found south of equator. FTE's have been found in both the magnetosheath and on the magnetosphere side of the magnetopause. The FTE's have a duration of 1 to 5 minutes. Their extent is about $1R_E$ in the direction normal to the magnetopause and about $2R_E$ in the tangential direction [Saunders et al., 1984; Rijnbeek et al., 1984]. The rate of occurrence of FTE's does not depend strongly on the sign of the IMF B_y component [Berchem and Russell, 1984].

11

P. E. Sandholt and A. Egeland (eds.), Electromagnetic Coupling in the Polar Clefts and Caps, 11–26.
© 1989 by Kluwer Academic Publishers.

Russell and Elphic [1979] first realized that FTE's were the result of patchy and inter-mittent reconnection and noted that the observations were consistent with isolated elbow shaped flux tubes passing over the spacecraft. They suggested that FTE's are created near the subsolar magnetopause and are convected to higher latitudes. Since the initial obser-vational papers there have been a number of models for flux transfer events. Lee and Fu [1985] presented an alternative model in which magnetic islands are formed by multiple x-line reconnection generated by the tearing mode in the magnetopause current layer. This causes twisted open flux tubes to form along the dayside magnetopause. Sonnerup pointed out that the differences between the Russell and Elphic and Lee and Fu models are signifi-cant. In the Lee and Fu model the large tearing mode island is a part of the magnetopause itself. In the Russell and Elphic model the FTE crosses the magnetopause boundary over a short segment. Galeev el al. [1986] suggested the possibility of spontaneous patchy re-connection related to the growth and overlapping of magnetic islands. Recently Scholer [1988] and Southwood et al. [1988] have suggested that the reconnection associated with FTE's occurs at a single location but is time dependent. In their models the rate of FTE formation changes as the amount of free energy in the system accumulates and dissipates.

There have been several attempts to build computational models of the FTE process. Fu and Lee [1985] used a two dimensional (2D) code to model the tearing mode in the stagnation region flow near a current sheet while La Belle-Harmer et al. [1988] included the effects of the Kelvin-Helmholtz instability. Shi et al. [1988] used a 2D incompressible magnetohydrodynamic (MHD) code to model multiple x-line reconnection. Sato et al. [1986] used a semi-global MHD model to demonstrate the creation of isolated magnetic flux tubes by repeated reconnection. Their model did not include a bow shock.

In this study we have used a time dependent and three dimensional global MHD code to model reconnection at the dayside magnetopause. In particular we studied the formation and dynamics of isolated magnetic flux tubes. We have considered two different IMF orientations. One in which the IMF has a large B_y component and one with a small B_y. For small B_y the results are consistent with subsolar point reconnection which forms twin isolated flux tubes which propagate away from the reconnection region. However for large B_y the results include twisted flux tubes with some of the features of the Lee and Fu model and some features not found in any of the other calculations. Initial results of our calculations can be found in Ogino et al., [1988].

2. The Simulation Model

We will briefly review the simulation model here since it has been described in detail elsewhere [Ogino, 1986; Ogino et al., 1985; 1986]. In order to study reconnection at the dayside magnetopause, we chose to use our highest resolution code and limited the calcu-lations to the northern hemisphere (Figure 1). This code solves the MHD and Maxwell's equations as an initial value problem by using the modified two step Lax-Wendroff scheme. The numerical diffusion terms in the calculation were chosen to be small enough that they had no significant influence on the global magnetospheric structure [Ogino, 1986]. The magnetic Reynolds number, which is the magnetic diffusion time divided by the Alfvén transit time, is $S = 200 \sim 2000$. It is small near the Earth and becomes larger in the distant magnetosphere.

In the simulation a solar wind with a number density $n_{sw} = 5\text{cm}^{-3}$, velocity $v_{sw} = 300\text{km}$

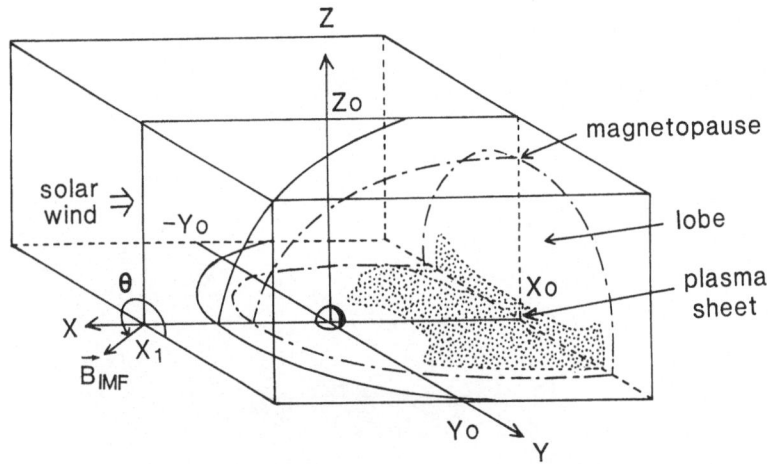

Figure 1. Coordinate system used in the simulation. The angle θ gives the direction of the interplanetary magnetic field. For southward and dawnward IMF $180° < \theta < 270°$.

s^{-1} and temperature $T_{sw} = 2 \times 10^{5°}K$ flows into the simulation box (Figure 1).

The IMF is given by $\mathbf{B_{IMF}} = (B_x, B_y, B_z) = B_{IMF}(0, cos\theta, sin\theta)$, where $B_{IMF} = 5nT$ and θ is defined counter-clockwise from dusk. Free boundary conditions, where the derivatives of all physical quantities are zero, were used at the top, back and sides of the box. The mirror boundary condition was used at $z = 0$ [Ogino et al.,1985; 1986]. Near the Earth a simple fixed ionospheric boundary condition was used [Ogino, 1986].

The MHD equations were solved on a $(N_x, N_y, N_z) = (96,96,48)$ point grid. The spatial mesh size was $\Delta x = \Delta y = \Delta z = 0.5R_E$ and the time step was $\Delta t = 1.87s$. The physical domain of the calculation is $-24.5R_E \leq x \leq 24.5R_E$, $-24.5R_E \leq y \leq 24.5 \ R_E$ and $0 \leq z \leq 24.5R_E$.

3. Simulation Results

At the start of the simulation we allowed a solar wind without an IMF to interact with a dipole which respresents the Earth's intrinsic magnetic field for 256 time steps or 8 minutes. Then we perturbed this system by gradually inserting an IMF at the upstream boundary ($x = x_1$ in Figure 1). The IMF was increased from zero to full value over 50 time steps.

3.1 THE CASE FOR LARGE IMF B_y ($\theta = 210°$)

Figure 2 shows the magnetic field configuration 16 minutes after the IMF was inserted. Two sets of field lines have been plotted in the the northern hemisphere. One set of field lines are IMF field lines which intersect the magnetopause and the other field lines are closed magnetospheric field lines. For $\theta = 210°$ the IMF field lines have a large dawnward component. A detailed examination of both the magnetic field and plasma data indicate that magnetic reconnection has not begun at this time. However the magnetospheric configuration already has been significantly distorted by the interaction. There is a dawn dusk asymmetry in the closed field lines with distorted field lines most evident in the northern

θ=210° t=16m

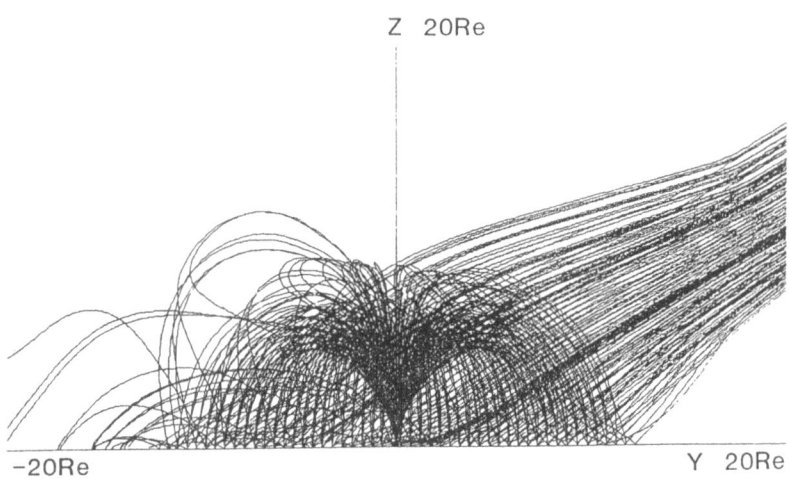

Figure 2. Magnetic field lines for the case with $\theta = 210°$ at $t = 16$m.

dawn region. These field lines connect to similarly distorted field lines in the southern dusk magnetosphere. The distortion results from the viscous convection of the IMF across the dayside magnetosphere.

About four minutes later reconnection began at high latitudes where the antiparallel field condition is satisfied between the IMF and the geomagnetic field [Ogino et al., 1986]. The magnetic configuration 24 minutes after the IMF insertion has been plotted in Figure 3a. Here we have removed the IMF field lines to more clearly show the magnetospheric configuration. The top panel shows the view from the north pole while the bottom panel shows the view from the Sun. That reconnection has begun can be seen from the sharply bent field lines in the polar magnetosphere on the northern dawn and southern dusk sides. These field lines have been recently reconnected. Distorted closed field lines also can be seen. These field lines are related to the viscous distortion displayed in Figure 2. Note that there is no evidence for reconnection at the subsolar point.

Four minutes later ($t = 28$m, Figure 3b), the configuration has evolved dramatically. At this time a twisted magnetic flux tube has started to form on closed field lines in a narrow channel connecting the two reconnection regions. There is no evidence of reconnection along the twisted flux tube. The twisted flux tube is fully formed by $t = 32$m (Figure 4a). The "flux tube" nature of this can be clearly seen in the top view.

Starting at about $t = 36$m (Figure 4b) the twisted flux tube has become open and has started to separate into northern and southern parts. The northern part convects north (i.e. tailward) and the southern part convects south. As the flux tubes convect the twist relaxes and the radius of curvature of the flux tubes increases. At later times these two flux tubes convect up over the polar caps (see Figure 1b of Ogino et al., 1988).

In Figure 5 we have plotted profiles of the magnetic field components, the velocity components, the plasma density and the plasma pressure in the z direction at the dayside magnetopause ($x = 12R_E$ and $y = -4R_E$) at $t = 32$m. The twisted flux tube is the isolated disturbance in B between $1R_E < z < 6R_E$. The wavelength of the disturbance (peak to peak) is about $5R_E$ in the z-direction and about $3R_E$ in the x-direction. The x-direction

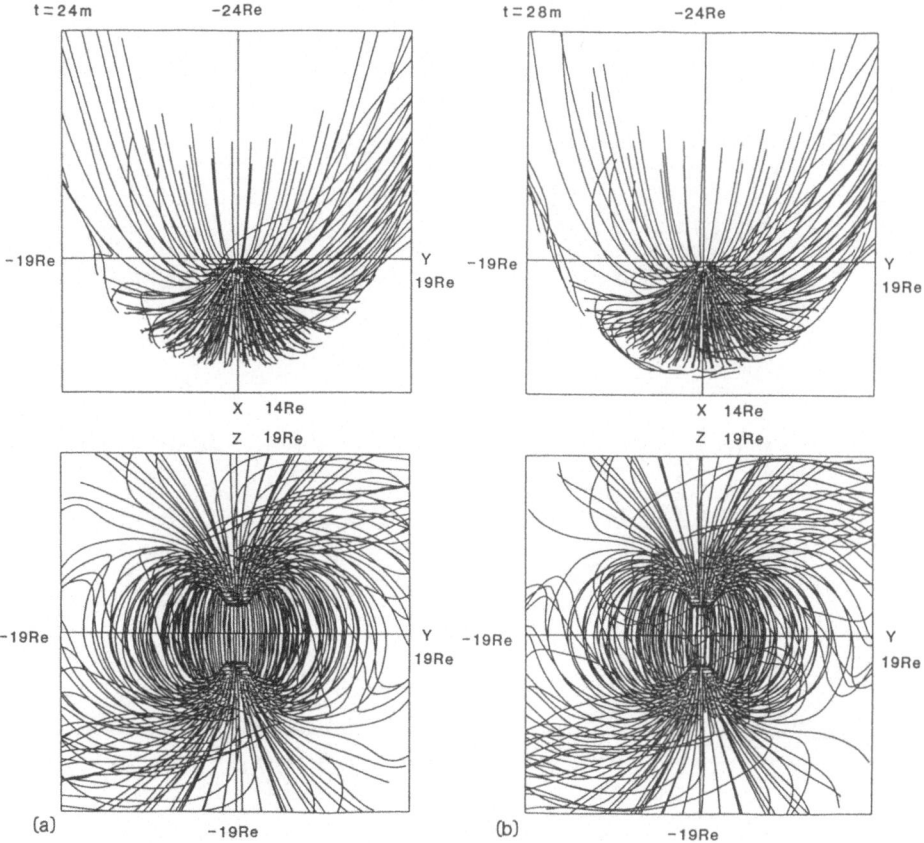

Figure 3. Magnetic field lines with at least one end attached to the Earth viewed from the north pole (top) and the Sun (bottom). Figure 3a is a snapshot at $t = 24$m and Figure 3b is at $t = 28$m.

is approximately normal to the magnetopause and B_x has a bipolar change with a peak to peak amplitude of 33nT. B_y decreases by 45nT and B_z increases by 12nT. The minimum in B_y occurs where the x-component passes through zero. This is the magnetic field configuration found in a magnetic flux rope in which the field aligned current confines a longitudinal magnetic field at the center [Elphic et al., 1986]. In the northern hemisphere the twisted flux tube moves dawnward and northward with a velocity of $(v_x, v_y, v_z) = (< -6, -35, 60)$ km s^{-1}. This velocity is close to the local Alfvén speed. The plasma pressure decreases in the flux tube region while the density is approximately constant. Beta $(\beta = 2\mu_0 p/B_y^2 = 0.35)$ is less than one at the center of the flux tube so the total pressure (plasma plus magnetic) increases across the twisted flux tube.

The maximum twist occurs in the antiparallel field region where the tube bends sharply. The field aligned current flows from dawn to dusk between the sharp bends and reverses between the bends and the ionosphere. The field aligned current density is $J_\parallel = 1.4 \times 10^{-9}$ Am^{-2} at the equator with a total parallel current of about $< J_\parallel > = 3.8 \times 10^5$A.

16

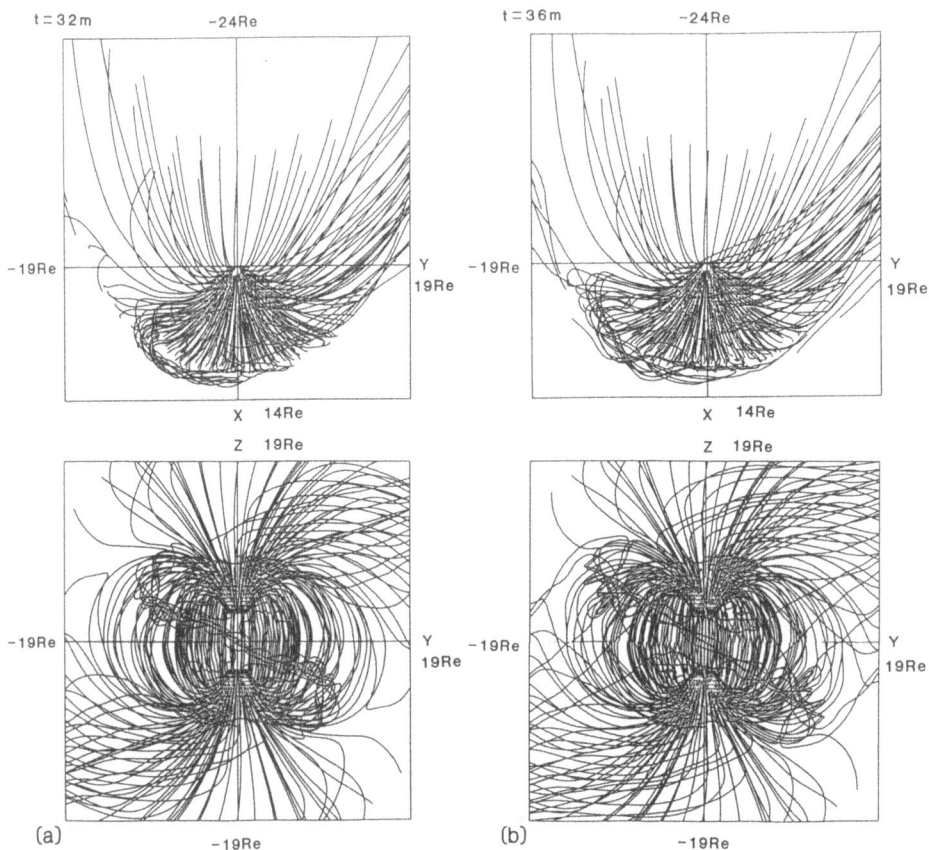

t≈32m   −24Re   t≈36m   −24Re

−19Re   Y 19Re   −19Re   Y 19Re

X 14Re   X 14Re

Z 19Re   Z 19Re

−19Re   Y 19Re   −19Re   Y 19Re

(a)   −19Re   (b)   −19Re

Figure 4. The same as Figure 3 for $t = 32$m and $t = 36$m.

All of the parameters of the model were evaluated at each grid point at each time. Figure 6 shows one way of presenting three-dimensional information in a two-dimensional format. The parallel vorticity ($\Omega = \nabla \times v$; $\Omega_{\parallel} = \Omega \cdot B/B$), the field aligned current ($J_{\parallel}$), the field aligned velocity ($v_{\parallel}$), the pressure (P) and the density (ρ) have been projected into the polar cap. The projected values were determined by calculating $\int (f/B)dl / \int (1/B)dl$ for each parameter, f, along the magnetic field lines. Only those parts of open field lines within the magnetosphere were included in the integrals. For $\Omega_{\parallel}$, $J_{\parallel}$ and $v_{\parallel}$ the contours for values parallel to the magnetic field have been shaded. For P and ρ the contours with the highest values have been shaded. In the upper left panel the region of open field lines has been shaded. Figure 6 gives the magnetospheric configuration in the absence of reconnection and gives us a basis with which to compare the values after the reconnection has started.

In Figure 6 there are two major convection cells [Ogino, 1986; Ogino et al., 1986]. The flow is tailward over the polar caps with return flow at lower latitudes. This is viscous convection. Between 70° and 75° the field aligned currents have a region 1 sense with current away from the ionosphere on the dusk side and towards the ionosphere on the dawn side. At higher latitudes the currents have the opposite sense.

Polar projections at $t = 24$m have been plotted in Figure 7. In these plots the region

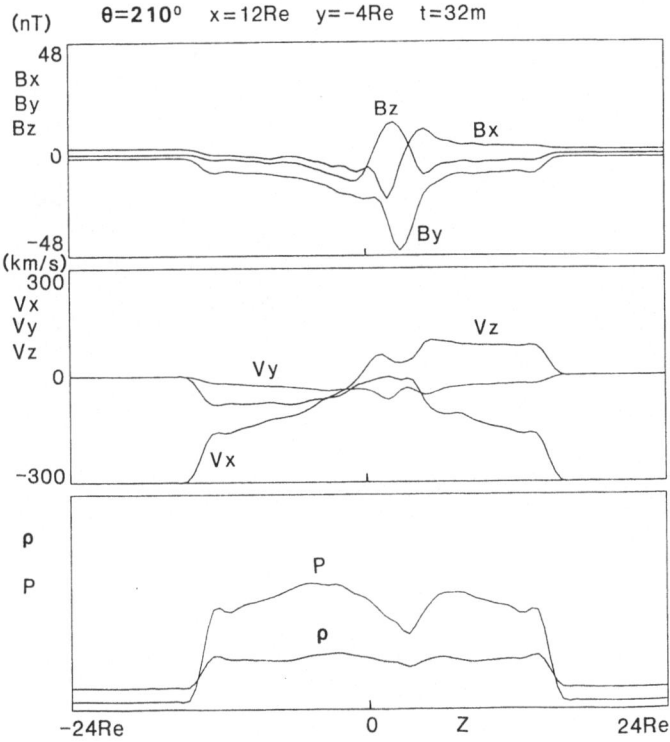

Figure 5. Plasma parameters versus position in z for $x = 12R_E$ and $y = -4R_E$ for $\theta = 210^\upsilon$ at $t = 32m$. The magnetic field is plotted in the top panel, the velocity is plotted in the middle panel and the density and pressure are plotted in the bottom panel.

of reconnection is a region of Earthward $v_\parallel$ between noon and about 10LT and between 80° and 82° (compare with Figure 6). The region of Earthward flow maps to a region of Earthward parallel current and to the region where the convection changes direction ($\Omega_\parallel$ panel). In Figure 8 the polar projections at $t = 36m$ have been plotted. The twisted flux tube in Figure 4b maps to the region of reversed parallel vorticity centered at about 9LT and just equatorward of 80^υ. That part of the twisted flux tube is still on closed field lines can be seen in the open-closed field line plot. At this time the parallel current pattern is complex and shows a layered structure around the cusp.

3.2 THE CASE FOR SMALL IMF B_y ($\theta = 240^\circ$)

In Figure 9 we have plotted the magnetic field lines at $t = 16m$ for the case when the IMF was in the $\theta = 240^\circ$ direction. As in the case for $\theta = 210^\circ$ the field lines on the dawn side have been distorted by the convection of the solar wind across the magnetopause. The field lines at $t = 32m$ are plotted in Figure 10a. In this case dayside reconnection occurred in the antiparallel field region near the subsolar point. The reconnected field lines have a helical twist which can be seen in the view from the north pole. By $t = 36m$ (Figure 10b) these twisted flux tubes have propagated over the poles. No additional twisted flux tubes had formed by the end of the run at $t = 40m$.

18

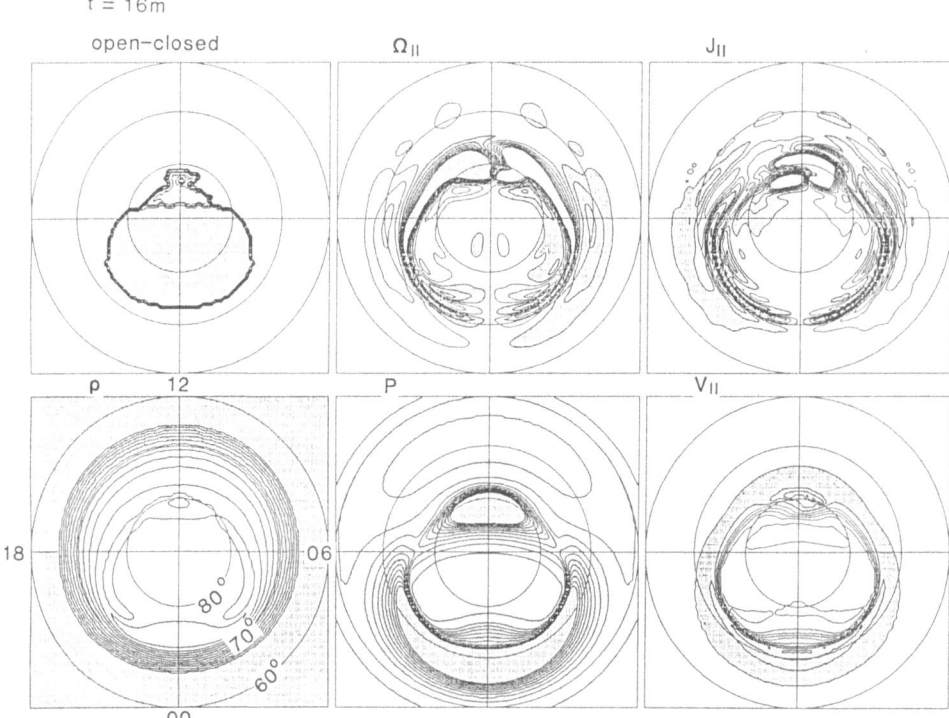

t = 16m

open–closed $\Omega_{\|}$ $J_{\|}$

ρ 12   P $V_{\|}$

18   06

80°
70°
60°

00

Figure 6. Plasma parameters projected onto the polar cap for $\theta = 210°$ at $t = 16$m. The parameters from top left are the region of open field lines, the parallel vorticity ($\Omega_{\|}$), the parallel current ($J_{\|}$), the density (ρ), the pressure (P), and the parallel velocity ($v_{\|}$).

Figure 11 shows the z profiles of the physical parameters at the magnetopause at $t = 32$m. The B_z component decreases at two locations $z \approx -5\text{R}_E$ and $z \approx 1\text{R}_E$. Both decreases in B_z are associated with bipolar changes in B_x with peak to peak amplitudes of 30nT and 22nT respectively. There is little change in B_y. The plasma flow velocity is $(v_x, v_y, v_z) = (-35, -63, -75)$ km s^{-1} for the southern dawn flux tube at $z \approx -5\text{R}_E$ and $(v_x, v_y, v_z) = (-6, 17, 23)$ km s^{-1} for the northern dawn flux tube at $z \approx 1\text{R}_E$. The southern flux tube is moving at approximately the local Alfvén velocity. Both flux tubes have wavelengths of about 6R$_E$. The field aligned current density in the magnetic flux tubes is $J_{\|} = (0.5 \sim 1.3) \times 10^{-9}\text{Am}^{-2}$ at the equator and the total current is about $< J_{\|} > = (1.4 \sim 3.4) \times 10^5\text{A}$. These values are a little less than those for the $\theta = 210°$ case.

The polar projections at $t = 36$m for the small B_y case are presented in Figure 12. The initial state is very similar to that in the $\theta = 210°$ case in Figure 6. There are only slight differences in the convection pattern because of the different IMF direction. In Figure 12 the region to which the open twisted flux tube maps extends to the dusk side (latitude less than 80° and local time earlier than 1400). As in the $\theta = 210°$ case the parallel current pattern is very complex at this late stage in the event. As before the region of the twisted flux tube corresponds to a flow reversal region in the parallel vorticity panel.

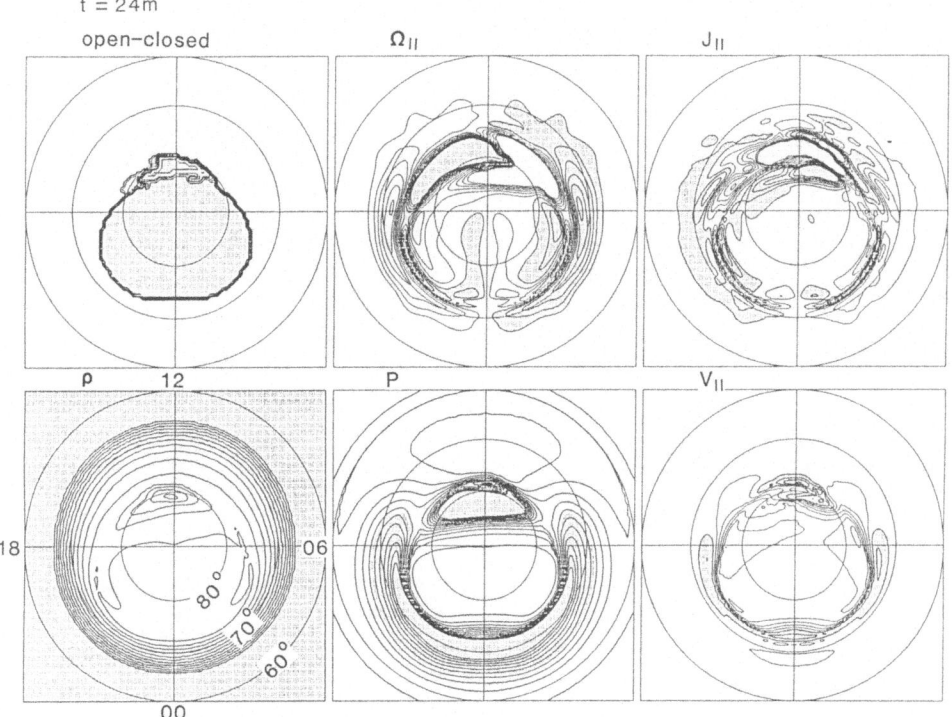

Figure 7. The same as Figure 6 for $t = 24$m.

4. Discussion and Conclusions

We have studied the formation and dynamics of isolated magnetic flux tubes at the dayside magnetopause by using a time-dependent and three-dimensional magnetohydrodynamic simulation model. For southward IMF we found that two types of magnetic flux tubes were formed depending on the size of the IMF B_y component. For large B_y a strongly twisted magnetic flux tube appeared across the dayside magnetopause. When the IMF B_y was small twin flux tubes were found on the dayside magnetopause. In both cases antiparallel magnetic reconnection was important for the generation of these flux tubes.

Although these flux tubes are near the minimum resolution of the numerical scheme, we believe they are significant. For example in the $\theta = 210°$ case the wavelength of the twisted flux tube is given by

$$\lambda_x \simeq 3R_E = 6\Delta x$$

$$\lambda_y \gg 11\Delta y$$

$$\lambda_z \simeq 5 \ or \ 6R_E \simeq 11\Delta z.$$

For three dimensional advection problems the key parameter for numerical accurary is $|\mathbf{k} \cdot \mathbf{C}|$ where $\mathbf{k}$ is the wave number and $\mathbf{C}$ is the advection velocity. It is important to have

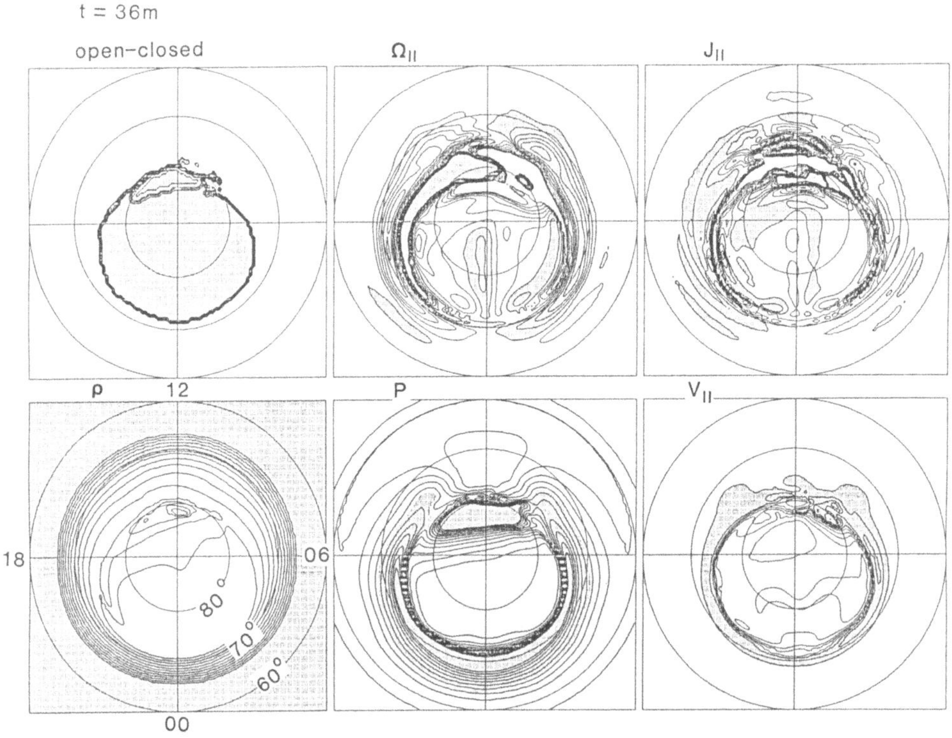

t = 36m

open-closed $\Omega_{||}$ $J_{||}$

ρ 12   P $V_{||}$

18   06

80°

70°

60°

00

Figure 8. The same as Figure 6 for $t = 36$m.

$\theta = 240°$ t=16m

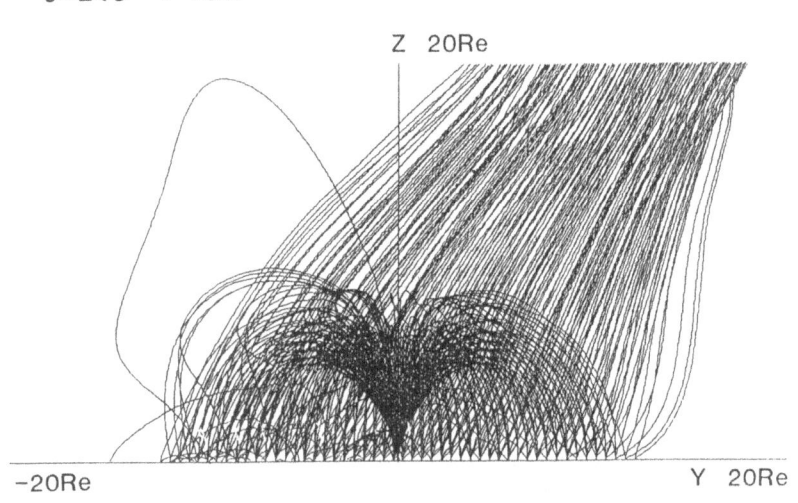

Z 20Re

−20Re   Y 20Re

Figure 9. The same as Figure 2 for $\theta = 240°$ at $t = 16$m.

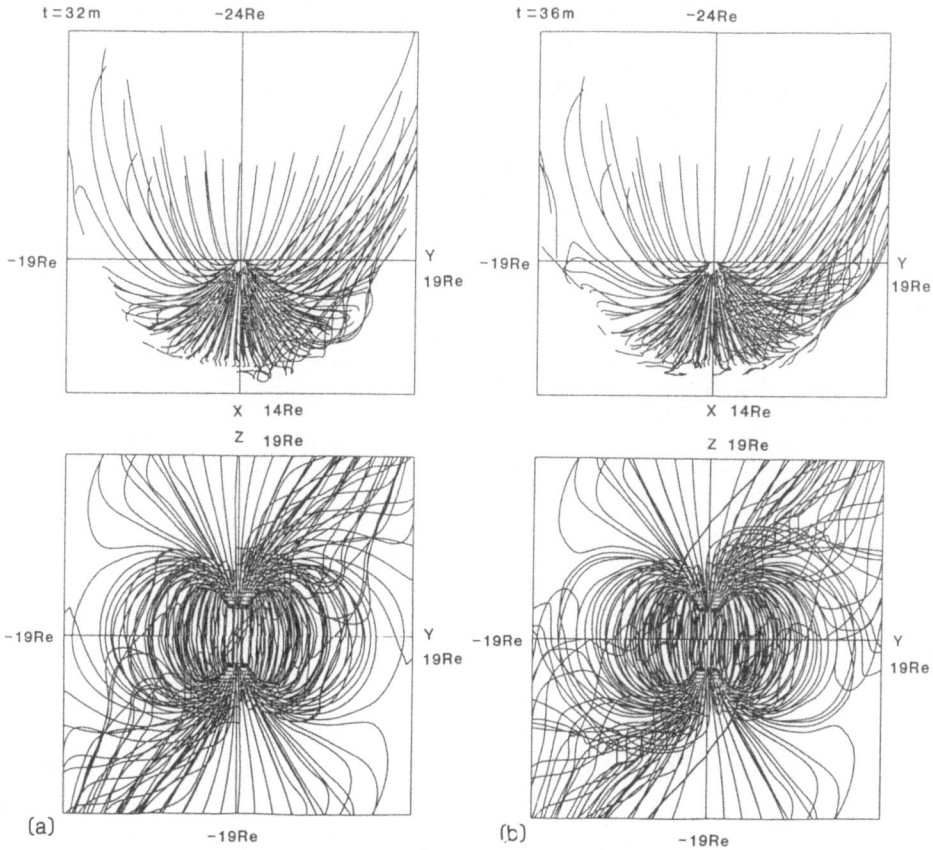

Figure 10. The same as Figure 3 for $\theta = 240°$ at $t = 32m$ and $t = 36m$.

sufficient grid points along any direction for which $|\mathbf{k} \cdot \mathbf{C}|$ is large. In our case we find that

$$k_x = 2\pi/\lambda_x = 2\pi/6\Delta x$$

$$k_y = 2\tau_i'\lambda_y \ll 2\pi/11\Delta y$$

$$k_z \cdot \quad /\lambda_z = 2\pi/11\Delta z$$

and

$$C_x < 6kms^{-1}$$

$$C_y = 35kms^{-1}$$

$$C_z = 60kms^{-1}$$

Thus we have $|k_x C_x| \ll |k_z C_z|$ and z is the critical direction for determining the numerical accuracy of the system. The 11 gird points per wavelength is sufficient to resolve the flux tubes.

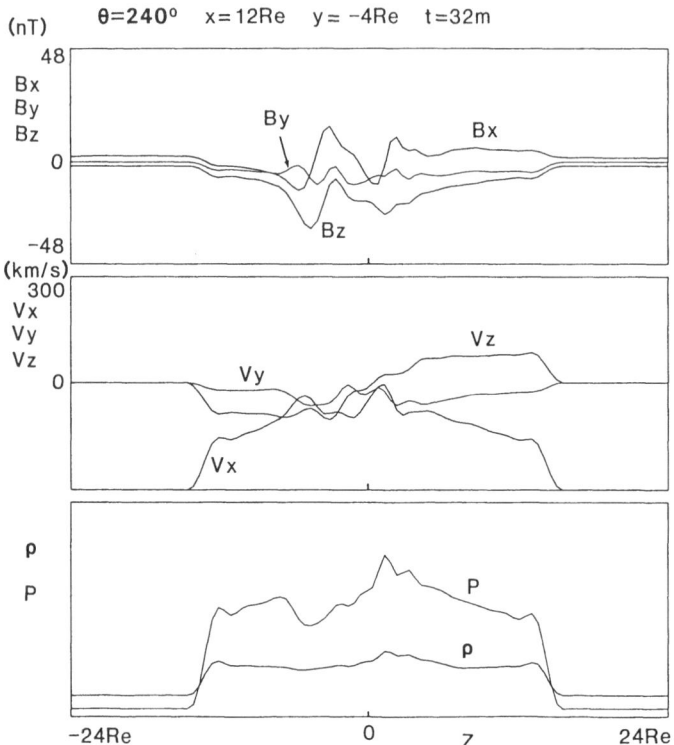

Figure 11. The same as Figure 5 for $\theta = 240°$ at $t = 32$m.

A cartoon showing the development and evolution of the twisted flux tube for the large B_y case is presented in panels a, b, and c of Figure 13. Before reconnection starts at the magnetopause viscous convection has distorted the magnetopause field lines. In panel a magnetic reconnection is just starting in the antiparallel field region where the IMF and the Earth's field are opposite (circled areas). The distortion of the field lines in the region between the two reconnection areas occurs because of a combination of the viscous convection and the reconnection. As reconnection proceeds the distorted field lines just Earthward of the reconnection region become twisted. The twisted flux tube is initally on closed field lines. That twisted flux tubes aren't found when the IMF is northward indicates that dayside reconnection is necessary for the generation of these twisted closed flux tubes. The reconnection enhances the distortion of the closed field lines by reducing the pressure in the reconnection areas.

In panel b the twisted flux tube has started to reconnect. The reconnection starts in both the northern and southern hemispheres when the twisted flux tube enters the antiparallel reconnection regions. When half of the flux tube has reconnected in both hemispheres the reconnection of the twisted flux tube stops as the newly reconnected flux tubes are convected tailward across the northern and southern magnetopauses (panel c). The convection velocity is 60 to 100 km s^{-1}.

The B_x component has a bipolar signature as a function of z and the B_y and B_z compot = 36m

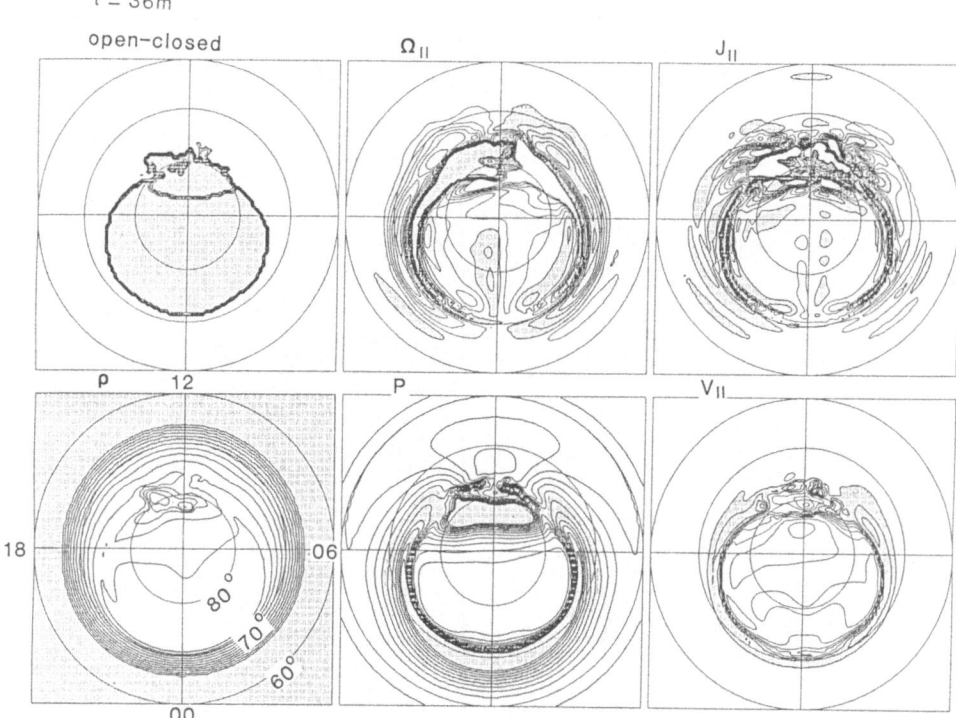

Figure 12 The same as Figure 6 for $\theta = 240°$ at $t = 36$m.

nents are intensified. In particular the change in B_y is largest where the B_x change is zero. This configuration is frequently referred to as a flux rope (see Elphic et al. [1986]). The plasma pressure decreases in the flux rope but the total pressure (plasma plus magnetic) increases. The field aligned current in the flux tube flows from dawn to dusk in that part of the flux tube between the reconnection regions. The current reverses in the region between the reconnection area and the ionosphere. A stationary spacecraft north of the equator will observe a positive then a negative change in the magnetic field component normal to the boundary as the flux tube convects past it. In the southern hemisphere the change will be opposite (i.e. negative then positive). It will take 3-5 minutes for the twisted flux tube to convect past the spacecraft. These simulation results are consistent with several features of satellite observations of flux transfer events [Russell and Elphic, 1978; 1979; Elphic and Russell, 1979; Berchem and Russell, 1984; Rijnbeek et al., 1984; Saunders et al., 1984].

The twisted flux tube in panel c is similar to that suggested by Lee and Fu [1985]. They suggested that component reconnection at multiple locations was responsible for the twisted structure. We have examined the flow and magnetic field pattern for evidence of component reconnection. Our examination indicates that while some component reconnection probably does occur, it is much less important for driving flows than is the antiparallel merging. In addition the twisted flux tube initially forms on closed field lines a feature not found in the Lee and Fu model.

$\theta = 210°$ $\theta = 240°$

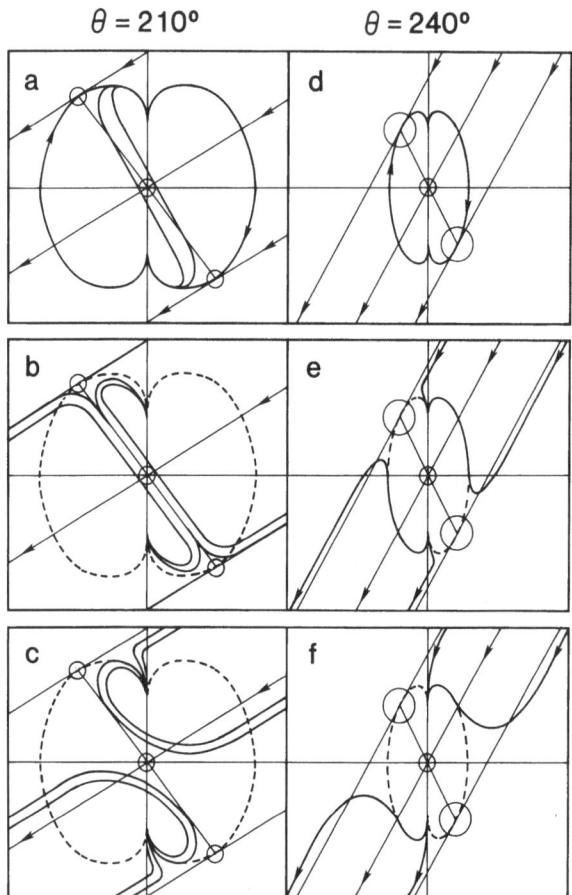

Figure 13. A sketch of the time evolution of the isolated flux tube for $\theta = 210°$ (panels a, b, and c) and for $\theta = 240°$ (panels d, e, and f).

The sequence of events which occurs when the IMF has a small B_y component are illustrated in panels d,e, and f of Figure 13. In this case the dayside reconnection occurs in the antiparallel field region (panel d). Twin magnetic flux tubes form episodically at the magnetopause and convect over the poles (panels e and f). Again the normal component of the magnetic field shows a bipolar change as the tube convects past a stationary spacecraft. However the B_y component has only small changes. We need to run the simulation longer to determine whether additional flux tubes form at the boundary and if so how often they form. In Figure 11 the plasma pressure increased in the southern tube and decreased in the northern tube. However the total pressure increased in both cases. Note that the slower northern tube is still near the reconnection region while the southern tube has convected away from it. The pressure increases as the tubes convect across the magnetopause.

If the IMF orientation has a duskward instead of a dawnward component we would expect that magnetic flux tubes would form at mirror image positions on the northern dusk and southern dawn magnetopause. In that case we would again expect that the field aligned current would flow from dawn to dusk and the changes in the normal component of the magnetic field would have the same signs as in the cases with dawnward IMF.

5. Acknowledgements.

We would like to acknowledge the support, encouragement and suggestions provided for the simulation studies by Maha Ashour-Abdalla. This work was supported by NASA Solar Terrestrial Theory Program grant NAGW-78 and NASA grant NGL-05-007-004 and by grants in aid from the Ministry of Education, Science and Culture. Computing support for our simulations was provided by the San Diego Supercomputing Center, by the Computer Center of Nagoya University and by the Computer Center of the Institute of Space and Astronautical Science.

6. References

Berchem, J., and C. T. Russell, Flux transfer events on the magnetopause: Spatial distribution and controlling factors, J. Geophys. Res., 89, 6689, 1984.

Elphic, R. C., and C. T. Russell, ISEE-1 and 2 magnetometer observations of the magnetopause, in Magnetospheric Boundary Layers,p. 43, ESA Scientific and Technical Publications, Noordwijk, The Netherlands, 1979.

Elphic, R. C., C. A. Cattell, K. Takahashi, S. J. Bame, and C. T. Russell, ISEE-1 and -2 observations of magnetic flux ropes in the magnetotail: FTE's in the plasma sheet?, Geophys. Res. Lett., 13, 648, 1986.

Fu, Z. F., and L. C. Lee, Simulation of multiple X lines in reconnection at the dayside magnetopause, Geophys. Res. Lett., 12,291,1985.

Galeev, A. A., M. M. Kuznetsova, and L. M. Zeleny, Magnetopause stability threshold for patchy reconnection, Space Sci. Rev.,44, 1, 1986.

Haerendel, G., G. Paschmann, N. Sckopke, H. Rosenbauer, and P. C. Hedgecock, The frontside boundary layer of the magnetosphere and the problem of reconnection, J. Geophys. Res.,83, 3195, 1978.

La Belle-Harmer, A. L., Z. F. Fu, and L. C. Lee, A mechanism for patchy reconnection at the dayside magnetopause, Geophys. Res. Lett., 15, 152, 1988.

Lee, L. C., and Z. F. Fu, A theory of magnetic flux transfer at the Earth's dayside magnetopause, Geophys. Res. Lett., 12, 105, 1985.

Ogino, T., A three dimensional MHD simulation of the interaction of the solar wind with the Earth's magnetosphere: The generation of field-aligned currents, J. Geophys. Res., 91, 6791, 1986.

Ogino T., R. J. Walker, M. Ashour-Abdalla, and J. M. Dawson, An MHD simulation of B_y-dependent magnetospheric convection and field-aligned currents during northward IMF, J. Geophys. Res., 90, 10835, 1985.

Ogino, T., R. J. Walker, M. Ashour-Abdalla and J. M. Dawson, An MHD simulation of the effects of the interplanetary magnetic field B_y component on the interaction of the solar wind with the Earth's magnetosphere during southward IMF, J. Geophys. Res., 91, 10029, 1986.

Ogino, T., R. J. Walker, and M. Ashour-Abdalla, A Magnetohydrodynamic simulation of the formation of magnetic flux tubes at the Earth's dayside magnetopause, Geophys. Res. Lett., (submitted), 1988.

Rijnbeek, R. P., S. W. H. Cowley, D. J. Southwood, and C. T. Russell: A survey of dayside flux transfer events observed by ISEE-1 and -2 magnetometers, J. Geophys. Res., 89, 786, 1984.

Russell C. T., and R. C. Elphic, Initial ISEE magnetometer results: Magnetopause observations, Space Sci. Rev.,22, 681, 1978.

Russell, C. T. and R. C. Elphic, ISEE observations of flux transfer events at the dayside magnetopause, Geophys. Res. Lett., 6, 33, 1979.

Sato, T., T. Shimada, M. Tanaka, T. Hayashi, and K. Watanabe, Formation of field twisting flux tubes on the magnetopause and solar wind particle entry into the magnetosphere, Geophys. Res. Lett., 13, 801, 1986.

Saunders, M. A., C. T. Russell, and N. Sckopke, Flux transfer events: Scale size and interior structure, Geophys. Res. Lett., 11, 131, 1984.

Scholer, M., Magnetic flux transfer at the magnetopause, Geophys. Res. Lett., 15, 291,1988.

Shi, Y., C. C. Wu, and L. C. Lee, A study of multiple X line reconnection at the dayside magnetopause, Geophys. Res. Lett., 15, 295, 1988.

Sonnerup, B. U. Ö., On the stress balance in flux transfer events, J. Geophys. Res., 92, 8613, 1987.

Southwood, D. J., C. J. Farrugia, and M. A. Saunders, What are flux transfer events?, Planet. Space Sci.,36, 503, 1988.

IMPULSIVE PENETRATION OF SOLAR WIND PLASMA IRREGULARITIES INTO THE MAGNETOSPHERE : RELEVANT LABORATORY EXPERIMENTS.

J. LEMAIRE *
*DESPA, Observatoire de Meudon,
France.*

ABSTRACT.   Before detailed high resolution and multi-point observations of the magnetopause region will become available in the 90's, it is interesting to reexamine existing Controlled Laboratory Plasma Experiments which are relevant to the study of solar wind - magnetosphere interaction. Two series of such laboratory experiments have been reviewed in this article. Their application to the solar wind - magnetosphere system has been presented and discussed. But other additional laboratory experiments exist; they should also be reconsidered and possibly extended.

1. What can we learn from laboratory experiments?

Before CLUSTER's detailed in-situ observations will become available to the science community in the 90's, it can be quite instructive to examine available Laboratory Plasma Experiments which are relevant to our topic.

The purpose of this paper is precisely to review two sets of laboratory plasma experiments which are mostly unknown or overlooked by a large fraction of our space science community. We will show below to which extent these basic plasma physics experiments are relevant for the study of solar wind-magnetosphere interaction.

---

* *On leave of absence from, IASB, 3, av. Circulaire, B-1180, Brussels, Belgium*

P. E. Sandholt and A. Egeland (eds.), Electromagnetic Coupling in the Polar Clefts and Caps, 27–42.
© *1989 by Kluwer Academic Publishers.*

The first set of these experiments is that of Baker and Hammel (1962,1965); it points out how the excess energy and momentum of plasmoids can be dissipated by Joule heating in the 'walls' of the plasma chamber (and similarly in the conducting ionosphere of the Earth). These experimental results will be discussed in section 2.

The second set of experiments are those of Demidenko *et al.* (1967,1969). They deal with plasma elements decelerated and accelerated in regions where the magnetic field intensity is non-uniform, i.e. where the gradient of B is either positive or negative. These largely unknown experimental results will be reviewed in section 3

2. Non-adiabatic deceleration of a plasmoid :
The Baker and Hammel's sets of experiments.

2.1. DESCRIPTION OF THE EXPERIMENTS.

The experiments by Baker and Hammel (1962,1965) consisted of injecting collisionless plasma stream in a vacuum chamber perpendicularly to the direction of a uniform magnetic field, B.

Their first set of experiments showed that the plasma stream conserves a constant velocity equal to its injection velocity provided that the wall of the tank are made of electrically non-conducting material : e.g. glass. When the walls are made of insulating material the plasma stream penetrates freely and unimpaired across magnetic field lines, whatever the value of the magnetic field intensity. The ability to penetrate magnetic field lines is then found to be independent of the value of 'Beta', the ratio between the plasma kinetic pressure and the magnetic energy density.

The ability to penetrate a uniform magnetic field is not either depending of the direction of the diamagnetic field carried by the laboratory plasma elements, at the contrary of what Schindler (1979) has inferred from a theoretical study of ideal MHD and 2-D current filaments of infinite length...

In other words, when the electrical Pederson conductivity transverse to the magnetic field direction is small everywhere along the field lines (i.e. even in the walls of the vacuum chamber), the plasma stream is not refrained from moving across magnetic field lines; in this case, 'magnetic field line tying' resulting from the frozen-in-field theorem (and illustrated by cartoons like that of fig. 1), is not supported by this first set of Baker and Hammel's experiments.

In another set of experiments Baker and Hammel showed that when the walls of the vacuum chamber are coated with good electrically conducting material the plasma stream is slowed down along the direction of injection; the plasma flow is then eventually deflected aside as illustrated in fig. 2. In other words as soon as the plasma gets on field lines which are rooted in conducting walls it is decelerated until its forward velocity eventually drops to zero.

(Note that there is no externally imposed electric field in none of
these experiments.)

Although, the local plasma density and magnetic field intensity
are similar in the vacuum chamber for both sets of experiments, the
plasma bulk velocity (or convection velocity) is constant in the
first case, while it is decelerated in the latter one; the different
conducts come from the different (non-local) boundary conditions : in
the first case the electrical conductivity transverse to the magnetic
field lines is everywhere nil or almost; in the second case there is
a 'weak place' (i.e. the walls) where conductivity is high.

2.2. THE PHYSICAL EXPLANATION

The explanation of the different behaviour of the flow velocities in
both sets of experiments is illustrated in fig.3, showing plasma
streaming across a uniform magnetic field. The electrons and ions are
deflected in opposite directions by the Lorentz force, $q \mathbf{v} \times \mathbf{B}$; as a
consequence, a net positive charge density builds up along the upper
boundary of the injected plasma stream, while a slight excess of
electrons forms a negative surface charge density on the opposite
side. These surface polarisation charges continue to grow until the
resulting charge separation electric field is strong enough to cancel
the deflecting effect the magnetic field; i.e. until the electric
field intensity measured in the frame of reference co-moving with the
plasma is equal to zero : $\mathbf{E}' = \mathbf{E} + \mathbf{v} \times \mathbf{B} = 0$.

In the laboratory frame of reference the polarisation electric
field, $\mathbf{E} = - \mathbf{v} \times \mathbf{B}$, is precisely equal to what is required for the
plasma inside the stream to keep its velocity constant, i.e. equal to
its injection velocity, $\mathbf{v}_0$. The plasma stream then continues to move
across the magnetic field with the well known $\mathbf{E} \times \mathbf{B}$ drift :
$\mathbf{v} = \mathbf{E} \times \mathbf{B} / B^2$. Since the plasma is collisionless, its transverse
Pederson conductivity is nil; there is then no way to discharge the
electric potential differences set up by the plasma dynamo via local
Pederson currents.

In the foremost section of the vacuum chamber where the plasma
stream moves across the non-conducting part of the vacuum tank (see
fig. 3) the transverse electric potential differences cannot be
discharged by electric currents within the walls since they are
insulating ones. Therefore, the convection electric field created by
the plasma itself is not short circuited, and consequently the stream
velocity remains constant.

This is not quite the case in the second portion of the vacuum
tank (see fig. 3) where depolarizing currents can flow within the
conducting walls; this short circuits the potential differences set
up in the plasma stream and distributed on the walls. Note that
magnetic field lines are almost equipotentials in collisionless
plasmas because the field-aligned conductivity is large, but the
local Pederson conductivity is very small within the plasma cloud
itself. The convection electric field being short circuited by
depolarisation currents in the walls, the plasma bulk speed is
reduced until its forward momentum vanishes, somewhat like that of a

rain drop falling into a viscous fluid. The initial kinetic energy
of the plasma element is dissipated into Joule heating in the walls;
its excess initial momentum is also transferred to the walls.

2.3. ADDITIONAL EFFECTS

This simple picture is subject to further refinements.
For example, the particles in the surface charge layers will
experience less electric force and consequently cannot keep up with
the drift of the main stream : an effect which has been described by
Dolique (1963). The 'peeled-off charges which are left behind can
provide a source of electric field for the plasma in the wake behind
the plasmoid.

    Another secondary effect is that the space charge repulsion in
the charge layers will spread the polarisation charges laterally
along the magnetic field lines, removing particles from the initial
stream (Baker and Hammel, 1965). Thus the plasma stream has a
tendency to spread along the field lines (Harris et al., 1957).

    This is also the place to mention that electrostatic double
layers form at the edges of plasmoids where magnetic field lines
traverse the surface of the plasmoid. This sets up the appropriate
electrostatic potential differences to avoid that the flux of
electrons running out of the cloud along magnetic field lines,
becomes larger than the escaping flux of the heavier positive
charges. Indeed, without such a retarding potential the plasmoid
would lose its electrons faster than its ions and become positively
charged.

    Furthermore, when the diamagnetic plasma stream penetrates the
external magnetic field, magnetic probes placed in the vacuum chamber
indicate a decrease of B (Baker and Hammel, 1962). The time variation
of the local magnetic field intensity as measured in the laboratory
frame of reference, depends on 'Beta', the ratio between the kinetic
and magnetic pressures inside the plasma. However, when Beta is
small the diamagnetic currents and the associated magnetic field
perturbations are small. In this case the local time derivatives of
the B are small also.

    Note that time derivatives of the local magnetic field
intensity generate induced electric fields in the vicinity of the
moving plasmoid. It is the importance of these induced electric
fields in high-beta plasma, that has been emphasized by Heikkila
(1982).

2.4. APPLICATION TO THE MAGNETOSPHERE

How do the Baker and Hammel series of experiments relate to the
problem of the solar wind interaction with the Earth's magnetosphere?
    The conducting Earth's Ionosphere corresponds to the walls in
the last section of the tank where plasma elements are slowed down;
the magnetic field lines at the interface between the conducting and
insulating sections in fig.3 correspond to the geomagnetic field
lines which are tangent to the magnetopause; on one side the magnetic

field lines are rooted in the dayside cusps ionosphere and have a relatively large integrated Pederson conductivity; while on the other side of the magnetopause they hang into the solar wind and are rooted in the distant solar atmosphere ; the integrated Pederson conductivity of interplanetary magnetic field lines is much lower.

The injected plasma elements of Baker and Hammel experiments correspond to impinging solar wind plasma irregularities (or plasmoids) which are almost always observed in the magnetosheath. From high resolution plasma and magnetic field measurements it can be inferred that similar plasma irregularities are also present in the solar wind beyond the bow shock. Radio scintillation observations from the corona and solar wind confirm also that during solar active periods the amplitude of these plasma density irregularities is enhanced.

Those plasmoids with larger momentum densities penetrate deeper into the magnetospheric boundary layer than those with a momentum smaller than the average solar wind plasma flow. While the former will be able to penetrate across closed geomagnetic field lines, the latter will be deflected aside in the middle of the magnetosheath without even having been able to reach the magnetopause surface.

The large peak in the ionospheric temperature distribution observed by Titheridge (1976) at the feet of dayside cusp field lines from below 400 km altitude up to 1000 km has been attributed to Joule dissipation of the solar wind plasmoids continuously raining into the magnetosphere (Lemaire and Roth, 1976).
Note however that part of this peak in the ionospheric temperature may also been attributed to elastic collisions of impulsively injected magnetosheath particles precipitating in the upper ionosphere along polar cusp field lines.

The observed poleward motion of ionospheric plasma over the polar caps (Goertz, et al., 1985), as well as recently observed poleward motion of polar cusp auroral features (Sandholt, et al., 1985) have been attributed to the momentum transfer of impulsively penetrating solar wind plasmoids in the frontside magnetosphere (Lemaire, 1977, 1987)
Heikkila et al. (1988) also emphasize that these moving structures should not be considered as unilateral evidence in favour of patchy reconnection or FTE's. Indeed, these same 'events' (or these same plasma structures) can indeed easily be explained in the framework of Impulsive Penetration Theories.

Since the solar wind is generally a high-beta plasma, the penetrating plasmoids produce large localized diamagnetic field perturbations similar to the magnetic field perturbations observed in Baker and Hammel's experiments.

A video film presented at this NATO Workshop by the author illustrates how diamagnetic fields carried by solar wind plasmoids change the local geomagnetic field distribution in a time dependent manner.

2.5. ADDITIONAL REMARKS

Another lesson that the experiments of Baker and Hammel tells us is that the local magnetic field intensity, the local plasma density distribution and the local pressure distributions within the vacuum chamber don't uniquely determine the bulk velocity of the plasma flow injected into it. Distant (non-local) boundary conditions in the walls (i.e. in the ionosphere) eventually determine the allowed plasma flow patterns, simply by imposing unavoidable constrains to the convection electric field.

Therefore, solving local hydromagnetic transport equations to determine the plasma flow in the distant magnetosphere and in the vicinity of an X-line, without specifying the boundary conditions in the 'walls' (or without giving the distribution of the integrated Pederson and Hall conductivities along magnetic field lines) appears a rather academic exercise, and has limited application to the dissipative magnetosphere-ionosphere system...

Unfortunately in many papers on magnetospheric convection, reconnection and merging, these non-local boundary conditions are not given explicitly; the implicit assumption being that the integrated Pederson conductivity is either equal to zero, or, in other cases (e.g. convection in the plasmasphere) that it is assumed to be infinitely large!...

3. Adiabatic deceleration of a plasmoid :
The Demidenko et al.'s experiments.

In the previous section we have examined laboratory experiments for which the external magnetic field distribution was uniform. In this section we recall the results obtained in the sixties when plasmoids are injected in a non-uniform magnetic field (Demidenko et al., 1966, 1969, 1972).

3.1. EXPERIMENTAL RESULTS

The magnetic field intensity is parallel to the Oz-axis and perpendicular to the Ox-axis along which the plasmoids are injected with a velocity v_{0x}. The magnetic field intensity varies as a function of x as indicated in fig.4. It increases from a small value at the entry of the vacuum chamber to a maximum value B_0 in the centre; beyond the central point the external magnetic field intensity decreases to a vanishinly small value at the opposite end of the drift tube. No electric field is applied from the outside; however, as in the experiments of Baker and Hammel a polarization electric field , $E_y(x)$, develops naturally within the plasma, as a result of the deflections in opposite directions of the electrons and ions in the magnetic field.

The experiments by Demidenko et al. have shown that the velocity inside the plasmoid decreases when it penetrates deeper in the B-field where the intensity becomes larger. The forward bulk velocity $v_x(x)$ satisfies the following relation :

$$v_x = (2/m_p)(W - \mu B_z)^{1/2} \qquad (1)$$

where m_p is the ion mass, μ is the mean magnetic moment of the particles and W their mean total energy when they penetrate in the non-uniform magnetic field region :

$$W = m_p v_{0x}^2/2 + \mu B_{0z} \qquad (2)$$

$$\mu = k T / B_{0z} \qquad (3)$$

Fig.5 shows that the square of v_x is indeed a linear function of the magnetic field intensity B_z, as indicated by equation (1). The slope of the solid line which is a least square fit across the experimental points; its value determines the value of the mean magnetic moment μ which is an adiabatic invariant of motion. The value of B where this straight line cuts the horizontal axis yields the critical field intensity, B_1, where v_x tends to zero.

The distribution of $v_x(x)$ is shown in fig.4 as a function of x. At $x=x_1$, where $B_z=B_1$, the bulk velocity of the plasmoid vanishes because its convection kinetic energy has been entirely transformed into thermal energy. Indeed, as a consequence of conservation of their magnetic moments, the energy associated with the gyro-motion of the particles increases proportionals to the value of B. The perpendicular thermal (gyro-motion) energy increases until the convection (or bulk motion) energy, $m_p v_x^2/2$, has become equal to zero; and this occurs at x_1 where

$$B_z = B_1 = W/\mu \qquad (4)$$

At this same distance x_1 the convection electric field

$$E_y = v_x B_z \qquad (5)$$

tends also to zero.

The convection electric field inside the plasmoid, $E_y(x)$ has a maximum value at $x=x_2$, where $B=B_2$:

$$B_2 = 2 B_1 / 3 = 2 W / 3 \mu. \qquad (6)$$

Furthermore, it has been shown by Demidenko *et al.*(1969) that, as a result of additional drifts in the crossed E_x and B_z fields, there is a variation in the transverse dimension (i.e. along the Oy-axis) of the plasma stream. The width of the stream, h, varies with x according to the relation :

$$h(x) \ E_y(x) = \text{constant}, \quad \text{or}$$

$$h \ v_x \ B_z = \text{constant} \tag{7}$$

The distribution of h as a function of x is illustrated in fig.4. Note that h(x) has a minimum at $x=x_2$, where the electric field has a maximum. When the front side of the plasmoid approaches the distance $x=x_1$, where v_x tends to zero, the width of the plasma stream tends to spread rapidly in the y-direction, somewhat like the section of a water droplet when it splashes against a solid surface.

From the conservation of particle flux it can also be deduced that the plasma density inside the plasmoid should increase proportionals to the value of B, when it penetrates in the higher field region and approaches x_1.

When the initial total energy W is large enough (i.e. when W > $\mu \ B_0$), the magnetic barrier is not quite high enough to stop the plasma element completely; beyond $x=x_0$, where the magnetic field intensity has a maximum, the plasmoid is accelerated adiabatically by the negative field gradient.

On the contrary, when the initial total energy W is smaller then $\mu \ B_0$, the plasma stream is stopped and should be reflected adiabatically. But the reflected plasma stream will then interact with the incoming stream; this leads to a cancellation of the polarization electric fields in the opposite directed plasma streams. The kinetic energy associated with the directed particle motion must then be transformed into gyro-motion, with the consequence that the perpendicular temperature increases.

In one dimensional flow system a backward propagating front should in principle appear. Behind this backward propagating wave front more and more heated plasma is piled up against the magnetic barrier. If the incoming plasma has a supersonic injection velocity, like the solar wind, the backward propagating wave front is a shock wave.

However, the distribution of laboratory magnetic fields (and of the geomagnetic field) are not 1-D, but 3-D ones. Therefore, instead of piling up indefinitely against the magnetic barrier, the stream is deflected aside along the flanks and finds a way where the gradient of the magnetic field intensity is smaller.

In all these Controlled Laboratory Experiments of Demidenko *et al.* where adiabatic deceleration and acceleration of plasmoids are investigated, the walls of the vacuum chamber were, of course, made out of glass which is an insulator; this avoids non-adiabatic deceleration resulting from depolarisation wall currents, as in the experiments of Baker and Hammel discussed above.

3.2. THE PHYSICAL EXPLANATION

Some of the experimental results discussed in the previous section had already been predicted theoretically by Schmidt (1960). The reader is conveyed to consult this original contribution, or the paper by Demidenko *et al.*(1967) where the equations (1)-(7) are also derived from first principles. There is no need to repeat, here, this straightforward derivation once more.

Let us just mention that this derivation is based on Poisson's equation and on the equation giving the polarisation current density in a plasma when the electric field intensity changes with time. The validity of this set of equations rests on the assumption that the dielectric constant of the plasma is much larger than unity; this is not only the case in all controlled laboratory experiments discussed above, but this is also always a valid approximation in magnetospheric and solar wind plasmas.

Note also that the results obtained in the first set of Baker and Hammel's experiments can be deduced from equation (1), when the external magnetic field is uniform. Indeed, when B_z is independent of x in equation (1), it can be seen that v^x is then also a constant, as precisely demonstrated experimentally by Baker and Hammel... at least when the walls of the vacuum chamber are coated with insulating material.

3.3. APPLICATION TO THE MAGNETOSPHERE

How do Demidenko *et al.*'s results apply to the study of impulsive penetration of solar wind plasmoids into the Earth's magnetosphere?

The magnetic field distribution corresponding to Demidenko *et al.*'s experiments increases from a small value to higher value without changing direction, just like the B-field at the front side magnetopause when the IMF is due Northward (NB_z IMF), or, along the surface of the magnetotail lobes when the IMF is Southward.

The magnetosheath corresponds to the region where the heated (compressed) solar wind plasma tends to pile up against the geomagnetic barrier, and where eventually it slips aside along the flanks of the 3-D magnetospheric cavity. The point of deepest penetration of an injected plasma element depends of its initial energy W; the value of the magnetic field intensity where its forward bulk velocity eventually vanishes is given by equation (4).

From the r^{-3} radial distribution of the geomagnetic field distribution it can be determined that the average solar wind plasmoids with an average momentum flux density should be stopped or deflected at about 10 Earth's radii in the sunward direction : i.E. at the average magnetopause position.

But a very large momentum flux density is required in the solar wind to obtain from equation (4), a value for B_1 which is equal to the geomagnetic field intensity at geostationary orbit. Therefore, it is only on very rare occasions that solar wind plasma elements will be able to penetrate as deep as 6.6 Earth's radii.

The plasmoids with the largest momentum flux density will penetrate deepest in the geomagnetic field, where they produce localized diamagnetic field perturbations, magnetosheath like plasma density enhancements, as well as localized electric field perturbations; these small scale and time-dependant electric fields perturbations produce convection in the ambient magnetospheric plasma along the flanks of the penetrating plasmoids as illustrated in fig.6. The external magnetospheric plasma is then pushed aside.

It has already been mentioned above that these small scale magnetospheric convection patterns driven by impulsively injected plasmoids map down in the dayside cusp ionosphere, where they have been observed as localized poleward moving ionospheric irregularities or auroral features (Goertz, et al., 1985, Sandholt, et al. 1985, Glassmeier, et al.,1988, Friis-Christensen, et al., 1988, Lanzerotti, 1989, Heikkila, et al. 1988).

Any plasma density irregularity which impacts on the Bow Shock with a momentum flux density smaller than the average solar wind value is not able to reach the average magnetopause position where 'average' plasmoids are stopped and deflected.

3.4. ADDITIONAL COMMENTS

Of course the IMF is rarely due Northward or Southward, generally it is at an angle between 0 and 180 degrees with respect to the direction of the geomagnetic field. In this case the eqs.(1) to (7) can be generalized and extended (see Lemaire, 1985). It would be most interesting to modify the magnetic field distribution from Demidenko et al.'s experiment to simulate this more general situation when the IMF is not aligned with the Northward geomagnetic field, i.e. when the magnetic field is sheared as usually near the magnetopause.

4. Discussion and conclusions.

There are several other plasma laboratory experiments whose results are published in technical Journals, and, which are relevant to our understanding of Solar Wind Magnetosphere Interaction (see for instance Fälthammar ,1988).

Let us mention for instance those results obtained by Bostick

(1956) with two interacting plasmoids moving in opposite directions against each others. These experiments may illustrate the dipole-dipole magnetic interaction between two plasmoids; indeed the current systems carried by plasmoids have dipole moments which can have a different orientation from the neighbouring one; consequently, as a result of the dipole-dipole interaction both plasmoids can be either attracted or repelled.

Bostick's type of experiments may be of great importance for those who are interested in the dipole-dipole interaction between interplanetary plasmoids and magnetospheres. Indeed, the magnetosphere is a huge plasmoid with a large dipole moment; it has been suggested by Lemaire (1985, 1987) that the magnetic dipole-dipole interaction between the Earth's dipole moment and that of the penetrating plasmoid (whose orientation is dependent of the ambient IMF direction) controls, to a certain extend, impulsive penetration of solar wind plasma elements into the magnetosphere (see also, Lemaire *et al.*, 1987).

Other experiments like that of Lindberg (1978) should also be mentioned as most relevant to the study of the solar wind magnetosphere interaction. Indeed, it may simulate solar wind plasma injected along the curved magnetic field lines of the dayside clefts.

These controlled laboratory plasma experiments could be reinstated and extended at relatively low cost. New types of experiments could also be imagined to enhance our current incomplete understanding of the interaction between the solar wind and the magnetosphere, e.g. the phenomenon of filamentation or breaking-off of plasmoids into separated entities when they penetrate into a non-uniform magnetic field like the geomagnetic field.

Moreover, experimental results always remain the test-bench for any theory even the most appealing and popular ones : the most careful referee will never replace a carefully conducted experimental test...

New breeds of controlled laboratory experiments could be most valuable to improve our current understanding of 'non-anomalous' interaction mechanisms between the **non-uniform and non**-steady state solar wind plasma and the Earth's Magnetosphere.

LIST OF REFERENCES

Axford W.I. and Hines C.O., Can. J. Phys., 39, 1433, 1961.

Baker, D.A. and Hammel, J.E., Phys. Fluids, 8, 713,1965.

Baker, D.A. and Hammel, J.E., Phys. Rev. letters, 8, 157-158, 1962.

Bostick, W.H., Phys. Rev., 10, 292-299, 1956.

Demidenko, I.I., Lomino, N.S., Padalka, V.G., Rutkevitch, B.N.,, and
        Sinel'niko, K.D., Soviet Physics - Technical Physics, 14, 16-
        22, 1969.

Demidenko, I.I., Lomino, N.S., Padalka, V.G., Soviet Physics -
        Technical Physics, 16, 1096-1101 , 1972.

Demidenko, I.I., Lomino, N.S., Padalka, V.G., Safranov, B.G., and
        Sinel'niko, K.D., J. Nucl. Energy, C, Plasma physics, 8, 433,
        1966.

Demidenko, I.L., Lomino,N;S., Padalka, V.G., Rutkevick, B.N. &
        Sinel'nikov, K.D., Soviet Phys. Techn. Phys., 11, 1354, 1967.

Demidenko, I.L., Lomino,N;S., Padalka, V.G., Rutkevick, B.N. &
        Sinel'nikov, K.D., Soviet Phys. Techn. Phys., 14, 16, 1969.

Dolique, J.M., Compt. Rendus, 256, 4170, 1963.

Dungey J.W., Phys. Rev. Lett., 6, 47, 1961.

Fälthammar, C.-G., Laser and Particle Beams, 3-3, 437-452, 1988.

Friis-Christensen, E., McHenry, M.A., Clauer, C.R. and Vennerstrom,
        S., Geophys. Res. Lett., 15, 253, 1988.

Glassmeier, K.-H., Hoenisch, M., and United, J.,
        J. Geophys. Res., 1988

Goertz, C.K., Nielsen, E., Korth, A., Glassmeier, K.H., Haldoupis,
        C., Hoeg, P., and Hayward, D., J. Geophys. Res., 87, 2147-
        2158, 1985.

Harris, E.G., Theus, R.B., and Bostick, W.N., Phys. Rev., 105, 46,
        1957.

Heikkila W.J., Geophys. Res. Lett. 9, 877, 1982.

Heikkila W.J., Planet. Space Sci., 26, 121, 1978.

Heikkila W.J., Jorgensen, T.S. and Lanzerotti, L.,
        A transient auroral event on the dayside, preprint, 1988.

Lanzerotti, L.J., Advances in Space Research, in press 1989.

Lemaire J. and Roth M., J. Atmospheric and Terr. Phys., 40 331-335,
        1978.

Lemaire J., Planet. Space Sci., 25, 887-890, 1977.

Lemaire, J., Interpretation of the Northward Bz (NBZ) Birkeland
        current system and polar cap convection patterns in terms of
        the impulsive penetration model, in Magnetotail Physics,
        Ed. A.T.Y. Lui, Johns Hopkins University Press, pp 83-90,
        1987.

Lemaire, J., J. of Plasma Physics, 33, part 3, 425-436, 1985.

Lemaire,J., Rycroft, M.J., and Roth, M., Planet. Space Sci., 27, 47-
        57, 1979.

Lindberg, L., Astrophys. and Space Sci., 55, 203-225, 1978.

Linhart, J.G., 'Plasma Physics', North Holland Publ. Co., Amsterdam,
        p 157-160, 1960.

Lundin, R. and Dubinin, E., Planet. Space Sci., 32, 745-755, 1984.

Lundin, R., Stasiewicz, K., and Hultqvist, B., J. Geophys. Res., 92,
        3214-3222, 1987.

Russell, C.T. and Elphic R.C., Space ScI. Rev., 22, 681, 1978.

Sandholt, P.E., Egeland, A., Lybekk, B;, and Deehr, C.S., Signature
        of flux transfer events in the dayside aurora?
        preprint, University of Oslo, 1985.

Schindler, K., J. Geophys. Res., 84, 7257-7263, 1979.

Schmidt, G., Phys. Fluids, 3, 961, 1960.

Sckopke, N., Paschman, G., Haerendel, G., Sonnerup, B.U.O., Bame,
        S.J., Forbes, T.G., Hones, E.W.,Jr., and Russell, C.T., J.
        Geophys. Res., 86, 2099, 1981.

Titheridge, J.E., J. Geophys. Research, 81, 3221, 1976.

40

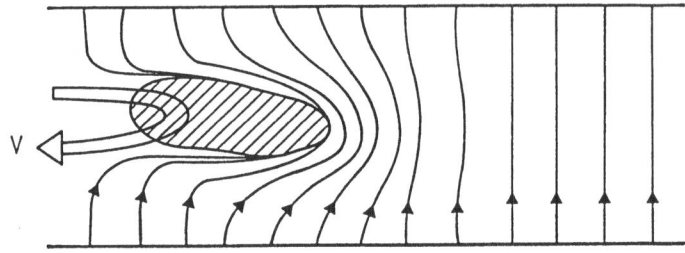

Fig. 1   Cartoon commonly drawn to illustrate magnetic field line tying, and reflection of a collisionless plasma element penetrating into a magnetic field.

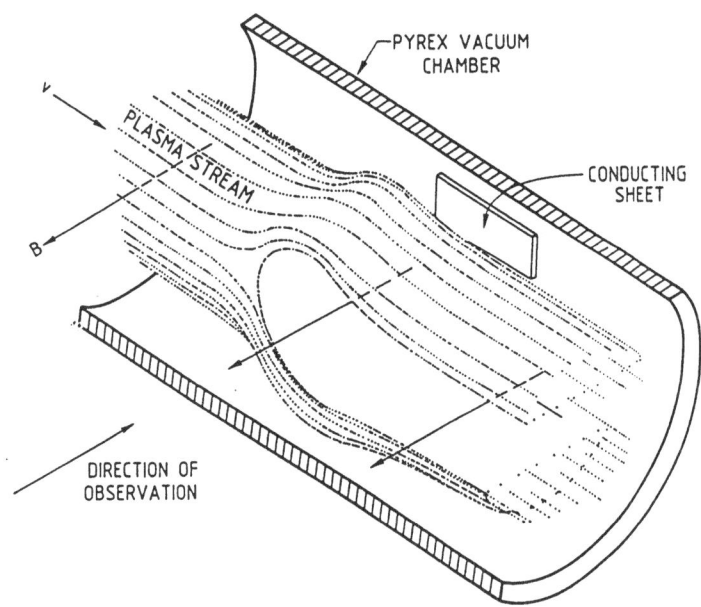

Fig. 2   Schematic diagram of a plasma flow around the cylindrical region of zero electric field created by a conducting strip located adjacent to the inside wall of the vacuum chamber. (after Baker and Hammel, 1962)

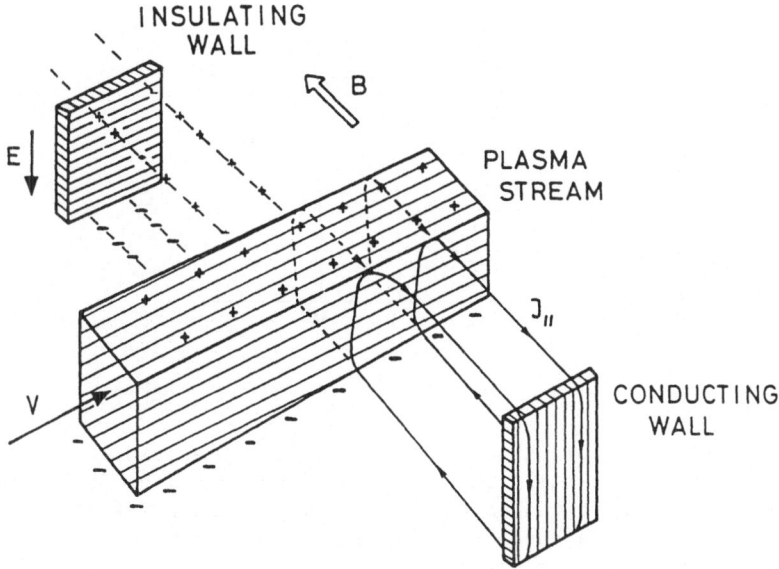

Fig. 3   Simplified model of a collisionless plasma flow
crossing a uniform magnetic field showing (i) the charging up effect
of an insulating wall and (ii) depolarizing effect of a conducting
wall. (after Baker and Hammel, 1965)

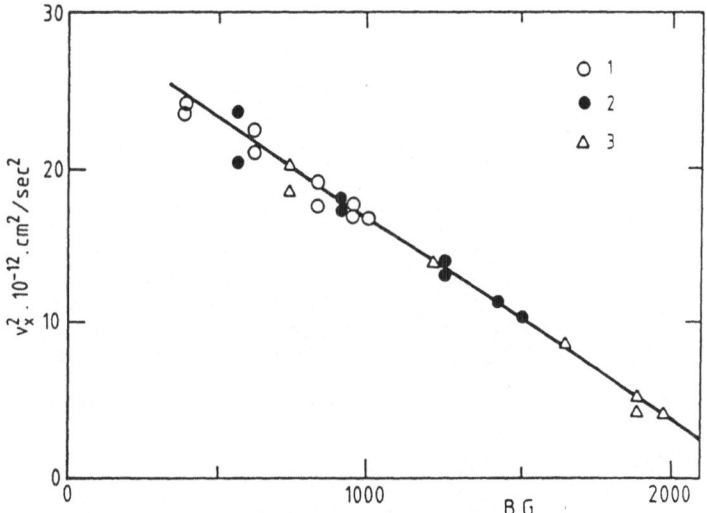

Fig. 4   Adiabatic deceleration of a plasmoid in a region of
increasing magnetic field intensity. Distribution as a function of x,
the distance along the injection axis of a plasmoid in a non-uniform
magnetic field whose intensity is parallel to the Oz-axis and whose
intensity is given by the curve labelled B(x).The bulk velocity, v_x,
and the width in the Oy direction, h(x), are also given. (after
Demidenko, *et al.*, 1972)

42

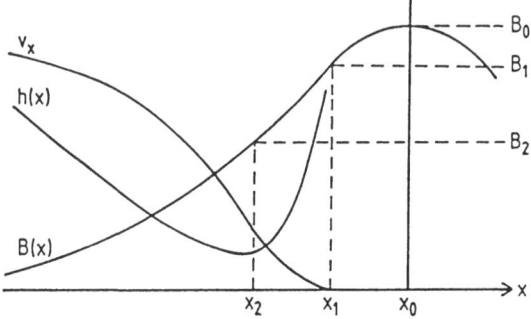

Fig. 5   Adiabatic deceleration of a plasmoid in a region of
increasing magnetic field intensity. The square of the bulk velocity
of the plasmoid is given as a function of the intensity of the
magnetic field.   The experimental points correspond to three different
magnetic field profiles for a peak magnetic field intensity of (1) B_o
= 1000 G ; (2) 1500 G ; (3) 2000 G. (after Demidenko, *et al*., 1969)

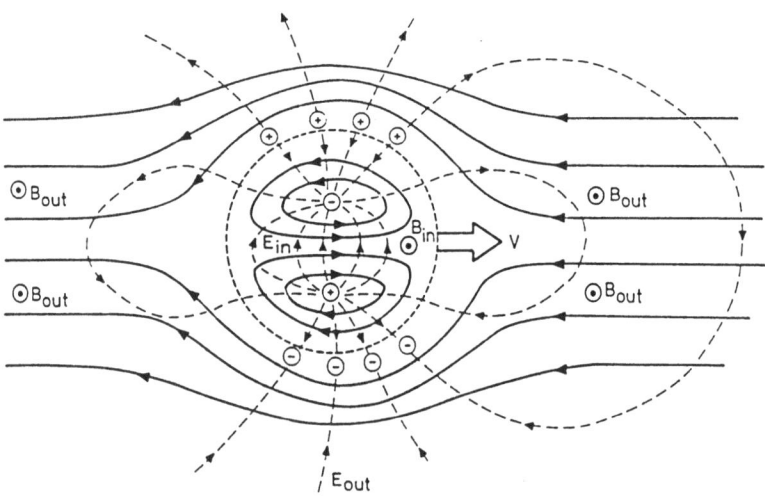

Fig. 6   Cartoon illustrating possible flow of external
plasma around a penetrating plasmoid. Possible plasma flow pattern
inside the plasmoid is also suggested.

POLAR RAIN AND THE QUESTION OF DIRECT PARTICLE ACCESS

M.S. Gussenhoven
Air Force Geophysics Laboratory
Hanscom AFB, MA 01731

ABSTRACT. Recent experimental findings on the weak, low-energy
electron population found in the tail lobes and the low altitude polar
regions to which they magnetically map are reviewed in light of two
different polar rain models. In one model the field aligned solar wind
electrons have direct entry to the magnetosphere in the distant tail
where the field lines are assumed to be open. In the other, the polar
rain is the potential-barrier-reflected electron component of the
magnetosheath population that enters the magnetosphere at the cusp, and
perhaps all along the magnetopause. The data do not fully support one
model over the other, and it is possible that both entry mechanisms
contribute depending on magnetopause position, on electron energy, and
on the size and extent of boundary and barrier potentials.

1. INTRODUCTION

Winningham and Heikkila (1974) first identified and named polar rain in
low altitude observations as a near-background, structureless, low-
energy electron population that precipitates over the 'unperturbed'
polar caps. Polar rain is of interest to the community of magnetospheric researchers because it is a distinct magnetospheric population
with entry, transport and precipitation mechanisms that require explanation, because it carries information about its source or sources, and
because it carries information about the distant tail magnetic field
configuration and the interplanetary magnetic field. Two models have
been proposed for the entry and transport of polar rain. In each the
solar wind is the ultimate source.

   In the 'direct entry' model, illustrated in Figure 1, the solar
wind electron population is assumed to enter the magnetosphere along
open field lines (field lines with one foot terminating in the Earth
and the other in the solar wind). It is then adiabatically transported
directly along these field lines to the low altitude polar caps. There
is no significant heating of the solar wind plasma in the vicinity of
the Earth as, say, in the magnetosheath. Electrons streaming away from
the sun will enter the northern (southern) polar cap when the IMF is
directed away from (toward) the sun while particles backstreaming

P. E. Sandholt and A. Egeland (eds.), Electromagnetic Coupling in the Polar Clefts and Caps, 43–60.
© *1989 by Kluwer Academic Publishers.*

toward the sun will enter the southern (northern) polar cap. An
analogy for this model is relativistic solar electron entry and
transport (Paulikas, 1974). Fairfield and Scudder (1985) quantitatively extended the model to the solar wind and polar rain energy
range.

The solar wind electron population that gains entry in this model
has two components, an isotropic core with average temperature and
density, 10 ± 2 eV and 10 ± 5 cm^{-3}, respectively, and a hotter halo population of 60 ± 9 eV and $.57\pm.23$ cm^{-3} (Feldman et al., 1975). The pitch
angle distribution of the halo population varies from isotropic to
highly peaked along the field line in the direction away from the sun
(Pilipp et al., 1987a,b). When the halo population is highly anisotropic it is referred to as the strahl because it is thought to proceed
virtually unscattered from the sun to the near-Earth region (Olbert,
1981). A cut in the measured solar wind electron distribution function
along the magnetic field, taken from Pilipp et al. (1987a), is shown in
Figure 2 for the case of a highly anisotropic strahl. In the energy
range 100-400 eV [$(6-12)10^5$ km/s] the ratio of fluxes in the direction
away from to that toward the sun is more than two orders of magnitude.
For direct entry this anisotropy will be carried unchanged into the
magnetosphere. In Figure 1 the tear-drop shapes represent a highly
anisotropic solar wind distribution function such that quite different
fluxes are incident upon the two hemispheres. The portions of the
distributions that are blocked by the magnetosphere have dashed lines.
Since the magnetic field increases by several orders of magnitude from
the magnetopause to the ionosphere, only particles in a loss cone of
$\sim1^\circ$ at the magnetopause will reach low altitudes before mirroring and
returning to the distant tail or plasma sheet. Thus in the direct
entry model, polar rain is only associated with the most field-aligned
population in the solar wind (Fairfield and Scudder, 1985; Gosling et
al., 1986, Baker et al., 1986).

The direct entry model predicts that for conditions of highly
anisotropic strahl the polar rain flux in the hemisphere that receives
electrons directed away from the sun (referred to throughout the
remainder of the paper as the 'preferred hemisphere') should have
stronger field alignment and greater intensity than in the opposite
hemisphere (the 'non-preferred' hemisphere). For conditions of
isotropic solar wind the fluxes in the two caps should be identical.
Pilipp et al. (1987b) found that a narrowly peaked strahl occurs in the
center of a sector coincident with the sector's high speed stream, that
a broadly peaked strahl occurs at the trailing edge of the high speed
stream and that the isotropic distribution occurs at a sector boundary.

The second model, called here the 'internal barrier' model and
illustrated in Figure 3, was put forth by Foster and Burrows (1976,
1977) to explain infrequent, intense fluxes of keV electrons that have
the same spatial smoothness as polar rain and no apparent counterpart
in the solar wind. In their model the source of 'normal polar rain' is
the magnetosheath. Magnetosheath electrons enter the magnetosphere
with ions in the region of the cusp. Most of the inflowing cusp
population mirrors and flows away from the Earth but remains within the

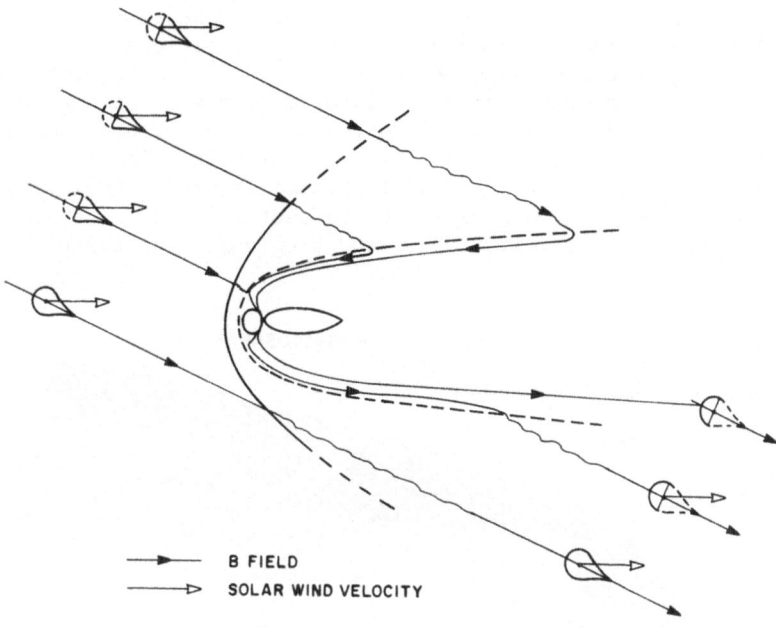

Figure 1.   Schematic diagram of the direct entry model for polar rain.

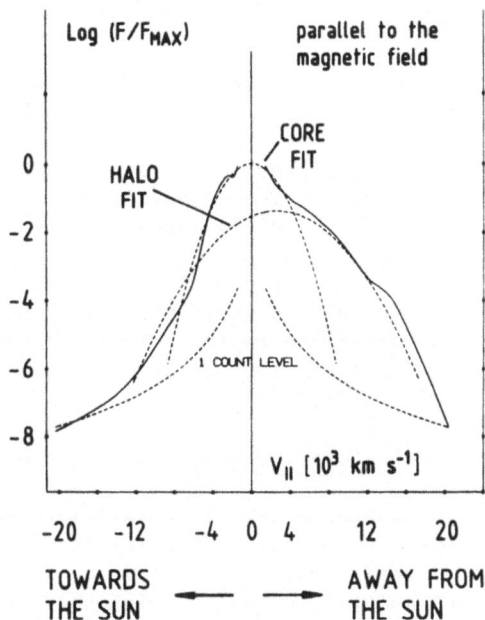

Figure 2.   Highly anisotropic solar wind distribution function.

magnetosphere due to the cross-cap convection electric field (Pilipp and Morfill, 1978). The outward flowing magnetosheath population is called the plasma mantle because it coats the tail lobe region just inside the magnetopause (Rosenbauer et al., 1975). This mantle population has been measured out to lunar distances. (See review by Sckopke and Paschmann, 1978). The extent to which it has expanded toward the plasma sheet at lunar distances is controlled by the IMF sector (Hardy et al., 1979a). Mantle plasma is preferentially seen at lunar distances in the morning (evening) sector in the northern (southern) hemisphere during away sectors and vice versa for toward sectors.

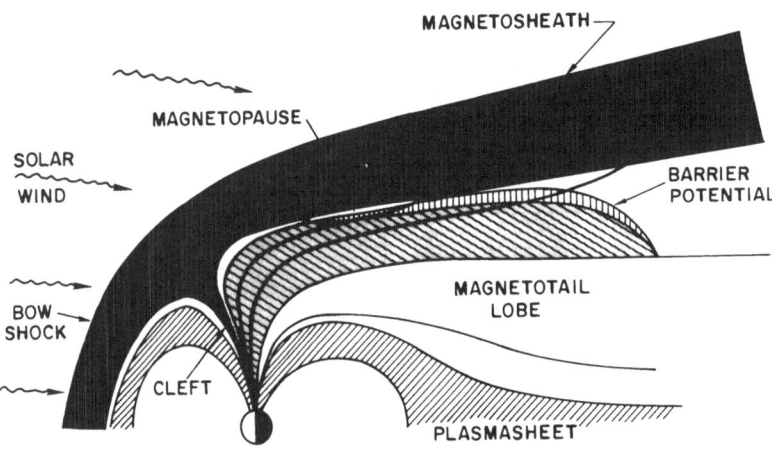

Figure 3. Schematic diagram of the internal barrier model of polar rain transport in the tail lobes.

In the Foster and Burrows picture an electrostatic barrier reflects a portion of the mantle electron population as polar rain. Occasionally the barrier is sufficiently large and widespread to accelerate distant tail lobe populations toward the cap to create intense, keV fluxes. The source for the electrostatic barrier is not identified. In the internal barrier model the polar rain distribution function is not directly comparable to the solar wind electron spectrum, but is a magnetosheath-like distribution modified by transport through a spatially-varying, field-aligned potential. Comparison of distribution functions in the near and distant tail lobes should reflect the existence of the total potential drop between the two positions.

In the following the characteristics of tail lobe electrons are reviewed in light of these two models. Measurements are organized by altitude starting with low altitudes, below several RE (RE = Earth radius), where the great majority of polar rain observations have been made; continuing through mid-range altitudes (5-60 RE); and ending with distant (>180 RE) tail lobe populations. The discussion will be confined, unless otherwise stated, to electrons in the energy range between approximately 50 eV and 1 keV. This is also the typical energy range for solar wind halo (strahl) electrons. Altitude comparisons of densities and temperatures will be made for Maxwellian fits to electron spectra in this energy range.

2. POLAR RAIN CHARACTERISTICS AT LOW ALTITUDES

Polar rain is identified at low altitudes by its position above the auroral oval, and its large-scale spatial/temporal homogeneity. Three types of electron precipitation are typically found in the polar caps: polar rain, polar showers and polar cap arcs. These and their ion counterparts are illustrated in Figure 4. Here the electron number flux (left hand side) and ion number flux (right hand side) from 100-1000 eV are shown for three pre-noon to pre-midnight DMSP passes over the auroral regions and polar caps at 840 km on three days in April, 1985. The hourly averaged values of the three interplanetary magnetic field (IMF) components and KP are listed to the right. In the top electron panel the region of slowly varying electron precipitation in the polar cap (above ∿75° MLAT) is polar rain. In the middle panel the electron precipitation in the highest latitude regions is designated polar showers and is characterized by high spatial (or temporal) variability. In the bottom panel the large blocks of intense electron precipitation extending to very high latitudes from the midnight region produce polar cap arcs. Polar rain most commonly occurs under conditions producing an active auroral oval, or when the IMF has a southward component; polar showers and polar cap arcs occur preferentially when the IMF has a northward component (Gussenhoven, 1982, Hardy, 1984).

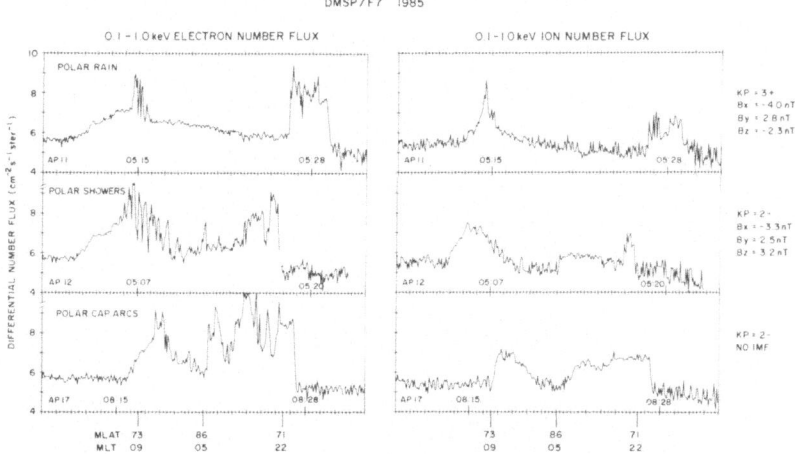

Figure 4. Precipitating electrons at low altitudes for three states of the polar cap.

48

The right hand panels in Figure 4 show the corresponding ion precipitation. In polar rain the ion precipitation is near background except
behind the cleft (or dayside boundary layer). For conditions of polar
showers and polar cap arcs significant ion precipitation is observed to
very high latitudes, particularly on the nightside.

The low-altitude spectral characteristics of polar rain were
quantified by Riehl and Hardy (1986) using 262 polar passes of DMSP
data. Figure 5 shows the distribution function of a typical polar rain
spectrum averaged over 45 seconds and fit to two Maxwellians in the
energy ranges below and above approximately 1 keV. The great majority
(70%) of the polar rain spectra examined by Riehl and Hardy did not
have a high-energy component. For their data sample the distribution
of polar rain temperatures fell mainly in the energy range 60-100 eV.
The average value was found to be 80±13 eV which is 40% higher than the
average solar wind halo population. The average density of the low
energy polar rain component was found to be .055±.038 cm^{-3}, an order of
magnitude smaller than the halo population. This comparison of average
values strongly suggests that the entry and transport of solar wind
electrons into and within the magnetosphere is not direct, but that
heating and redistribution is part of the process. The pitch angle
distributions of polar rain electrons at low altitude have not been
studied systematically. Early observations suggest that it is isotropic over the down-coming hemisphere (Winningham and Heikkila, 1974).

Low altitude studies comparing the northern and southern polar
rain flux levels on a daily basis show that there is almost always a
higher flux in the preferred hemisphere for direct entry of fluxes
streaming away from the sun: northern (southern) hemisphere for IMF
away (toward) sectors (Fennell et al., 1975, Meng and Kroehl, 1977,

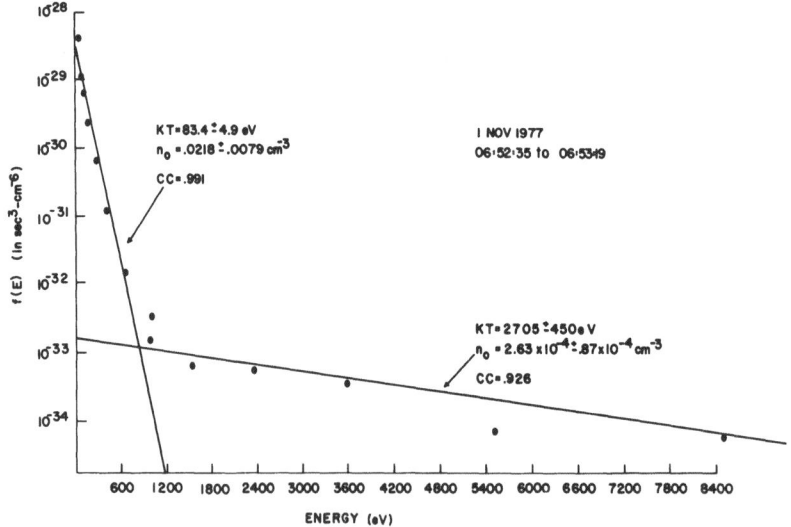

Figure 5. Typical two-Maxwellian polar rain distribution function at
low altitude (840 km).

Makita and Meng, 1987). Figure 6 was prepared using solar wind data
(King, 1979) and DMSP particle data. It shows the variation in the
solar wind velocity, ion density, ion temperature, and field magnitude
(top four panels) and a daily value for the north and south polar rain
energy flux and number flux (bottom two panels) for three full sectors
(marked T and A and separated by vertical dashed lines) following the
vernal equinox in 1979. The polar rain fluxes are taken within 5° of
either magnetic pole. The preferred hemisphere clearly alternates from
one pole to the other with sector change. This asymmetry is one of the
strongest arguments for the direct entry model.

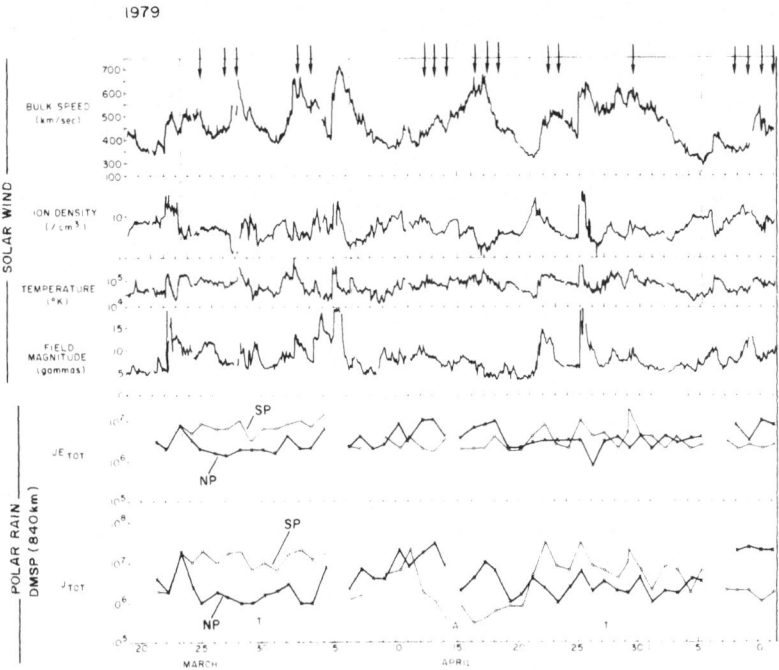

Figure 6. Variation of solar wind parameters and low altitude polar
rain energy flux, JE_{TOT} in keV/(cm^2 s sr), and number flux, J_{TOT} in
(cm_2 s sr)$^{-1}$, from 20 March to 11 May, 1979.

The two-dimensional morphology of polar rain in the preferred and
non-preferred cap was statistically determined using DMSP electron data
at 840 km by Gussenhoven et al. (1984). Maps were constructed in
magnetic latitude and local time of the average polar rain characteristics using one year of data and separating the data by KP and by IMF
sector. This study confirmed that the hemispheric, sector-dependent
asymmetry in flux levels is also found in the average sense. When
taken across the entire polar cap the total number flux input to the
preferred hemisphere was approximately 2 times that in the non-preferred hemisphere.

Gussenhoven et al. also found that within a given cap there is a large statistical variation from high to low flux along a prenoon to premidnight axis of symmetry. This variation (also evident in the top electron panel of Figure 4) is, on average, larger than the hemispheric asymmetry. Table 1 illustrates both the average day-night variation within a given cap and the hemispheric asymmetry. To construct Table 1 a Maxwellian fit was made to the average spectra shown in Gussenhoven et al. (1984) at two positions along the noon-midnight axis for the preferred and unpreferred caps. The left-hand comparison is for the dayside at 79° MLAT directly behind the cusp. The right-hand comparison is on the nightside at 75° MLAT above the midnight oval. On the dayside the average difference in density between hemispheres is a factor of 3, on the nightside it is a factor of 7. Furthermore, the day-night difference in density in the preferred cap is much less (a factor of 2.4) than that in the unpreferred cap (a factor of 6).

TABLE 1. Density (n) and temperature (kT) of polar rain along the noon-midnight axis, KP = 3-5+

	79° (noon)		75° (midnight)	
	$n(cm^{-3})$	kT(eV)	$n(cm^{-3})$	kT(eV)
Preferred	.22	79	.086	90
Unpreferred	.073	55	.012	81

Gussenhoven et al. (1984) also identified a pronounced seasonal variation in polar rain flux. Not only is the peak flux higher in the summer than winter, but the region of high number flux extends much further across the caps in summer.

The direct entry mechanism does not easily provide a mechanism for cross-cap variation and seasonal dependence of polar rain flux. On the other hand, the internal barrier model is expected to produce a field-aligned potential drop having strong latitude and altitude dependencies.

Although a clear modulation of polar rain intensity has been shown to occur in a given hemisphere and between hemispheres with IMF sector, there has been little success in relating polar rain characteristics to other solar wind parameters (Riehl and Hardy, 1986; Makita and Meng, (1987). Figure 6 also demonstrates that no obvious relationship exists between the polar rain flux and solar wind parameters. Pilipp et al. (1987b) have shown that narrowly peaked strahl occurs in high speed solar wind streams which tend to occur in the middle of well-formed sectors. Such streams are often characterized by low solar wind density (see also Fairfield and Scudder, 1985). When the strahl is narrowly peaked the hemispheric asymmetry in polar rain flux should be greatest. Arrows at the top of Figure 6 mark instances where the hemispheric difference in polar rain number flux is an order of magnitude or greater. These occur frequently, but not exclusively in high speed streams and periods of low density. This lack of a one-to-one correlation supports entry and transport mechanisms for polar rain in which some modification of the solar wind plasma occurs.

3. POLAR RAIN AND THE MANTLE POPULATION FROM 5-60 RE.

The electron populations in the lobes, which magnetically map to the polar caps, are far less systematically studied than the polar cap populations at low altitudes. Two particle populations are found at mid-altitudes in the lobe. One is the boundary population, including the plasma mantle. It is characterized by tailward streaming ions of moderately high density ($\sim.5$ cm^{-3}). The other is the plasma that is measured when the boundary plasma is not present. In this plasma ions above ~50 eV are not detected. Weak electron fluxes >50 eV are measured, and it is these fluxes that are associated with low altitude polar rain.

One of the most interesting features of the 'polar rain' measurements beyond 5 RE is that while plasma mantle populations are explicitly excluded from study, no distinction is made between polar rain, polar shower and polar cap arc electron populations. In other words, no requirement on spatial/temporal smoothness or on IMF Bz conditions are imposed to identify polar rain at mid- to high altitudes.

Shortly after the identification of polar rain in low altitude data Yeager and Frank (1976) studied the >250 eV electron population in the northern lobes from 3-7 RE using Imp 5 data. Their 'polar cap' population was distinguished from interplanetary, magnetosheath, polar cusp, and plasma mantle electron populations. Their polar cap electron spectra are similar to the low altitude polar rain spectra and range in density from .0015-.045 cm^{-3}. They are isotropic within factors of 3 to 4. The variation of electron fluxes averaged over each northern polar cap crossing for ten months in 1970 showed systematic increases in away sectors. The enhancement between low and high fluxes in the energy range 305-510 eV varied between 2 and 50. There was no strong local time dependence in the fluxes. Ions between 80 eV and 50 keV were below detectability, putting an upper limit on density of .05 cm^{-3} in this energy range.

More recently Fairfield and Scudder (1985), using ISEE 1 data, examined the electron population in the northern tail lobe between 10-23 RE. They excluded from their study magnetopause boundary layer and plasma mantle populations. At times they found extended periods for which the fluxes of electrons above ~100 eV were significantly greater both parallel and antiparallel to the magnetic field than perpendicular to the field indicating bidirectional streaming. An example of their data is shown in Figure 7 (taken from Greenspan et al., 1986). The ISEE 1 detector response during an intense polar rain event is given for various energy channels as a function of pitch angle. Dashed lines indicate background levels. In 399 hours of such data the anisotropy, defined as the ratio of the parallel to the perpendicular flux, was

52

found to be more pronounced during away sectors. In the 180 eV channel
it was greater than 1.1 more than 90% of the time. For toward sectors
it was greater than 1.1 only 13 % of the time. At this energy the
ratio of the average flux in away sectors to that in toward sectors was
2.4. The authors concluded that the electron characteristics in the
preferred hemisphere, namely, the bidirectional anisotropy and the
enhanced flux, were clear signatures of the direct entry of an anisotropic strahl population in the solar wind and its adiabatic transport to low altitudes.

It should be noted that most of the cases reported by Fairfield and Scudder correspond to unusually intense polar rain events. These events are more readily compared to the events discussed by Foster and Burrows (1976, 1977) than to normally occurring polar rain. At low altitudes the keV electrons in intense events were also found to have a strong field alignment. Foster and Burrows (1977) successfully postulated a fluctuating magnetotail barrier potential to explain the measured pitch angle anisotropy. Recall that in the Foster and Burrows event there was no evidence of enhanced keV electron fluxes in simultaneous solar wind measurements.

In order to examine whether or not a potential drop exists from ISEE 1 altitudes to low altitudes, Greenspan et al. (1986) compared northern hemisphere electron spectra in intense polar rain events at ISEE 1 to those at DMSP. An example which typifies the DMSP-ISEE 1 comparisons, taken from Greenspan et al. is shown in Figure 8 for May 30, 1978. The angles marked are the pitch angles at which the ISEE 1 data were taken. For the most part the two spectra show very good agreement.

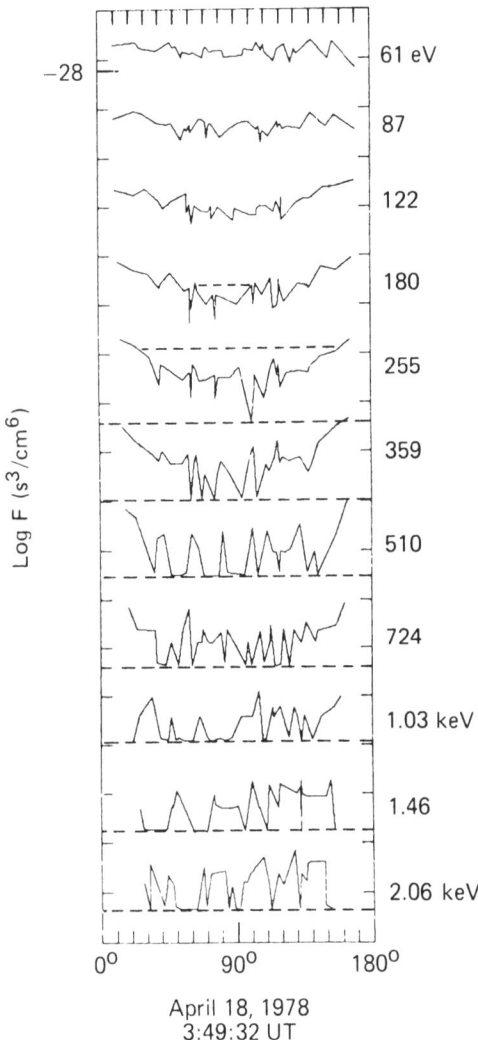

Figure 7. Polar rain pitch angle variations at ISEE 1 altitudes.

April 18, 1978
3:49:32 UT

There is no constant offset of one spectrum with respect to the other
indicative of a potential drop between the two measurements. The
tendency for high energy values of the distribution function at DMSP

(low altitude) to be higher than those at ISEE 1 (mid-altitude) was found in almost all cases. The suggested explanation for this is that the ISEE 1 detector does not adequately sample the strongly peaked loss cone population which maps adiabatically to DMSP altitudes.

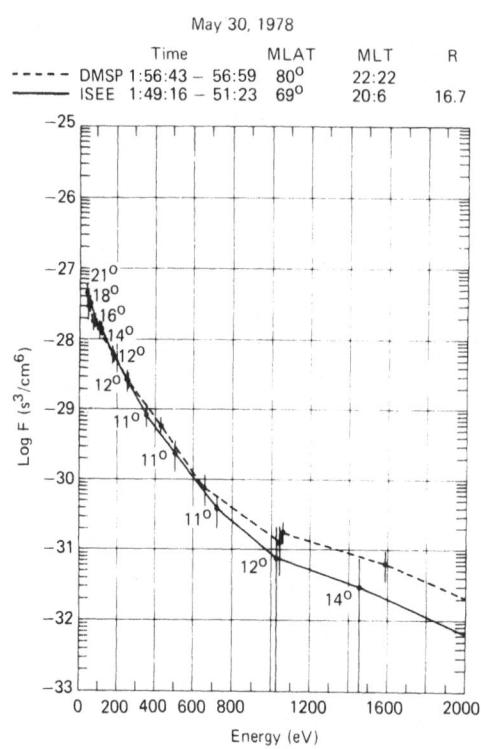

May 30, 1978

Time	MLAT	MLT	R
----- DMSP 1:56:43 – 56:59	80°	22:22	
——— ISEE 1:49:16 – 51:23	69°	20:6	16.7

Figure 8. Comparison of low and mid-altitude polar rain spectra.

Before turning to high altitude tail lobe measurements some comments should be made concerning the electrons accompanying the tailward streaming ions in the mid-altitude boundary population. In the near-cusp mantle the electrons have magnetosheath characteristics, and are somewhat cooler and several orders of magnitude denser than the polar rain measurements of Yeager and Frank. The plasma mantle electrons do not show the strong velocity filter effects of ions and maintain reasonably constant characteristics well inward from the magnetopause where the ions cool systematically and fall below the energy range of the detector (Sckopke and Paschmann, 1978). At the moon the electrons accompanying the mantle ions are also hotter and denser than those of the polar rain population (Hardy et al., 1979b). Thus there is a clear bimodal distribution of tail lobe electrons in the altitude range of 0 to 60 RE. The bimodality is also evident in the statistical study of Zwickl et al. (1984) using ISEE 3 data from transfer orbit periods. Between 0-60 RE the electron density has two strong, near-equal peaks in occurrence frequency at .01 cm^{-3} and .4 cm$^-$^{3}. The temperature is singly peaked at ∿80 eV but has a broad tail at lower values. On average the electrons are found to be weakly streaming away from the Earth.

Of major importance here is that the two tail lobe plasma populations share magnetic field lines. The denser boundary plasma fans out down-tail, occupying field lines that when mapped back to lower altitudes contain only the near-background polar rain electrons. Thus the populations at 60 RE on a given field line are not those adiabatically mapped to 1-10 RE on the same field line. They are constrained from reaching those altitudes by convection and down-tail streaming. The distance at which the electron pressure on a field line begins to substantially deviate from polar rain values is apparently greater than

10-23 RE since out to these distances the Greenspan et al. (1986) comparisons showed good agreement.

4. TAIL LOBE ELECTRONS AT HIGH ALTITUDES

The positioning of ISEE 3 in the distant tail made possible the collection of a large body of data on the plasmas and fields found there. One of the principal findings from these data is that the tail lobes and the plasma sheet are well-defined to distances exceeding 200 RE.

In comparing tail lobe electron characteristics at high altitudes and low altitudes, both statistically and on a case by case basis, it was found that the high altitude populations are much denser. The statistical study of Zwickl et al. (1984) showed that for >180 RE the frequency of occurrence of densities between .05 and 3 cm^{-3} is almost constant, and the peak at low densities (.01 cm^{-3}), associated with polar rain, is missing. The temperature profiles from low to high altitudes cool slightly from peak occurrence at 80 eV to peak occurrence at 54 eV. Baker et al. (1987) compared tail lobe electron spectra from ISEE 3 near 200 RE to DMSP polar rain spectra. A sample comparison for northern and southern hemispheres on 24 January, 1983 is

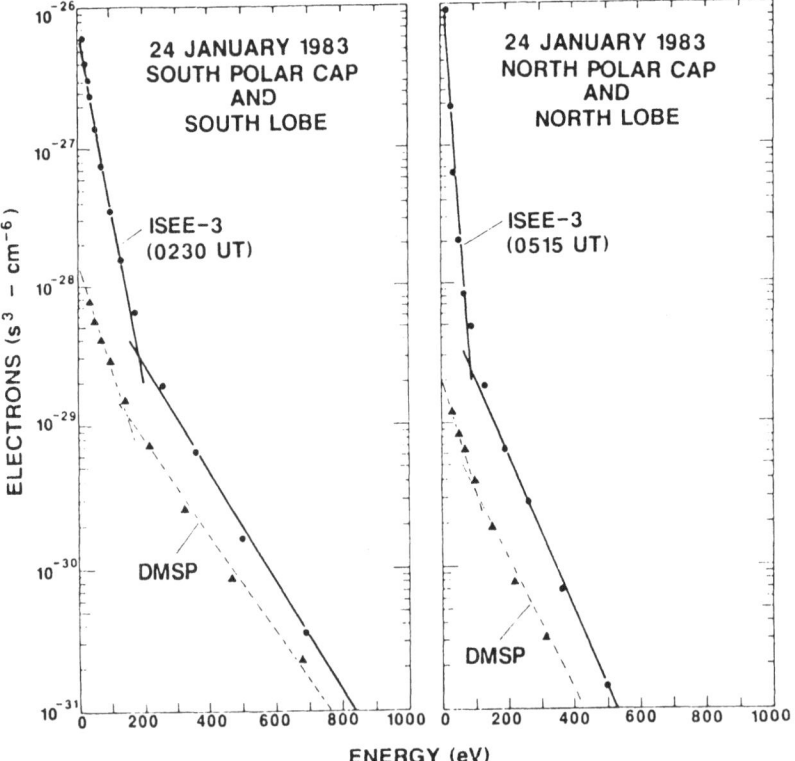

Figure 9. Comparison of low and high altitude polar rain spectra.

shown in Figure 9. The ISEE 3 distributions were measured as close as possible to field-alignment under relatively steady conditions. The DMSP data were taken near the magnetic pole and were long term averages. Because of a clear break-point in the ISEE 3 data the distributions were fit to two Maxwellians above and below 200 eV. The low energy portion of the DMSP data is quite different than that at ISEE 3, being both far less intense (by orders of magnitude) and warmer. The temperature of the high energy component is nearer that measured at ISEE 3 but again the intensity is much lower. Because the displacement of the ISEE 3 high energy distribution function value is fairly uniform by an amount 100-150 eV the comparison of the two spectra is suggestive of a potential drop between the two by that amount.

Figure 9 also shows that the electron population in the tail lobes has a greater intensity in the preferred hemisphere as does the polar rain at low altitudes. Gosling et al. (1985) studied 21 rapid crossings from the tail lobe in one hemisphere to the tail lobe in the other hemisphere and found that the morning sector density was 3-10 times higher in the northern (southern) hemisphere for IMF away (toward) sectors. The preferred hemisphere was in the opposite sense in the one evening side passage for which there were IMF data.

Baker et al. (1984) found not only that the preferred hemisphere for tail lobe fluxes was the same as for low altitude polar rain, but that the higher fluxes exhibited strong bidirectional streaming or field-aligned pitch angle anisotropies similar to that found in the mid-altitude polar rain data of Fairfield and Scudder (1985). Furthermore, examining cases where ISEE 3 passed rapidly from the tail lobe to the magnetosheath, they found that the spectrum for electrons streaming away from the sun in the magnetosheath was nearly identical to that for electrons streaming toward the Earth in the tail lobe. Figure 10 shows their example of this. On the left are two examples of electron pitch angle distributions for a variety of energy channels. The top, left panel is taken in the magnetosheath, and the first set of peaked count rates is in the field-aligned direction and away from the sun. The bottom left panel is in the tail lobe some ten minutes later, and the first set of peaked count rates is field-aligned and toward the Earth. The most field-aligned spectra are compared on the right. The two spectra are consistently separated from each other by at most 10 eV, which the authors consider the extent of the potential barrier in crossing from the magnetosphere to the magnetosheath. They feel that this comparison rather convincingly proves direct entry. One notes, however, that the tail lobe plasma flowing from the Earth (reflected plasma in the direct entry picture) is more intense than that flowing from the source toward the Earth, and that while the antisunward streaming magnetosheath plasma may be interpreted as having easy entry into the magnetosphere, the reflected, tailward streaming magnetosheath plasma does not get out.

To conclude this section we compare electron spectra taken in the magnetosheath and solar wind to those at low altitude in the polar cusp and polar cap on 9 October, 1984. The comparison is shown in Figure 11. The data outside the magnetopause were taken aboard the AMPTE/UKS

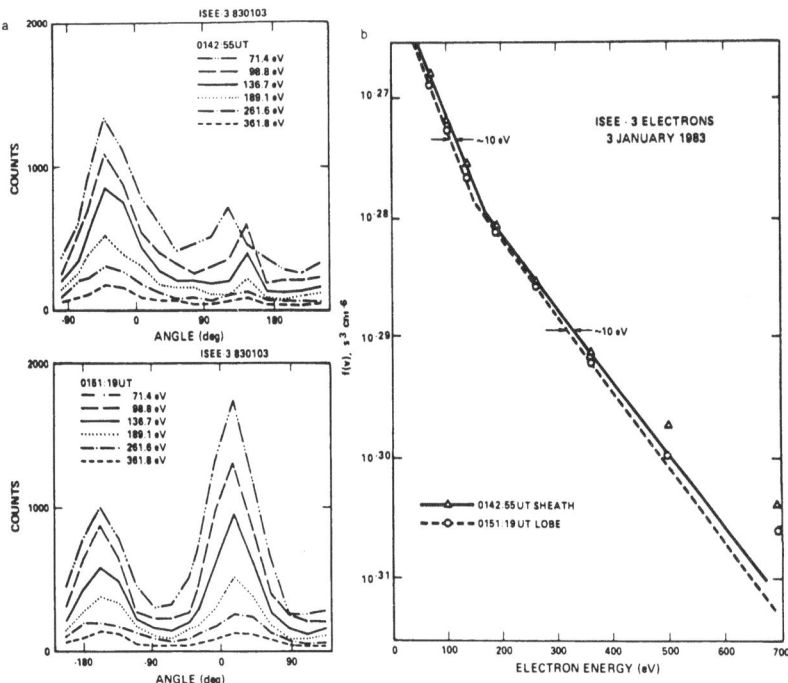

Figure 10. Magnetosheath and tail lobe pitch angle variations (left) and comparison of the two spectra at the most field aligned position.

satellite (David Hall, private communication, 1988) at low inclination on the dayside. The spectra were constructed from 20 minute averages of data. For the solar wind, average spectra parallel, antiparallel and perpendicular to the IMF are shown. A single magnetosheath spectrum is given because it is nearly isotropic. The low altitude precipitating electron data were taken aboard the DMSP satellites. The cusp spectrum is taken near local noon during the time of the AMPTE/UKS magnetosheath measurement and is averaged for 1 minute. The polar rain spectra are constructed from daily averages of data taken near the magnetic pole in either hemisphere.

The following spectral comparisons can be made: Above 50 eV the magnetosheath spectrum bears little resemblance to the solar wind spectra. It is more intense and falls off much faster from its peak value. Below 50 eV, however, the comparison is good. The cusp and magnetosheath spectra have the most favorable comparison, both being quasi-thermal with temperatures near 100 eV. The cusp spectrum is lower in intensity than the sheath spectrum below 500 eV and higher above. The solar wind and the polar rain spectra compare favorably in shape above 300 eV, but the polar rain intensity is at least an order of magnitude less in intensity. In the 100 eV range the polar rain

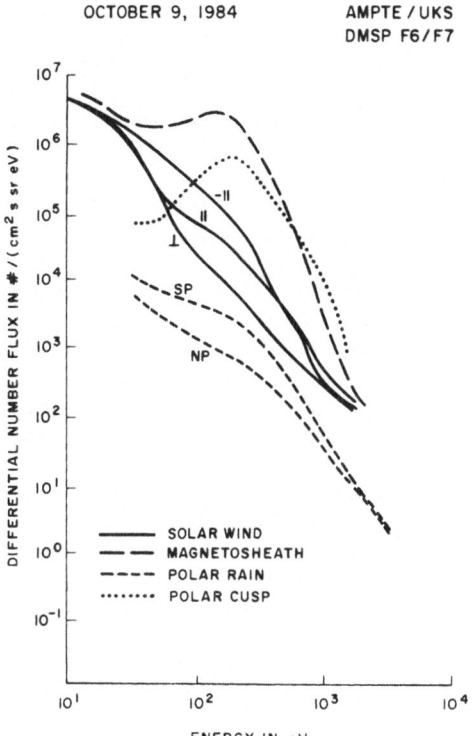

OCTOBER 9, 1984   AMPTE / UKS
                       DMSP F6 / F7

Figure 11.   Comparison of solar wind, magnetosheath, polar cusp and
polar rain spectra on 9 October, 1984.

spectra are considerably warmer than the solar wind spectra.   In sum,
the processes required to produce any one of these spectra from any
other one must be complex.   No straight-forward process, such as direct
entry or field-aligned acceleration can account for the spectral
differences.

5.   CONCLUSIONS

Within the magnetosphere the weakest plasma occurs in the most simply
configured magnetic field region, namely the tail lobes.   The apparent
simplicity of the tail lobes tempts us to ignore internal lobe processes and use the lobes as pass-through regions for information about
more complicated processes such as the solar wind-magnetosphere
interactions or those in the solar corona.   The direct entry model of
the solar wind plasma allows us to do this, and is highly attractive
because of it.   Recent measurements in the near to far tail lobes indicate that such a simple picture is premature.
        Statistical studies show that the tail lobes become increasingly
more dense with increasing distance downtail.   The extremely weak

electron populations are apparently only common at low and mid-altitudes. Even at mid-altitudes the boundary plasma in regions of strong
convection fills much of the lobes. In the distant tail it is rare to
find regions where the electron density is extremely weak. Plasma
measurements there indicate that the magnetopause is leaky and rather
efficiently lets in magnetosheath plasma. Thus a single polar cap
field line could conceivably pass through a low altitude region of weak
polar rain electrons, a comparatively dense, somewhat warmer plasma
mantle population, another rarified region beyond the plasma mantle,
and then into an increasingly denser magnetosheath type population
before exiting the magnetosphere. The pressure changes along the field
line in this scenario would most certainly be accompanied by field-
aligned potential differences.

   The preferred hemisphere dependence on IMF for low and mid-
altitude polar rain intensity and for pitch angle anisotropy is
convincing evidence for direct entry of the solar wind strahl. A
remaining concern is whether or not the hemispheric difference in the
polar rain fits the characteristics of the backstreaming solar wind
electron population. On average the difference in intensity between
polar rain in the preferred and non-preferred hemispheres is considerably less than that found in the anisotropic strahl measurements. It
also remains to be shown that hemispheric polar rain differences match
solar wind strahl anisotropies on a case by case comparison. Occurrence conditions for highly anisotropic strahl have not been shown to
be the same as those for either intense polar rain events, or for polar
rain events with large hemispheric differences.

   Tail lobe plasma electrons beyond some undetermined altitude
apparently cease to change character as a function of IMF Bz. At lower
altitudes one can nearly predict the sign of Bz by the change from
polar rain to polar showers or polar cap arcs. Furthermore, strong
local time and seasonal dependencies seem to be much more pronounced at
low altitudes than at high altitudes. One should be cautious here
because of the difficulty in obtaining sufficient data to make these
comparisons in the distant tail. Nevertheless, these dependencies are
far more suggestive of a gating mechanism operating in the lobes that
allows more or less plasma through from high altitudes to low altitudes. Such a mechanism could be one or more field-aligned potential
drops, each driven separately. Thus there are indications of the
existence of field-aligned potential differences at high latitudes, but
as yet no clear method for producing them has gained acceptance.

   On the one hand, we have a simple and elegant model for polar
rain, direct entry, which cannot explain many of the polar rain
characteristics in detail. On the other, we have the internal barrier
model which requires quantification in order to gain any foothold. In
reality a combination of both models may be required to fully explain
tail lobe dynamics. Progress can be made by considering the following
questions: What are the characteristics of the ion lobe population as
a function of altitude? Are the ion and electron populations as we
understand them hydromagnetically self-consistent? How important is
the convection electric field in the distant tail and what is the

source of the IMF sector dependence of the lobe plasma density there?
If there are field-aligned potential differences, how do they couple to
the ionosphere? And if there is direct entry of the strahl, why does
it, unlike the rest of the solar wind plasma, escape perturbation by
the bow shock? It is incumbent upon us, therefore, to be dissatisfied
with excessively simple pictures that lead us away from the opportunity
to understand a complex situation, even if the complex situation is one
of the easier ones presented to us in magnetospheric physics.

ACKNOWLEDGEMENTS

I wish to express my appreciation to David S. Hall of the Rutherford
Appleton Laboratory for providing the AMPTE/UKS data and to Gary Mullen
and David Hardy at the Air Force Geophysics Laboratory for continuing
discussions on polar cap dynamics and insightful comments on the
manuscript.

REFERENCES

Baker, D.N., S.J. Bame, W.C. Feldman, J.T. Gosling, R.D. Zwickl, J.A.
    Slavin, and E.J. Smith, J. Geophys. Res., 91, 5637, 1986.
Baker, D.N., S.J. Bame, J.T. Gosling, and M.S. Gussenhoven, J. Geophys.
    Res., 92, 13,547, 1987.
Fairfield, D.H., and J.D. Scudder, J. Geophys. Res., 90, 4055, 1985.
Feldman, W.C., J.R. Asbridge, S.J. Bame, M.D. Montgomery and S.P. Gary,
    J. Geophys. Res., 80, 4181, 1975.
Fennell, J.F., P.F. Mizera, and D.R. Croley, Proc. Int. Conf. Cosmic
    Rays 14th, 4, 1267, 1975.
Foster, J. C., and J. R. Burrows, J. Geophys. Res., 81, 6016, 1976.
Foster, J. C., and J. R. Burrows, J. Geophys. Res., 82, 5165, 1977.
Gosling, J.T., D.N. Baker, S.J. Bame, W.C. Feldman, and R.D. Zwickl,
    J. Geophys. Res., 90, 6354, 1985.
Gosling, J.T., D.N. Baker S.J. Bame, and R.D. Zwickl, J. Geophys. Res.,
    91, 11352, 1986.
Greenspan, M.E., C.-I. Meng and D.H. Fairfield, J. Geophys. Res., 91,
    11,123, 1986.
Gussenhoven, M.S., J. Geophys. Res., 87, 2401, 1982.
Gussenhoven, M.S., D.A. Hardy, N. Heinemann, and R.K. Burkhardt, J.
    Geophys. Res., 89, 9785, 1984.
Hardy, D.A., J. Geophys. Res., 89, 3883, 1984.
Hardy, D.A., H.K. Hills, and J.W. Freeman, J. Geophys. Res., 84, 72,
    1979a.
Hardy, D.A., P.H. Reiff, and W.J. Burke, J. Geophys. Res., 84, 1382,
    1979b.
King, J.H., Rep. NSSDC 79-08, NASA Goddard Space Flight Center,
    Greenbelt, MD., 1979.
Makita, K. and C.-I Meng, J. Geophys. Res., 92, 7381, 1987.
Meng, C.-I. and H.W. Kroehl, J. Geophys. Res., 82, 2305, 1977.

60

Olbert, S., _Eur. Space Agency Spec. Publ._, SP-161, 135, 1981.

Paulikas, G.A., _Rev. Geophys. Space Phys._, 12, 117, 1974.

Pilipp, W.G., H. Miggenrieder, M.D. Montgomery, K.-H. Muhlhauser, H. Rosenbauer and R. Schwenn, _J. Geophys. Res._, 92, 1075, 1987a.

Pilipp, W.G., H. Miggenrieder, K.-H. Muhlhauser, H. Rosenbauer and R. Schwenn, and F.M. Neubauer, _J. Geophys. Res._, 92, 1103, 1987b.

Pilipp, W.G., and G. Morfill, _J. Geophys. Res._, 83, 5670, 1978

Riehl, Kevin B. and David A. Hardy, _J. Geophys. Res._, 91, 1557, 1986.

Rosenbauer, H., H. Grunwaldt, M.D. Montgomery, G. Paschmann, and N. Sckopke, _J. Geophys. Res._, 80, 2723, 1975.

Sckopke, N., and G. Paschmann, _J. Atmos. Terr. Phys._, 40, 261, 1978.

Yeager, D.M. and L.A. Frank, _J. Geophys. Res._, 81, 3966, 1976.

Winningham, J.D. and W.J. Heikkila, _J. Geophys. Res._, 79 949, 1974.

Zwickl, R.D., D.N. Baker, S.J. Bame, W.C. Feldman, J.T. Gosling, E.W. Hones, Jr., D.J. McComas, B.T. Tsurutani, and J.A. Slavin, _J. Geophys. Res._, 89, 11007, 1984.

AURORAL OVAL CONFIGURATION DURING THE QUIET CONDITION

CHING I. MENG
Applied Physics Laboratory
The Johns Hopkins University
Johns Hopkins Road
Laurel, Maryland 20707-6099

ABSTRACT. The auroral oval configuration during the quiet condition is examined in this paper. The quiet condition can be defined as the very low K_p or AE index (i.e. $K_p \leq 1$ and AE $\leq$ 100 nT) or the northward interplanetary magnetic field. The previous investigations are also reviewed here. On the basis of the auroral electron precipitations, the quiet time auroral oval is characterized by the spatially extended continuous electron precipitations poleward to above 80° geomagnetic latitude from the normal equatorial boundary of the oval. This extended oval configuration is evident both statistically and on the case study basis. The instantaneous quiet time auroral oval was imaged and the optical auroral oval is consistent with the particle precipitation configuration. Without any doubt that the polar cap size diminishes and the auroral oval widens during the quiet condition, it is believed that the so-called "sun-aligned polar cap arcs" are the manifestation of the auroral oval arcs in the latitudinally expanded dawn or dusk part of the oval. The optical emission intensity and the energy flux of the oval are much weaker during the quiet time.

1. Introduction

It has been well established, on the basis of various types of satellite observations, that the earth's magnetosphere is open, namely, some of the geomagnetic field lines from the two polar regions extend into interplanetary space and connect with those of solar origin. Variations of the solar wind and the interplanetary magnetic field (IMF) undoubtedly affect the configuration of the terrestrial magnetosphere. The importance of the IMF in the solar-terrestrial interaction has been recognized, but the energy coupling process (or processes) is not well understood yet. The most obvious consequences of the dynamic interaction of IMF and the magnetosphere are the variations of the auroral oval configuration, which is the interface between the open and closed geomagnetic field regimes, the polar cap size, and the global auroral display. The polar cap is defined here as the region poleward of the instantaneous auroral electron

61

precipitations and presumably the region of opened geomagnetic field lines.

Extensive observations of low-energy auroral particle precipitations have been made by low-altitude satellites, such as Injun 5 [Frank and Ackerson, 1971], Ogo 4 [Hoffman and Burch, 1973], ISIS 1 and 2 [Heikkila and Winningham, 1971; Winningham et al., 1975], Cosmos 261 [Fedorova et al., 1971], Esro 1/Aurorae [Hultqvist, 1974], the series of DMSP spacecraft [Hardy et al., 1985], and others. Morphologically, electron precipitations over the polar auroral region have been described in terms of two distinct types, namely, soft discrete precipitations located in the poleward part of the auroral oval (called the "boundary plasma sheet"), and the broad, harder, structureless precipitation situated in the equatorward region of the auroral oval (called the "central plasma sheet") [Hoffman, 1972; Winningham et al., 1975].

Comparing these precipitation features with the morphology of the visual auroras, shows that the region of the poleward discrete electron precipitations corresponds to the bright discrete auroral region in the poleward part of the optical auroral oval, and the structureless hard electron precipitation is associated with the diffuse aurora in the equatorward part of the auroral oval. Such relationships between electron precipitation characteristics and the different auroral forms have been confirmed by coordinated ground-based and satellite observations as well as simultaneous auroral display and precipitation measurements from a single satellite, such as Injun 5, ISIS 2, and DMSP [Ackerson and Frank, 1972; Winningham et al., 1973; Deehr et al., 1976; Meng, 1976, 1978; Lui et al., 1977].

Morphologically, averaged observations of the auroral electron precipitations have generally corresponded to moderately active geomagnetic conditions (during magnetospheric substorms) when the auroral optical display and electron precipitations are very intense and dynamic. Therefore, previous studies of auroral electron precipitations primarily represent the active and moderately active magnetosphere. During periods of relative magnetic quiescence, the global auroral display observed by imagers on board ISIS 2 and DMSP satellites showed that quiet auroral arcs extend along the contracted auroral oval, and, at times, the luminosity of the auroral oval is even below the detector threshold of these satellites [Akasofu, 1974]. Because of the rarity of extremely quiet magnetospheric conditions and the nondynamic nature of the auroral display, the morphology of the very quiet auroral oval has not yet been well established [Hoffman and Burch, 1973; Winningham et al., 1975; Lui et al., 1976; Meng, 1981a; Murphree et al., 1982; Makita and Meng, 1984; Makita et al., 1988; Lassen et al., 1988]. The purpose of this paper is to present some of the morphology of the auroral oval and the polar cap during the very quiet condition. The first part is a review of quiet time electron precipitations. The second part is a report of the optical auroral observation during a very quiet period with simultaneous particle precipitation observations. The final part is a comparison with previous ground-based observations and the implication of the very quiet time auroral oval configuration.

2. Quiet Time Electron Precipitations

A typical example of the electron precipitations observed by a dawn-dusk polar orbiting satellite during the quiet condition is shown in Figure 1 (October 6, 1978).   The panel illustrates the auroral electron integral number flux, energy flux, and electron average energy.   The satellite trajectory, given at the top of the diagram, is in corrected geomagnetic latitude local time coordinates.   As the satellite traveled from the evening to the morning sector, it passed through the approximate center of the polar cap, at about 85° geomagnetic latitude along the midnight meridian [Holzworth and Meng, 1975].   In this event, the hourly value of the AE index was less than 50 nT from 2000 UT on October 5 to 0300 UT on October 6.   The observation was made at ~0140 UT on October 6, about 5.5 hours after the recovery of a weak substorm of ~300 nT occurred at about 1830 UT on October 5 and ended at about 2000 UT.   The auroral electron precipitations were detected over most of the polar region above 70° geomagnetic latitude extending from 70° to 85° geomagnetic latitude in the dusk sector and 70° to 86° in the dawn sector.   The equatorward and poleward boundaries of the region of auroral electron precipitation were determined at the edge of the extended precipitation.   The polar cap covers from ~85° geomagnetic latitude in the eveningside and to ~86° geomagnetic latitude in the morningside with a dawn-dusk diameter of only ~10° wide.   On the basis of characteristics of electron precipitations, two distinctly different precipitation regions of the auroral oval can be recognized on both the dawn and the dusk sectors.   The average energy in the lower latitude part is generally higher than 500 eV, corresponding to a region of hard electron precipitation that maps to the central plasma sheet.   The poleward part of the precipitation region has an average electron energy lower than 500 eV, corresponding to a region of soft precipitation or the boundary plasma sheet as defined by Winningham et al. [1975].   The boundary between these hard and soft regions is defined here as the transition boundary.   The selection of 500 eV is somewhat arbitrary since it merely represents a change in precipitation characteristics.   The latitudinal characteristics of the precipitation also change near the transition boundary from somewhat smooth and continuous in the lower latitude to highly fluctuating burst type in the higher latitude.   In this example, the region of high-average-energy electron precipitation is rather narrow, about 3° in the dusk sector and about 6° in the dawn sector, compared with the width of the entire auroral oval.   The region of low-average-energy electron precipitation is very wide, with widths of about 12° in the dusk sector and about 10° in the dawn sector.

The quiet time electron precipitations in the noon and midnight sectors are illustrated by the next example (Figure 2).   The hourly value of the AE index was continuously low at the background noise level from 0000 to 2400 UT on December 25. This example was observed during the extended quiet condition. The region of high-average-energy electron precipitation extends from 69° to 70° geomagnetic latitude in the midnight sector and 72° to 76° in the noon sector. The region of low-average-energy electron precipitation extends from 70° to 88° in the midnight sector and 76° to 80° in the noon sector. The polar cap is only ~12° in dimension in the noon-midnight orientation. It again indicates a very small polar cap in the quiet times.

It is important to know whether the polar cap region defined by using the poleward boundary of the soft extended auroral electron precipitations is a reasonable one. One way to check is to examine the polar cap boundaries over both the northern and southern hemispheres. On the basis of the definition of the polar cap as the area of opened geomagnetic field lines, a near symmetry is expected. A small polar cap region was detected over both the northern and southern hemispheres, similar to previous results reported by Meng [1981b].

3. Average Location of Quiet Time Precipitation Boundaries

From the examples illustrated, it is clear that the polar cap is rather small in all of the local time sectors and also that, in the quiet time, the soft electron precipitation region is extremely wide in latitude. The statistically spatial distribution of the polar cap and the auroral electron precipitation regions based on the latitudes of the poleward, transition, and equatorward boundaries during quiet geomagnetic conditions was examined from 1.5 years of DMSP observations [Makita and Meng, 1984]. The geomagnetic quiescence was defined by continuous AE ≤ 50 nT extending over 6 hours. We found 434 polar cap crossings. Over the northern hemisphere, there were 114 winter observations (October to March) and only 41 summer ones (April to September). For the southern hemisphere observations, there were 62 winter passes (April to September) and 217 summer passes (October to March). The average electron precipitation patterns were grouped by the local season irrespective of the hemisphere. These data were subgrouped into four magnetic local time (MLT) sectors, namely, from 0300 to 0900, 0900 to 1500, 1500 to 2100, and 2100 to 0300 MLT.

Figure 3 illustrates the average locations of the poleward, transition, and equatorward boundaries of the precipitation region in the winter and summer hemispheres during geomagnetically quiet periods. Because of the limited coverage of the sun-synchronous DMSP orbits, observations from 1200 to 1800 and 0000 to 0300 MLT zones were not abundant compared with other sectors. Therefore, the average locations of boundaries in the noon sector (0900 to 1500 MLT), the evening sector (1500 to 2100 MLT), and the midnight sector (2100 to 0300 MLT) were determined by using the data mainly from 0900 to 1200, 1800 to 2100, and 2100 to 0000 MLT, respectively. One of the interesting results was that the average location of the poleward boundary in both winter and summer polar regions was at very high geomagnetic latitudes in all local time sectors, at about $82°$ to $84°$ in the morning, noon, and evening sectors, and at about $81°$ to $82°$ in the midnight sector. The average size of the polar cap of the very quiet magnetosphere was only about $12°$ to $14°$ in diameter in contrast to >$30°$ nominal [Makita et al., 1983].

It is important to point out that in various local time sectors, the average locations of the auroral equatorward and transition boundaries were nearly identical between the winter and the summer hemisphere, except in the noon sector (0900 to 1500 MLT). The average location of the equatorward boundary was about $70.0°$ to $70.2°$, $68.6°$ to $68.9°$, and $69.5°$ to $69.7°$ in the evening, midnight, and morning sectors, respectively. The transition boundary was at about $73.4°$ to $73.5°$, $71.3°$ to $71.6°$, and $75.4°$ to $75.5°$ in the evening, midnight, and morning sectors, respectively. In the noon sector, there was a slight difference between the winter and the summer hemisphere in the averaged equatorward and transition boundary locations. The locations were at $70.0°$ and $76.7°$ in the winter hemisphere and $70.8°$ and $77.5°$ in the summer hemisphere, perhaps indicating that, in the noon sector, the region of electron precipitations in the summer hemisphere was located about $1°$ higher than it was in the winter hemisphere.

4. Auroral Electron Precipitation During Northward IMF

It is also generally known that various kinds of disturbances in the auroral region become weak during northward IMF periods. Akasofu [1975] examined DMSP auroral images and corresponding IMF data and showed that the auroral oval becomes very dim and less active during periods of northward IMF. Berkey et al. [1976], Lassen and Danielson [1978], Ismail and Meng [1982], and Gussenhoven [1982] examined the auroral morphology at very high geomagnetic latitude and concluded that discrete auroral arcs were observed mainly when the IMF was northward and geomagnetic activity was low along the auroral oval. Thus, it is interesting to examine the auroral electron precipitations during the northward IMF. Makita et al. [1988] studied the average precipitation pattern during small (B_z < 2.2 nT) and large (B_z > 5.2 nT) IMF B_z > 0 cases while AE was very low (<76 nT). Only data with these two criteria, continuous over 5 hours, were used in that analysis.

Since there are limitations on the availability of IMF and electron data for consecutive orbits, the number of examples selected on the basis of the above-mentioned criteria was very limited. On the basis of 17 months of DMSP electron data from August 1978 to December 1979, we found only seven periods of weak IMF events and ten periods of strong northward IMF events. For the weak IMF events, the average B_t is 3.2 nT and B_z is 0.9 nT, the solar wind velocity is 321 km/s, and the density is 10.0 cm^{-3}. For the strong northward cases, the average B_t and B_z are 10.8 and 7.8 nT, respectively, and the solar wind velocity and density are 371 km/s and 8.2 cm^{-3}, respectively. The corresponding AE values in the weak IMF case are slightly lower than those for the strongly northward IMF cases. Indeed, the average AE, including all weak IMF periods, is only 34 nT and it is 48 nT for the ten strongly northward IMF periods. The average Kp values during the weak and strongly northward IMF periods are 0 and 1, respectively. There were 108 polar passes during the weak IMF periods and 102 polar passes during the strongly northward IMF events, and they were obtained over both hemispheres. On the basis of these data, the average latitude of the poleward, transition, and equatorward boundaries for the two types of IMF conditions were determined; the statistical results are shown in Figure 4. The average latitudes of precipitation boundaries for the northern and southern hemispheres were not very different; the deviation of average latitude is generally less than 1°. Thus, to make the average precipitation pattern statistically more significant, the electron precipitation data observed in both hemispheres were combined. The locations of the average poleward, transition, and equatorward boundaries were grouped in four magnetic local time sectors (0300-0900, 0900-1500, 1500-2100, and 2100-0300 UT).

The left panel of Figure 4 shows the average location of three boundaries of electron precipitation during the weak IMF period. The hatched line and dot shaded areas indicate the average region of hard ($E_{ave} \gtrsim 500$ eV) and soft ($E_{ave} < 500$ eV) precipitation. The poleward edge of the soft region may be interpreted as the equatorial boundary of the average polar cap region that is void of intense auroral oval precipitation but may have uniform polar rain. In the weak IMF cases, the poleward boundary of the soft electron precipitation region is located at about 82° to 83° in the morning, noon, and evening sectors, whereas it is at 75.8° in the midnight sector. The transition boundary between the soft and hard region is at 78.3° in the dayside, at about 75° in the morning and evening sectors, and at 71.9° in the night sector. The latitudinal extent of this soft region is about 7° to 8° in both the evening and morning sectors, whereas the width of the soft region in the nightside and dayside reduces to about 4° to 5°. The equatorward boundary of the auroral oval is at about 71° to 72° in the day and evening sectors, whereas it is at 69.6° in the night and morning sectors. The latitudinal extent of the hard region in the morning and dayside is about 5° to 6°, and its width is narrower, about 4° in the evening sector. The width of the hard region in the midnight sector is even narrower, only about 2°. Statistically, the soft region in the midnight sector is about 4° wide. However, it sometimes disappears; there were 20 polar passes (~20% of the total) with no soft precipitation region in the midnight sector.

The right panel of Figure 4 shows the locations of the three averaged boundaries during the periods of very strong northward IMF. The equatorward boundary is at about 68° in the evening, midnight, and morning sectors, whereas in the dayside, it is slightly higher, at about 70°. The transition boundaries in the noon and morning sectors are located at 76.5° and 74.1°, respectively. In the evening and midnight sectors, the boundaries are at slightly lower latitudes, about 73.0° and 71.9°, respectively. The latitudinal extent of the hard precipitation region is about 7° in the morning and dayside and about 4° to 5° in the evening and midnight sectors. The width of this region expands by about 1°, and the equatorward boundaries shift to the lower latitudes by about 1° to 3° during the strongly northward IMF period in comparison with those in the weak IMF period. The poleward boundaries of the soft precipitation are seen at latitudes higher than ~ 85° in the morning, noon, and evening sectors. In the midnight sector, the soft precipitation location is at latitudes higher than ~83°.

It is difficult to determine the poleward boundary of the soft region for 47 of the polar passes (46% of all events). Burst-type soft electron precipitation is observed continuously throughout the entire pass, even at the highest-latitude point of satellite orbit above 85°. Near the very high latitude (>80°) region in the dayside, the burst-type soft electrons are frequently observed over the entire area. Therefore, in such cases, the highest-latitude value of satellite trajectory was used in calculating the average poleward boundary. The average poleward boundary shown in Figure 4 gives only the lower bound. It is greater than ~85° in the morning, noon, and evening sectors and higher than ~83° in the midnight sector. From these statistical results, it is clear that the size of the polar cap is very small for periods of both weak and strongly northward IMF. This is very consistent with statistical analysis using the AE index shown in Figure 3.

The relationship of average locations of the quiet time poleward, transition, and equatorward boundaries of auroral precipitation with the IMF B_z component is examined statistically in the following. The data base is the same as that used in Figure 3. Figure 5 shows scatter diagrams of the three precipitation boundaries subgrouped into four MLT sectors (0900-1500, 1500-2100, 2100-0300, and 0300-0900 UT). Within each diagram, boundary locations are plotted according to the concurrent hourly average of the IMF B_z value. Since DMSP electron data used here were selected during the quiet period, the concurrent hourly average value of IMF B_z was mostly (> 80%) northward. The poleward, transition, and equatorward boundaries are represented by circles, triangles, and solid dots, respectively. The linear regression was calculated for each type of boundary by using the least-squares-fit method, and the correlation coefficients are shown on the figure. There are large scatters with respect to the linear regressions; however, some general trends of these boundaries can be stated.

The responses of the poleward, transition, and equatorward boundaries to the IMF B_z are similar in all local time sectors. Generally, the latitudinal position of the poleward boundary of auroral electron precipitation was proportional to the magnitude of the northward IMF B_z component. The average displacement rate of the poleward boundary in the noon, evening, midnight, and morning sectors was about 0.14°, 0.37°, 0.43°, and 0.26° per 1nT change of the IMF B_z, respectively. This indicates a decrease of the polar cap radius by ~0.3° per 1 nT of northward IMF B_z. The average polar cap size for large B_z is about 10° in diameter, consistent with Figure 4. The effect of IMF B_z on the transition and equatorward boundaries during quiet times was quite different from that of the poleward boundary. The transition boundary moves only slightly equatorward with the increasing northward IMF component in all local time sectors, whereas the equatorward boundary did not show any significant variation with the northward change of IMF B_z magnitude as reported by Hardy et al. [1981]. These results indicate that, in all local time sectors, the width of the region of low-average-energy electron precipitation (between the poleward and transition boundaries) expanded both poleward and equatorward with increasing northward IMF B_z intensity; however, the width of the region of high-average-energy electron precipitation (between the transition and equatorward boundaries) slightly reduced its latitudinal width with the increasing northward IMF component. The poleward shift of the poleward precipitation boundary with increasing northward IMF implies the reduction of the polar cap size with the increasing magnitude of positive B_z. Such a gradual poleward movement of the poleward edge of the auroral oval is very consistent with the ground-based observation on the distribution of discrete auroras in different IMF conditions reported by Lassen and Danielsen [1978], since the discrete arcs occur in the poleward part of the auroral oval.

5. Optical Aurora During The Very Quiet Period

It is interesting to examine the characteristics of optical aurora and the auroral distribution in the very quiet period. Makita and Meng [1984] searched DMSP auroral images associated with the quiet time electron precipitation boundaries. Several images with distinguishable auroral display were found during very quiet conditions from the enormous DMSP auroral images data set, since the faint diffuse aurora is very difficult to discern with the earth albedo and moonlight background present. Furthermore, the DMSP auroral imager has a very broad spectral passband that is dominated by broadband backgrounds more than the narrow auroral emissions. Faint stable discrete auroral arcs were occasionally detected in the evening sector and located near the equatorward part of the electron precipitation region (i.e., in the region of high average energy of ~1 keV), in contrast to their normal location in the poleward soft precipitation region during moderate activity. Thus, the DMSP auroral images are not suitable for the investigation of the quiet time weak auroral display.

Using data from the narrow band pass auroral scanner at 5577 and 3914 Å aboard the ISIS 2 satellite, Murphree et al. [1982] reported auroral observations during the strong northward IMF ($B_z > 3.5$ nT) condition (i.e., corresponding to the geomagnetic quiescence). They reported that the distinction between the auroral oval auroras and the polar cap auroras is no longer clear. Diffuse weak emission can fill in the region between the auroral oval and the region of polar cap arcs. The merging between auroral oval auroras and the polar cap auroras was observed. Murphree et al. [1982] concluded that the region of closed geomagnetic field lines can expand poleward to occupy much of the high-latitude region; this supports the suggestion of the poleward widening of the auroral oval during very quiet times, made by Meng [1981b] on the basis of electron precipitation observations. In this section, the very quiet time auroral display is examined by using the auroral images in far ultraviolet and 3914Å emissions obtained by the Polar Bear satellite from 1000km altitude, and the nearly simultaneous DMSP auroral particle precipitations.

The main advantage of using far ultraviolet auroral emissions is that the auroral display can be monitored in both dark and sunlit hemispheres [Meng and Huffman, 1984] while the earth albedo is no longer transparent in the atmosphere for those wavelengths. The UV/VIS imaging instrument is the advanced version of that flown on the HILAT satellite [Schenkel and Ogorzalek, 1984]. It can simultaneously image the global distribution of auroral and atmospheric emissions at four different wavelengths, two in the FUV via a spectrometer and two in the near UV and visible via a filtered photomultiplier tube through the same telescope [Meng and Huffman, 1987]. The details of the instrumentation are described elsewhere [Schenkel and Ogorzalek, 1987].

The very quiet auroral event was observed on January 30, 1987, between ~0300 to 1200 UT, while K_p was 0_+, 0_0, 0_+ with a daily ΣK_p of only 7_0. The K_p between 0000 and 0300 UT was 2 corresponding to a very weak substorm of about 100 nT at ~0130 UT. Within this period, the IMF was mainly positive. The Polar Bear satellite was in the nearly noon-midnight orbit over the northern polar region and recorded the quiet time auroral display over most of the oval at about 0320 and 0510 UT, and that of the dayside oval at about 0710, 0850, 1030, and 1220 UT. The auroral distributions were imaged at 1356±15, 1595±15 and 3914 ± 5 Å. In this section the analysis is centered on two consecutive passes of 0320 and 0510 UT with the extended oval coverage. Figure 6 illustrates these two passes in false color presentation; in each pass only images of 1356 and 3914 Å are presented. The 1595 Å auroral images were not presented because of the low emission intensity and the very low count rate recorded by the detector. The geomagnetic latitude contours of every 10° and local time meridian of every 3 hours are drawn together with the coastlines below 80° geographic latitude. The monochromatic intensity of the aurora can be inferred from the color bar at the right of each image.

At ~0320 UT, the auroral display was not very active; weak discrete-type auroras were seen at the poleward edge of the midnight auroral oval. Most of the oval was characterized by very weak but spatially extended diffuse auroras. These features occurred in images of both wavelengths. The dawn and dusk parts of the oval were very wide in latitude from about 70° to 85° geomagnetic latitude at dawn and from <68° to 80° at dusk. (The very bright red and yellow horizontal features on the top of the 3914 Å image are associated with the scattered light into the optical train near the spacecraft sunrise.)

At the next pass (~0510 UT), the auroral oval configuration is very similar to the previous orbit; note the disappearance of the brighter discrete auroral features near midnight. The auroral oval was very diffused, and the extended dawn and dusk parts of the oval were over 10° to 15°; the only discrete feature was a very long (>2000 km) noon-midnight aligned discrete aurora near the poleward edge of the dusk. The auroral intensity of 1356 Å along the oval was 300 R. At 3914 Å, the existence of the auroral oval was not distinct at all, indicating an upper limit of the oval emission brightness of ≲1000 R.

To understand the very quiet auroral oval configuration, the auroral particle precipitations were also examined. Two DMSP satellites carrying both electron and ion detectors traversed the northern oval near the times of two Polar Bear auroral oval observations; two dawn-dusk particle observations were "simultaneous" with two images. Figure 7 shows two color energy-time spectrograms of these dawn-dusk oval crossings corresponding to auroral displays in Figure 6. DMSP-F6 first entered the dawn oval at 0510 MLT and exited the dusk oval at 1730 MLT associated with the 0320-UT image, and at 0600 and 1750 MLT, respectively, for the 0515-UT image. The precipitating auroral particle measurements indicate that the dawn oval extended over 15° in latitude from about ~65° to 82° geomagnetic latitude and from ~85° to 69° in the dusk side during the first image at 0320 UT. The precipitated energy flux was only a fraction of 1 erg/cm^2.s.sr. For the auroral image at 0515 UT, the particle precipitation extended from ~68° to 85.5° and from ~86° to 66° over the dawn and dusk oval, respectively. The energy flux was ~0.2 to 0.4 ergs/cm^2.s.sr in the dawnside and only ~0.04 to 0.1 ergs/cm^2.s.sr in the dusk side. The spatial distribution of auroral particle precipitations is very consistent with the optical auroral distribution in images.

The data from the Polar Bear satellite were only available in the real time mode, while the spacecraft was within the field of view from a dedicated ground receiving station. Thus, the nightside auroral oval was not imaged after the 0515-UT pass during this quiescent period. However, two DMSP satellites provided a continuous monitoring of auroral particle precipitations over both hemispheres in the dawn-dusk and near 10 am to 10 pm orbits. Figure 8 illustrates the loci of auroral precipitations detected from about 0100 to 1200 UT in the geomagnetic latitude and local time coordinates over both polar regions. The heavy line indicates the region with equatorward harder smooth precipitations (i.e. the central plasma sheet type), whereas the thin line is the poleward softer burst precipitation region (i.e., boundary plasma sheet type). It is important to point out that after about 0230 UT (i.e., start of the quiet period), the latitudinal extent of the auroral oval was very wide (≳10°), especially in the dawn and dusk part of the oval. If we define the polar cap as the void region of auroral particle precipitations poleward of the instantaneous auroral oval (see Figure 7), these polar plots reveal a very small polar cap of about only 5° to 8° in diameter. Such a tiny polar cap is distinct in both hemispheres, even though perfect conjugate distributions are difficult to establish.

The quiet time distributions of auroral oval imaged by the Polar Bear satellite and determined from DMSP precipitating auroral particle measurement consistently indicate that the instantaneous auroral oval widens drastically poleward, especially the dawn and dusk part,whereas the polar cap has a very small dimension. This observation is consistent with the quiet auroral oval distribution proposed by Meng [1981b].

6. Comparison With Ground-Based Observations

Recently, Lassen et al. [1988] studied the distribution of quiet time electron precipitation boundaries and compared it with various ground-based optical observations. The following is the summary of our results. The DMSP auroral electron observations over Greenland during six midwinter months in 1978 and 1979 seasons, were used to determine the quiet time average electron spatial distribution. Figure 9 gives the locations of boundaries of each pass and the average curves for the equatorward, transition, and poleward boundaries.

Figure 9(d) shows the average configuration of these three boundaries in polar magnetic coordinates. The electron precipitation extends over a wide belt around the pole. The higher energy part of the belt (dotted region) is about 2° wider in the dawn sector than in the dusk sector. In the afternoon sector the DMSP measurements, although scarce, indicate a considerable narrowing of the belt. The low-energy belt (hatched area) forming the poleward part of the auroral electron precipitation region is quite narrow in the midnight sector but broadens rapidly in about three hours of local time before and after midnight. Also the polar cap proper has a teardrop shape, as seen in a few ISIS 2 auroral images [Murphree et al., 1982] and Dynamic Explorer images [Frank, private communication, 1987]. Poleward of the low-energy belt, the total flux rapidly decreases to a minimum level below the detectable 10^6 electrons $cm^{-2}s^{-1}sr^{-1}$ (Figures 1 and 2). This region is the polar cap region.

The distribution of 6300 Å emission over the northern polar region during quiet conditions was reported by Shepherd [1979] on the basis of one ISIS 2 pass on 16 December 1971, at 0500 UT during a very quiet period (K_p = 0+, AE = 24 nT). Figure 10, a reproduction of Shepherd's plot, gives an instantaneous distribution of 6300 Å emission together with the DMSP electron precipitation poleward and equatorward boundaries from the individual orbits for the K_p = 0 condition. The 200-R contours pass smoothly from dayside to nightside. The equatorward boundary (solid dots) of the electron precipitation follows the equatorward 200-R contour, whereas the poleward boundary (open dots) clusters along the poleward 200-R contour.

Therefore, we can conclude that the quiet time electron precipitation pattern from the DMSP observation is nearly identical to the spatial distribution of quiet time 6300 Å auroral emission observed by ISIS-2. The soft electron precipitation region (i.e., poleward part of the quiet time oval) coinciding with the 6300 Å emission region certainly is a natural consequence, since the 6300 Å emission is generally produced by electrons of energy less than 300 eV [Shepherd and Shepherd, 1979]. In the higher energy electron belt (i.e. equatorward part), the 6300 Å emission may be produced by an intense low-energy component of the total auroral electron spectrum. The similarity between electron precipitation and 6300 Å emission regions indicates that the spatial configuration is probably a consistent pattern during quiet times. As pointed out by Shepherd [1979], the dayside and nightside maximum regions both follow constant (but different) latitude circles. Figure 10 further demonstrates that these two regions are separated in latitude by the transition boundary. Thus, the nightside 6300 Å maximum region is wholly situated in the higher energy precipitation, whereas the dayside maximum belongs entirely to the low-energy part of the auroral electron region.

The Danish Meteorological Institute has made continuous recordings of auroral displays from a network of all-sky cameras in Greenland for about three decades. It is interesting to compare the statistical space-based observations with statistical ground-based auroral observations. Figure 11 [from Lassen, 1972] shows the average distribution of the occurrence frequency of discrete and diffuse auroras observed during quiet intervals (K_p = 1) of the years 1964-1969. Also included in Figure 11 are the smoothed equatorward and transitional boundaries as well as the individual polar boundary locations observed by DMSP for K_p = 1. There is a good agreement between the distribution of these particle boundaries and the (in part, interpolated) isofrequency contours of the ground-based auroral observation.

7. Conclusion

During the quiet time, the statistical distribution of precipitating auroral electrons does not have a well-known oval configuration. It has the shape of an irregular annular belt encircling a rather small polar cap. Optically, the auroral particle precipitation belt is characterized by a diffuse background of 6300-, 5577-, and 3914-Å emissions with embedded discrete forms. The high-latitude, sun-aligned arcs in the dawn and dusk sector reported by Lassen and Danielsen [1978] are in the region of the poleward expanded auroral oval.

Comparing the quiet time observations with previous established auroral features, we believe that the bright discrete auroras forming the classical oval are located in the poleward (most intense) part of the higher energy belt; the high-latitude dayside auroral arc-forms and sun-aligned so-called "polar cap arcs" are embedded in the high-latitude, low-energy belt. The dayside auroral forms are located in the equatorward (most intense) part of this belt, several degrees poleward of the oval maximum. The high latitude sun-aligned arcs are observed poleward of the dayside oval, including both dawn and dusk parts. There is no obvious signature difference in the electron precipitation between the two high-latitude subgroups. This observation is in agreement with the findings of Lassen and Danielsen [1978], Meng [1981b], and Murphree et al., [1982], that the high-latitude daytime and sun-aligned arcs together form one quiet time pattern (the so-called "polar cap arc pattern"). Therefore, it is probable that the sun-aligned arcs are signatures of a poleward widening of the dayside auroral maximum during quiescence, as suggested by Meng [1981b]. Figure 12 is a schematic of quiet time particle precipitations and auroral distribution.

The poleward area without significant precipitation (except the polar rain), encompassed by the electron precipitation belt is believed to be the polar cap proper. We find that the instantaneous polar cap, thus defined, is very small and sometimes may be pear-shaped (or teardrop-shaped), with its largest extension in the noon-midnight direction, as seen occasionally in satellite images. The poleward boundary of the expanded oval can be much higher than shown, and there is a dynamic variation of its shape and location, probably depending on the instantaneous orientation and magnitude of the interplanetary magnetic field.

The optical emission intensities of the quiet time oval are rather low, about a few hundred Rayleighs or less at 1356 Å and below 1000R at 3914-Å, corresponding to the observed auroral electron energy flux of about 10^{-2} to 10^{-1} ergs/cm^2.s.sr.

8. Acknowledgement

This work was supported by the Air Force Office of Scientific Research Grants AFOSR 88-0101 and 86-0057 and the Atmospheric Sciences Division, National Science Foundation Grant ATM-8713212 to The Johns Hopkins University. Polar Bear Satellite Program and subsequent data analysis are supported by the Atmospheric Effects Division of the Defense Nuclear Agency under Contract N00039-87-C-5301 to the Johns Hopkins University, Applied Physics Laboratory.

9. References

Ackerson, K. L., and Frank, L. A. (1972) 'Correlated satellite measurements of low-energy electron precipitation and ground level observations of a visual arc', *J. Geophys. Res.*, **77**, 1128.

Akasofu, S.-I. (1974) 'A study of auroral displays photographed from the DMSP-2 satellite and from the Alaska meridian chain of stations', *Space Sci. Rev.*, **16**, 617.

Akasofu, S.-I. (1975) 'The roles of the north-south component of the interplanetary magnetic field on large-scale auroral dynamics observed by the DMSP satellite', *Planet. Space Sci.*, **23**, 1349.

Berkey, F. T., Cogger, L. L., Ismail S., and Kamide, Y. (1976) 'Evidence for a correlation between sun-aligned arcs and the interplanetary magnetic field', *Geophys. Res. Lett.*, **3**, 145.

Deehr, C. S., Winningham, J. D., Yasuhara, F., and Akasofu, S.-I. (1976) 'Simultaneous observations of discrete and diffuse auroras by the ISIS-2 satellite and airborne instruments', *J. Geophys. Res.*, **81**, 5527.

Fedorova, N. I., Temny V. V., and Galperin Yu. I. (1971) 'Morphology of auroral electron energetic and angular distributions according to "Cosmos 261" measurements', *J. Atmos. Terr. Phys.*, **33**, 731.

Frank, L. A., and Ackerson, K. L. (1972) 'Local-time survey of plasma at low altitudes over the auroral zones', *J. Geophys. Res.*, **77**, 4116.

Gussenhoven, M. S. (1982) 'Extremely high latitude auroras', *J. Geophys. Res.*, **87**, 2401.

Hardy, D. A., Burke, W. J., Gussenhoven, M. S., Heineman N. H., and Holeman, E. (1981) 'DMSP/F2 electron observations of equatorward auroral boundaries and their relationships to the solar wind velocity and the north-south component of the interplanetary magnetic field', *J. Geophys. Res.*, **86**, 9961.

Hardy, P. A., Gussenhoven, M. S., and Holeman E. (1985) 'A statistical model of the auroral electron precipitation', *J. Geophys. Res.*, **90**, 4229.

Heikkila, W. J., and Winningham, J. D. (1971) 'Penetration of magnetosheath plasma to low altitudes through the dayside magnetic cusps', *J. Geophys. Res.*, **76**, 883.

Hoffman, R. A. (1972) 'Properties of low energy particle impacts in the polar domain in the dawn and dayside hours', in *Magnetosphere-Ionosphere Interactions*, edited by K. Folkestad, p. 117, Universitetsforlaget, Oslo.

Hoffman, R. A., Burch, J. L. (1973) 'Electron precipitation patterns and substorm morphology', *J. Geophys. Res.*, **78**, 2867.

Holzworth, R. H., and Meng, C.-I. (1975) 'Mathematical representation of the auroral oval', *Geophys. Res. Lett.*, **2**, 377.

Hultqvist, B. (1974) 'Rocket and satellite observations of energetic particle precipitation in relation to optical aurora', *Ann. Geophys.*, **30**, 223.

Ismail, S., and Meng, C.-I. (1982) 'A classification of polar cap auroral arcs', *Planet. Space Sci.*, **30**, 319.

Lassen, K. (1972) 'On the classification of high-latitude auroras', *Geofys. Publ.* **29**, 87, Universitetsforlaget, Oslo.

Lassen, K. and Danielsen, C. (1978) 'Quiet time pattern of auroral arcs for different directions of the interplanetary magnetic field in the Y-Z plane', *J. Geophys. Res.*, **83**, 5277.

Lassen, K., Danielsen C., and Meng, C.-I. (1988) 'Quiet-time average auroral configuration', *Planet. Sci.*, **36**, 791.

Lui, A. T. Y., Venkatesan, D., Anger, C. D., Akasofu, S. I., Heikkila W. J., Winningham, J. D., and Burrows, J. R. (1977) 'Simultaneous observations of particle precipitations and auroral emissions by the ISIS-2 satellite in the 19-24 MLT sector', *J. Geophys. Res.*, **82**, 2210.

75

Lui, A. T. Y., Akasofu, S. I., Hones, E. W., Jr., Bame, S. J., and McIlwain, C. E. (1976) 'Observations of the plasma sheet during a contracted oval substorm in a prolonged quiet period', J. Geophys. Res., **81**, 1415.

Lui, A. T. Y., Akasofu, S.-I., Hones, Jr., E. W., Bame, S. J., and McIlwain, C. E. (1978) 'Observations of the plasma sheet during a contracted oval substorm in a prolonged quiet period', J. Geophys. Res., **83**, 5277.

Makita, K., Meng, C.-I., and Akasofu, S.-I. (1988) 'Latitudinal electron precipitation patterns during large and small IMF magnitudes for northward IMF conditions', J. Geophys. Res., **93**, 97.

Makita, K. and Meng, C.-I. (1984) 'Average electron precipitation patterns and visual auroral characteristics during geomagnetic quiescence', J. Geophys. Res., **89**, 2861.

Makita, K., Meng, C.-I., and Akasofu, S.-I. (1983) 'The shift of the auroral electron precipitation boundaries in the dawn-dusk sector in association with geomagnetic activity and interplanetary magnetic field', J. Geophys. Res., **88**, 7967.

Meng, C.-I. (1978) 'Simultaneous observations of low-energy electron precipitation and optical auroral arcs in the evening sector by the DMSP-32 satellite', J. Geophys. Res., **81**, 2771.

Meng, C.-I. (1978) 'Electron precipitations and polar auroras', Space Sci. Rev., **22**, 223.

Meng, C.-I. (1981b) 'The auroral electron precipitation during extremely quiet geomagnetic conditions', J. Geophys. Res., **86**, 4607.

Meng, C.-I. (1981b) 'Polar cap arcs and the plasma sheet', Geophys. Res. Lett., **8**, 273.

Meng, C.-I., and Huffman, R. E. (1984) 'Ultraviolet imaging from space up to the aurora under full sunlight', Geophys. Res. Lett., **11**, 315.

Meng, C.-I., and Huffman, R. E. (1987) 'Preliminary observations from the auroral and ionospheric remote sensing imager', Johns Hopkins APL Techincal Digest, **8**, 303.

Murphree, J. S., Anger, C. D., and Cogger, L. L. (1982) 'The instantaneous relationship between polar cap and oval auroras at times of northward interplanetary magnetic field', Can. J. Phys., **60**, 349.

Schenkel, F. W., and Ogorzalek, B. S. (1984) 'The HILAT vacuum ultraviolet auroral imager', Johns Hopkins JHU Technical Digest, **5**, 131.

Schenkel, F. W., and Ogorzalek, B. S. (1987) 'Remote sensing imager auroral images from space: Imagery, spectroscopy, and photometry', Johns Hopkins JHU Technical Digest, **8**, 308.

Shepherd, G. G. (1979) 'Dayside cleft aurora and its ionospheric effects', Rev. Geophys., **17**, 2017.

Shepherd, M. M., and Shepherd, G. G. (1979) 'Comments on the low-altitude optical signatures of the magnetospheric boundary layers', Proc. Magnetospheric Boundary Layers Conf., Alpbach, 11-15 June 1979 (ESA SP-148, August 1979).

Winningham, J. D., Akasofu, S.-I., Yasuhara, F., and Heikkila, W. J. (1973) 'Simultaneous observations of auroras from the south pole station and of precipitating electrons by ISIS-1', *J. Geophys. Res.*, **78**, 6579.

Winningham, J. D., Yasuhara, S.-I., Akasofu, and Heikkila, W. (1975) 'The latitudinal morphology of 10 eV to 10 keV electron fluxes during magnetically quiet and disturbed times in the 2100-0300 MLT sector', *J. Geophys. Res.*, **80**, 3148.

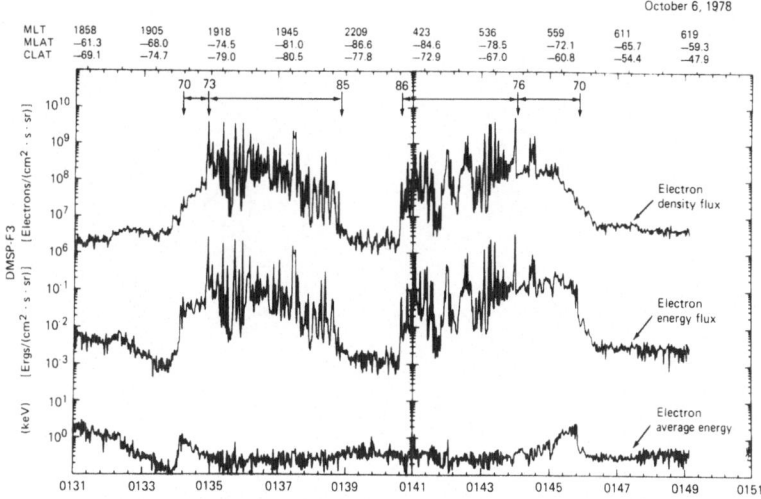

Figure 1. An example of the auroral electron precipitation in the dawn-sector during a quiet period. Panel consists of the electron total number flux, energy flux, and average energy observed by the DMSP-F3 satellite. The satellite trajectory in the corrected geomagnetic latitude local time coordinates as given at the top of the figure.

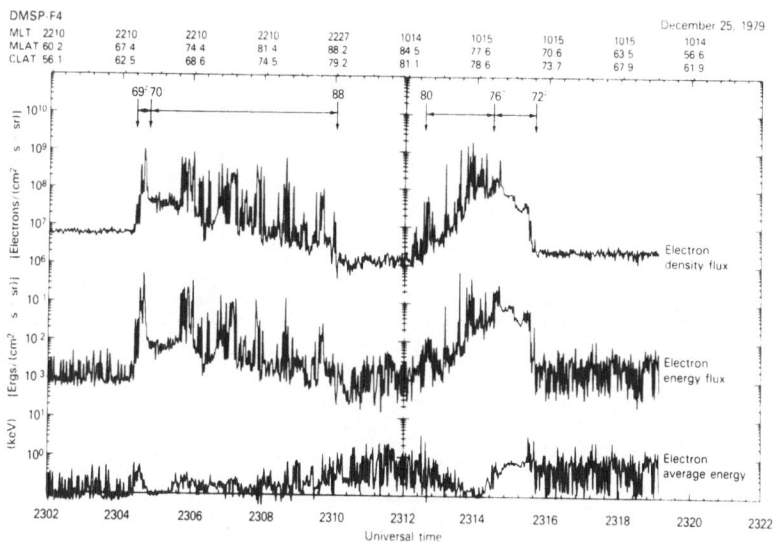

Figure 2. An example of the auroral electron precipitation in the noon-midnight sectors during quiet periods observed by the DMSP-F4 satellite. The format of the presentation is the same as that of Figure 1 but for December 25, 1979, ~2310 UT.

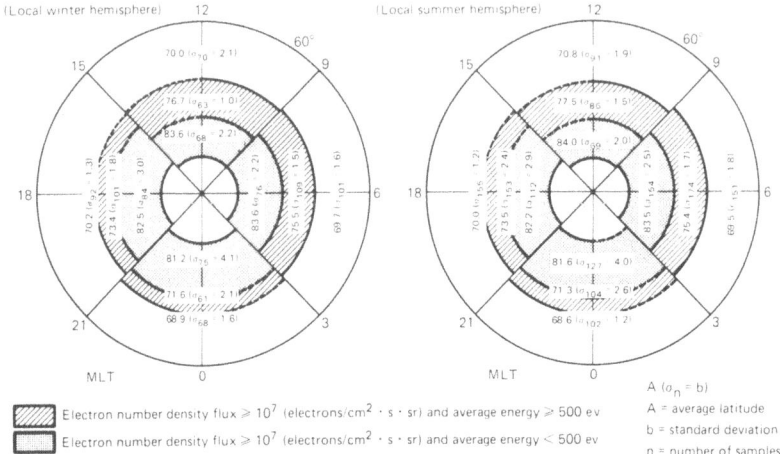

Figure 3. Average distribution of auroral electron precipitation on the local winter and the local summer hemispheres during geomagnetic quiescence. The lightly shadowed region corresponds to the poleward low-average-energy auroral electron precipitation region, and the heavily shadowed region corresponds to the equatorward high-average-energy auroral electron precipitation. Note that the boundaries of poleward precipitation are statistically located at very high latitude, above 80° in all local time sectors during very quiet periods.

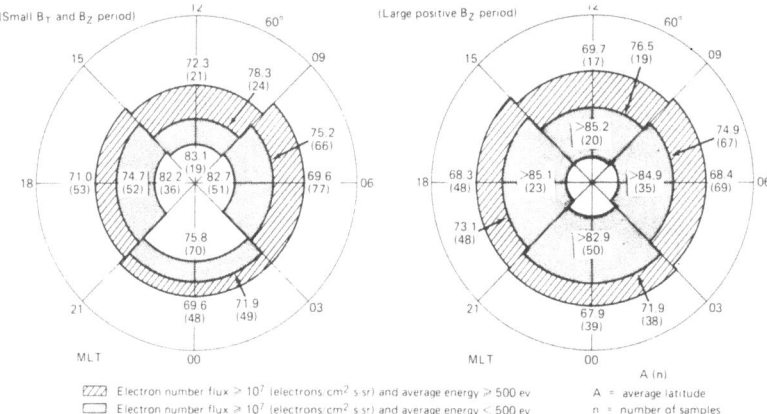

Figure 4. The left and right panels show the average electron precipitation pattern during the period of weak IMF and strongly northward IMF, respectively. The hard electron region ($E_{ave} \geq 500$ eV) in the left panel is located at a slightly higher latitude than that in the right panel. (Conversely, the poleward boundary of the soft electron region ($E_{ave} < 500$ eV) in the left panel is located at a lower latitude than that in the right panel.) Notice that the poleward boundary of the soft region in the right panel shows the minimum average value.

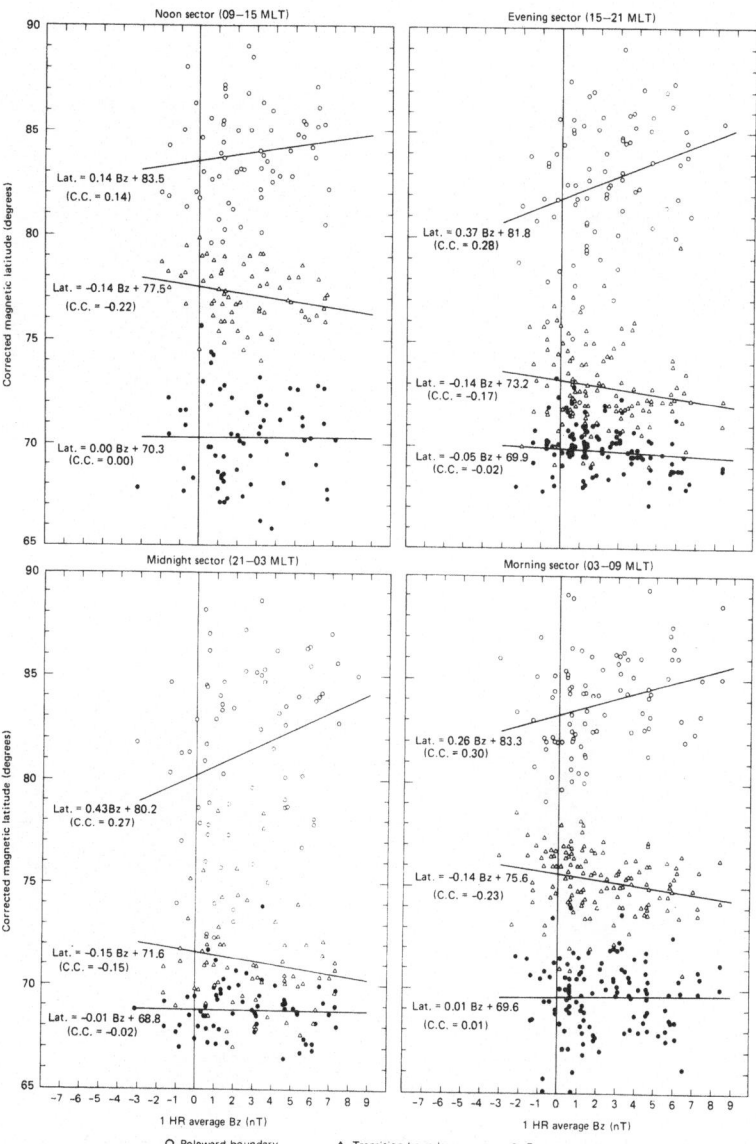

Figure 5.   The scatter plot of the poleward, transition, and equatorward boundaries of the auroral electron precipitation in the four MLT sectors (09–15, 15–21, 21–03, and 03–09).   The poleward, transition, and equatorward boundaries are represented by circles, triangles, and solid dots, respectively.   The linear regression was calculated for each type of boundary by using the least-squares-fit method; the correlation coefficients (c.c.) are also shown.

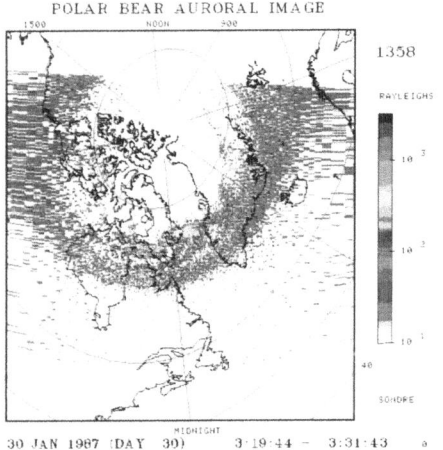
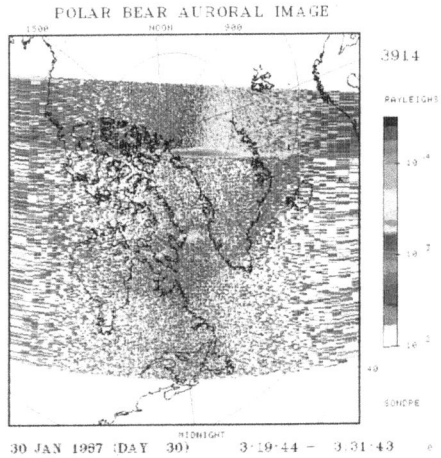

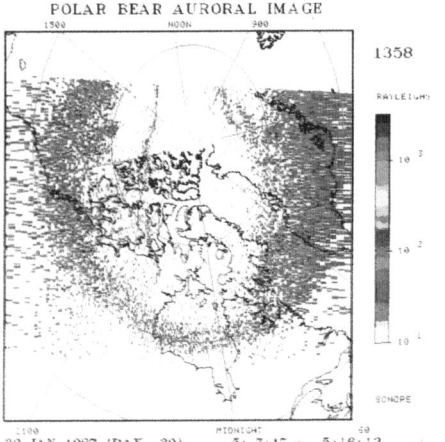
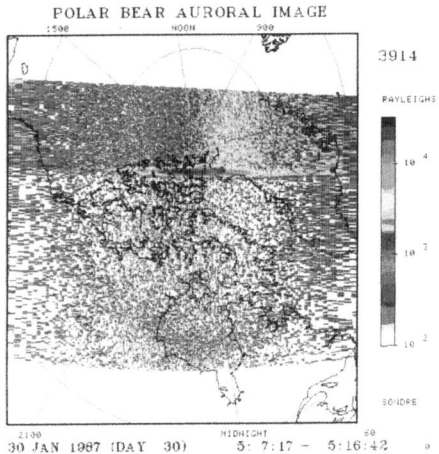

Figure 6. Quiet time auroral oval configuration observed during two consecutive orbits in false color. Images are presented in geomagnetic local time latitude coordinates. They were obtained by the Polar Bear satellite at 1356 ± 15 Å and 3914 ± 5 Å. The diffused oval with very weak brightness was obvious in both wavelengths. (The intensity of auroral emission is color coded by the color bar at right). A localized weak auroral activity was seen near the poleward edge of the midnight oval. The bright horizontal bands in 3914 Å is the artifact of the scattered sunlight near the spacecraft sunrise. At ~0515 UT, the emissions of quiet time oval are so weak near the imager sensitive threshold. Both dawn and dusk oval extends over 10° wide in latitude.

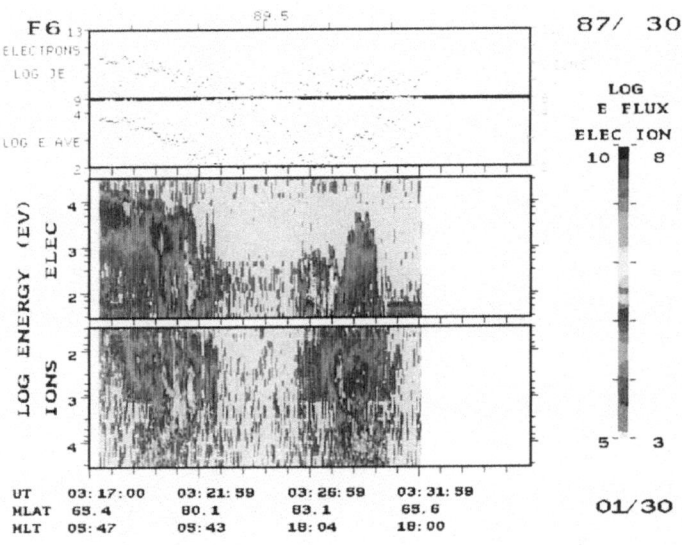

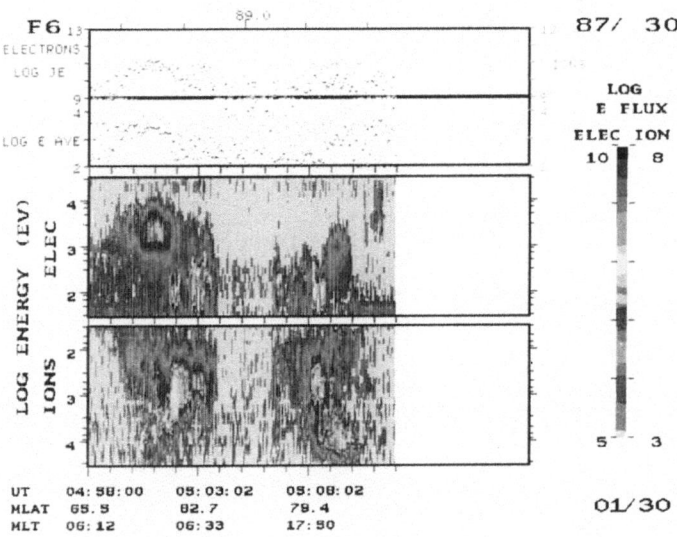

Figure 7. Auroral particle precipitations observed along the dawn-dusk meridian associated with Figure 6 two optical auroral distributions by DMSP-F6 satellite. The particle precipitation is presented in energy-time color spectrogram. The small void region near the middle of the polar crossing corresponds to the polar cap. The DMSP-F6 traversed the center of the polar region. The format of E-t spectrogram is given in a paper by Newell and Meng in this book.

82

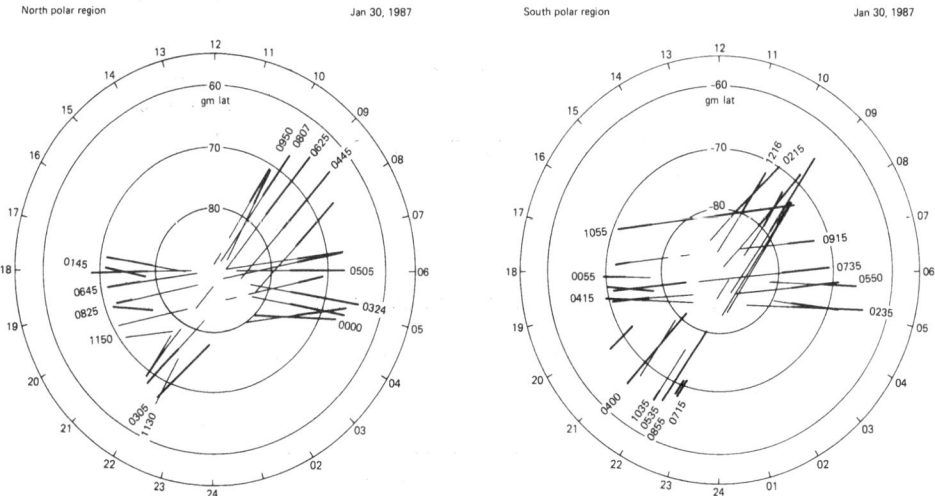

Figure 8. Electron precipitation regions observed by DMSP–F6 and F7 between 00 to 12 UT on January 30, 1987 corresponding to quiet time auroral ovals shown in Figure 6. The heavy and thin lines represent the harder and softer electron regions. Note the extended auroral electron precipitation region between 03 to 12 UT. The oval was wider than 10° to 15° gm latitude in both dawn and dusk sectors. The center polar cap region is very small as seen in both northern and southern hemispheres.

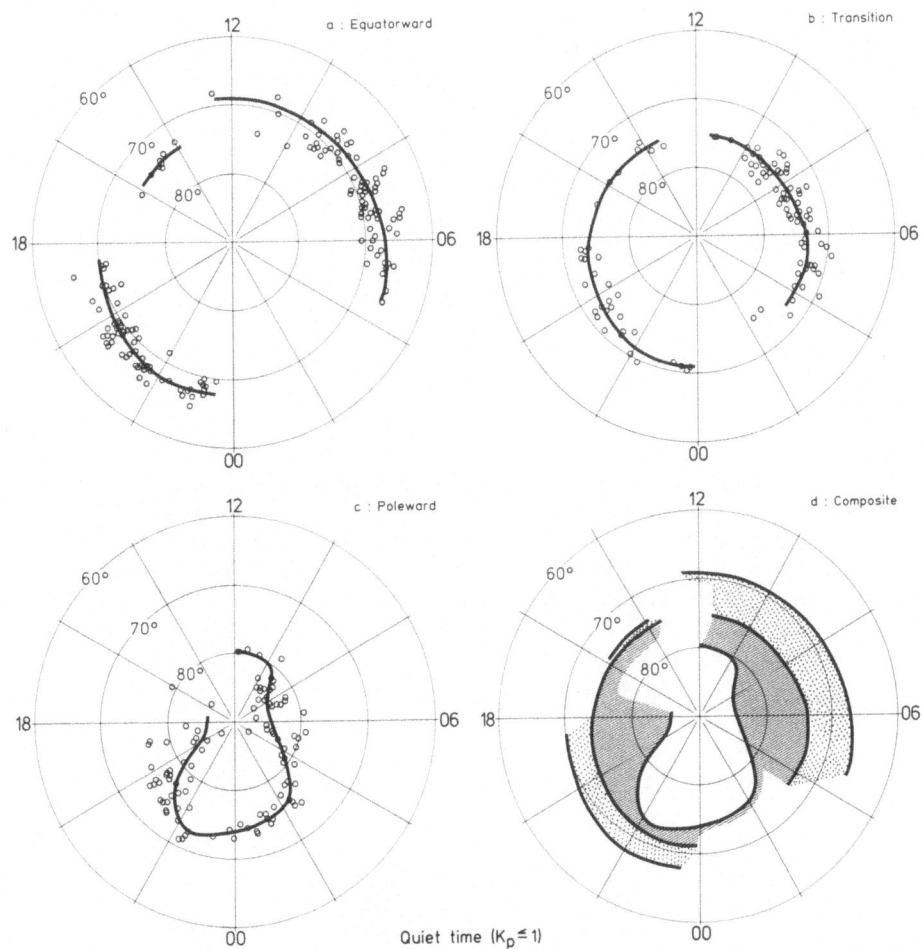

Figure 9. Position of observed electron precipitation boundaries in a corrected geomagnetic polar coordinate system, K_p = 0 or 1. (See text). (a) Equatorial, (b) transition, (c) poleward boundary and (d) smoothed boundaries from (a), (b) and (c).

84

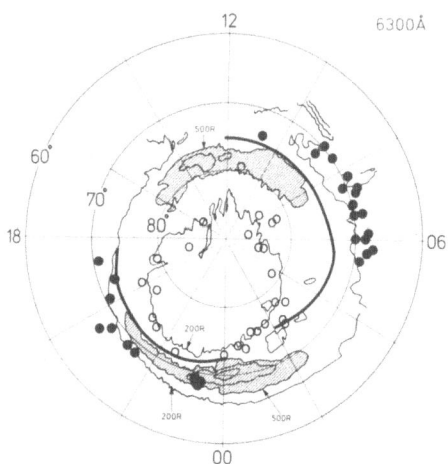

Figure 10. Position of observed electron precipitation boundaries in a corrected geomagnetic coordinate system, $K_p = 0$, superposed on a polar projection of contours of constant 6300 Å emission rate obtained from one ISIS-2 pass (Shepherd, 1979). Open circles denote poleward boundaries, and filled circles denote equatorward boundaries. Contours of 200 R, 500 R, 1 kR and 2 kR are shown. Also shown is the average transition boundary for orbits with $K_p = 0$. Hatching indicates emission rates greater than 500 R.

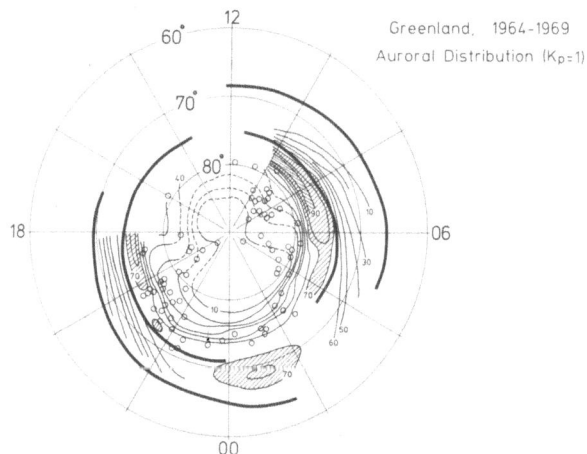

Figure 11. Position of observed poleward boundaries (open circles) of electron precipitation in a corrected geomagnetic coordinate system, $K_p = 1$, superposed on a graph showing average distribution of frequency of occurrence of auroras observed by the network of all-sky cameras in Greenland 1964-1969, $K_p = 1$ (From Lassen, 1972). Also shown are the average transition and equatorward boundaries of the electron precipitation from Fig. 9(d). Hatching indicates auroral frequencies above 70%.

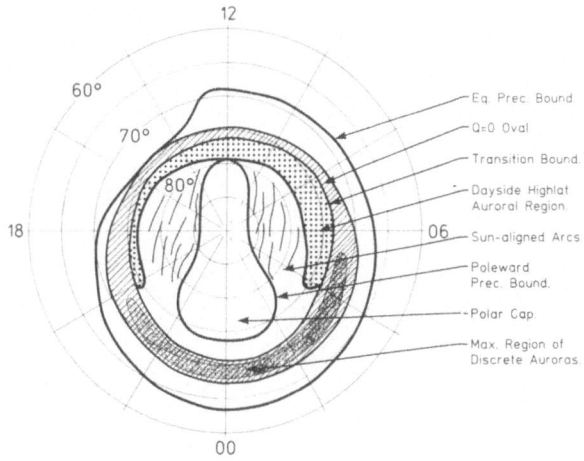

Figure 12. Model showing schematically in geomagnetic coordinates the position of the quiet auroral precipitation belt, together with the distribution of visual auroras: the Feldstein-Starkov auroral oval (hatched), the high-latitude daytime auroras (dotted), and the sun-aligned arcs.

ON QUANTIFYING THE DISTINCTIONS BETWEEN THE CUSP AND THE CLEFT/LLBL

PATRICK T. NEWELL AND CHING-I. MENG
The Johns Hopkins University
Applied Physics Laboratory
Laurel, Maryland 20707

Abstract: The distinction between the cusp (the region of fairly direct entry of magnetosheath plasma to low altitudes) and the cleft (the ionospheric signature of the magnetospheric boundary layer) is placed on a firm quantitative foundation. Case examples illustrating the difference are shown; when both regions are seen on a given pass the cusp lies poleward of the cleft, generally with a sharp boundary between the regions. The statistical differences are developed, for example the ion number flux in the cusp is approximately 4 times higher than in the cleft. The different responses of the cusp and the cleft to the interplanetary magnetic field B_z are documented; the cusp (cleft) ion fluxes increase (stay nearly unchanged) and the statistical local time width increases (decreases) when B_z changes from northward to southward. A brief review of the various previous attempts at distinctions between the cusp and the cleft is given, as is a brief summary of all known differences between the two regions as observed at low altitude.

1. Introduction

Particles with energies similar to, but usually somewhat above, magnetosheath values precipitate in a band which essentially covers the dayside. This band of precipitation is easily distinguishable from the several keV electron precipitation associated with the drifting of plasma sheet particles from the nightside; indeed spectrograms generally show a latitudinally sharp boundary between these regions. Early work established that the softer roughly magnetosheathlike band of particles on the dayside covered a low-altitude latitudinal extent of about 4° MLAT [Burch, 1973] and appeared to extend to at least the dawn-dusk meridian. The low energies and high flux levels strongly suggested that the ultimate source of this dayside precipitation was the magnetosheath, leading Heikkila and Winningham [1971] to name the band the "cusp", suggesting a funnel shaped region of direct shocked solar wind plasma entry into the magnetosphere. The longitudinally and latitudinally extended shape of the band soon led Heikkila [1972] to propose an alternate

87

P. E. Sandholt and A. Egeland (eds.), Electromagnetic Coupling in the Polar Clefts and Caps, 87–101.
© *1989 by Kluwer Academic Publishers.*

terminology, the "cleft"; and for several years thereafter the two terms were used interchangeably.

However external magnetospheric measurements indicate a much smaller region of direct magnetosheath entry [Paschmann et al., 1976; Haerendel et al., 1978], suggesting that it is unlikely that the entire low altitude band corresponds to a direct entry region. Formisano [1980] used Heos-2 electron data to distinguish between boundary layer and cusp electrons at low altitudes based on average energy; a distinction which was reported to match that between the magnetosheath and boundary layer population in the equatorial plane. The cusp as defined by Formisano was narrower than hitherto; but still very extended in local time.

More recently, Heikkila [1985] and others have suggested that the cusp be regarded as a limited subset of the cleft near noon, although it has not been clear how to make the distinction for a given low altitude satellite pass. Gussenhoven et al. [1985] identified the cusp with the minimum in electron average energy near noon in statistical maps of electron precipitation binned by magnetic latitude and local time, although it is difficult to discern a unique and consistent cusp in such maps. Because the cusp position is so variable, no bin position in a MLT/MLAT map very closely resembles the cusp as observed on individual passes. Most recently Lundin [1988] has used Viking data to make a distinction between the cusp proper and the cleft based (in addition to somewhat vague average energy criteria) on the isotropy or anisotropy of the electrons, with the former condition characterizing the cusp.

We propose the following conceptual definition: *"The low-altitude cusp is the dayside region in which the entry of magnetosheath plasma to low altitudes is most direct. Entry into a region is considered more direct if more particles make it in (the number flux is higher) and if such particles maintain more of their original spectral characteristics."* The present work will demonstrate that simple criteria, based primarily on the average energy of ion and electron precipitation, identifies a localized cusp distinct from the remainder of the cleft consistent with the conceptual definition given above. The resulting algorithm is applied to three year's worth of DMSP F7 data, and certain statistical results (for example, the probability of observing each region as a function of MLT) are reported. The present work is partly an extension of a previous study which was based on one year's data [Newell and Meng, 1988]; particularly this is true of Section 2, which deals with identifying the cusp on a case basis; and Section 3, which has certain basic statistical results. Some new effects as how the cusp differs from the cleft in responding to changes in the IMF are presented in Section 4; a critical discussion of various previous definitions of the cusp proper is given in Section 5; and finally, a summary of all known differences between the cusp and cleft are given in Section 6. It is the purpose of the present work to establish the distinction between the cusp and cleft as clearly and objectively as possible.

2. Distinguishing the Cusp and the Cleft in the DMSP F7 Data Set

The SSJ/4 package on the DMSP F7 satellite uses electrostatic analyzers to measure electrons and ions from 32 eV to 30 keV in 20 steps each. Hardy et al. [1984] have described the instrument and its calibration in detail. The detector apertures always point towards local zenith, which at the latitudes of interest here mean that only precipitating particles are observed. DMSP F7 is in a nearly circular sun-synchronous polar orbit at about 838 km altitude in the prenoon-premidnight local time meridian.

We start by showing, in Figure 1(a), a pass near 12 noon which typifies a clean "cusp" signature. Notice that the ion energy scale is inverted. The arrows denote the region of fairly direct magnetosheath entry; the poleward low energy ion "plume" results from the simple fact that the slow moving ions have sufficient time from entering the magnetosphere until precipitating to undergo significant poleward ExB deflection [Shelley et al., 1976; Reiff et al., 1977]. Within the arrows, the cusp can be seen to have a strong characteristic peak in ion energy flux at about 1 keV, a value comparable to that of the magnetosheath bulk flow and thermal velocities combined. The electron average (about 100 eV) and spectral peak (about 40-70 eV) energies are also quite low; and the number fluxes are high. In contrast, Figure 1(b) shows the cleft, or boundary layer, near 0830 MLT on December 1, 1983. A clear transition from the equatorward hard keV electron spectra which indicates a nightside plasma sheet origin and the poleward softer boundary layer/cleft fluxes can be observed at about -75.2° MLAT. However although the boundary layer fluxes are softer than the plasma sheet fluxes, they are still harder than the cusp fluxes of Figure 1(a). The number fluxes, particularly for the ions, are reduced by a factor of about 4-5; and the spectral peaks at magnetosheath energies are less pronounced. Discrete bursts of acceleration can be observed in the electrons, some resulting in energy fluxes large enough to correspond to visible aurora. Thus a comparison of Figures 1(a) and 1(b) makes clear the distinction between the cusp and boundary layer, at least for these cases (whether the same differences hold statistically is a question which will be addressed in the next section).

The question which now arises is whether there is one continuously varying region of roughly magnetosheathlike precipitation which gradually varies from looking like Figure 1(a) to looking like Figure 1(b); or whether there are in fact two morphologically distinct regions, the cusp and the cleft/BL. Figure 1(c), from a dayside high latitude pass near 1000 MLT on December 5, 1983, is an example illustrating that the latter is the case. Three distinct regions can be observed in Figure 1 moving from right (equatorward) to left (poleward). First is a region of a few keV electron precipitation (from 72-76° MLAT), these are plasma sheet particles which have drifted from the nightside. Immediately poleward, from 76-79° MLAT, is a region of a few hundred eV electrons and several keV ions; based on the spectral characteristics it is likely that this region maps to

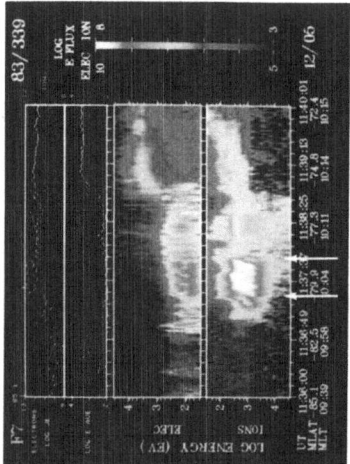

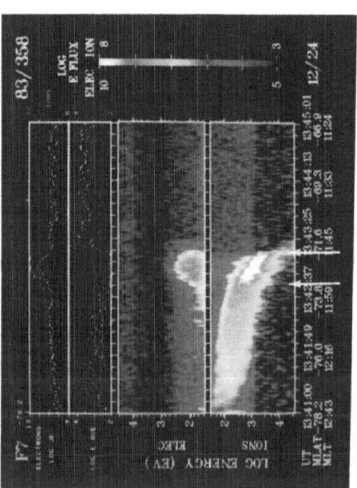

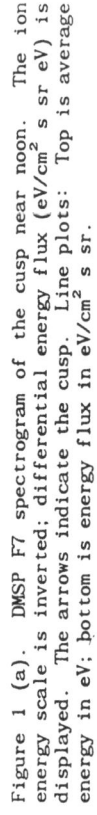

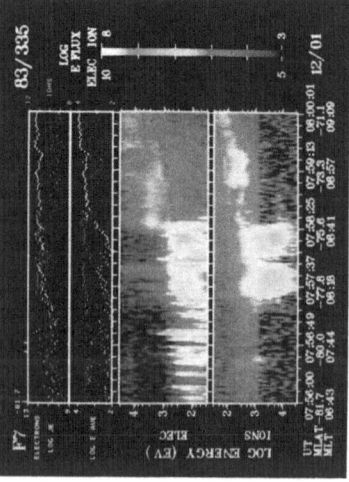

Figure 1 (a). DMSP F7 spectrogram of the cusp near noon. The ion energy scale is inverted; differential energy flux (eV/cm^2 s sr eV) is displayed. The arrows indicate the cusp. Line plots: Top is average energy in eV; bottom is energy flux in eV/cm^2 s sr.

Figure 1 (b). A cleft (boundary layer) pass near 0830 MLT. The precipitation is still roughly magnetosheathlike, but the number and energy fluxes are smaller and the average energy higher than near noon. The arrows indicate the cusp proper.

Figure 1 (c). A DMSP F7 pass near 1010 MLT. The cusp and the cleft/LLBL are both present at the same MLT. There is considerable variation within the cleft, but the number fluxes tend to be lower and the average energy higher than in the cusp.

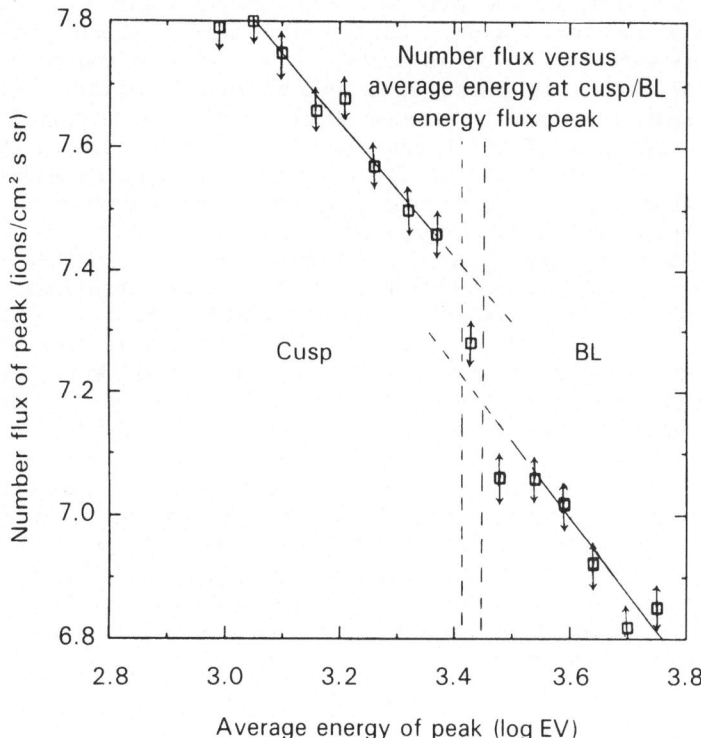

Figure 2. For each DMSP F7 dayside pass, the point of peak ion energy flux was recorded irrespective of region (cusp or boundary layer) and binned by average energy. Plotted is the (3 yr) average number flux versus log energy bin. Lower average energy passes have higher number fluxes; and the clear break which occurs at about 3000 eV corresponds to the transition between regions classified as cusp and those classified as cleft.

the boundary layer in the external magnetosphere (in this case, the low latitude boundary layer, LLBL). The third region, which lies furthest poleward and which is marked by arrows, is the cusp proper. This identification is based on a number of characteristics. The electron energy flux peak is around 50 eV and the ion energy flux peak is below 1 keV, both of which are closer to true magnetosheath values than is the distinct boundary layer region immediately equatorward. Note also that there is a strong flux enhancement moving across the fairly sharp cusp–LLBL boundary; the ion number flux in the cusp region is about 5 times that in the boundary layer region, a value which agrees with magnetosheath/boundary layer crossings in the external magnetosphere [Sckopke et al., 1981]. Note also the high energy ion tail (10s of keV) in the boundary layer, presumably of magnetospheric origins [e.g., Williams et al., 1988] which drops out moving into the cusp; a behavior which again mimics that of a crossing from the boundary layer into the magnetosheath. Thus based on fluxes and spectral properties the poleward region marked with arrows in Figure 1(c) is more magnetosheathlike, and so fits our conceptual definition of the low–altitude "cusp". Of course if only a few examples are studied there is the possibility of being misled by unusual conditions, or perhaps differing magnetospheric conditions. The systematics of the occurrence of regions such as that shown in Figure 1(a) (the cusp) and Figure 1(b) (the cleft/boundary layer) are discussed in the next section.

3. The Systematics of the Cusp and Cleft/BL

The peak number flux observed on a dayside pass and the average energy of the ions at that peak are negatively correlated quantities. Figure 2 illustrates this relationship. In this Figure, the peak number flux observed on each pass was binned by average energy (regardless of whether the region was cusp or boundary layer), and the number fluxes in each bin were averaged over a three year period. It can be clearly seen that regions with lower average energies (which are closer to the original magnetosheath values) are also the regions of higher number flux (i.e., more particles are entering). Figure 2 also evidences a clear break in the average energy/number flux relationship indicating the transition from cusplike to cleftlike regions. The transition energy is about 3 keV; which experience also indicates works well as a dividing point between cusp and cleft/boundary layer.

The plot analogous to Figure 2 for electrons also shows lower average energies corresponding to higher number fluxes; but the sharp break between cusp and cleft is not evident. Experience shows that making the distinction between cusp and cleft based solely on the electron average energy gives less clear and consistent results. However a useful rule of thumb is that cusp average electron energies lie below about 200 eV (the criteria advocated, although not always strictly adhered to, by Meng [1983] and coworkers).

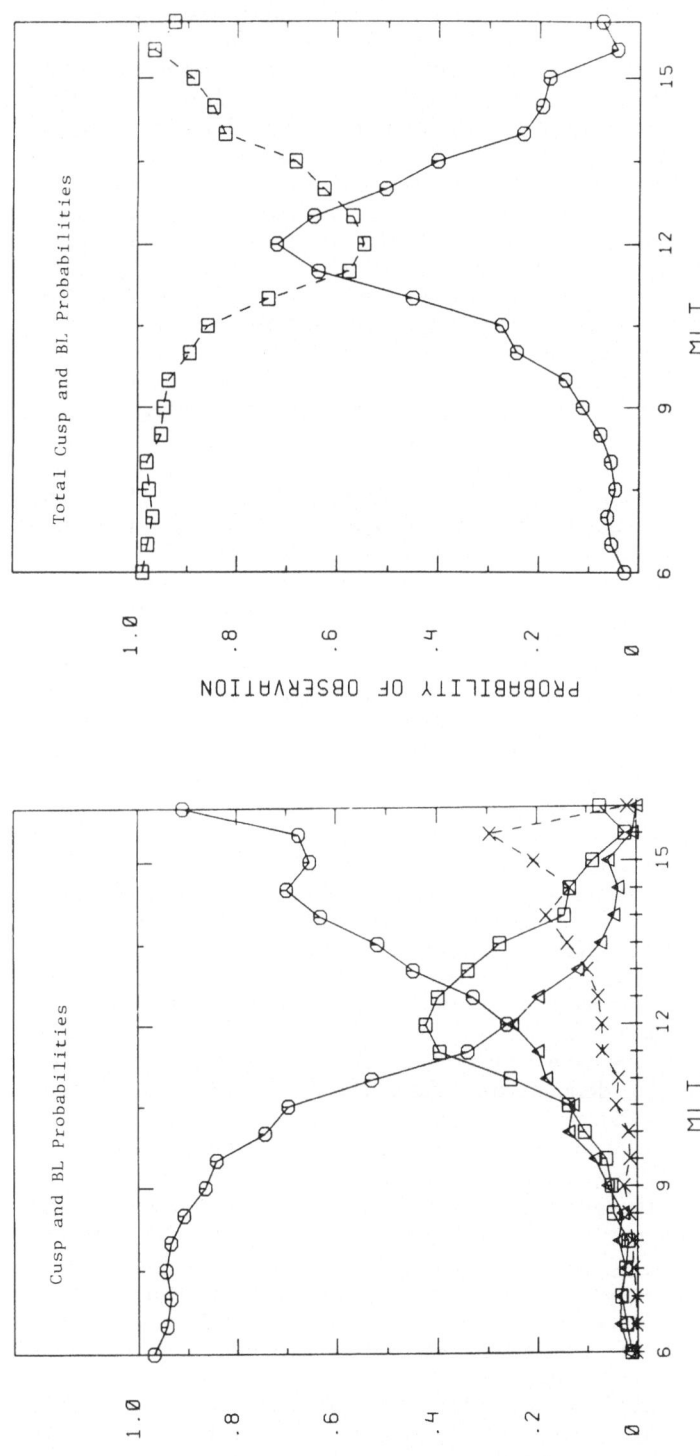

Figure 3 (a). The probability as a function of MLT (in half hour bins) of observing on a given DMSP F7 dayside pass (1) The cusp alone; (2) The cleft (BL) alone; (3) The cusp poleward of the cleft; (4) MLT changing too rapidly to allow a determination. The probabilities here sum to 1.

Figure 3 (b). The normalized probabilities of observing the cusp and the cleft. Since both regions can be observed on a single pass, the probabilities here sum to more than one.

In order to conduct a large scale statistical investigation of the distinction between the cusp and the cleft, the following simple criteria were used to identify the regions in an automated fashion. (i) If the energy flux of the ions (electrons) is less than 10^{10} eV/cm^2•s•sr (6×10^{10}) the region is neither cusp nor boundary layer; (ii) If the energy flux in either the 2 or 5 keV electron channel is greater than 10^7 eV/cm^2•s•sr•eV the region is neither cusp nor boundary layer, since such fluxes would indicate plasma sheet presence; (iii) If the first two criteria are met, the region is boundary layer if *either* 3000 eV < E_i < 6000 eV or 220 eV < E_e < 600 eV where E_i and E_e are the average electron and ion energies respectively; (iv) If the first two criteria were met and both 300 < E_i < 3000 eV and E_e < 220 eV the region was identified as cusp.

Slightly more than three years of DMSP F7 dayside polar passes, from December 1983 to December 1986, were categorized according to the above criteria, amounting altogether to about 12600 passes (previously Newell and Meng reported statistical results from one year's data). Figure 3(a) shows the probability, as a function of MLT, of observing each of four possible types of DMSP F7 dayside passes: (i) The cusp but not the boundary layer was observed; (ii) The boundary layer/cleft but not the cusp was observed; (iii) the cusp was observed poleward of the cleft; (iv) The MLT changed by more than an hour from entering a structure to exiting it, thus the pass is not useful for sorting.   The probability of observing a "pure" cusp signature -- that is, one without a boundary layer associated -- peaks at noon, where the boundary layer in the external magnetosphere is known to be thinnest. Incidentally, Figure 3(a) answers statistically the question of how typical the pass shown in Figure 1(c) is: such a pass can be expected at 10 MLT about 20% of the time.

Figure 3(b) presents the probability of observing the cusp and cleft/boundary layer as a function of MLT. The probabilities in this Figure sum to more than 1, since both can be observed at a given MLT; the probabilities of Figure 3 have been normalized after discarding the passes in category (iv) (rapid change in MLT) of Figure 3(a).   The cusp is observed to have its peak at noon.   The cleft/boundary layer has its minimum of occurrence at noon, with an essentially certain chance of observing some type of boundary layer (probably in some cases PSBL rather than LLBL) well away from noon.

Figures 3 (a) and (b) demonstrate that the practical (automated) definition of the cusp does indeed give a region of precipitation generally confined near noon.   Integrating the area under the curve gives a statistical average local time extent of 2.6 hours.   The energies within the limited region are closer to magnetosheath by definition (the identification algorithm is based on the average energies). Finally the ration of number flux in the limited region to that in the boundary layer is about 4.5 (the exact number depends somewhat on how the averaging is done).   Thus the criteria of the conceptual definition of the cusp is met: There does exist a spatially limited region in which magnetosheath plasma entry is more direct in the sense that more particles are entering, and their average energies are closer to the original magnetosheath values.   We

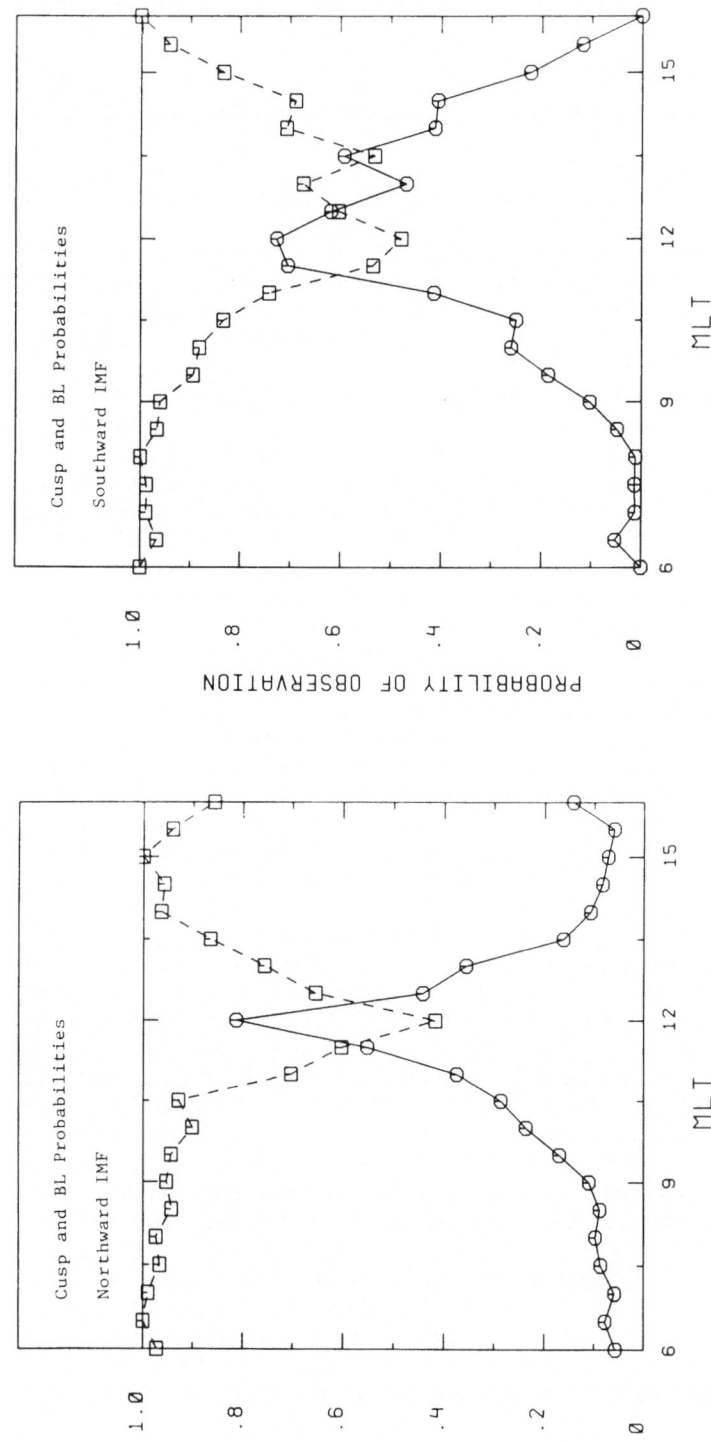

Figure 4 (a).   The same as Figure 3(b), except restricted to northward B_z cases.

Figure 4 (b).   The same as Figure 3(b), except for B_z southward.

conclude that the practical and conceptual definitions of the cusp given above coincide.

4. Differences in the Response of the Cusp and the Cleft to IMF B_z

This section compares the differing responses of the cusp and the cleft/boundary layer to the interplanetary magnetic field. For example, quantities associated with various phenomena are calculated for B_z southward and for B_z northward, and the ratios compared. The results prove to agree fairly well with intuition, and to support the belief that the cusp and cleft are topologically distinct regions.

Figures 4(a) and 4(b) show how the probabilities of observing the cusp and the cleft differ for positive and negative IMF B_z. Figure 4(a), which is calculated from the 2446 passes for which IMF data was available and B_z was positive, shows that the cusp is very localized near noon these conditions. Integrating the area under the curves for which there are enough passes (20) in each bin to make the statistics reasonably significant (6 to 15 MLT) we find that the cusp statistical local time width for northward B_z is 2.1 hours, while the cleft is 8.2 hours. It should be noted that our present fairly simple algorithm is not capable of distinguishing the PSBL from the LLBL; and some type of boundary layer (presumably mostly PSBL) is always seen at local times well away from noon. The LLBL is known to be thinnest at noon; and indeed is apparently often thin enough that no low altitude signature can be identified (our program requires a minimum .3° MLAT width).

For southward B_z, Figure 4(b) shows the cusp to be statistically wider in local time extent: integration over the same 6 to 15 MLT gives 2.8 hours. Interestingly, the probability of observing the cusp precisely at noon is actually less than for the B_z northward case. The reason proves to be the effect of the IMF B_y component, which controls the cusp latitudinal position more strongly for southward than for northward B_z [Newell et al., 1989]. For southward B_z the probability of observing the boundary layer declines, primarily near noon: the usual integration gives 7.6 hours. We thus conclude that the cusp has a greater statistical local time extent for southward than for northward B_z; whereas the boundary layer is more likely to be observed for northward B_z.

Another question concerning IMF response is the change in particle flux levels. Candidi and Meng [1984] have reported that the ratio of cusp electron precipitation for southward to northward B_z is "roughly a factor of two." They did not distinguish between the cusp and the cleft. It is of interest to investigate the question with higher precision, and considering the cusp and cleft separately (as well as including the ions). We computed the southward IMF to northward IMF ratio of particle precipitation in the boundary layer and the cusp. The results were that the ion (electron) number flux ratio $B_z(-)/B_z(+)$ for the cusp is 1.75 ± 12 (1.59 ± .10). For the boundary layer the same ratios were for ions and electrons respectively, 1.18 ± .05 (1.36 ± .08). Hence there is a clear difference in response to IMF in that the ion cusp fluxes are significantly larger for southward IMF than for northward IMF; but for the boundary layer no such result holds.

5. A Review of Selected Cusp Definitions

In this section we evaluate various cusp identifications that have been proposed elsewhere, in terms of their appropriateness or accuracy given the present research results.   We will primarily consider workers who sought to make a distinction between the cusp proper and the rest of the cleft.   However it is worth discussing briefly the apparent correspondence between the average electron energy criteria used here and that advocated by Meng [1983] and co-workers.   Meng [1983], who was not seeking to distinguish the cusp from the cleft, defined the cusp as a region of electron precipitation with an average energy $\lesssim$ 200 eV and number flux around 10^9 eV/cm^2-s-sr.   This is similar to the criteria presently advocated, although the earlier work [e.g., Meng 1983; Candidi and Meng 1984; Carbary and Meng 1986a,b] did not actually adhere to this criteria very strictly, and indeed often identifies as cusp regions we would now consider to be boundary layer. The more recent inclusion of ion detectors on the DMSP satellites has allowed for a much clearer distinction between the cusp and the cleft/boundary layer.

Potemra et al. [1977], studying several cases of dayside auroral oval passes by AE-C and AE-D, argued for the separation of the regions of roughly magnetosheath energy into equatorward boundary layer fluxes and poleward true "cusp" passes.   In some cases their identifications agree with those we have been led to by the present study.   However Figure 6 of Potemra et al. shows a cases where the equatorward region has by several orders of magnitude the more intense fluxes, and average energies quite compatible with the cusp.   The more poleward region has only a drizzle of very low energy particles, and we would identify this as the mantle.   Apparently because of the location of the "cusp" current system, the poleward portion was identified as the cusp despite the comparative dearth of particles.   Based on recent work [e.g., Erlandson et al., 1988] it seems reasonable to conclude that the "cusp" current system lies at the poleward portion of the cusp or in the mantle.   In any case, we regard an identification based on particle characteristics to be primary.

Heikkila [1985] has advocated this definition of the cusp and cleft: "The cleft is the low altitude region around noon of about 100 eV electron precipitation associated with 6300 A$^\circ$ emission, but containing also structured features of higher energy.   The cusp is a more localized region near noon with the cleft characterized by low energy precipitation only, having no discrete auroral arcs, but often displaying irregular behavior, presumably associated with the magnetic cusp."   Our statistical results tend to agree with the morphology implied by the spirit of this definition.   The chief difficulty with the definition as it stands is that it is too vague to be much practical   use   in   differentiating   the   cusp   and   the   cleft.

Incidentally, as this definition implies, an enhancement of 6300 A°
emissions are to be expected in both the cusp and the cleft, so that
the 6300/5577 ratio is of little value in making this distinction.
Even worse is to define the cusp as the region of 6300 A° emission
without considering the 5577 A° line at all [e.g., Eather, 1985];
since essentially all aurora include 6300 A° emissions.

Formisano [1980] used HEOS-2 electron observations to attempt a
distinction between the cusp and boundary layer at high and
mid-altitudes based on the electron spectra. This study has some
similarities to our own; the chief difference (aside from the shear
size of the present study) is that based on the electrons alone it is
much harder to develop a consistent picture of a cusp localized near
noon with higher fluxes; this is because of the variability of the
electron population in the boundary layer.

Finally we wish to briefly comment on the identification of the
"cusp proper" by the Viking team. We have found that their
identifications do indeed meet the criteria developed here for
distinguishing the cusp from the cleft, so that on a practical basis
we are in agreement. However there is some discrepancy between the
energies shown in the spectrograms and textual statements. For
example, Lundin [1988] states that the cusp electron precipitation is
"typically less 100 eV" and the ion precipitation is "typically less
than 1 keV". Figure 1 of Lundin shows that the electron population in
the cusp proper (as he defines it) reaches at least 200 eV, and there
are also significant ion fluxes above 1 keV. It would be more
accurate to state that within the cusp the spectral peaks satisfy the
criteria given by Lundin [1988].

6. The Difference Between the Cusp and the Cleft/Boundary Layer

It is the purpose of this section to summarize as clearly and
succinctly as possible the various differences between the cusp and
the cleft/boundary layer as observed at low altitudes.

The average energies, and especially the spectral peaks, are closer
to the original magnetosheath values in the cusp than in the cleft.
For reasons given above and elsewhere the average energies are
significantly higher than the spectral peaks. Within the cusp there
is a clear spectral peak in the ion flux around approximately 1 keV;
although the average energy can be as high as 3 keV. The electron
average energy is generally around 140 eV; although it can be as high
as 220 eV. Usually there is a electron flux spectral peak below 100
eV. Values higher than these are cleft, not cusp.

The ion number flux in the cusp is about 4-5 times larger than in
the cleft. The probability of observing the cusp peaks at 12 MLT;
whereas the cleft is thinnest and least likely to be observed at noon
[Newell and Meng, 1988]. Judging from the examples which have been
presented elsewhere, ion conics and the "cleft ion fountain", seem to
be indeed a feature of the cleft and not the cusp [Peterson et al.,
1988; Lundin, 1988]. (This is a happy coincidence, since Lockwood et
al. [1985] and Horwitz and Lockwood [1985]) were not distinguishing

between the cusp and the cleft). Possibly related is the fact that inverted Vs and discrete bursts of precipitation are often observed in the cleft, but not in the cusp [cf Figures 1(a)-(c)].

The cusp and the cleft have different responses to the IMF B_z component. When B_z turns southward, the statistical local time width (defined as the area under the probability of observation curve) of the cusp increases from 2.1 hours to 2.8 hours; whereas the probability of observing the cleft decreases, particularly near noon. The cusp ion number flux is much larger for southward than northward B_z (the ratio is 1.75 ± .12); whereas the cleft ion flux is little changed by the sign of B_z (1.18 ± .05).

When both the cusp and the cleft are seen on a given dayside pass, the cusp lies poleward of the cleft. Often high energy ions, presumably of magnetospheric origin, ($\sim$ 30 keV) can be seen in the cleft but not the cusp (as in Figure 1(c)). Often a velocity filter effect can be seen in the cusp ions, with lower energy ions being convected poleward of the cusp injection point (as in Figure 1(a)). A word of caution however: these last two characteristics are neither sufficient nor necessary conditions for identifying the dayside region that they are typical of, for higher energy ions can sometimes be observed in the cusp (as indeed they can in the external magnetosheath) and poleward convection can sometimes be observed in the boundary layer (since, after all, the LLBL is anti-sunward convecting).

Finally, cusp precipitation tends to be smoother (more homogeneous spatially and temporally) than does cleft precipitation. Particularly this is true of electron precipitation, which can be quite bursty and erratic in the cleft.

7. **Acknowledgements**

The DMSP F7 particle data were provided by AFGL (D. Hardy) through the World Data Center A in Boulder, Colorado. The IMF data was provided by Goddard Space Flight Center for which we thank R. Lepping and J. King. This work was supported by the Atmospheric Sciences division, National Science Foundation grant ATM-8713212, and by the Air Force Office of Scientific Research grant 88-0101 to the Johns hopkins University.

8. **References**

Burch, J. L. (1973) 'Rate of erosion of dayside magnetic flux based on a quantitative study of the dependence of polar cusp latitude on the interplanetary magnetic field', *Radio Science*, 8, 955-961.

Candidi, M., and Meng, C.-I. (1984) 'The relation of the cusp precipitating electron flux to the solar wind and the interplanetary magnetic field', *J. Geophys. Res.*, 89, 9741-9751.

Carbary, J. F., and Meng, C.-I. (1986a) 'Relations between the interplanetary magnetic field B_z, AE index, and cusp latitude', J. Geophys. Res., 91, 1549-1556.

Carbary, J. F., and Meng, C.-I. (1986b) 'Correlation of cusp latitude with B_z and AE(12) using nearly one year's data', J. Geophys. Res., 91, 10047-10054.

Eastman, T. E., Popielawska, B., and Frank, L. A. (1985) 'Three-dimensional plasma observations near the outer magnetosphere boundary', J. Geophys. Res., 90, 9519-9539.

Eather, R. H. (1985) 'Polar cusp dynamics', J. Geophys. Res., 90, 1569-1576.

Formisano, V. (1980) 'HEOS 2 observations of the boundary layer from the magnetopause to the ionosphere', Planet. Space Sci., 28, 245-257.

Gussenhoven, M. S., Hardy, D. A., and Carovillano, R. L. (1985) 'Average electron precipitation in the polar cusps, cleft, and cap', in The Polar Cusp, J. A. Holtet and A. Egeland (Eds), pp. 85-97, D. Reidel, Hingham, Mass.

Hardy, D. A., Schmitt, L. K., Gussenhoven, M. S., Marshall, F. J., Yeh, H. C., Shumaker, T. L., Hube, A., and Pantazis, J. (1984) 'Precipitating electron and ion detectors (SSJ/4) for the block 5D/flights 6-10 DMSP satellites: Calibration and data presentation', Rep. AFGL-TR-84-0317, Air Force Geophys. Lab., Hanscom Air Force Base, Mass..

Haerendel, G., Paschmann, G., Sckopke, N., Rosenbauer, H., and Hedgecock, P. C. (1978) 'The Frontside boundary layer of the magnetosphere and the problem of reconnection', J. Geophys. Res., 83, 3195-3216.

Heikkila, W. J., and Winningham, J. D. (1971) 'Penetration of magnetosheath plasma to low altitudes through the dayside magnetospheric cusps', J. Geophys. Res., 76, 883-891.

Heikkila, W. J. (1972) 'The morphology of auroral particle precipitation', in Space Research 12, 1343-1355, Akademie-Verlag.

Heikkila, W. J. (1985) 'Definition of the cusp', in The Polar Cusp, J. Holtet and A. Egeland, (eds), 387-395, D. Reidel Publishers, Hingham, Mass.

Horwitz, J. L., and Lockwood, M. (1985) 'The cleft ion fountain: A two-dimensional kinetic model', J. Geophys. Res., 90, 9749-9762.

Lockwood, M., Chandler, M. O., Horwitz, J. L., Waite, J. H., Moore, T. E., and Chappell, C. R. (1985) 'The cleft ion fountain', J. Geophys. Res., 90, 9736-9748.

Lundin, R. (1988) 'Acceleration/heating of plasma on auroral field lines: Preliminary results from the Viking satellite', Annales Geophysicae, 6, 143-152.

Meng, C.-I. (1983) 'Case studies of the storm time variation of the polar cusp', J. Geophys. Res., 88, 137-149.

Newell, P. T., and Meng, C.-I. (1988) 'The cusp and the cleft/boundary layer: low altitude identification and statistical local time variation', *J. Geophys. Res.*, 93, 14549-14556.

Newell, P. T., Meng, C.-I., Sibeck, D. G., and Lepping, R. (1989)"Some low altitude cusp dependencies on the interplanetary magnetic field', *J. Geophys. Res.*, 94, 1989.

Paschmann, G., Haerendel, G., Sckopke, N., Rosenbauer, H., and Hedgecock, P. C. (1976) 'Plasma and magnetic field characteristics of the distant polar cusp near noon: The entry layer', *J. Geophys. Res.*, 81, 2883-2899.

Peterson, W. K., Andre, M., Crew, G. B., Persoon, A. M., Engebretson, M., and Pollock, C. (1988) 'Heating of thermal oxygen ions near the equatorward boundary of the mid-altitude polar cusp: Dynamics Explorer Observations', *EOS*, 69, 1379.

Reiff, P. H., Hill, T. W., and Burch, J. L. (1977) 'Solar wind plasma injection at the dayside magnetospheric cusp', *J. Geophys. Res.*, 82, 479-491.

Shelley, E. G., Sharp, R. D., and Johnson, R. G. (1976) 'He^{++} and H$^+$ flux measurements in the dayside cusp: Estimates of convection electric field', *J. Geophys. Res.*, 81, 2363-2370.

Sckopke, N., Paschmann, G., Haerendel, G., Sonnerup, B. U. O., Bame, S. J., Forbes, T. G., Hones, E. W., and Russell, C. T. (1986) 'Structure of the low-latitude boundary layer', *J. Geophys. Res.*, 86, 2099-2110.

Williams, D. J., Mitchell, D. G., Frank, L. A., and Eastman, T. E. (1988) 'Three-dimensional magnetosheath plasma ion distributions from 200 eV to 2 MeV', *J. Geophys. Res.*, 93.

HEATING OF THERMAL IONS NEAR THE EQUATORWARD BOUNDARY OF THE MID-ALTITUDE POLAR CLEFT

W. K. Peterson
Lockheed Palo Alto Research Laboratory
Palo Alto, California 94304

M. André
Swedish Institute of Space Physics
University of Umeå
Umeå S-901, 87 Sweden

G. B. Crew
Center for Space Research
MITCambridge, MA 02139

A. M. Persoon
Department of Physics & Astronomy
University of Iowa
Iowa City, Iowa 52242

M. J. Engebretson
Augsberg College
Minneapolis, MN 55454

C. J. Pollock
U. S. National Research Council
Marshall Space Flight Center
Huntsville, AL 35812

M. Temerin
Space Science Laboratory
University of California
Berkeley, CA 94720

ABSTRACT. Intense, energetic oxygen ions are frequently observed to be heated near the equatorward boundary of the mid-altitude polar cleft on the Dynamics Explorer -1 satellite. These observations confirm reports from the VIKING satellite. In this paper we present examples of heating of oxygen ions equatorward of the 'central' cusp/cleft region and use simultaneously obtained low-frequency electric and magnetic field observations to quantitatively test resonant and non-resonant heating processes. We find that, within observational and model uncertainties, the heating rates inferred from observed electric field spectra are large enough to produce the observed perpendicular oxygen temperatures in time scales determined by the measured poleward drift rate of thermal oxygen ions. The examples presented provide a good comparison of basic plasma theory and observation.

Introduction

The entry of intense fluxes of shocked solar wind (magnetosheath) plasma into the polar cusp region creates plasma waves and instabilities which heat and help extract significant quantities of plasma from the ionosphere. The wide variety of plasma waves

P. E. Sandholt and A. Egeland (eds.), Electromagnetic Coupling in the Polar Clefts and Caps. 103–113.
© *1989 by Kluwer Academic Publishers.*

observed in the cleft and cusp regions, both from the ground and in-situ, indicate that several different plasma processes are involved in heating ionospheric plasma. To understand this aspect of electromagnetic coupling in the mid-altitude cusp/cleft region, it is necessary to both understand how energy is transferred to ionospheric plasma by the different processes and to determine the relative geophysical importance of various microphysical processes.

Recent reports from the VIKING satellite (Andre et al., 1988) have revealed unexpected, intense heating of thermal ions at mid-altitudes near the equatorward edge of the cusp/cleft region, not in the region of the most dense magnetosheath plasma. Andre et al. found that the ion heating region was closely associated with a strong spatial gradient in the low-frequency (i.e. in the range of ion gyrofrequencies) electric field turbulence. The ion heating regions were identified by angular distributions that show peaks not aligned with the local magnetic field.These ion distributions, which are commonly called ion conic distributions, were first described by Sharp et al. (1977). Andre et al. suggested that 1) the observed ions were oxygen and 2) that they were resonantly heated by the simultaneously observed broad-band, low-frequency plasma waves that have a very sharp gradient in intensity equatorward of the cusp/cleft region. The process suggested by Andre et al. is that thermal oxygen equatorward of the cusp/cleft region drifts poleward into the region of intense low-frequency waves and is heated. This process could be geophysically important because it results in a new, high altitude source of magnetospheric energetic O+ ions. The geophysical conditions (geometry) in the cusp also provide an excellent laboratory in which the predictions of a resonant heating mechanism (Chang et al, 1986) and a non-resonant heating mechanism (Temerin and Roth, 1986) can be tested with in-situ data.

In this paper we demonstrate that data obtained by the Dynamics Explorer (DE) -1 satellite qualitatively confirm the model presented by Andre et al. (1988). We then present examples of mid-altitude (r/Re ~4) heating of oxygen ions equatorward of the cusp/cleft region obtained from the DE -1 satellite. We then use simultaneously obtained low-frequency electric and magnetic field observations to quantitatively test the resonant and the non-resonant heating process.

Observations

The DE -1 Satellite was launched in September 1981 into a 4.7 r/Re by 400 km, 90 degree inclination orbit. The satellite has a complement of plasma instruments including two ion mass spectrometers (Shelley et al. 1981, Chappell et al. 1981), a multi-function plasma wave instrument (Shawhan et al., 1981) and a high resolution magnetometer (Farthing et al. 1981). The ion distributions and plasma wave data presented here were obtained in the satellite spin-plane which lies in the orbit-plane. The basic measurement interval of 6 seconds is set by the satellite spin period.

The cusp/cleft region can be identified by its unique signatures in each of the plasma instruments. Peterson (1985) has presented and discussed the unique cusp signature in the Energetic Ion Composition Spectrometer (EICS) data. We have used this characteristic signature to identify intervals when the satellite passes through the cusp/cleft region and have found that intense energetic oxygen conic distributions were frequently observed equatorward of the most intense downward fluxes of energetic H+ and He++ ions as suggested by Andre et al. (1988). Figure 7 in Peterson (1985) presents a typical example.

We also occasionally detected intense H+ and He+ ion conic distributions equatorward of the central cusp/cleft region.

To quantitatively test the resonant heating mechanism suggested by Andre et al. (1988) we have assembled high resolution data from the relevant DE -1 instruments for three cusp crossings in 1984 where oxygen ion conics were also observed. We required that several complete energy-mass-angle ion distributions be acquired between the region of most intense downward flowing H+ ions and the equatorward onset of the observed oxygen ion conics. This requirement led us to consider only cusp passes above r/Re ~3.5; at these altitudes the local oxygen ion gyrofrequency is ~1 Hz.

In addition to data from two ion mass spectrometers, plasma wave data in the frequency range 0 to 8 Hz was obtained from the Fourier transform of the 16 sample per second DC electric field measurements (Shawhan et al., 1981). In one case three dimensional magnetic field spectra over the same frequency range were available. Figures 1 and 2 present an overview of the observed ion mass-energy-angle spectra and electric field data from two of the three events studied in detail. Note that the time scales are not uniform. Universal time is indicated below the third, fourth, and fifth panels. Altitude in units of r/Re, McIlwain L parameter, magnetic local time (MLT) and geomagnetic latitude (MLAT) are indicated only below the fifth panel.

Figure 1 presents data obtained on August 10, 1984 when the spacecraft was crossing the cusp into the polar cap near noon magnetic local time. The top two panels show EICS energy-time spectrograms for hydrogen (top) and oxygen for a 7 minute interval starting near 01:20. The observed count rate, in units of counts per second which is proportional to number flux, is encoded using the gray bar on the right. During this interval the EICS instrument sampled H+ and O+, at 15 logarithmically spaced energy steps from 10 eV to 17 keV at 24 pitch angles in 24 seconds (4 spin periods). The display illustrates that energy sampling is not contiguous by showing gaps between energy bins. The data obtained during each instrument cycle have been sorted into energy and pitch angle for presentation resulting in the apparent 24 second spin period shown in the pitch angle vs.time trace in the third panel. The characteristic hydrogen cusp signature begins in the instrumental cycle starting at ~01:23:30. This butterfly pattern has been discussed at length by Peterson (1985) and Burch et al. (1982). The region of most intense downward flowing hydrogen ions, and therefore downward flowing cusp electrons,(Burch, 1985) was observed after 01:27. On this crossing of the cusp, only one plasma injection region was observed. Typically more than one injection event such as that beginning at 01:23:30 is observed (see, for example, Peterson, 1985).

Intense, relatively local, heating of the oxygen ions is indicated by the conic-type angular distributions that appear in the instrument cycle starting just before 01:24 in the second panel. Because of the way that the data are sampled, it is possible that oxygen heating began during the previous instrumental cycle that included the onset of the cusp/cleft ion signature. The oxygen angular distributions shown do not all peak at the same angle; they peak at smaller angles at lower energies. Klumpar et al. (1985) have noted the frequent occurrence of this type of angular distribution and have called them bi-modal. They pointed out that bi-modal distributions could be produced by both parallel and perpendicular acceleration mechanisms occurring together or in series on the same magnetic field line. Temerin (1986) and Chang et al. (1986) have shown that transverse heating over an extended altitude range also produces bi-modal conic distributions.

The bottom panel shows the observed oxygen ion counting rate from the Retarding Ion Mass Spectrometer (RIMS) instrument for the interval from 01:10 to 1:34. The RIMS

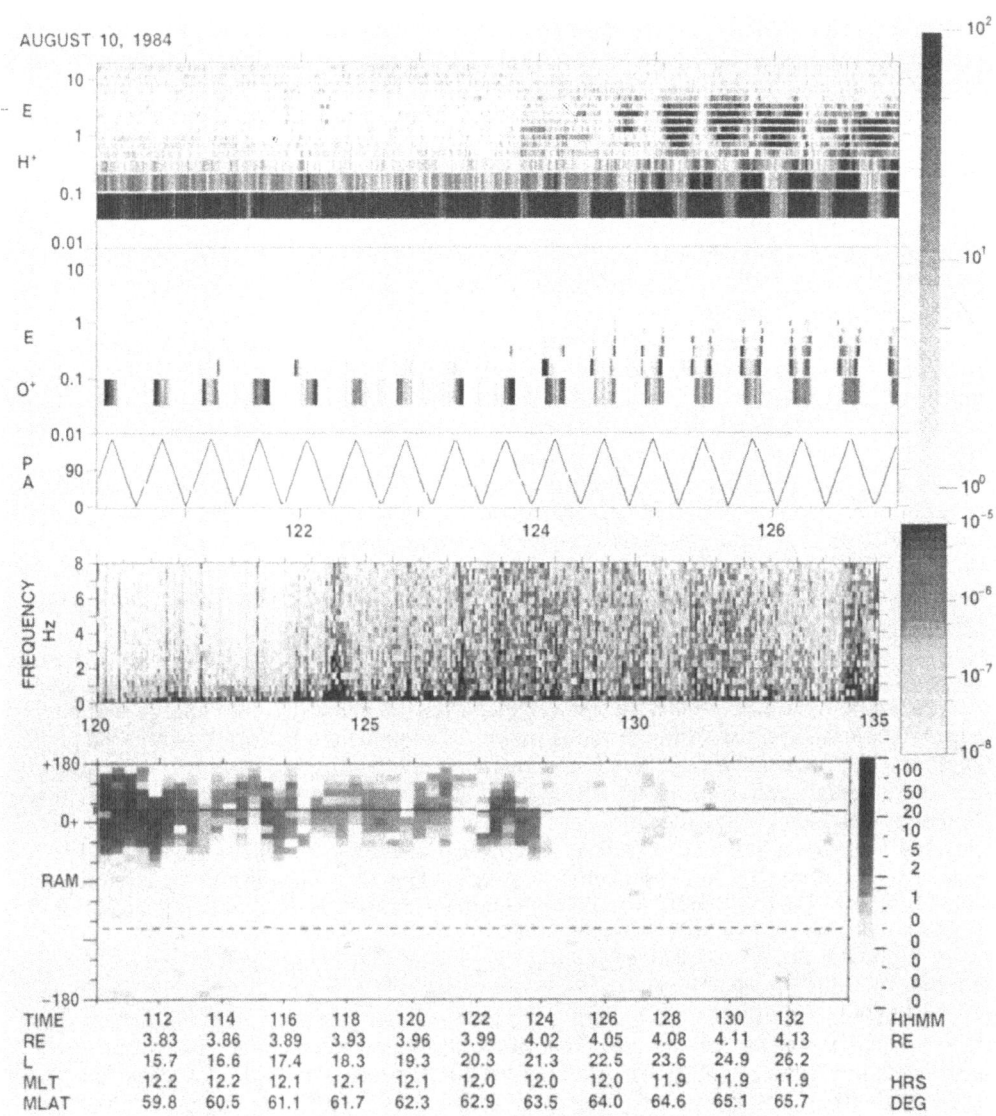

Figure 1. Ion and plasma wave data obtained from Dynamics Explorer -1 on August 10, 1984. Note that the time scale is different for the bottom two panels.

count rate, which is proportional to the number flux of ions with energies above the spacecraft potential (~1 eV) and below ~50 eV, is encoded using the gray bar shown on the right. The RIMS data are presented in an angle-time spectrogram format. The solid and dotted lines indicate directions of the magnetic field. The RIMS data show a strong flux of thermal oxygen ions flowing up the magnetic field line before ~01:24. Upflowing thermal oxygen ions are not observed after ~01:24. Comparison with the energetic oxygen spectra in the second panel suggests that the thermal oxygen ions are energized out of the RIMS energy range after ~01:24.

The fourth panel in Figure 1 presents the electric field power-spectral density derived from the 16 sample per second DC electric field measurements obtained from the long wire antenna. The power-spectral density in units of $V^2/M^2/Hz$ has been encoded using the gray bar on the right. These measurements were made in the plane of the spacecraft and therefore are an average of the electric field spectrum both parallel and perpendicular to the local magnetic field. The data are presented with 1/3 Hz frequency and 3 second time resolution. Near 01:21 and after ~01:23:30 the wave power is more intense.

Detailed examination of the energetic oxygen ion distributions from 01:23 to 01:25 show that they are characterized by drifts in the spacecraft frame of ~20 km/s upward and ~10 km/s poleward and perpendicular to the local magnetic field in the satellite spin-plane. Since the satellite motion is 3km/sec poleward, the poleward drift of the ions is 10 + 3 = 13 km/s. Near 01:23 the RIMS oxygen distribution has a density of ~1 cm^{-3} and a temperature on the order of a few eV. Near 01:25 the EICS oxygen distribution has a density of ~1 cm^{-3} and a bulk temperature perpendicular to the local magnetic field of ~150 eV.

The data in Figure 1 are consistent with the model suggested by Andre et al. (1988). Specifically, thermal oxygen is present equatorward of the cusp/cleft and equatorward of a relatively sharp boundary in the low-frequency wave field; thermal oxygen is drifting poleward and is energized perpendicularly to the magnetic field poleward of the low-frequency wave boundary. To quantitatively check this model we proceed as follows: we compare the measured oxygen temperature perpendicular to the magnetic field obtained at a point poleward of the low-frequency wave boundary with the expected temperature inferred from the product of a calculated heating rate and an estimated time for ions to move from the wave boundary to the point where the perpendicular temperature is determined. To estimate this 'exposure' time we note that the distance the satellite travels from the time it crosses the wave boundary at 01:23:30 to 01:25 where we have determined the oxygen perpendicular temperature is ~270 km (90 s x 3 km/s). If the wave boundary is stationary with respect to the earth's magnetic field and perpendicular to the direction of satellite motion, then it takes plasma approximately 20 seconds to drift this distance (20 s ~270 km /(10+3)km/s). If the boundary was oblique to the satellite path or was moving poleward or equatorward the real 'exposure' time could be longer or shorter than 20 seconds. The rate of cusp/cleft movement at radial distances of r/Re ~ 4 is not expected to exceed ~1 km/s, however, even during quite active periods (Eather et al. 1979).

Using the exposure time of 20 seconds inferred from the simplest possible geometry we see that a heating rate of ~7 eV/s ((150-10)eV / 20 s)would be required to energize the thermal plasma observed before 01:23:30 to the 150 eV energy observed at 01:25. We now use the average power-spectral density in the electric field data from 01:24 to 01:25 and either a resonant heating mechanism proposed by Chang et al. (1986) or a non-resonant heating mechanism proposed by Temerin and Roth (1986) to calculate heating rates.

The observed average power-spectral density from 01:24 to 01:25 in the vicinity of the oxygen gyrofrequency (~0.85 Hz) is ~1 x 10^{-5} $V^2/M^2/Hz$. Using equation 4 from Chang et al. (1986) we obtain a heating rate of 30 eV/s. The product of the estimated heating rate and exposure time is ~4 times the observed oxygen ion heating of ~150 eV which is consistent with the observations, given the expectation that much of the wave power is in modes which do not participate in ion heating. To calculate the non-resonant heating rate using equation 4 from Temerin and Roth (1986) we use the power-spectral density measured at half the oxygen gyrofrequency (~3 x 10^{-5} $V^2/M^2/Hz$), assume a perpendicular wavelength of 100 km, and a bandwidth of 0.5 Hz. The 100 km perpendicular wave length was chosen as a typical cusp/cleft scale size; smaller wavelengths would correspond to higher non-linear ion heating. These assumptions lead to a heating rate of ~0.9 eV/s. The non-resonant heating rate is sensitive to the assumptions made on both the average and peak electric field strength and this leads to greater uncertanties in our estimate. If the electric field power spectral density has been thus undervalued, the non-resonant heating mechanism can account for the observations. We defer a discussion of uncertainties in the above comparisons until we have presented the data from a second example.

Figure 2 displays data acquired on March 7, 1984 in the same format as Figure 1 except that the RIMS ions displayed in the bottom panel are He+ rather than O+. During the interval on March 7, 1984, the flux of O+ ions below ~50eV was below the RIMS instrumental threshold. Note also that on March 7 the satellite direction was from the polar cap region through the cusp/cleft and into the magnetosphere which is opposite to that shown in Figure 1. The cusp/cleft region on this pass was highly disturbed, with multiple injection boundaries detected, consistent with a relatively high level of magnetic activity. (The Kp index was 4- and the hourly average AE index was 409.) One of the several cusp injection events can be identified by the characteristic signature of precipitating hydrogen ions near 16:34. Note that near 16:34, in the top panel of Figure 2, there is an onset of downflowing energetic (~5 keV) hydrogen above a strong flux of low energy hydrogen, that disperses to lower energies (~1 keV) poleward, i.e. backward in time. The time 16:34 also corresponds to a very sharp boundary in the electric field power- spectral density (fourth panel), and the disappearance of low energy upflowing He+ ions. Poleward of 16:34 energetic bi-modal oxygen conics (second panel) are clearly present; equatorward are beams of energetic upflowing oxygen.

Detailed examination of the energetic oxygen distributions obtained poleward of 16:34 at 16:33:45 and equatorward at 16:35 show that they are both characterized by bulk upward velocities of ~40 km/s and densities of ~0.25 cm^{-3}. During the interval from 16:33:30 to 16:35 the drift of the plasma perpendicular to the magnetic field and in the satellite spin-plane fluctuated but had an average poleward velocity of ~20 km/s. The oxygen distribution had a characteristic perpendicular temperature of ~340 eV at 16:33:45. After 16:34 (i.e. equatorward of the region of intense low-frequency waves) the parallel and perpendicular temperatures of the oxygen distribution were approximately equal (~40 eV). The satellite traveled ~45 km from the time of the last complete measurement cycle in the intense low-frequency waves to the boundary at 16:34 (15 s x 3 km/s). If we assume that the wave boundary is stationary with respect to the magnetic field and perpendicular to the orbit-plane, approximately 3 seconds (~45 km / (20-3) km/s) are required for the

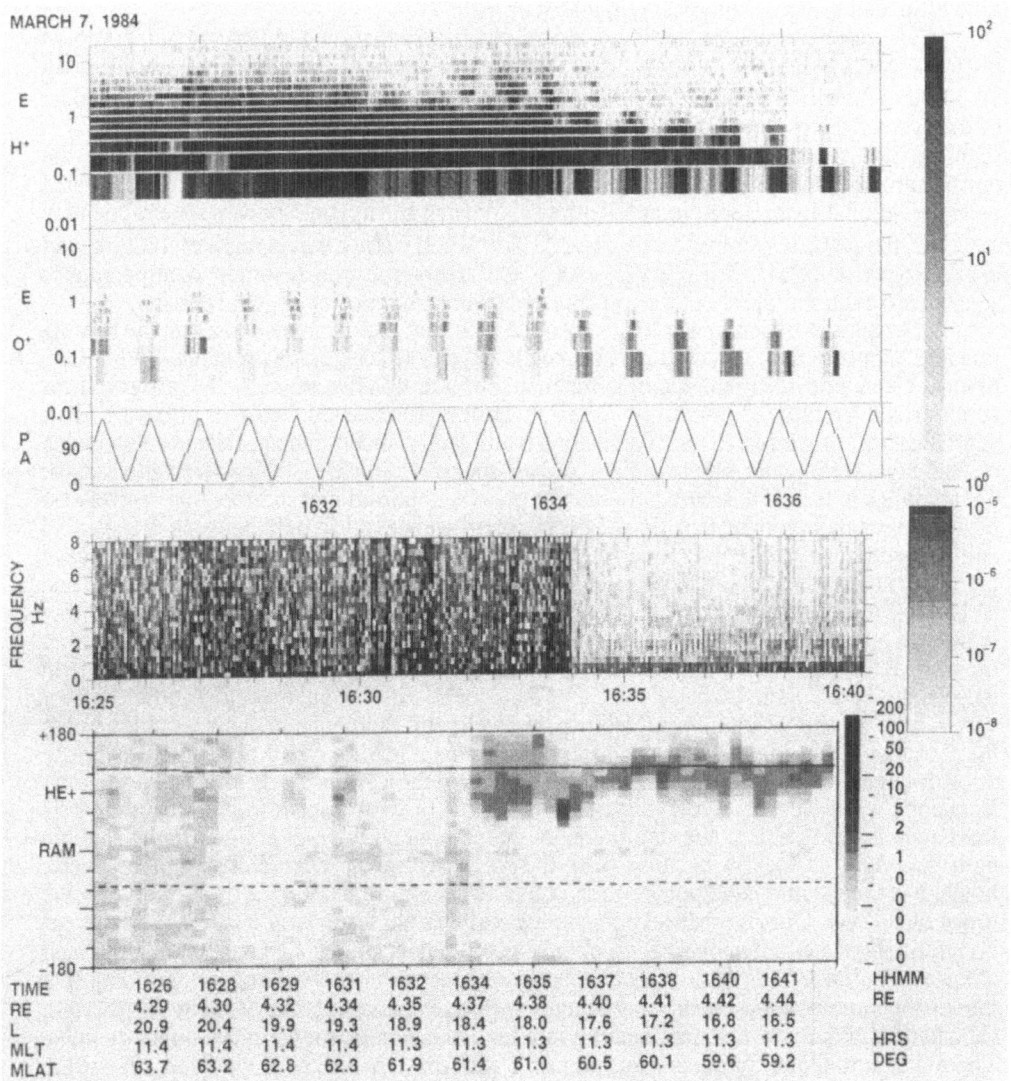

Figure 2. Ion and plasma wave data obtained from Dynamics Explorer -1 on March 7, 1984. Note that the time scale is different for the bottom two panels.

plasma to drift from the boundary to the observation point. Thus a heating rate of 100 eV/s (~(340-40)eV /3 s) would be required to energize the plasma in the 3 second exposure time estimated using the simplest possible geometry.

The resonant oxygen heating rate obtained from equation 4 in Chang et al. (1986) is 60 eV/second using the observed average power-spectral density from 16:33:45 to 16:34:00 of ~2 x 10^{-5} V^2/M^2/Hz near the oxygen gyrofrequency (~0.63 Hz). The product of the estimated exposure time (3 seconds) and heating rate is 180 eV which is slightly less than the 300 eV observed heating, but still consistent with the observations within the uncertainties of the model and the measurements. The non-resonant heating rate calculated using equation 4 from Temerin and Roth (1986) using the average power-spectral density near half the oxygen gyrofrequency of ~5 x 10^{-5} V^2/M^2/Hz, a wavelength of 100 km, and a bandwidth of 0.3 Hz is ~2 eV/s. As noted above the non-resonant heating rate is sensitive to assumptions about the average and peak power power-spectral density.

Because of uncertainties in the absolute values of the parameters used in the heating rate and exposure time calculations above, the oxygen heating rates predicted by both the resonant and non-resonant heating mechanisms are consistent with the observations summarized in Figures 1 and 2. The ion temperatures used and the poleward drift velocities are known to ± 30%. There are slightly larger uncertainties in the absolute value of the measured wave electric field power-spectral densities. However the largest uncertainties in the comparison between the observations and measurement are introduced by 1) the uncertainty in the partition of the observed wave electric field between that parallel and perpendicular to the magnetic field; 2) the assumption that the low-frequency wave boundary is aligned with the local magnetic field and perpendicular to the satellite path; and 3) lack of any information on the polarization of the low-frequency electric field. Only wave energy perpendicular to the local magnetic field is effective in heating ions by either the resonant or non-resonant mechanism. Under the assumption that the wave electric field is isotropic in the plane perpendicular to the magnetic field, the actual electric field power would be three times that inferred from a single electric field antena. This would increase the resonant heating rate 3 times and the non-resonant heating rate 9 times the previously quoted values. The exposure time used in the calculation is made assuming that the low-frequency wave electric field is aligned and stationary with respect to the local magnetic field and perpendicular to the satellite path. If the boundary were inclined to the satellite path generally or locally because of a surface wave in the cusp/cleft/magnetospheric boundary, the calculated exposure time could be several times longer or shorter than the times calculated. Only right-hand polarized waves will not resonantly heat ions and there is no information about the polarization of the electric field available. Chang et al. (1986) and Crew et al. (1988) have found reasonable agreement with observations by assuming that the left-hand polarization near the oxygen ion frequency was on the order of 10 per cent. Calculation of the non-resonant heating rate involves significant additional uncertainties. These are introduced by the assumption of a 100 km perpendicular wavelength and the sensitivity of the mechanism to the electric field strength. The non-linear nature of the mechanism means that local intensification of the power-spectral density would provide significantly more heating than the same power averaged over a longer interval. For these reasons we find that even though the non-resonant heating rates calculated from the time averaged spectra are significantly lower,the mechanism might, under conceivable circumstances contribute a substantial fraction of the heating needed to form the observed distributions. Unfortunately, the uncertainties in the observations do not allow a quantitative determination of the relative contribution.

On March 7, 1984 there are several additional observations that are consistent with the model proposed by Andre et al. (1988) which we are testing. Specifically the RIMS instrument observed singly charged helium ions after ~16:34, and three dimensional magnetic field spectra are available over the frequency range 0 to 8 Hz. We can use the average electric field spectra to estimate the helium ion heating rates for both the resonant and non-resonant mechanisms. The resonant mechanism gives a helium heating rate of ~70 eV/s; the non-resonant mechanism gives a rate ~0.4 eV/s. If we use three times the average power-spectral density at half the helium gyrofrequency, the non-resonant helium heating rate is ~3 eV/s. We do not have observations of the energetic helium spectra so we can not verify that helium ions acquired enough energy to move them out of the RIMS energy range (0-50 eV) poleward of 16:34; so we can not rule out the possibility that there is no energetic helium poleward of 16:34. On March 7 the magnetometer was in its high sensitivity mode and the three-dimensional, 0-8 Hz magnetic field power spectra obtained from the 16 vector samples per second were above the instrumental noise limit. Examination of these spectra revealed no measurable elliptical or circular polarization perpendicular to the magnetic field at frequencies near the oxygen gyrofrequency. Only the resonant heating mechanism requires left-hand polarized waves. The lack of a measurable polarization is possible if the waves responsible for the energization have a broad wavelength spectrum. Using the electric and magnetic field power-spectral density near 16:34 we calculate an Alfvén velocity of ~3 x 10^3 km/s which is consistent with the value estimated using the measured mass density and magnetic field intensity. In addition field aligned currents observed from 16:31 to 16:40 were significantly less than those observed in the region of most intense cusp ion precipitation before 16:31.

Discussion

We have examined the energetic ion composition data for many transits of the mid-altitude cusp/cleft region by the Dynamics Explorer -1 satellite and have found that oxygen ions frequently are heated well equatorward of the region of most intense downward fluxes of magnetosheath plasma. These observations are consistent with those reported by Andre et al. (1988) and confirm the existence of an unexpected mechanism for energizing oxygen at mid-altitudes in the cusp/cleft.

We have used the data summarized in Figures 1 and 2 to quantitatively confirm the model proposed by Andre et al. (1988) to explain this unexpected heavy ion heating region. We have found that the heating rates inferred from resonant (Chang et al. 1986) and non-resonant (Temerin and Roth, 1986) heating mechanisms are large enough to produce the observed perpendicular oxygen temperatures in time scales determined by the poleward drift rate of thermal oxygen ions found well equatorward of the cusp/cleft region. We have also discussed the uncertainties in the measurements and geometry of the the model and the measurements.

The examples presented provide an unusually clean comparison of theory and observation. As noted by several investigators (e.g. Peterson et al., 1988) unambiguous comparisons between space plasma data and theory have proven to be extremely difficult. We discuss below two of the ambiguities in the data reported here that are perhaps inconsistent with the simple model we have tested. First there is a feature in the data for the August 10 event (Figure 1) that can be interpreted as an electrostatic shock which could be responsible for some of the oxygen heating observed. Secondly we consider the source and nature of the sharp boundary in the low-frequency wave field. We also note that we expect that other heating mechanisms are operative in the mid-altitude cusp/cleft.

The intensification in the low-frequency wave power near 01:24:30 in Figure 1 is associated with a ~10s interval where the DC electric field ramps up to over 20 mV/m and back down. This feature can be interpreted as a perpendicular electric field shock and as such could energize oxygen with energies greater than ~10 eV perpendicular to the magnetic field as discussed by Lennartsson (1980) and Borofsky (1984). Since the oxygen conic ion angular distribution after 01:24 is strongly peaked in angle and not exactly at ninety degrees to the magnetic field, the perpendicular heating must have occurred over a rather limited altitude range below the satellite. However, the source altitudes of oxygen heating regions inferred from detailed examination of individual angular distributions are not consistent with a single source altitude. This fact does not prove that the oxygen ions on August 10, 1984 are not heated as they pass through an electrostatic shock, but suggests that a series of shocks is required to produce the observed angular distributions. The feature near 01:24:30 in Figure 1 is the only one that could be interpreted as a perpendicular electrostatic shock found in the three cusp crossings examined in detail. It is possible to measure the perpendicular oxygen temperature at several times after 01:24 to determine if the observed increase indicated in Figure 1 is consistent with the resonant or non-resonant wave heating rates inferred. The software to perform this comparison is under development. This comparison will help resolve the ambiguity in the interpretation of Figure 1.

The source and nature of the strong gradient in low-frequency electric field found in the cases examined here and the results summarized by Andre et al. (1988) are not well known. Gurnett et al. (1984) have reported correlated low-frequency electric and magnetic noise along cusp, cleft and auroral field lines in the altitude range from 1.1 to 3 r/Re. Gurnett et al. (1984) were unable to determine if the observed low-frequency wave spectra, which are similar to those reported here, were due to static electric fields embedded in the ionosphere or electromagnetic embedded waves propagating from the magnetosphere into the ionosphere. It should also be noted that Temerin and Parady (1980) have reported low-frequency electric and magnetic field spectra similar to those reported here at ionospheric heights from a rocket. Their report suggests the waves could be caused by static electric fields embedded in the ionosphere.

Finally we note that the simple geometry in the cusp/cleft region where cool ions drift into a region of intense waves provides a particularly well defined test of an aspect of basic plasma theory using observations from spacecraft. Crew et al. (1988) have shown that the resonant heating mechanism also explains to a comparable degree of accuracy the bi-modal oxygen distributions reported by Klumpar et al. (1985).

Acknowledgements

The work at Lockheed was supported by NASA Contract NAS5-28710 and internal funds; MIT by NASA Grant NAGW-1532 and the Air Force Office of Scientific Research (AFSC) Contract F49620-86-C-0128; at Augsberg College by NASA Grant NAG5-529, and NSF Grant ATM-86-06388; at Iowa by NASA Grant NAG5-310; and at the University of California by The Office of Naval Reseach Contract N0014-81-C-006.

References

Andre, M., H. Koskinen, L. Matson, and R. Erlandson, Local transverse ion energization in and near the polar cusp, *Geophys. Res. Lett. 15*, 107, 1988.

Borofsky, J.E., The production of ion conics by oblique double layers, *J. Geophys. Res., 89*, 2251, 1984.

Burch, J.L., R.H. Reiff, R.A. Heelis, J.D. Winningham, W.B. Hanson, C. Gurgiolo, J.D. Menietti, R.A. Hoffman, and J.N. Barfield, Plasma injection and transport in the mid-altitude polar cusp, *Geophys. Res. Lett. 9*, 921, 1982.

Burch, J. L., Quasi-neutrality in the polar cusp, *Geophys. Res. Lett. 12*, 469, 1985.

Chang, T., G.B. Crew, N. Hershkowitz, J.R. Jasperse, J.M. Retterer, and J.D. Winningham, Transverse acceleration of oxygen ions by electromagnetic ion cyclotron resonance with broad band left-hand polarized waves, *Geophys. Res. Lett, 13*, 636, 1986.

Chappell, C.R., S.A. Fields, C.R. Baugher, J.H. Hoffman, W.B. Hanson, W.W. Right, H.D. Hammack, G.R. Carignan, and A.F. Nagy, The retarding ion mass spectrometer on Dynamics Explorer -A, *Sp. Sci. Instrum. 5*, 477, 1981.

Crew, G.B., T. Chang, J.M. Retterer, W.K. Peterson, D.A. Gurnett, and R. L. Huff, Ion conics: Detailed comparison of theory and observation, (Abstract), *EOS, 69*, 1374, 1988.

Eather, R.H., S.B. Mende, and E.J. Weber, Dayside aurora and relevance to substorm current systems and dayside merging, *J. Geophys. Res. 84*, 3339, 1979.

Farthing, W.H., M. Sugiura, B.G. Ledley, and L.J. Cahill, Jr., Magnetic field observations on DE-A and -B,*Sp. Sci. Instrum. 5*, 551, 1981.

Gurnett, D.A., R.L. Huff, J.D. Menietti, J.L. Burch, J.D. Winningham, and S.D. Shawhan, Correlated low-frequency electric and magnetic noise along the auroral field lines, *J. Geophys. Res. 89*, 8971, 1984.

Klumpar, D.M., W.K. Peterson and E.G. Shelley, Direct evidence for two-stage (bimodal) acceleration of ionospheric ions, *J. Geophys. Res. 89*, 10779, 1985.

Lennartsson, W., On the consequences of the interaction between the auroral plasma and the geomagnetic field, *Planet. Space Sci., 28*, 135, 1980.

Peterson, W.K., E.G. Shelley, S.A. Boardsen, D.A. Gurnett, B.G. Ledley, M. Sugiura, T.E. Moore, and J.H. Waite, Jr, Transverse ion energization and low-frequency plasma waves in the mid-altitude auroral zone: A case study, *J. Geophys. Res, 93*, 11405, 1988.

Peterson, W.K., Ion injection and acceleration in the polar cusp, in *The Polar Cusp*, J.A. Holtet and A. Egeland Editors, Reidel Publishing Co. 1985.

Shawhan, S.D., D.A. Gurnett, D.L. Odem, R.A. Helliwell, and C.G. Park, The plasma wave and quasi-static electric field instrument (PWI) for Dynamics Explorer -A, *Sp. Sci. Instrum. 5*, 535, 1981.

Sharp, R.D., R.G. Johnson, and E.G. Shelley.Observations of an ionospheric acceleration mechanism producing energetic (keV) ions primarily normal to the geomagnetic field direction, *J. Geophys. Res., 82*, 3324 1977.

Shelley, E.G., D.A. Simpson, T.C. Sanders, E. Hertzberg, H. Balsiger, and A. Ghielmetti, The energetic ion composition spectrometer (EICS) for the Dynamics Explorer -A, *Sp. Sci. Instrum. 5*, 443, 1981.

Temerin, M. and B. Parady, Observations of ULF electric field fluctuations in the dayside auroral oval, *J. Geophys. Res. 85*, 2925, 1980.

Temerin, M., Evidence for a large bulk ion conic heating region, *Geophys. Res. Lett., 13*, 1059, 1986.

Temerin, M., and I. Roth, Ion heating by waves with frequencies below the ion gyrofrequency, *Geophys. Res. Lett. 13*, 1109, 1986.

TWO-DIMENSIONAL MAPPING OF DAYSIDE CONVECTION

J. C. Foster, H.-C. Yeh, J. M. Holt
M.I.T. Haystack Observatory
Westford, MA 01886
U.S.A.

D. S. Evans
NOAA/SEL
Boulder, CO 80302
U. S. A.

ABSTRACT

Millstone Hill radar azimuth scans have been used to map the large-scale features of the ionospheric convection pattern in the vicinity of the cusp and cleft. Each scan covers 5 hours of MLT and 20° of invariant latitude, Λ, with 30 minute temporal resolution Individual "snapshots" of the convection pattern for disturbed conditions on 31 January 1982 span the entire region of convection convergence near noon and compare favorably with average model representations of the dayside region. The characteristic features of ion and electron precipitation observed during satellite overflights of the radar field of view are used to identify the cusp and cleft and to relate the location of these magnetospheric features to the pattern of ionospheric convection electric field. Cusp precipitation is seen at 70°Λ and 09 MLT at the sunward/anti-sunward convection reversal immediately after a sudden turning of interplanetary magnetic field (IMF) B_y from -5 nT to +5 nT while IMF B_z was -10 nT.

1. INTRODUCTION

Under proper conditions a cusp forms at the dayside magnetopause associated with the merging of the solar and geomagnetic fields and the direct entry of magnetosheath particles. The high-altitude cusp projects down field lines to the dayside ionosphere to a region which is, at best, difficult to identify and localize. Whereas a discrete cusp is formed at the magnetopause, the related phenomena at ionospheric heights appear spread over many hours of local time and several degrees of latitude into a broader region referred to as the cleft. Precipitating particle signatures involving both low energy electrons and ions appear to offer an operational technique for identifying and differentiating the cusp and cleft in order to facilitate the study of these important regions of the magnetosphere and ionosphere (eg. Reiff et al., 1977; Newell and Meng, 1989).

Ionospheric convection at high latitudes reflects the large-scale circulation and dynamics of the magnetosphere which is driven by its interaction with the solar wind (Foster, 1984). The ionospheric convection pattern in the vicinity of the cleft and cusp has been investigated through the synthesis of large numbers of satellite overflights (eg Heelis, 1984) or through patterns built up as a ground-based observatory rotated across the dayside region (eg Foster and Doupnik, 1984). The large-scale characteristics of the noontime convection pattern respond strongly to the interplanetary

115

P. E. Sandholt and A. Egeland (eds.), Electromagnetic Coupling in the Polar Clefts and Caps, 115–125.
© 1989 by Kluwer Academic Publishers.

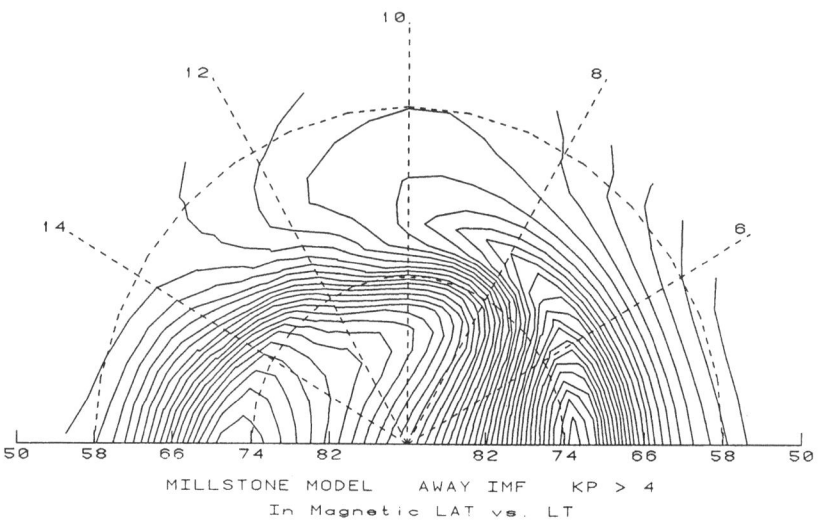

MILLSTONE MODEL   AWAY IMF   KP > 4
In Magnetic LAT vs. LT

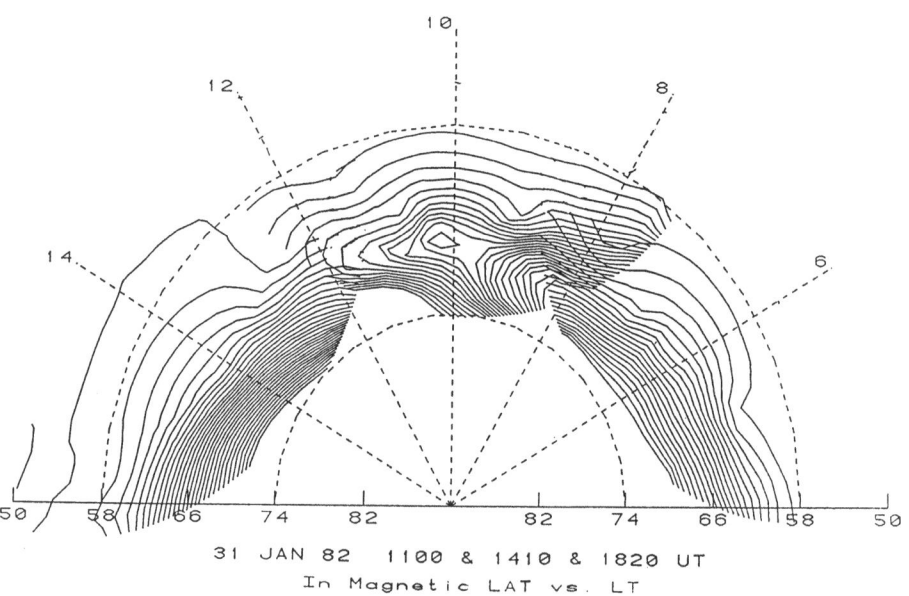

31 JAN 82   1100 & 1410 & 1820 UT
In Magnetic LAT vs. LT

Figure 1. (top) Average ionospheric convection electric field pattern derived as an average of Millstone Hill radar observations for inferred IMF AWAY sector conditions with Kp >4 (after Holt et al., 1987). Equipotential contours with 2 kV spacing are shown. Only the dayside portion of the convection pattern is presented with 10 MLT at the top of the figure.
(bottom) Individual scan maps of the convection pattern observed from Millstone Hill during the 31 January 1982 event compare closely with the average convection pattern shown above.

magnetic field (IMF) orientation and strength. Average patterns of dayside convection electric field for disturbed geomagnetic conditions (Kp > 4) have been prepared from the large synoptic data base obtained with the Millstone Hill incoherent scatter radar (Holt et al., 1987). The pattern for inferred IMF away sector conditions (IMF By positive) is presented at the top of Figure 1. Strong westward and poleward-directed convection velocity characterizes the dayside pattern for this IMF orientation.

The radar azimuth scan technique has been used at Millstone Hill since 1978 to provide wide latitude and local time coverage of the mid and high-latitude ionosphere with approximately 30 minutes temporal resolution. By scanning the monostatic radar beam in azimuth at low elevation angle, Doppler returns indicating the line-of-sight component of the F-region plasma motion are observed over a large range of latitude and longitude. Making the assumption that the observed plasma motion is due solely to E x B forces associated with an electrostatic field, the electric field pattern can be derived across the region observed by a single scan (Holt et al., 1984). This technique provides two-dimensional mapping of the ionospheric convection pattern over a range of latitude and longitude and has been used in a number of studies to investigate the local time extent and variation of the auroral-latitude convection pattern (eg Foster et al., 1989a). Azimuth scan experiments performed with the Sondrestrom, Greenland incoherent scatter radar span 2 hours of local time and have been used to investigate the dayside convection pattern in the vicinity of the cleft (Foster et al., 1985).

2. DAYSIDE PLASMA CONVECTION: RESPONSE TO CHANGING IMF

Two-dimensional maps of the dayside pattern of ionospheric plasma convection spanning 5 hours of MLT and 20°Λ along the Millstone Hill meridian (75° W longitude) were derived from radar azimuth scans during a period of strong magnetic disturbance on 31 January 1982. A large geomagnetic disturbance (Kp ranged from 4 to 6) took place on 31 January 1982 while the Millstone Hill radar surveyed the daytime ionosphere as a part of the MITHRAS (de la Beaujardiere et al., 1984) program. The regions of cusp and cleft precipitation and the dayside convection convergence moved to relatively low latitudes (70°Λ) during this event (Foster et al., 1989b), well within the radar field of view. High plasma density in the topside F region, which characterizes solar cycle maximum, further enhanced the radar's effective range and sensitivity and extended its high-latitude coverage to 75°Λ for spectral measurements and to 80°Λ for densities. Individual radar "snapshots" of the ionospheric convection pattern are superimposed in the lower half of Figure 1 providing a two-dimensional mapping of portions of the overall dayside pattern at different times during the event.

The level of activity and the configuration of the convection pattern changed dramatically during the course of the event in response to changes in the orientation of the interplanetary magnetic field (IMF). Figure 2 presents the variations of the IMF components and one-hour averaged AE index. The IMF was observed 255 R_e upstream from the earth and a 1 hour propagation delay has been applied to those data prior to plotting the figure. Local noon is at 17:00 UT on the Millstone Hill meridian and the interval of interest for our investigation of the dayside convection pattern extends from 10:00 UT to 20:00 UT. The 1-hour AE index reached a maximum of 1000 nT and Kp was 6 during the disturbed interval before 15:00 UT. IMF B_z turned strongly southward before 11:00 UT reaching -10 nT around 14:00 UT. B_y was negative for 2 hours before turning sharply positive at approximately 13:10 UT. B_z remained negative until after 18:00 UT.

Convection maps representative of the dayside region on this disturbed day are presented in Figure 3. The scan at 14:10 UT found the convection convergence region centered at 09 MLT for IMF B_y positive conditions and found some 50 kV potential difference across a span of 3 hours local time at 70°Λ. The configuration observed at this time is well represented by the average convection pattern for IMF B_y positive conditions as seen in Figure 1. Rapid convection from the post-noon F

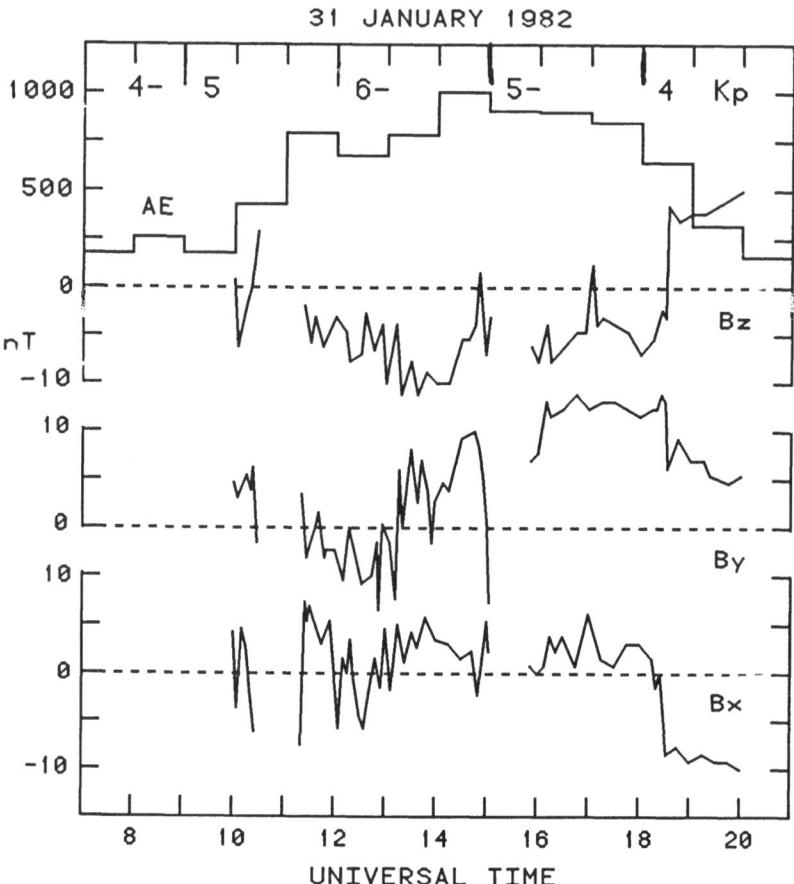

Figure 2. Interplanetary magnetic field from ISEE-3 and 1-hour averaged AE index for the 31
January 1982 storm. A one hour propagation time correction has been applied to the
IMF data.

region carried solar-enhanced plasma to high latitudes in the polar cap (Foster et al., 1989b).
During the 40 minutes which separated the two scan maps presented in Figure 3, the IMF B_z
component turned positive and B_y turned negative (followed by a gap in the available IMF data).
The convection pattern at 14:45 UT, shown in the lower panel of the figure, shifted briefly to be
more consistent with the IMF towards sector average pattern presented by Holt et al.(1987).

During the 31 January 1982 radar experiment, ion temperatures exceeding 3500°K were
observed immediately equatorward of the convection reversal in the region of strong sunward
plasma flow. The close relationship between the convection reversal and a narrow region of strong
ion heating is seen in the upper panel of Figure 3. The elevated ion temperatures, indicated by the
shaded region, lie equatorward of the region of ion precipitation in the cusp (as identified in Section
3, below). The high speed ion flow and rapidly changing convection near the reversal lead to a large
difference velocity between ions and neutrals and thus to strong frictional heating and elevated ion
temperatures. Ion velocities in excess of 1 km/sec were observed by the radar in this region.

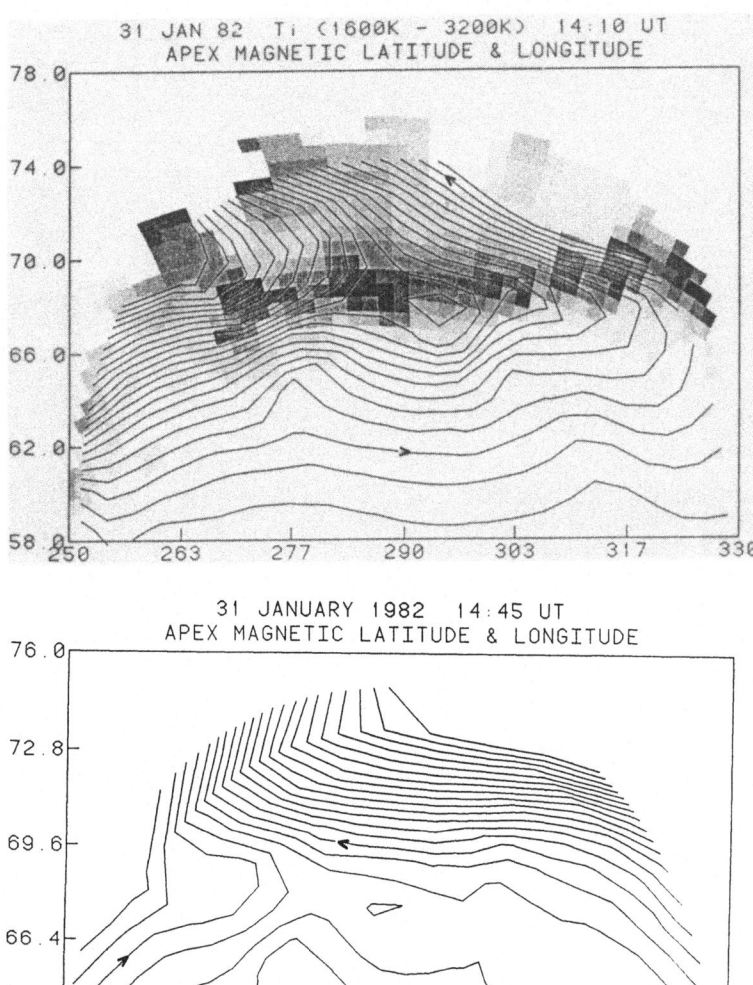

Figure 3. (top) A narrow band of enhancemented ion temperatures was observed of 31 January 1982 associated with frictional heating at the cusp/cleft convection reversal (see text). (bottom) The large-scale dayside convection pattern shifted to a TOWARD sector configuration at 14:45 UT in response to a negative turning of IMF B_y.

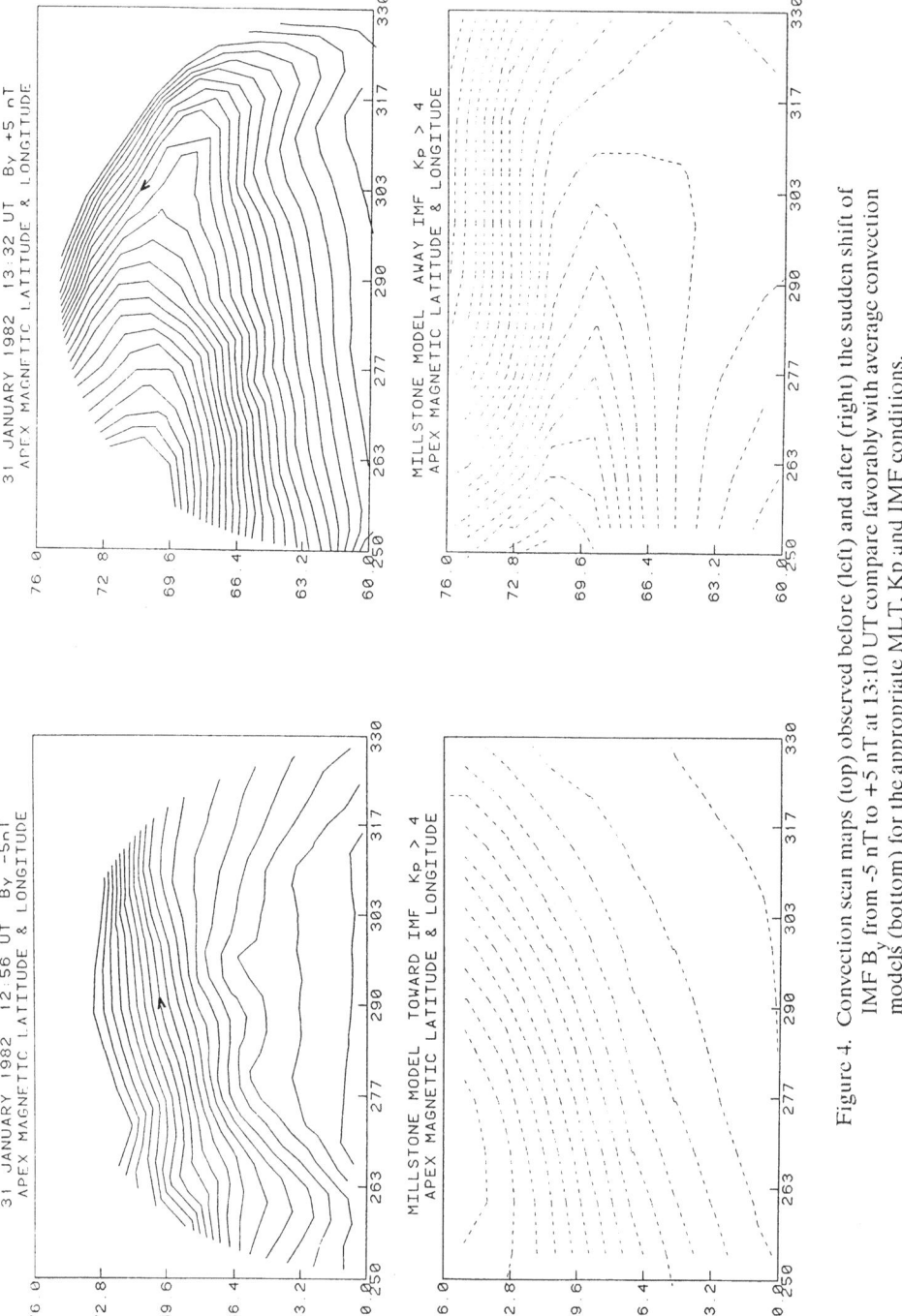

Figure 4. Convection scan maps (top) observed before (left) and after (right) the sudden shift of IMF B_y from -5 nT to +5 nT at 13:10 UT compare favorably with average convection models (bottom) for the appropriate MLT, Kp and IMF conditions.

Throughout the event the large-scale dayside convection pattern seen with the radar experiment closely followed changes in the IMF orientation. Figure 4 presents "snapshots" of the convection pattern observed with the Millstone Hill azimuth scan technique for times immediately before and after the sharp positive turning of IMF B_y at 13:10 UT. The radar sampling technique and the equipotential contour fit to the observed data smooth over features in the convection pattern less than 100 km - 200 km scale size. For comparison, Millstone electric field models (Holt et al., 1987) for the appropriate MLT, IMF, and Kp conditions are displayed in the lower portion of each figure. The convection pattern changes promptly and dramatically in response to the sudden change in IMF B_y. Whereas convection was directed poleward and eastward toward noon for B_y negative conditions, a distinct convection reversal at 70° is observed for B_y positive. In both cases the average model electric field pattern closely matches the large-scale observations.

3. LOW-ALTITUDE SIGNATURES OF THE CUSP AND CLEFT

Away from noon the ionospheric convection reversal between sunward plasma velocities at auroral latitudes and anti-sunward flow in the polar cap is situated in the region of closed magnetic field lines (McDiarmid et al., 1978; Heelis et al., 1980) and has been associated with the low latitude boundary layer (LLBL) where convection is driven by a viscouslike interaction with the shocked solar wind plasma of the magnetosheath (Eastman et al., 1976). In the dayside sector, precipitation associated with the LLBL constitutes the ionospheric cleft, a region which spans several hours of MLT around noon. A narrower region in which magnetosheath particles have more direct entry to the ionosphere is thought to mark the low altitude signature of the cusp on the magnetopause (Heikkila, 1985). Following on the work of the many authors who have discussed the signatures of the cleft and the cusp at low altitude, Newell and Meng (1989) provided a set of criteria for identifying and differentiating the cleft and the cusp based on a statistical study of precipitating ions and electrons in these regions. They found that boundary layer, or cleft, precipitation is characterized by ion energies greater than 3000 eV and a total number flux some 3.5 times less than that observed in the cusp proper, where average ion energies are less than 3000 eV. Both regions are associated with precipitating electrons with average energy less than 200 eV. The cusp is found with 70% probability near the noon meridian, but with rapidly decreasing probability at local times away from noon. The cleft is found with high probability away from noon and occurs equatorward of the cusp when the two occur simultaneously at the same local time.

Precipitating ion and electron measurements were made by the total energy detector experiment (Fuller-Rowell and Evans, 1987) on the NOAA-6 satellite at 850 km altitude during overflights of the Millstone Hill azimuth scan field of view at the time of the sudden positive turning of the IMF B_y component. In Figure 5 the ionospheric footprint of the satellite is projected onto the radar convection pattern and the average energy and total energy flux of ions and electrons are presented (the energy range of the NOAA-6 instrument is 300 eV to 20,000 eV). Ion precipitation with 1500 eV average energy and electron precipitation with energy <300 eV was observed poleward of 69.5°Λ, nearly coincident with the convection reversal. Equatorward of 69.5°Λ the low energy ion precipitation terminated abruptly and 2 to 10 keV electron precipitation, typical of the plasma sheet, was observed across the region of strong sunward convection. The particle signatures seen between 69.5°Λ and 71°Λ at 08 MLT are indicative of the cusp as defined by Newell and Meng (1989) while little evidence of the cleft is seen. Statistically, those authors found only a 10% probability of observing the cusp at this local time, but a 100% likelihood of finding LLBL precipitation and the cleft. Our observations of cusp precipitation at this early local time suggest that merging and the high-altitude cusp extended far into the dawn sector in response to the sudden positive turning of IMF B_y.

NOAA-6 overflew the edge of the convection pattern observed by the radar during the interval of negative IMF B_y at 14:45 UT as shown in Figure 6. The ion precipitation observed at this

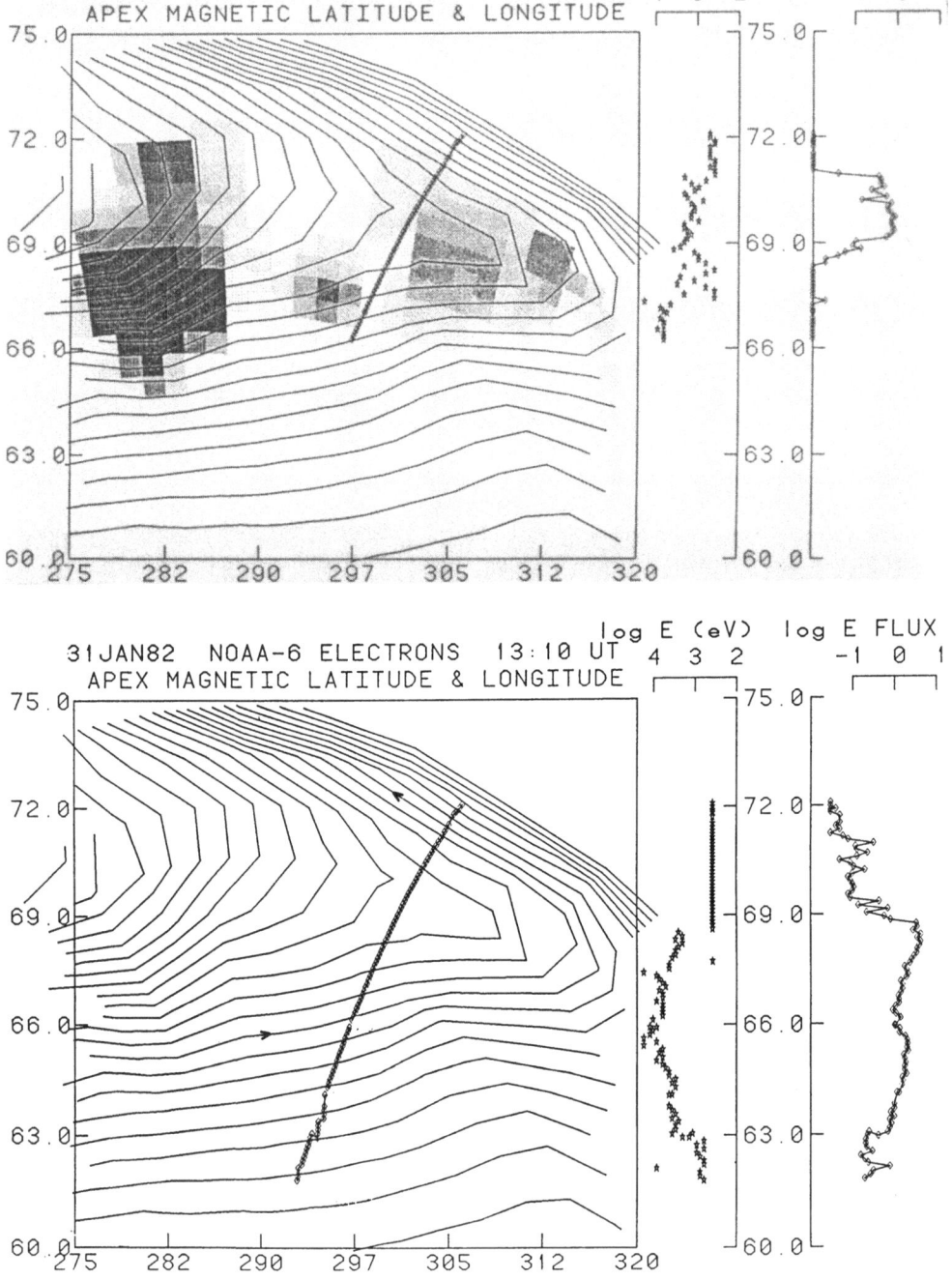

Figure 5. Precipitating protons (top) and electrons (bottom) observed as the NOAA-6 satellite overflew the radar field of view at 13:10 UT locate the cusp at 70°Λ, concident with the convection reversal at 09 MLT.

time was more energetic than seen during the pass at 13:10 UT (average energy 4250 eV) and began abruptly equatorward of 67°Λ, coincident with the region of <300 eV electron precipitation. The number flux of the precipitating ions was approximately a factor of 5 lower in this region than observed during the earlier pass through the cusp. The characteristics of the precipitation between 65°Λ and 67°Λ closely match the criteria specified by Newell and Meng (1989) for the cleft/LLBL region and we note that the cleft signature lies immediately equatorward of the convection reversal at this time. Electron temperatures are enhanced in the cleft associated with the intense low energy electron precipitation in that region.

4. SUMMARY

Radar azimuth scans from Millstone Hill have produced "snapshots" of the large-scale dayside convection pattern during disturbed conditions. These patterns agree closely with the average models derived for the appropriate IMF and activity conditions. A narrow-latitude region of elevated ion temperatures lies immediately equatorward of the dayside convection reversal and the region of cusp ion and electron precipitation. These are associated with convection velocities in excess of 1 km/sec. Precipitating ion and electron signatures differentiate the cleft/LLBL from the cusp and locate these magnetospheric features with respect to the two-dimensional convection pattern. On 31 January 1982 we observe the cusp to be coincident with the sunward/anti-sunward convection reversal at 09 MLT and 70°Λ during very disturbed conditions with IMF B_y positive. Cleft/LLBL precipitation is seen later in the event immediately equatorward of the pre-noon convection reversal as B_y turns negative.

ACKNOWLEDGEMENTS

Radar observations at Millstone Hill were supported by the U. S. National Science Foundation. The analyses leading to this report were supported by the U. S. Air Force Office of Scientific Research through grant AFOSR-86-0023 and by National Science Foundation Cooperative Agreement ATM-88-08137 with the Massachusetts Institute of Technology.

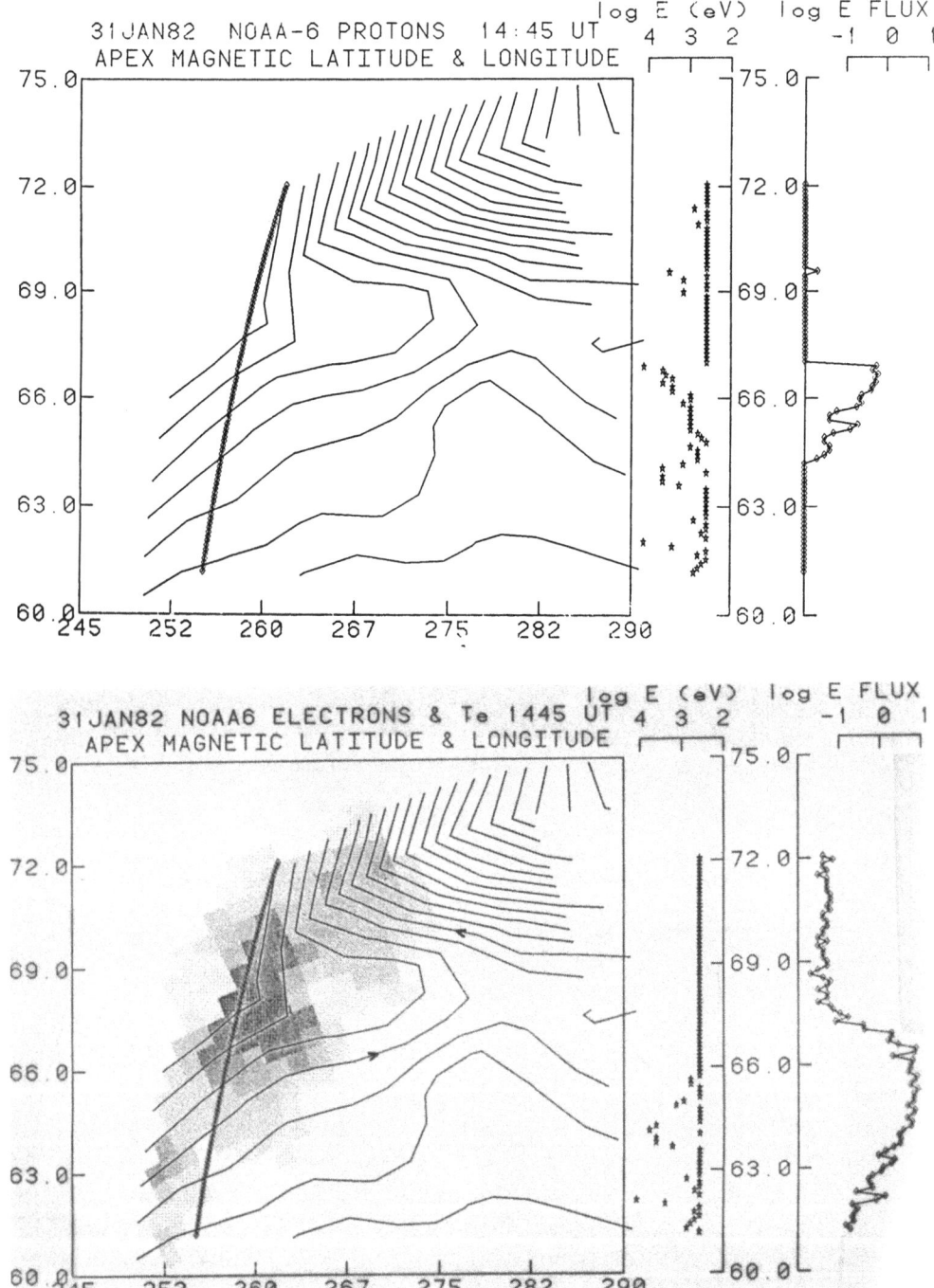

Figure 6. Precipitating protons (top) and electrons (bottom) observed by NOAA-6 at 14:45 UT locate the cleft/LLBL at 67°Λ, immediately equatorward of the convection reversal.

REFERENCES

de la Beaujardiere, et al., MITHRAS: a brief description, Radio Science, 19, 665, 1984.

Eastman, T. E., E. W. Hones, Jr., S. J. Bame, and J. R. Asbridge, The magnetospheric boundary layer: site of plasma, momentum and energy transfer from the magnetosheath to the magnetosphere, Geophys. Res. Lett., 6, 685, 1976.

Foster, J. C., Ionospheric signatures of magnetospheric convection, J. Geophys. Res., 89, 855, 1984.

Foster, J. C. and J. R. Doupnik, Plasma convection in the vicinity of the dayside cleft, J. Geophys. Res., 89, 9107, 1984.

Foster, J. C., J. M. Holt, J. D. Kelly, and V. B. Wickwar, High resolution observations of electric fields and F-region plasma parameters in the cleft ionosphere. The Polar Cusp, J. A. Holt et and A. Egeland (eds.), D. Reidel Pub., Dordrecht, 349, 1985.

Foster, J. C., T. Turunen, P. Polari, H. Kohl, and V. B. Wickwar, Multi-radar mapping of the auroral ionosphere, Adv. Space Res., in press, 1989a.

Foster, J. C., Plasma transport through the dayside cleft: a source of ionization patches in the polar cap, in Electromagnetic Coupling in the Polar Clefts and Cap, A. Egeland and P. E. Sandholt (eds.), this volume, 1989b.

Fuller-Rowell, T. J., and D. S. Evans, Height-integrated Pedersen and Hall conductivity patterns inferred from the TIROS-NOAA satellite data, J. Geophys. Res., 92, 7606, 1987.

Heelis, R. A., The effects of interplanetary magnetic field orientation on dayside high-latitude ionospheric convection, J. Geophys. Res., 89, 2873, 1984.

Heelis, R. A., J. D. Winningham, W. B. Hanson, and J. L. Burch, The relationships between high-latitude convection reversals and the energetic particle morphology observed by Atmospheric Explorer, J. Geophys. Res., 85, 3315, 1980.

Heikkila, W. J., Definition of the cusp, The Polar Cusp, J. A. Holtet and A. Egeland (eds.), D. Reidel Pub., Dordrecht, 387, 1985.

Holt, J. M., R. H. Wand, and J. V. Evans, Millstone Hill measurements on 26 February 1979 during the solar eclipse and formation of a midday F region trough, J. Atmos. Terr. Phys., 46, 251, 1984.

Holt, J. M., R. H. Wand, J. V. Evans, and W. L. Oliver, Empirical models for the plasma convection at high latitudes from Millstone Hill observations, J. Geophys. Res., 92, 203, 1987.

McDiarmid, I. B., J. R. Burrows, and M. D. Wilson, Comparison of magnetic field perturbations at high latitudes with charged particle and IMF measurements, J. Geophys. Res., 83, 681, 1978.

Newell, P. T., and C.-I. Meng, The cusp and the cleft/LLBL: low altitude identification and statistical local time variation, J. Geophys. Res., 94, in press, 1989.

Reiff, P. H., T. W. Hill, and J. L. Burch, Solar wind plasma injection at the dayside magnetospheric cusp, J. Geophys. Res., 82, 479, 1977.

IONOSPHERIC CONVECTION IN THE POLAR CAP AS SEEN BY OPTICAL IMAGING

J. H. Doolittle, S. B. Mende, G. R. Swenson, R. M. Robinson

Lockheed Palo Alto Research Laboratory, Dept. 91-20, Bldg. 255, 3251 Hanover Street, Palo Alto, California 94304

ABSTRACT.

An all-sky imager operated at Sondre Stromfjord, Greenland has been used to make auroral measurements in correlation with data from the incoherent scatter radar. The auroral images were recorded on video tape for later display as time lapse sequences. Reviewing the images as sequences allows the eye to discern subtle features which are not readily apparent in the individual frames. One such feature which is the subject of this study is the faint 6300Å airglow background seen in the presence of much brighter auroras. Previously considered to be uniform, the time lapse observations show that it is in fact structured and that the structures exhibit a coherent drift in the image field. The direction and magnitude of this drift is different from the motion of the auroras present in the field. The motion of the airglow structures was analyzed from the images and compared to available radar data which provided a measure of the ion drift velocities. Excellent agreement was found in this comparison, which shows that the airglow structures are non-uniformities in the ionosphere where the 6300Å emissions are produced by recombination. The observation of the motion of these structures could provide a routine measurement of the ionospheric convection speeds in places where the radar is not available.

1. INTRODUCTION

Anti-sunward moving airglow enhancements in the polar cap F region were described first by *Weber, et al.* [1984]. They have shown that these 6300 O(1D) airglow enhancements were produced by recombination in patches of enhanced ion density. They showed that these patches occur preferentially at altitudes greater than 200 km and from simultaneous satellite measurement they also showed that the airglow enhancements were not caused by local precipitation. The airglow enhancements were observed mainly at the time of negative B_z and were observed to drift anti-sunward, presumably driven by ionospheric convection field. The source mechanism of the ionospheric enhancements is either dayside solar photo ionization or dayside cusp particle precipitation. Since the motion of the associated airglow enhancement shows the direction and magnitude of the ionospheric plasma drift, detection and systematic mapping of the airglow enhancement

P. E. Sandholt and A. Egeland (eds.), Electromagnetic Coupling in the Polar Clefts and Caps, 127–136.
© *1989 by Kluwer Academic Publishers.*

could be a significant tool in determining the instantaneous convection morphology.

Monochromatic all sky cameras are routinely used for studying faint aurora and airglow enhancements [*Mende, et al.* 1976, *Mende, et al.,* 1977, *Weber, et al.* 1978]. Starting in the winter of 1985, such a camera was used at Sondrestromfjord, Greenland to make coordinated optical measurements with the Sondrestrom radar. Several imporovements were incorporated into the instrument prior to the February, 1988 CEDAR campaign which was in support of the National Science Foundation Aeronomy program entitled Coupling, Energetics and Dynamics of the Atmospheric Regions. Simultaneous radar data was also obtained to support the campagin observations.

The imager data was recorded on video tape and when replayed in a speeded up mode, it was observed that the airglow enhancements could be distinguished which showed systematic drift motions. In this paper we present the analysis of these observations, compare the drifts with the radar observations and discuss the significance of these observations in the framework of the current understanding of auroral electric fields.

2. OPTICAL OBSERVATIONS

The optical observations presented in this paper were obtained with an all sky monochromatic imaging camera. The original camera was essentially the same as described by *Mende, et al.,* [1977], but was modified over the years to include several improvements. In January 1988 the original SEC vidicon was replaced by a cooled, intensified CCD detector. The instrument in its final configuration consisted of a 15 mm focal length large format (75 mm film format) fish eye lens with an eight position double filter wheel set. The system had the conventional tele-centric lens elements (see *Mende, et al.,* [1977]) to assure that the chief ray from all parts of the image are parallel to the optic axis when it passes through the filter. In the new configuration the detector consisted of a Varo 25 mm inverter image intensifier and a fiber-optically coupled Fairchild 3000 type line transfer CCD. The whole system including the image intensifier, the fiber-optics reducing taper and the CCD detector were placed inside a thermoelectrically cooled housing. This housing maintained the temperature of the detector at about -5° C. The CCD was further cooled by an additional small thermoelectric cooling element attached to it inside the main cooler. This new type of imaging camera can take exposures which are several minutes long and can detect extended sources of a few Rayleighs in intensity.

In Figure 1 we are presenting a single frame of video data on which radar drift velocity vectors are superimposed. In this presentation, generated by an upward viewing all sky imager, geographic north is at the top of the figure, east is on the left and west is on the right. At Sondrestrom magnetic north is 27 degrees west of true geographic north. The solar direction appropriate to the local time of the image is also shown toward the west. In the lower left half of the image there is a bright feature which is identified as a discrete arc. This arc represents the poleward boundary of the auroral precipitation. The boundary is aligned approximately in a

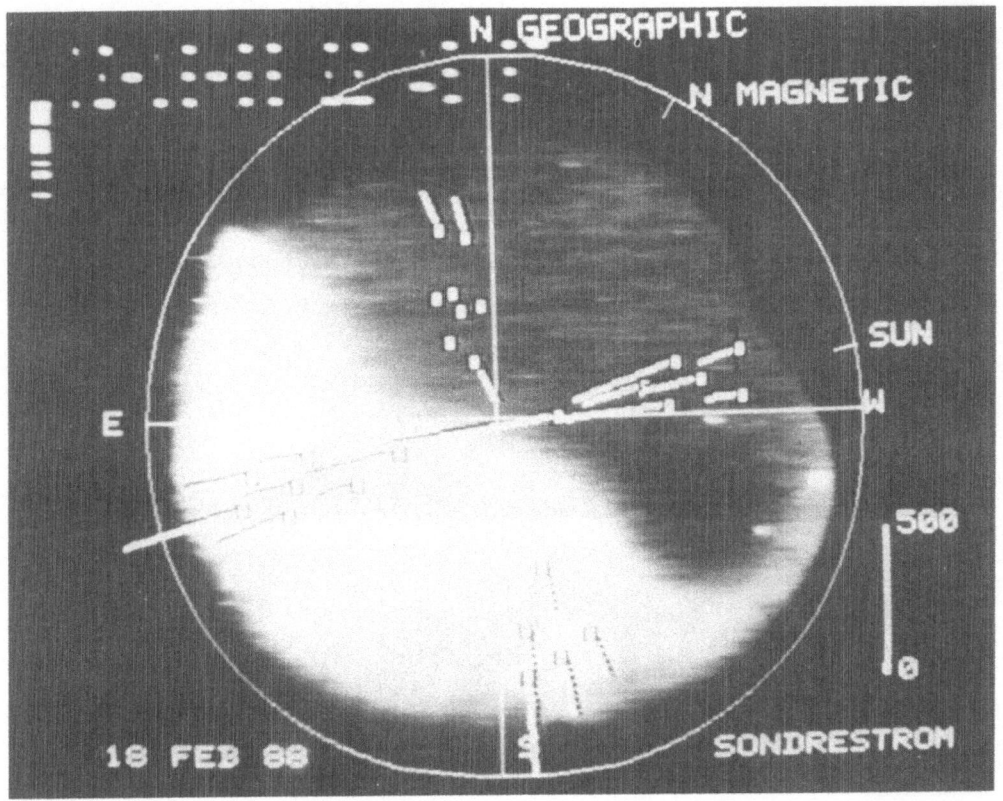

Figure 1. The all sky 6300 Å imager data with superimposed radar drift velocity vectors. Geographic north is at the top, east is to the left and west is to the right. Geomagnetic north and the position of the sun for the time of the image are shown. The origin of the radar vectors were found by projecting the position of the range gate to 300 km altitude along the magnetic field line.

magnetic east-west direction and is located slightly south of the station zenith. The region extending from the center of the image to the top right, or approximately in the magnetic north direction, is dark and apparently uniform.

The data were recorded by two video tape recorders. One of these recorded only the 6300 filter data and produced a continuous video record which, when played back at normal recording speed, provided a presentation which was speeded up approximately by a factor of 60. The examination of these speeded up video records showed that the dark region representing the airglow in the polar cap in fact contained coherent structures which were perceived by the eye to be moving.

SONDRESTROM   18 FEB 1988

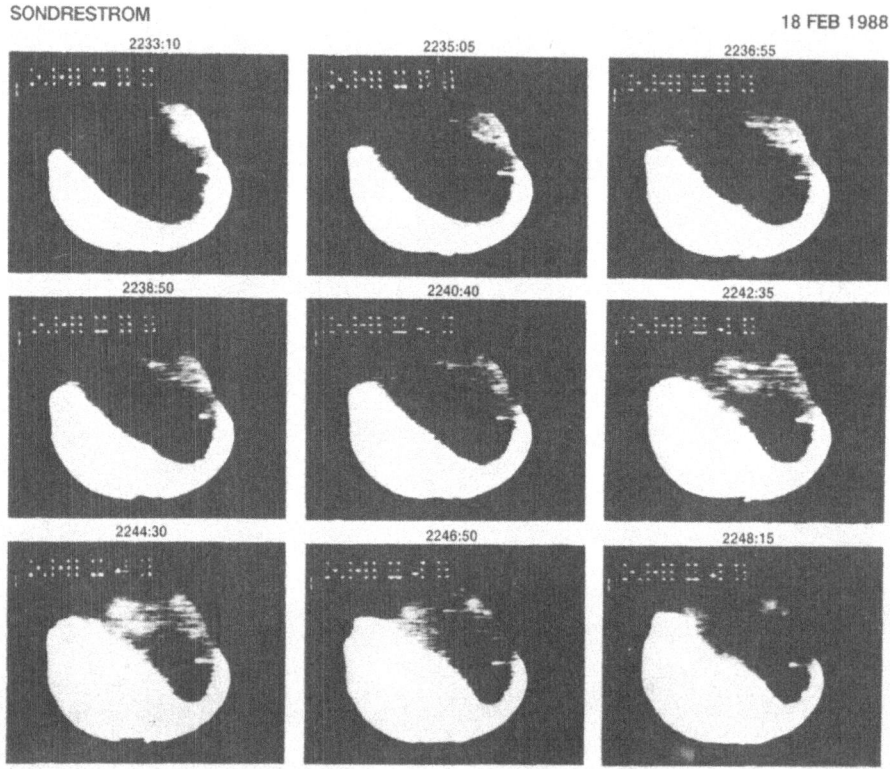

Figure 2 Collage of the all sky images taken between
22:23:10 UT and 22:48:15 UT showing the intensity enhanced airglow patches.

In order to show the data in a printed form, video enhancement techniques
were used to stretch the intensity range of the airglow region until the desired features would stand out. The result of this enhancement is shown in a collage of
the enhanced images in Figure 2. In these enhanced images the bright aurora is
completely saturated but, the faint airglow enhancement patch is perceptible.

From Figure 2 we can detect the appearance of an airglow enhancement travelling from the right (magnetic north-west) to the left (magnetic south-east). The
collage shows that the enhancement is convecting in the anti-sunward direction towards the aurora and it appears to flow directly into the aurora where it is masked by
the bright auroral luminosity. From these images it cannot be established whether
the patch continues its motion through the aurora. The examination of the collage further shows that, while the patches are actually moving in the anti-sunward
direction, the aurora is in fact expanding and moving in the opposite direction.

Using an intensity color coding technique of image processing, it was possible

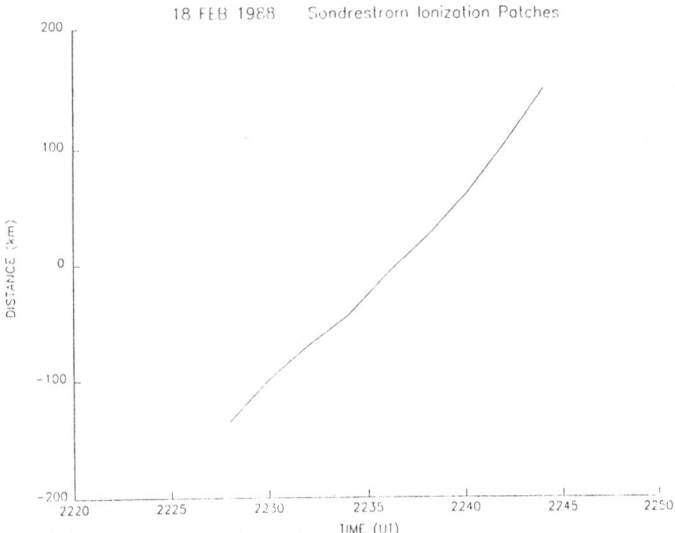

Figure 3. The position of one of the patches as a function of time.

to locate the advancing contour of an airlglow patch associated with an intensity of about 400 Rayleighs. The locations were mapped into geographic coordinates relative to Sondre Stromfjord, assuming an emission altitude of 300 km. Interpolation was then used to place the 400 Rayleigh contour at locations corresponding to two minute increments, as shown in Figure 3. From this figure we can see that the anti-sunward velocity of a patch was about 300 m/sec in the polar cap and accelerated to about 500 m/sec as it approaced the auroral boundary.

3. RADAR MEASUREMENTS

During the optical observations the Sondrestrom radar was also obtaining data in a north-south and east-west meridian scanning mode. As we have shown previously in Figure 1, the radar data Doppler line-of-sight velocity vectors were superimposed on the auroral images. The foot of each vector was placed at the point where the projection of the range gate intersected the 300 km altitude region. The projection was performed along the magnetic field line in the up or down direction.

The radar data was also reduced to ion density contour plots. These are shown in Figure 4. The top panel shows the north-south scan and the bottom panel shows the east-west scan. Note that in Figure 4 we have followed the more frequently adopted convention in which east is shown on the right. Both plots show clearly the presence of the auroral arc as an intense ionization feature which extends down to 95 km altitude. The high altitude ionization is also shown on the ion density plots at 300 km altitude. The radar data also shows some structure in the ionization and the presence of some of the so called ionization "blobs" can be discerned.

In summary, the radar data confirms the presence of enhanced ionization

132

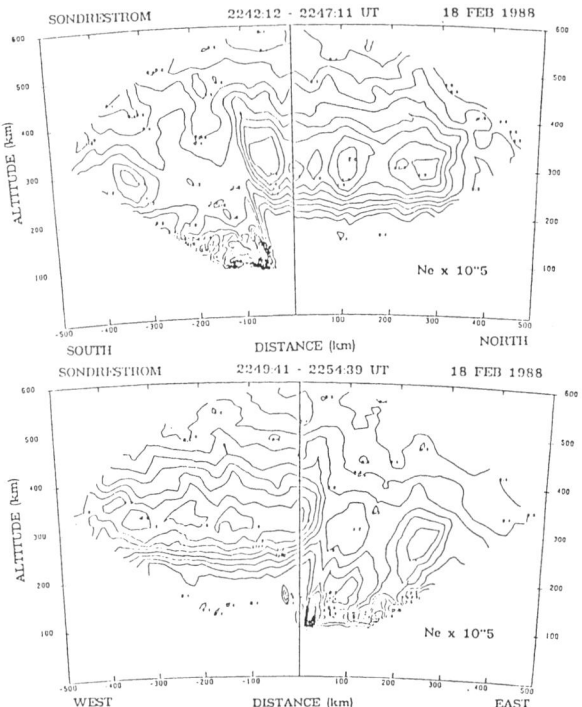

Figure 4. Ionospheric electron density contours for the
"north-south" (top) and "east-west" (bottom) scan.

patches in the altitude range 250-350 km. Because of the high altitude of these
ionization enhancements, it is very unlikely that the observed optical enhancements
would be produced by auroral electron precipitation. The radar data also shows
that the bulk velocity of the ionization is mainly in the anti-sunward direction and
its magnitude is several hundred meters per second. The radar data corroborates
our interpretation that the airglow patches are drifting ionization enhancements of
the type described by *Weber, et al.,* [1984].

4. DISCUSSION

As we have seen, the faint airglow observed northward of the aurora
showed structure and these structures were recognized as drifting ionization
patches driven by the anti-sunward convection in the polar cap. The radar
data confirmed the existence of the high altitude ionization enhancements and
that they were drifting equatorward at the velocity predicted from the optical data. The patches were seen to drift anti-sunward towards the discrete auroral arc, while the arc was moving in the opposite, or sunward direction. Thus any electric field which may be responsible for the motion of
the aurora is different from the anti-sunward convection driving the ionosphere.

As the drifting patches are not discernable when they drift above the brighter
auroras, we cannot determine optically whether the drift is continuous across the

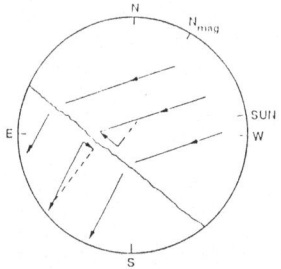

Figure 5. Schematic representation of the drift vectors
when split into components which are parallel and perpendicular to the arc.

auroral arc. Looking at the radar vector data (Figure 1) it appears that the ion
drift motion transverse to the arc could be continuous. The drift vectors perpendicular to the arc tend to be discontinuous and the observed tangential velocities are
consistent with the vector scheme illustrated in Figure 5. At the top of Figure 1 the
magnetic east-west scan shows a velocity discontinuity which could be interpreted
to occur at the poleward boundary of the arc. Therefore it is possible that the
poleward boundary of the arc is colocated with a discontinuity in the drift velocity
component tangential to the aurora. In Figure 5 we have illustrated schematically
the drift vectors as two components, one parallel and the other perpendicular to
the arc. The drift velocity can be represented by an electric field perpendicular
to the drift motion. If we represent the drift vector component parallel to the arc
as electric fields, then the electric field would be in the sense as to point towards
the boundary from both directions and therefore would drive a Pederson current
consistent with an upward flowing field aligned current at the field discontinuity.
Upward flowing aligned current is the correct sense for an enhancement of downward accelerated electrons. This situation is illustrated in Figure 6. The electric
field points inward towards the poleward boundary of the arc from both directions
and the continuity of the current generated by the electric field requires the upward
Birkeland current.

So far the observations are consistent with the currently accepted ideas regarding the mechanism responsible for Birkeland current and auroral arc formation.
However, in the case described by our observational data the boundary where the
field aligned current generation occurs is not on an equipotential since the bulk of
the plasma flow is perpendicular to the boundary. Accordingly, our data shows that
an auroral arc is formed in a region where there is substantial electric field parallel
to the arc.

In Figure 7 we have illustrated the observation geometry on a segment of a
map of the auroral oval. The observation occured around 20:30 local time in the
evening sector. According to a simplified view the most poleward arc should be on

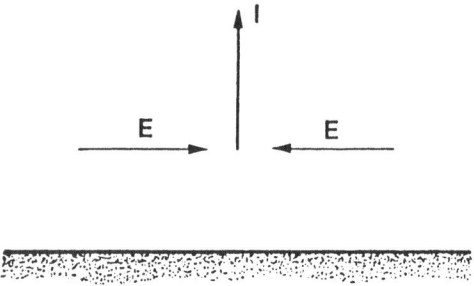

Figure 6. Schematic representation of the electric fields and currents in a cross-sectional view of the auroral arc.

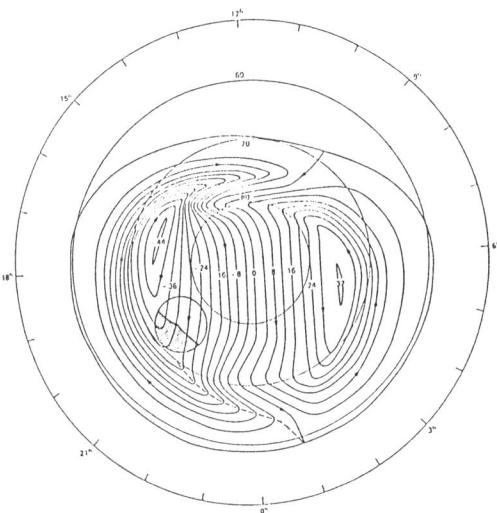

Figure 7. Model convection electric field showing a global view of the auroral oval (after *Heppner*, 1977) with the field of view of the all-sky imager located at Sondre Stromfjord at 2030 local time.

the boundary of polar cap and should be located close to the convection boundary separating the polar cap anti sunward drift from the lower auroral zone sunward drift. However our data shows that the predominant ionospheric drift is essentially perpendicular to the arc and the boundary of the polar cap. The picture shown by our data is consistent with a main feature of the ionospheric convection models such as proposed by *Burch et al.*, [1985] or derived by *Heppner* [1977]. These convection models show strong anti-sunward drift in the late evening near midnight sector for B_z negative. However this region is also the location of the nightside auroral oval with predominantly east-west aligned arcs. Thus the convection models show drifts perpendicular and across the arcs. Our data shows a single case study in which the convection is largely perpendicular to the auroral forms.

5. CONCLUSION

From all sky video records of the 6300Å emission at Sondre Stromfjord and coordinated incoherent scatter radar data, it was found that the airglow intensity and the ionization density in the region poleward of the auroral oval was structured and was drifting in an anti-sunward direction. These airglow enhancement patches are probably the same as the ones previously detected in the polar cap in the absence of oval auroras by *Weber, et al.*, [1983]. From the comparison with the radar data we could deduce that the drifting airglow enhancements were produced by high altitude ionization structures and were not produced by local precipitation. The Doppler measurements of the ion drift velocity by the radar agreed with the airglow drift velocity measurements both in direction and magnitude. Using the radar measurements, it was shown that the electric field pattern is consistent with a strong electric field parallel with the arc causing the ion drift to be predominantly perpendicular to the arc, while still allowing a small gradient in the electric field perpendicular to the arc. The sense of the fields at either side of the gradient is consistent with the generation of an upward field aligned current near or within the arc. This is of course in agreement with the idea that the electron acceleration in the discrete arc is caused by upward Birkeland current flow. The data presented here shows the existence of a strong electric field parallel to the arc which demonstrates that the arc is not produced along an ionospheric equipotential. It was particularly interesting to note that the auroral arc moved sunward during the period while the F region ionization patches were observed to drift anti-sunward.

REFERENCES

Burch, J. L., P. H. Reiff, J. D. Menietti, R. A. Heelis, W. B. Hanson, S. D. Shawhan, E. G. Shelley, M. Sugiura, D. R. Weimer, and J. D. Winningham, 'IMF By Dependent Plamsa Flow and Birkeland Currents in the Dayside Magnetosphere 1. Dynamic Explorer Observations'. J. Geophys. Res. **90,** 1577-1593, 1985.

Heppner, J. P., 'Empirical Models of High-latitude Electric Fields', J. Geophys. Res., **82,** 1115, 1977.

Mende S. B., R. H. Eather, and E. K. Aamodt, 'Instrument for the Monochromatic Observation of All-sky Auroral Images', Applied Optics, **16,** 691, 1977.

Mende S. B., and R. H. Eather, 'Monochromatic All-sky Observations and Auroral Precipitation Patterns', J. Geophys. Res., **81,** 3771, 1976.

Weber E. J., J. A. Klobuchar, J.Buchau, H. C. Carlson, R. C. Livingston, O. De La Beaujardiere, and M. McCready, J. G. Moore, G. J. Bishop, 'Polar Cap F Layer Patches: Structure and Dynamics', J. Geophys. Res., **91,** 12121, 1986.

Weber E. J., J.Buchau, J. G. Moore, J. R. Sharber, R. C. Livingston, J. D. Winningham, and B. Reinisch, 'F Layer Ionization Patches in the Polar Cap', J. Geophys., Res., **89**, 1983.

Weber E. J. and J. Buchau, R. H. Eather, S. B. Mende, 'North South Aligned Equatorial Airglow Depletions'. J. Geophys. Res. **83,** 712, 1978.

BALLOON OBSERVATIONS OF THE ELECTRIC FIELD OVER SOUTH POLE: CONVECTION PATTERNS

Edgar A. Bering, III, James R. Benbrook, Gregory J. Byrne,
Danqing Liang and Zhong-Min Lin
Physics Department
University of Houston
Houston, Texas 77204-5504, U.S.A.

ABSTRACT. During the 1985-86 austral summer, eight balloon flights were launched from Amundsen-Scott Station, South Pole, Antarctica by the University of Houston in order to measure the ionospheric electric field in the vicinity of the polar cusp and the polar cap. One objective of this work was to determine the average daily ionospheric plasma convection patterns near the polar cusp as measured by the balloon experiment. The solar wind and interplanetary magnetic field conditions during the campaign period have been obtained using IMP 8 satellite data. IMP 8 was in the solar wind during four of the balloon flights. The balloon data have been binned separately as functions of Kp, interplanetary magnetic field (IMF) B_z and IMF B_y. The intent is to determine if there are any differences between data obtained in this way and data obtained from other techniques. The overall patterns show the expected two-cell convection pattern. In detail, the various individual patterns show subtle but interesting departures from previous results.

1. Introduction

This paper presents ionospheric electric field data from the 1985-86 South Pole Balloon Campaign (Bering *et al.*, 1987) in the form of equivalent ionospheric convection patterns. The data from the whole campaign are presented in the form of averages over magnetic local time and various geophysical parameters such as K_p and the interplanetary magnetic field direction. In addition, the data from a single day will be shown in order to illustrate the limitations of the averaging technique.

The polar cusp region is important to the study of the electrodynamics of coupling between solar wind, magnetosphere, and ionosphere. Understanding of this region requires knowledge of the ionospheric convection patterns at high-latitude. Measurements can be made by earth orbiting satellites (Heppner, 1972; Heelis *et al.*, 1983; Heppner and Maynard, 1987) and ground based radars (Foster, 1983; Foster *et al.*, 1986) or other techniques such as balloon measurements. The balloon-borne technique is one that uses a roughly earth-fixed platform, that has a higher time resolution than radar techniques and that requires no assumptions about temporal stationarity, spatial uniformity or pattern recurrence. Thus, it permits concentrated study of a single region and provides less ambiguous separation of temporal and spatial variations. However, the single point nature of the measurement means that large scale flow patterns can only be inferred in some average sense.

P. E. Sandholt and A. Egeland (eds.), Electromagnetic Coupling in the Polar Clefts and Caps, 137–150.

The task of understanding a complex system must begin with an accurate picture of the dynamics of the system. The system of interest in this paper is the Earth's magnetosphere, which is a very large, optically transparent system comprising a rarefied plasma. Since this system is tens of earth radii in size and the techniques for remote sensing the flow of very transparent plasma are expensive and of limited range, it has proven very difficult to obtain reliable pictures of magnetospheric dynamics, particularly in snapshot form. *In situ* point measurements of the plasma flow velocities in the magnetosphere can be made directly or inferred from measurements of the electric field on satellites or balloons. The plasma flow in the ionosphere can be sensed remotely by a few specialized radars. None of these techniques provide snapshots, and for each of them there is a particular set of assumptions that must be made in order to infer global patterns. For satellite measurements of either plasma drift or electric field, the basic assumptions are that the pattern is stationary for the 15 min it takes to make a pass across the auroral zone and that the same pattern recurs fairly often during periods of similar solar wind conditions. Aliasing of the pattern by short period temporal variations, probe work function variations and accurate subtraction of $\vec{v} \times \vec{B}$ also present serious problems to the use of satellite data. For radar measurements, the basic assumptions are that the field is sufficiently uniform in space and stationary in time to allow a vector to be inferred from multi-point line-of-sight velocity measurements and that the pattern is more or less stationary for the 24 hours it takes for any one radar to rotate under it. For balloon measurements, one needs to make only the second of these two assumptions.

One consequence of the difficulties in measuring magnetospheric convection is that there are persistent controversies over the details of the flow in certain regions. For example, the exact value of the width of the region of plasma entry into the polar cap through the cusp is quite controversial, presumably because of the foregoing differences in basic assumptions (Heelis *et al.*, 1983). The data that have fueled this controversy were mostly obtained by satellites and radars, since few balloons had been flown near the cusp prior to 1985. The problem has arisen because the cusp is a region where the basic assumptions discussed above are not, in fact, particularly valid. One major purpose of the 1985-86 South Pole balloon campaign was to add a third viewpoint to the picture by obtaining a large amount of balloon electric field data from the vicinity of the polar cusp.

Another consequence of the difficulties in measuring ionospheric convection is that very little attention has been paid to the possibility that there may be hemispheric differences in convection patterns arising from the off-center character of the earth's magnetic field. In particular, all of the radars that have been used to study ionospheric drifts lie in the northern hemisphere. After correcting for a known hemispheric asymmetry in IMF B_y dependence (Heppner, 1972), satellite studies appear to mix the hemispheres indiscriminantly. However, theorists modeling the thermosphere have concluded that there should be some difference in hemispheric convection patterns and feel that their models have reached a level of detail where some data on this issue has become necessary (Sojka and Schunk, 1986). Another major purpose of the South Pole Balloon Campaign was to provide some Southern Hemisphere convection pattern data.

This paper will present convection patterns inferred from the data from the balloon campaign both in average form and in a specific example. The paper will then make a preliminary attempt to address the questions posed in the previous two paragraphs.

1.1 DATA SOURCES

The electric field data were acquired during the 1985-86 South Pole Balloon Campaign (Bering *et al.*, 1987) in which eight balloon payloads carrying three-axis double-probe electric field detectors and X-ray scintillation counters were launched sequentially from South Pole Station, Antarctica, at an invariant latitude of 74.5°. The noise level of the electric field instrument was ~0.4 mV m^{-1}, the digitization increment was 0.1 mV m^{-1}, and the data were sampled at 8 Hz. Balloon payload attitude was determined from an on-board magnetometer. A total of 468 hours 30 minutes of data were obtained under a wide range of magnetic conditions. Solar wind plasma and IMF data used to organize the data averaging process were obtained from instruments on the IMP 8 spacecraft. The K_p index values for the period were obtained from Coffey (1986a&b). The location of the various particle precipitation boundaries in the auroral zone and polar cap were determined using particle data from the SSJ detector packages on the DMSP F-6 and F-7 satellites. Cusp and cleft locations have also been inferred from the DMSP particle data (P. Newell, private communication, 1988).

2. Results

2.1 15 JANUARY 1986

The most impressive feature of the electric field data that were obtained by the balloon campaign is the amount of variability in the data. Before presenting the data in average form, it is worthwhile to examine the data from a single day in order to understand the limitations of long term averages as a representation of the data. Two-minute averages of the balloon data obtained on 15 January 1986 are shown in Figure 1. The Southern Hemisphere clockdial format of this figure will be used throughout the paper. In the figures, the electric field data are plotted as inferred $\vec{E} \times \vec{B}/B^2$ ionospheric drift velocities. The data are plotted in corrected geomagnetic latitude and magnetic local time coordinates. The UT values corresponding to the average corrected geomagnetic longitude of the balloon during the day are shown on the inner dial. The figure shows a very typical daily convection pattern, with sunward flows on the morning and evening flanks and antisunward flows near noon and midnight.

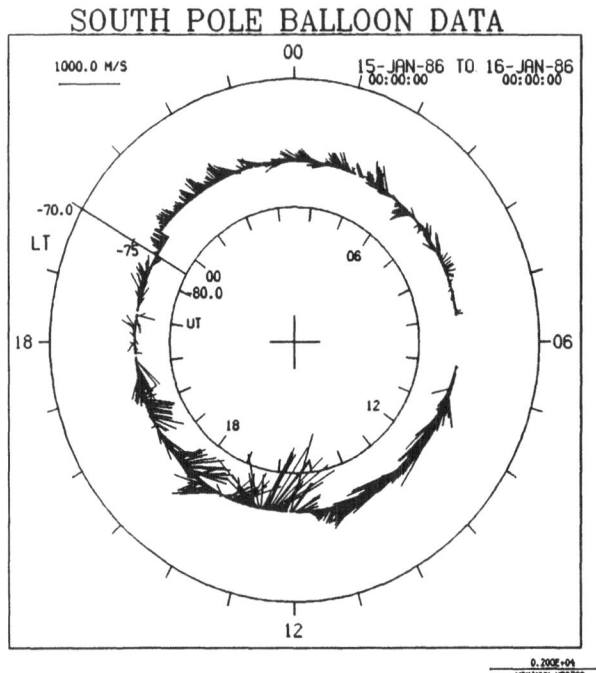

Figure 1. Two minute average ionospheric drift velocites for 15 January 1986. The vectors have been plotted in a clockdial format in the earth-fixed reference frame. The base of each flow vector is plotted at the balloon position in magnetic local time - corrected geomagnetic latitude coordinates.

It is obvious from Figure 1 that there is a lot of variation superimposed on this overall pattern. In order to understand these variations, one needs to know where the balloon was relative to the auroral oval, and what the interplanetary magnetic field was doing during the day. The DMSP satellite data have been used in an attempt to place the balloon relative to the oval. Figure 2 shows an example of the process. This figure shows 12 hours of data centering around magnetic local noon at South Pole on 15 January 1986. All DMSP orbits that cross 75° invariant latitude within 4 hours of local time of the balloon location have been plotted. Five particle precipitation features are shown on each orbit, if present: the equatorward boundary (E), the transition between central plasma sheet and boundary plasma sheet (T), the poleward boundary (P), the cleft, and the cusp. The numbered circles show the balloon location at the times corrsponding to the DMSP passes shown. This figure shows that the balloon was between the T and P boundaries most of the time. The cusp extended from ~0900 to at least 1300 MLT. Over most of this range, the cusp was somewhat poleward of the balloon position. However, the cusp appears to have extended over the balloon position from ~1130 MLT onward. On the nightside, a similar analysis (figure not shown) indicates that the balloon was poleward of the P boundary and therefore in the polar cap from the start of the day until 0430 MLT.

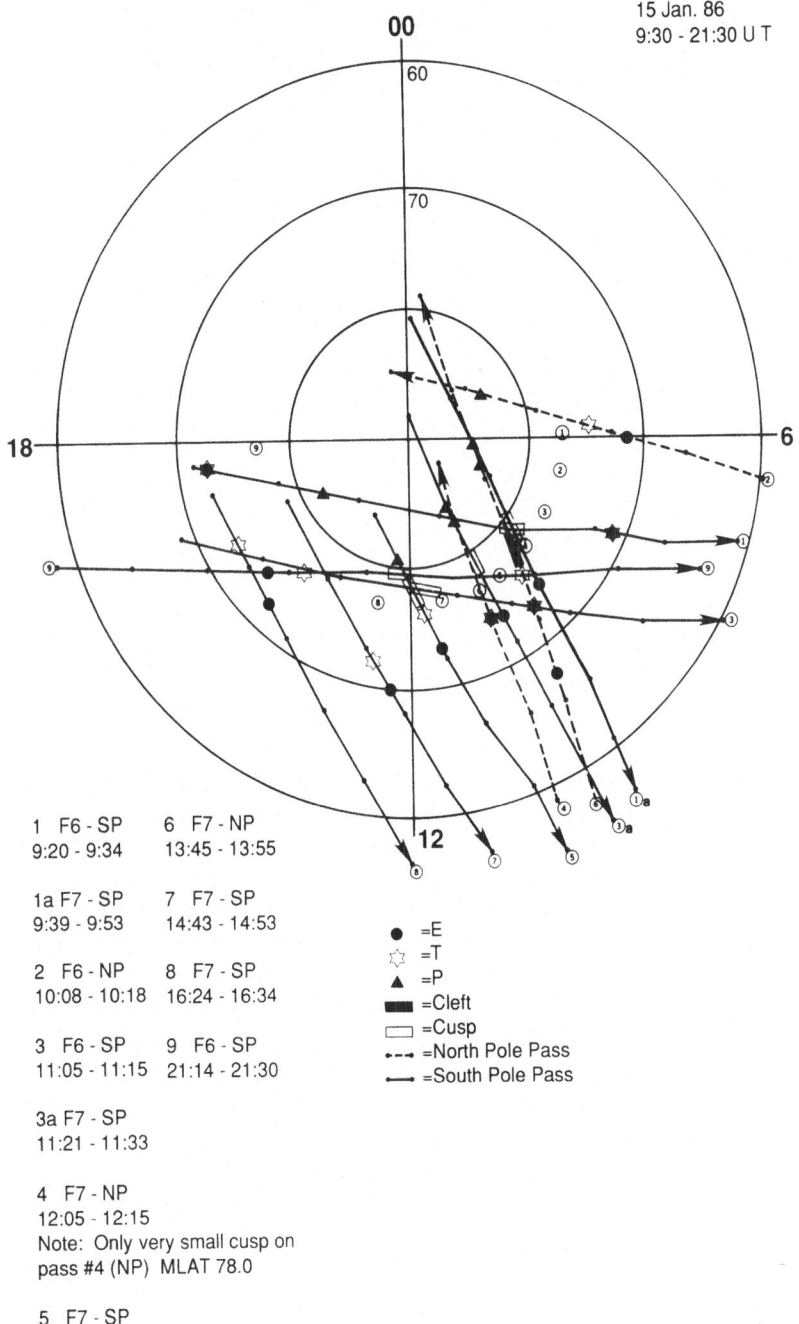

Figure 2. DMSP F-6 and F-7 ground tracks and particle boundaries on 15 January 1986.

It would appear therefore that the region of anti-sunward flow shown on the nightside

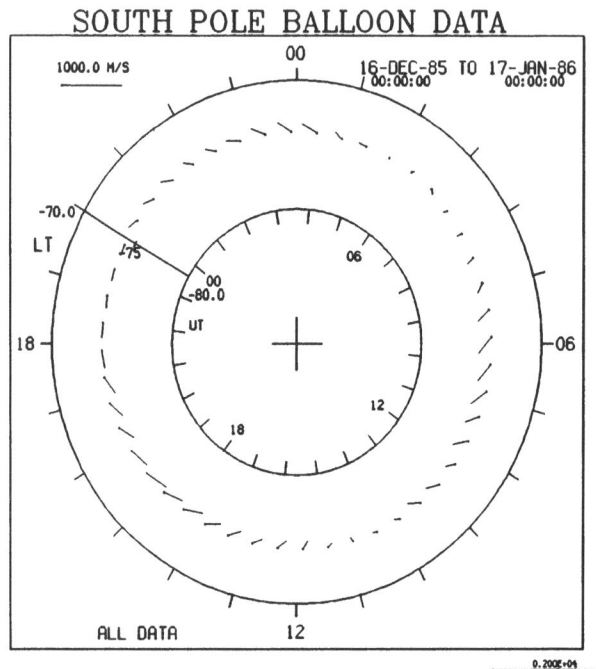

Figure 3. One-half hour averages of the data from the entire campaign binned as a function of magnetic local time and plotted in the same format as Figure 1.

in Figure 1 corresponds to the time that the balloon was in the polar cap. The fluctuations that we see during this period possibly can be attributed to temporal variations in the polar cap flow and not to boundary crossings. In the morning sector, the strong sunward flow that is seen appears to be taking place just equatorward of the cusp as seen in DMSP particle data and the auroral oval as seen in DE image data. The convection reversal that takes place just prior to magnetic noon seems to coincide with the cusp location indicated in Figure 2. We suggest therefore that this convection reversal is associated with a boundary crossing. About one half hour later, the IMF turns southward and stays that way for just over two hours. This southward turning appears to be responsible for the intensification of poleward flow seen during that time. Passage out of the cusp region and northward turning of the IMF occur essentially simultaneously at ~1400 MLT. Several of the "sloshes" in the overall sunward flow that appear during the next several hours also appear to be associated with short southward turnings of the IMF.

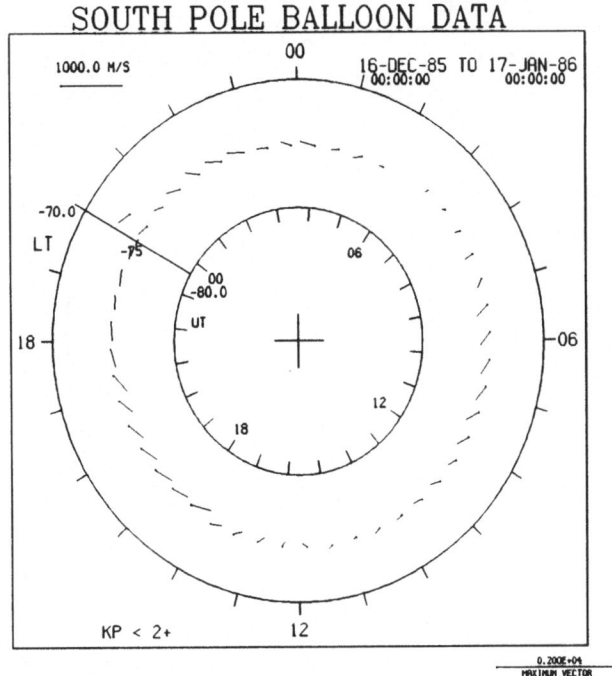

Figure 4a. Averages similar to those shown in Figure 3, under the conditions that $K_p \leq 2\mathrm{o}$.

2.2 AVERAGES

2.2.1 *Global Average.* One of the ways by which we can begin to separate ephemeral temporal effects from the underlying global structure is to take long term averages of the data. Two minute average data such as those shown in Figure 1 have been averaged over the whole campaign, binned in $\frac{1}{2}$ hour intervals of magnetic local time. The first step was to average the entire data set in this fashion, as shown in Figure 3. The purpose of showing Figure 3 is to establish that the balloon detectors over South Pole were responding properly to the well-known average two-cell morphology of ionospheric convection (Heppner, 1977; Foster *et al.*, 1986). The expected two-cell pattern is clearly evident in Figure 3, which shows sunward flow on both flanks and a broad region of flow with an antisunward component centered on 1100 MLT. On the nightside, we see anti-sunward flow from 2200 to 0200 MLT. The only noteworthy feature in the figure is the fact that the dawn cell is significantly smaller than the dusk cell. In fact, we did not observe much flow out of the nightside into the dawn cell. This anomaly reflects two things: that the basic assumption of 24 hour stationary patterns is incorrect and that the balloon flights were concentrated in one sector of the IMF.

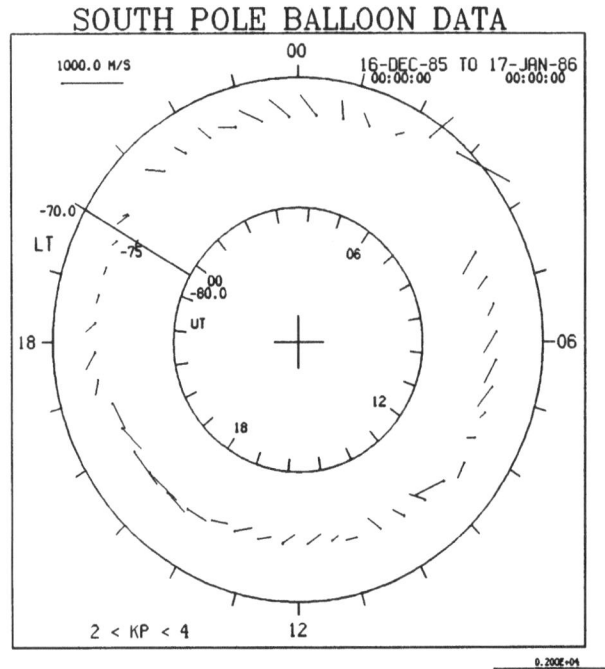

Figure 4b. Same as Figure 4a, for $2+ \geq K_p \leq 4-$.

2.2.2 K_p *dependence*. The balloon data set is extensive enough to permit us to examine the dependence of the convection patterns on magnetic activity level and IMF
orientation. We shall begin by presenting the dependence of the cusp latitude Southern
Hemisphere convection pattern on the level of internal magnetic activity as expressed in
the K_p index. The results of averaging the data binned as a function of K_p are shown
in Figures 4a, 4b, and 4c. The averages shown in these figures were computed from an
average of 11.4 half-hour data blocks per bin in 4a, 3.75 data blocks per bin in 4b and
3.12 data blocks per bin in 4c. The interval from 0400 to 0800 is somewhat undersampled
in the two latter figures. The quiet-time pattern is shown in Figure 4a. This pattern is
very similar to Figure 3 in overall outline, with two significant differences. First, the flow
speed is slightly slower overall, particularly in the dawn sector. Second, in the pre-noon
region of eastward flow on the dayside, the sun-tail component remains sunward until just
30 minutes prior to the east west reversal, which occurs just after noon. This pattern is
a contrast to the earlier east-west reversal and wider region of anti-sunward flow seen in
Figure 3. The moderate activity pattern is shown in Figure 4b. In this Figure, the east-
west reversal on the dayside has moved to an earlier local time, 1100 MLT, the regions
of antisunward flow have widened on both the dayside and the nightside and the overall
speed of the flow has increased. The active-time pattern is shown in Figure 4c. The region
of anti-sunward flow on the dayside now covers $7\frac{1}{2}$ hours in local time, running from 0700
to 1430 MLT. The east-west reversal takes place at 1000 MLT. The day side flows are still
larger in speed than in the previous two figures. The nightside average flows, by contrast,
are small and disorganized. In the pre-midnight sector, these small flows probably reflect

the fact that the balloons spent nearly equal amounts of time on either side of the Harang discontinuity with the result that the average flow is nearly zero. After midnight, the turbulent nature of the flow in the auroral zone in the post-break-up pre-dawn situation probably accounts for the disorganized character of the averages.

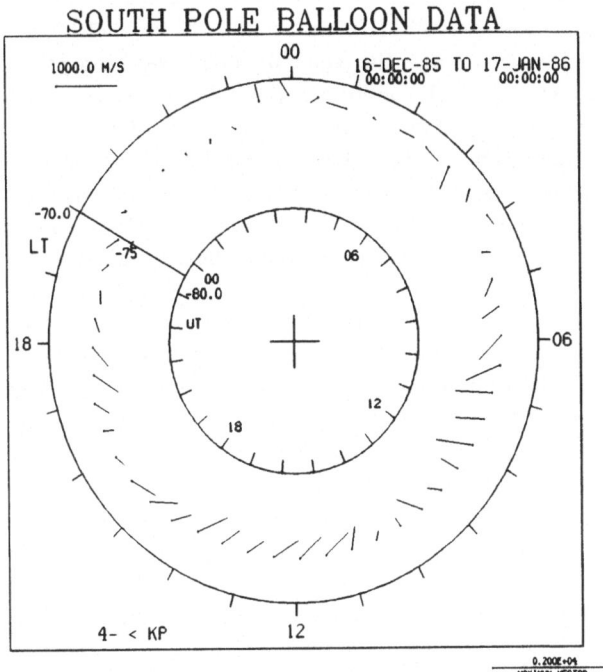

Figure 4c. Same as Figure 4a, for $4o \leq K_p$.

2.2.3 *IMF orientation dependence.* The response of the large-scale convection pattern at high-latitude to changes in the interplanetary magnetic field has been studied extensively by other methods (Heppner, 1972, 1977; Heelis, 1984; Foster *et al.*, 1986; Heppner and Maynard, 1987). In order to investigate influence of the IMF orientation on the daily average convection patterns from our data, we have used high time resolution solar wind plasma and IMF data from the IMP 8 spacecraft. The spacecraft-ionosphere delay time was computed using measured solar wind velocity and density data (Etemadi *et al.*, 1988) and assuming a 220 s magnetopause-ionosphere propagation time (Clauer and Friis-Christensen, 1988). The balloon data were binned using weighted one hour average IMF B_y and IMF B_z values. Separate averages were computed for each two minute balloon data point, using a Rayleigh function with a 22 minute time constant (Clauer and Friis-Christensen, 1988) as a weighting function. The results of averaging the data binned as a function of IMF orientation are shown in Figures 5a, 5b, 5c and 5d. The averages shown in these figures were computed from an average of 1.99 half-hour data blocks per bin in 5a, 1.44 data blocks per bin in 5b, 1.60 data blocks per bin in 5c and 0.62 data blocks per bin in 5d. Since the interval from 0000 to 0800 is seriously undersampled in all the figures, between 3 and 6 days were sampled in 5a around magnetic noon, between 2 and 3 days in 5b, between 2 and 4 days in 5c and between 1 and 3 days in 5d. The invariant latitude variations are due to differences in stratospheric wind direction during the different flights. Figure 5a shows the convection pattern for B_y and B_z positive. The point of interest in this figure is the presence of two separate regions of anti-sunward convection on the dayside, from 0800 to 0900 MLT and from 1130 to 1330 MLT. Because of a UT dependence in IMP 8 data coverage, the region of antisunward convection on the nightside was not sampled under these IMF conditions. Figure 5b shows the convection pattern for B_y negative and B_z positive. Two regions of antisunward flow can still be seen on the dayside. However, the pre-noon region has expanded to the interval from 0800 to 1100 MLT and the post-noon region has contracted to the interval from 1300 to 1400 MLT. The presence of two convection reversals sandwiching a region of sunward flow near noon is consistent with the presence of four convection cells during periods of northward IMF (Reiff, 1982).

Figures 5c and 5d show the two B_y cases for B_z negative. Both Figures show that when B_z was negative (southward), two-cell patterns were observed. In addition, when B_y was postive, we observed that the dayside/dawn cell flow velocities had greater magnitude than in the other cases. Furthermore, the dayside convection reversal rotates further into the prenoon sector toward dawn than it does in the case where B_y was negative shown in Figure 5d. (Note that the episode of sunward flow near noon shown in Figure 5d is derived from just one flight, and may not therefore represent the average situation.) On the nightside, antisunward flow is seen after midnight in Figure 5c and before midnight in 5d. This B_y dependence is in agreement with previous results that the sign of the IMF-Y component shifts the convection pattern in local time (Heppner, 1972; Heelis, 1984).

3. Discussion

There was considerable discussion at the workshop where this paper was initially presented concerning the validity and utility of averages such as the ones presented above. It is certainly the case that the balloons were near major ionospheric boundaries such as the Harang discontinuity during much of the campaign, as shown in Figure 2. Because of this proximity, significant physics may be lost in the averaging process. Figure 1 also shows that there is a lot of temporal variation and detailed dynamics to be found in the data that will be lost. Third, Figure 1 also makes clear that actual snapshots are very unlikely to resemble the averages in anything but the crudest fashion. Nonetheless, we have proceeded to compute and present these averages for three reasons. First, averages exist (Vasyliunas, 1988) and are well-defined well-understood statistics. Second, averages are easy outputs for modelers to produce for validation purposes. Third, we have not yet determined what the statistics and estimators should be used in characterizing the dynamic variability of our data. Simple presentation of twenty days of clockdial plots, regardless of how interesting they may be, does not give modelers much that can be used later. One possible approach to this dilemma involves fitting detailed quantitative models (e.g. Heelis et al., 1982; Moses et el., 1988) to each day separately and then examining

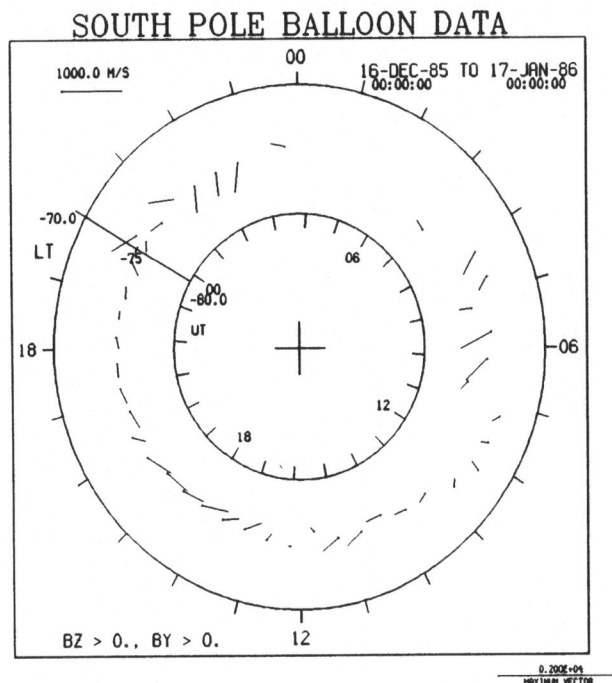

Figure 5a. Averages similar to those shown in Figure 3, under the conditions that IMF $B_y > 0$. and $B_z > 0$..

how the parameters of these models behave as a function of solar wind conditions and magnetic activity level.

4. Conclusions

Examination of the Southern Hemisphere ionospheric convection patterns observed by balloons at 75° invariant latitude near the south polar cusp has shown the overall convection patterns to be two-celled, especially during quiet days. On average, the dusk cell is larger than the dawn cell and extends into the dawn side near both magnetic noon and midnight. The dayside convection reversal region appears to rotate towards dawn as the level of magnetic activity increases.

The results have also shown dependence of the convection patterns on the orientation of the interplanetary magnetic field. When IMF $B_y < 0$, it seems that the convection pattern in the Southern hemisphere shifts toward dusk and when $B_y > 0$ the pattern shifts toward dawn. When B_z is northward, two dayside convection reversals may occur, with a region of sunward flow in between.

The averaging process used in this paper and others like it necessarily supresses a great deal of interesting physics that is expressed in the daily patterns. It is almost certainly the case that the average patterns never actually exist in an instantaneous snapshot sense.

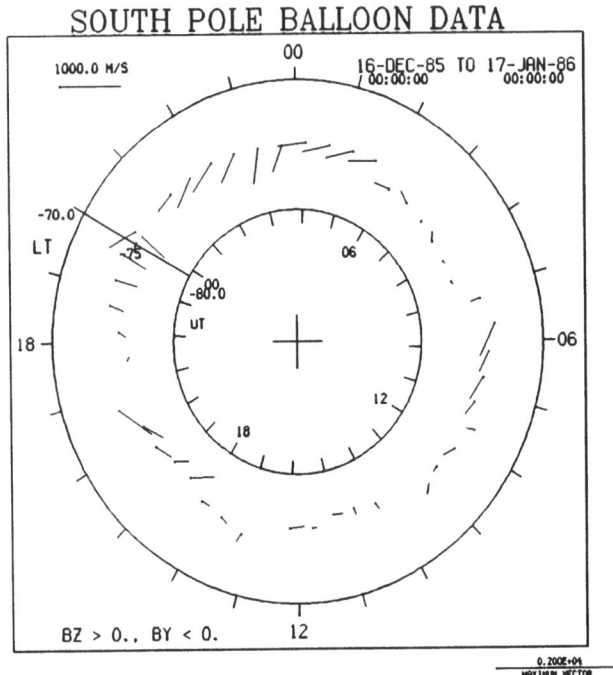

SOUTH POLE BALLOON DATA

Figure 5b. Averages similar to those shown in Figure 3, under the conditions that IMF $B_y < 0$. and $B_z > 0$..

The real challenge that these data pose is finding ways to quantify the dependence of the variability and detailed structure of the daily patterns on the IMF and activity level.

Acknowledgments

We wish to thank Dr. Susan Gussenhoven for the DMSP particle data, Dr. Ron Lepping for the IMP 8 magnetometer data, Dr. Al Lazarus for the IMP 8 plasma data, and Dr. Patrick Newell for the DMSP cusp crossing identifications. We also wish to thank Dr. Nelson Maynard and Dr. Rod Heelis for helpful discussions. This work was supported in part by National Science Foundation Grant DPP-8614091.

References

Bering, E. A., III, J. R. Benbrook, J. M. Howard, D. M. Oró, E. G. Stansbery, J. R. Theall, D. L. Matthews and T. J. Rosenberg (1987) 'The 1985-86 South Pole Balloon Campaign', *Proceedings of the Nagata Symposium on Geomagnetically Conjugate Studies and the Workshop on Antarctic Middle and Upper Atmosphere Physics Which Were Held at*

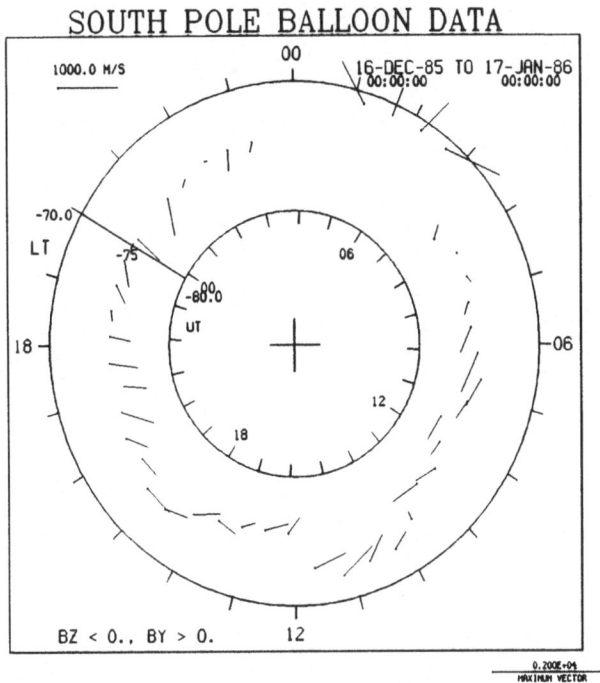

Figure 5c. Averages similar to those shown in Figure 3, under the conditions that IMF $B_y > 0$. and $B_z < 0$..

SCAR XIX, Memoirs of the National Institute of Polar Research, Japan, Special Issue No. **48**, 313–317.

Clauer, C. R., and E. Friis-Christensen (1988) 'High-latitude dayside electric fields and currents during strong northward interplanetary magnetic field: observations and model simulation', *J. Geophys. Res.*, **93**, 2749.

Coffey, H. E. (1986a) 'Geomagnetic and Solar Data', *J. Geophys. Res.*, **91**, 4609.

Coffey, H. E. (1986b) 'Geomagnetic and Solar Data', *J. Geophys. Res.*, **91**, 7161–7162.

Etemadi, A., S. W. H. Cowley, M. Lockwood, B. J. I. Bromage, D. M. Willis, and H. Lühr (1988) 'The dependence of high-latitude dayside ionospheric flows on the north-south component of the IMF: A high time resolution correlation analysis using EISCAT "polar" and *AMPTE UKS* and *IRM* data', *Planet. Space Sci.*, **36.**, 471.

Foster, J. C. (1983) 'An empirical electric field model derived from Chatanika radar data', *J. Geophys. Res.*, **88**, 981.

Foster, J. C., J. M. Holt, R. G. Musgrove, and D. S. Evans (1986) 'Ionospheric convection associated with discrete levels of particle precipitation', *Geophys. Res. Lett.*, **13**, 656.

Heelis, R. A. (1984) '*J. Geophys. Res.*, **89**', *Geophys. Res. Lett.*, **13**, 2873.

Heelis, R. A., J. C. Foster, O. de la Beaujardiere, and J. Holt (1983) 'Multistation measurements of high-latitude ionospheric convection', *J. Geophys. Res.*, **88**, 10111.

Heelis, R. A., J. K. Lowell, and R. W. Spiro (1982) 'A model of the high-latitude ionospheric convection pattern', *J. Geophys. Res.*, **87**, 6339.

150

SOUTH POLE BALLOON DATA

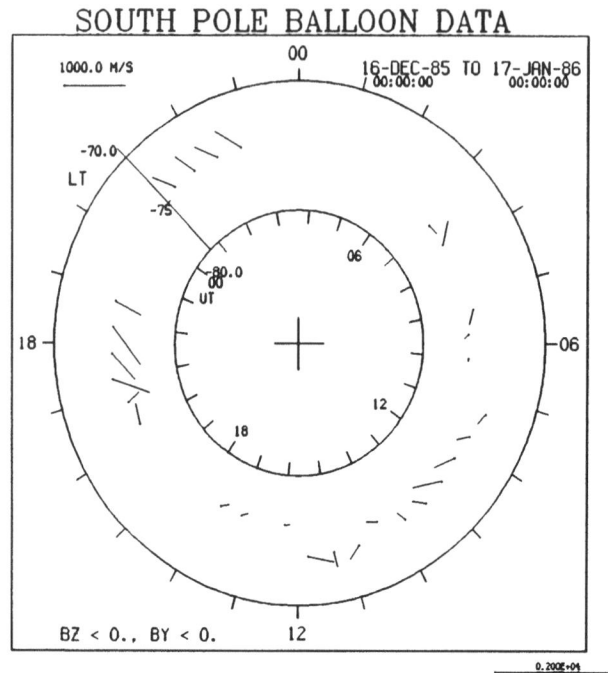

Figure 5d. Averages similar to those shown in Figure 3, under the conditions that IMF $B_y < 0$. and $B_z < 0$..

Heppner, J. P. (1972) 'Polar cap electric field distributions related to the interplanetary magnetic field direction', *J. Geophys. Res.*, **77**, 4877.

Heppner, J. P. (1977) 'Empirical models of high-latitude electric fields', *J. Geophys. Res.*, **82**, 1115.

Heppner, J. P., and N. C. Maynard (1987) 'Empirical high-latitude electric field models', *J. Geophys. Res.*, **92**, 4467.

Moses, J. J., G. L. Siscoe, R. A. Heelis, and J. D. Winningham (1988) 'A model for multiple throat structures in the polar cap flow entry region', *J. Geophys. Res.*, **93**, 9785.

Reiff, P. H. (1982) 'Sunward convection in both polar caps', *J. Geophys. Res.*, **87**, 5976.

Sojka, J. J. and R. W. Schunk (1986) 'Asymmetries in the plasma characteristics of the conjugate high-latitude ionospheres', paper presented at the *International Symposium on Large Scale Processes in the Ionosphere-Thermosphere System*, Boulder, Colorado.

Vasyliunas, V. M. (1989) 'Electrodynamics of the ionosphere/magnetosphere/solar wind system at high latitudes', paper presented at the *NATO Advanced Research Workshop on Electromagnetic Coupling in the Polar Clefts and Caps*, Lillehammer, Norway.

GROUND MAGNETIC PERTURBATIONS
IN THE POLAR CAP AND CLEFT:
STRUCTURE AND DYNAMICS OF IONOSPHERIC CURRENTS

E. FRIIS-CHRISTENSEN
Danish Meteorological Institute
Division of Geophysics
100 Lyngbyvej
DK-2100, Copenhagen
Denmark

ABSTRACT.   Ground magnetic perturbations in the polar cap and cleft are very closely related to variations in the solar wind parameters through the magnetic field-lines which map to the boundary layers of the magnetosphere. The magnetic variations in the polar cap respond almost immediately to variations in the interplanetary magnetic field (IMF), although it takes some time for the large-scale ionospheric currents to adapt to a sudden change in the IMF. Dense arrays of magnetometers provide a possibility to resolve the structure of the ionospheric currents, which are very complex around magnetic local noon. The cleft is thought to map to the low-latitude boundary layer which is generally located on closed field-lines. In this region small-scale features known as magnetic impulse events have been shown to correspond to tailward moving ionospheric current vortices which are probably related to sudden changes in the solar wind dynamic pressure. But also other magnetopause processes may have ionospheric signatures which are seen in the ground magnetic records.

1. Introduction

The magnetic perturbations in the polar cap and polar cleft are caused by currents which are intimately related to processes occurring at the boundary layers of the magnetosphere. These processes are determined by the conditions in the solar wind, in particular the interplanetary magnetic field (IMF) which is the major cause of the large-scale polar magnetospheric plasma convection. A number of studies have shown that magnetospheric convection responds immediately to changes in the IMF (Clauer et al., 1984; Rishbeth et al., 1985). This indicates a direct electrical coupling between the interplanetary electric field and the high-latitude ionosphere via field-aligned currents.

In a companion paper (Friis-Christensen, 1989) the average relationship between the IMF and the electric fields and currents at polar latitudes is reviewed . In the present paper we will emphasize features which are too localized or too short lived to systematically affect the average patterns of polar cap and cleft ground magnetic perturbations. We will see that these features can only be sufficiently described using closely spaced measurements of the magnetic variations, and that they therefore are difficult to observe, and especially to interpret using single point data. Nevertheless, the small-scale ground-based magnetic perturbations contain information about structure and dynamics which is important for the understanding of the interaction between the solar wind and the coupled magnetosphere-ionosphere system.

P. E. Sandholt and A. Egeland (eds.), Electromagnetic Coupling in the Polar Clefts and Caps, 151–165.
© *1989 by Kluwer Academic Publishers.*

In section 2 we will first investigate how the response of changes in the IMF is seen in the polar cap and cleft magnetic perturbations. In section 3 we will consider the structure of the ionospheric currents and their development in time. Section 4 deals with a particular aspect of ground magnetic perturbations in the cleft known as impulsive magnetic events which have been proposed to be the ionospheric signatures of dayside reconnection events. Recent work, however, calls for caution in identification of ground-based magnetic signatures of flux transfer events.

2. Ground Magnetic Response to Changes in the IMF

The reconfiguration of the magnetosphere to a new state corresponding to a reorientation of the IMF is a process which involves time delays. Furthermore, the time delays are different at different locations in the magnetosphere. The ground magnetic response, which is a combined effect of local ionospheric currents and distant ionospheric and field-aligned currents, will therefore show a spatial distribution which is different from the patterns obtained from average statistical models.

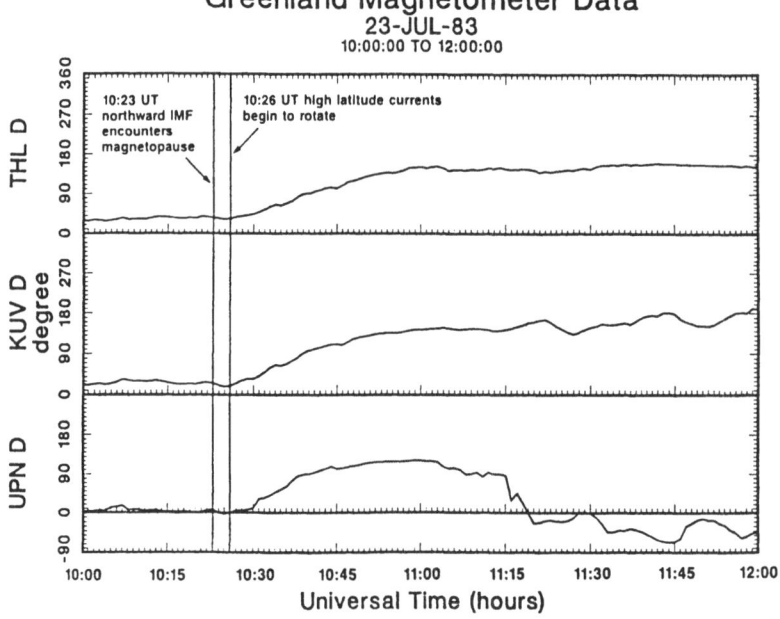

Figure 1. Horizontal magnetic field declination variations measured relative to the average declination at three polar cap stations on the West Coast of Greenland. At 1023 UT the northward IMF is estimated to encounter the magnetopause, and at 1026 UT the field at THL and KUV begins a 120° rotation. (After Clauer and Friis-Christensen, 1988)

In Figure 1 taken from Clauer and Friis-Christensen (1988) is shown the rotation of the horizontal magnetic perturbation vector observed at three locations in the polar cap and cleft. A northward turning of the IMF is estimated to encounter the magnetopause at

SIMULATED ION VELOCITY

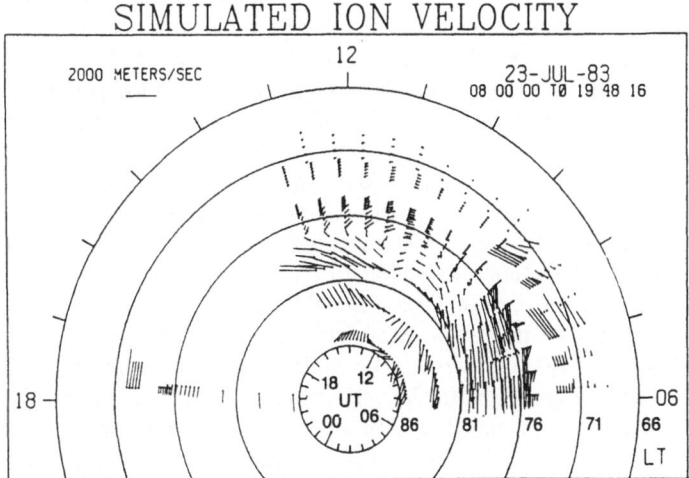

Figure 2. Polar plot showing simulated F region ion convection velocity vectors as a function of latitude and local time for July 23, 1983. Local time is indicated around the outer circumference of the plot, and universal time is indicated around the inner circumference. Local magnetic noon (1400 UT) is at the top of the plot. The three complete rows of highest-latitude vectors are convection velocities inferred from ground magnetometer data. The other vectors in the plot were obtained using the Søndre Strømfjord, Greenland incoherent scatter radar. (After Clauer and Friis-Christensen, 1988)

MEASURED ION VELOCITY

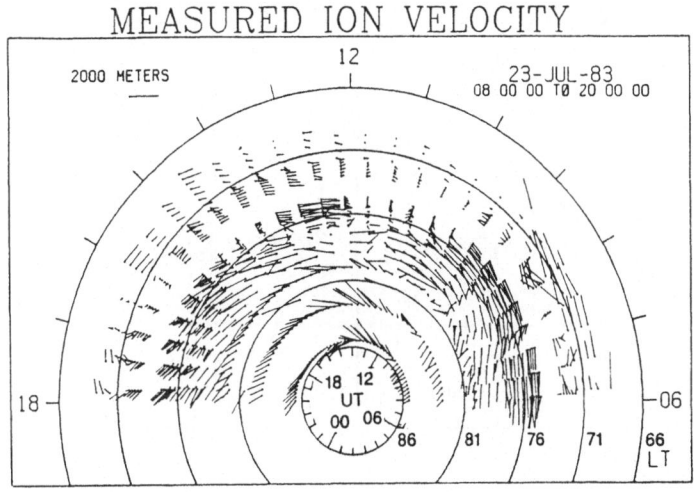

Figure 3. Measured F region ion convection velocity vectors for July 23, 1983, shown in the same format as Figure 2. (Clauer and Friis-Christensen, 1988)

1023 UT . Three minutes later all three stations located at invariant latitudes from 79.9° to 85.5° start to respond. The delay time was concluded to correspond to the estimated

154

propagation time for an Alfven wave to travel from the dayside magnetopause to the ionosphere. It is also seen that the large-scale reconfiguration of the currents appears to evolve
over approximately 22 minutes. This time delay is close to the time delay of 30 minutes
(including transit time from the spacecraft position to the magnetopause) which statistically gives the best correlation between IMF variations and ground magnetic perturbations
(Friis-Christensen et al., 1985).

Clauer and Friis-Christensen (1988) simulate the ionospheric plasma convection velocities and magnetometer observations using a simple electrodynamic model in which the
dayside high-latitude field-aligned current configuration is determined by the strength and
orientation of the IMF. The model is similar to that proposed by Banks et al. (1984) and
the main idea is based upon the assumption that it is possible to decompose the high-latitude
field-aligned current systems into two systems which are separately controlled by the B_y and
B_z components of the IMF. Comparing the simulated time development of the polar cap
convection velocities in Figure 2 with the velocities measured with the Søndre Strømfjord
incoherent scatter radar supplemented with equivalent convection velocity directions inferred from the highest latitude magnetometer stations on the West Coast of Greenland
(Figure 3), they conclude that the simple representation of the currents controlled by the
measured IMF reproduces the majority of features seen in the observed plasma convection.
There are, however, also features which are not explained by the simplified model.

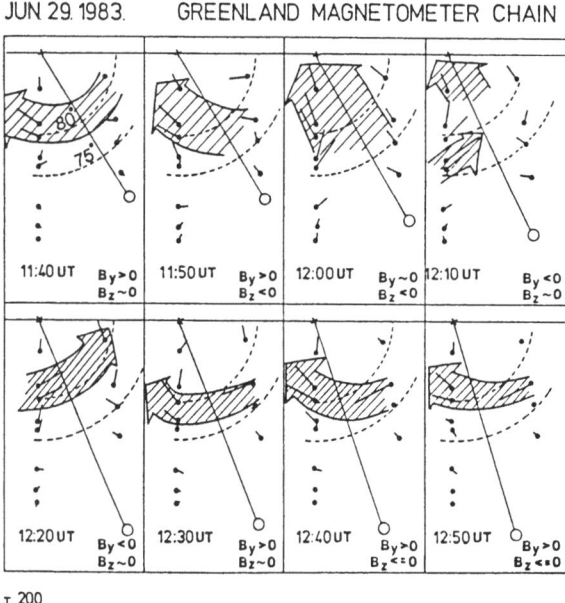

Figure 4. Time development of the spatial distribution of convection velocity vectors
inferred from the magnetometer chain in Greenland. The magnetic pole is indicated
by a cross and the meridian magnetometer chain on the West Coast of Greenland
is placed on a vertical line. The East Coast stations are placed according to the
position in an invariant latitude-magnetic local time diagram where magnetic local
noon is indicated by a line connecting the pole and the position of the sun. The
main areas of convection are emphasized by hatching.

Comparing magnetometer observations on the West Coast and on the East Coast of Greenland Friis-Christensen (1986) shows the spatial distribution of the time variations of the polar cap and polar cleft magnetic perturbations. This is illustrated in Figure 4 in the form of convection velocity directions inferred every 10 minutes during 80 minutes during which the IMF B_y component changes sign several times. As it is seen in several of the panels, the response to the IMF does not take place simultaneously at all stations. The change in B_y at 1210 UT for instance, when B_y becomes negative, is first seen in the prenoon sector corresponding to the start of a poleward plasma convection probably from closed to open field-lines in agreement with the build up of a merging-cell of convection (Reiff and Burch, 1985).

3. Structure of the Convection Currents — DPY1 and DPY2

In the preceding section we have seen that the magnetic perturbations in the polar cap and cleft respond immediately to the changes in the IMF. The response is primarily, at least in the dayside sunlit ionosphere, due to the ionospheric Hall currents since the effect from the field-aligned currents is to a large extent cancelled by the effect from the Pedersen currents assuming that only small conductivity gradients exist in the ionosphere.

In the most simple approximation the ionospheric plasma convection consists of an antisunward convection in the polar cap and a sunward return flow along the auroral oval in the electrojet regions. Using the Søndre Strømfjord incoherent scatter radar Clauer et al. (1984) and Jørgensen et al. (1984) show that the plasma convection around noon is predominantly east-west directed, at least during intervals of a significant IMF B_y component. This is consistent with the earlier results regarding the ionospheric currents (Friis-Christensen and Wilhjelm, 1975), and confirms that the ground-based magnetic perturbations are indeed caused by the ionospheric Hall currents.

Unfortunately the incoherent scatter radar in Søndre Strømfjord at 74° invariant latitude is not able to look very far into the polar cap, but is limited to approximately 80° invariant latitude. Latitude profiles along a magnetic meridian show, however, often fine structures in the currents, which indicate that the polar cap convection consists of separate cells. In an attempt to describe the Hall current distribution and hence the plasma convection in a quantitative way Friis-Christensen and Vennerstrøm (1989) model the high-latitude ionospheric currents. Using a model consisting of a number of linear ionospheric sheet currents, they found that very often the ground magnetic variations can be sufficiently well modeled by three east-west directed currents sheets. Namely the auroral electrojet (westward in the morning and eastward in the afternoon) and two additional currents located poleward of this. These two currents which normally have the same direction, form the B_y controlled DPY-current (Friis-Christensen and Wilhjelm, 1975). We therefore call them the $DPY1$ and the $DPY2$ currents, $DPY1$ being the poleward of the two.

In Figure 5 is shown an example of the time variations of the latitudinal extent of these three currents calculated every 10 minutes during three hours from 09 to 12 UT corresponding to 07 to 10 MLT. The lower panel shows 10 minute average values of the IMF B_y and B_z components. B_y is negative during most of the interval, but the B_z component changes sign several times. Comparing the latitudinal variations of the ionospheric currents with the IMF variations it is seen that consistent with a time delay of about 20 minutes the $DPY1$ and $DPY2$ currents vary systematically as a function of the IMF components.

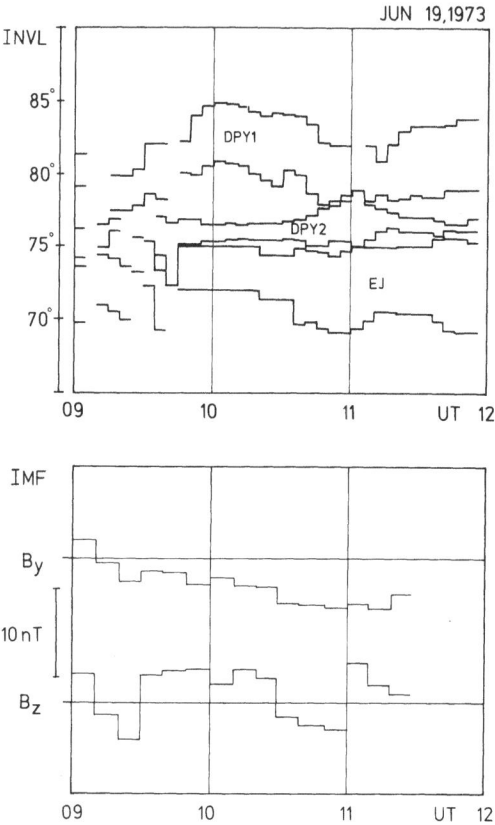

Figure 5. The upper panels show the poleward and the equatorward boundaries of the $DPY1$, the $DPY2$, and the electrojet currents and their development in time from 09-12 UT on June 19, 1973. The lower panel shows 10 minute average values of the B_y and B_z components of the IMF.

When B_z becomes positive the $DPY1$ current moves poleward while the width of the $DPY2$ current decreases. Oppositely, when B_z becomes negative the width of the $DPY2$ current increases, as does the electrojet, whereas the $DPY1$ current moves equatorward and merges with the $DPY2$ current. Since the $DPY1$ current seems to be related to positive B_z conditions and the $DPY2$ current to negative B_z conditions the $DPY1$ and $DPY2$ currents could correspond to the lobe-cell (L-cell) and merging-cell (M-cell) in the convection model of Reiff and Burch (1985). In their model, however, the merging-cell and the lobe-cell are parallel in the morning sector for $B_y > 0$ and in the afternoon sector for $B_y < 0$ in the Northern Hemisphere. It has not yet been possible to find a similar symmetry in the magnetometer observations.

The separation of the $DPY1$ and $DPY2$ currents is expected to be associated with a corresponding fine structure of the field-aligned currents. Friis-Christensen and Lassen (1989) find that statistically the region 1 field-aligned currents consist of two parts, the region 1a and region 1b currents which have different relationships to the IMF B_z component. Although a number of low-altitude satellite observations do reveal fine structures in

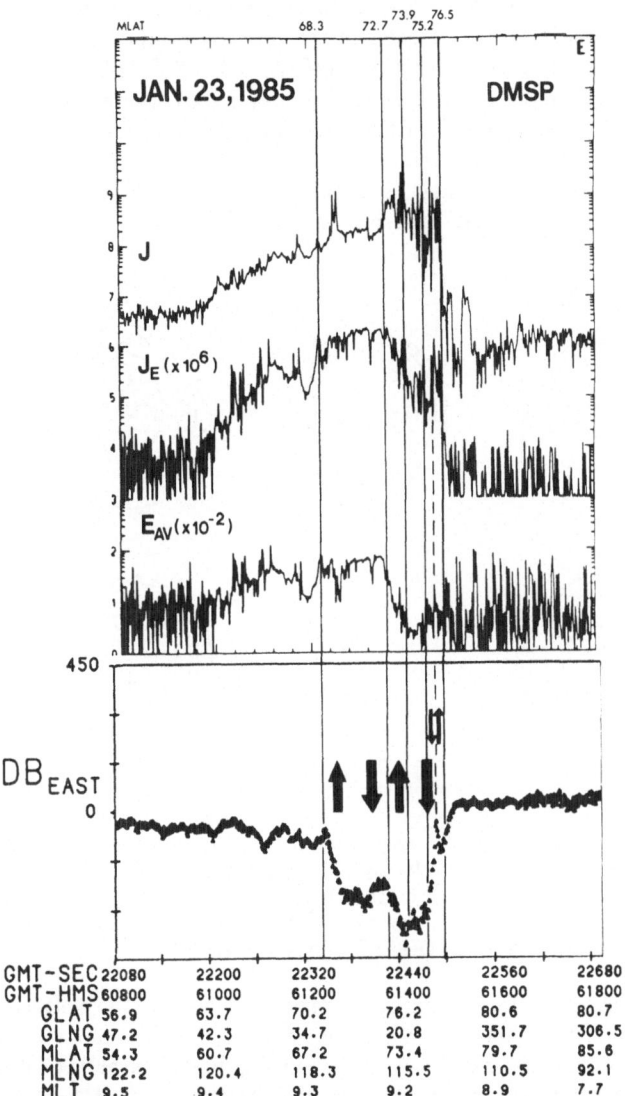

Figure 6. Electron precipitation (particle flux, energy flux, average energy) and eastward magnetic deflection during a DMSP-F7 pass (061435 UT) on Jan 23, 1985. Vertical lines mark different zones of particle precipitation and Birkeland current (current direction marked by arrows). (After Sandholt et al., 1988)

the field-aligned current patterns, no systematic relationship to the IMF has been reported. Sandholt et al. (1988) report on an example of a multiple field-aligned current structure which is reproduced here in Figure 6. They interpret the equatorward pair of field-aligned currents as the region 1/region 2 system. This system is seen to be associated with particles characteristic of the plasma sheet. Poleward of this, a second system is located in a region populated with low-latitude boundary layer particles. Friis-Christensen and Vennerstrøm (1989) find that particularly during changes in the IMF B_z component, there is a larger probability of the simultaneous presence of the $DPY1$ and the $DPY2$ system. The observation by Sandholt et al. (1988) shown in Figure 6 does indeed take place during a drastic change in the IMF B_z component (Friis-Christensen and Vennerstrøm, 1989). This is consistent with the idea of Friis-Christensen and Lassen (1989) that the equatorward pair of currents, including the region 1b current, is associated with $B_z < 0$ and that the poleward pair, including the region 1a current, is associated with $B_z > 0$ and with the low-latitude boundary layer.

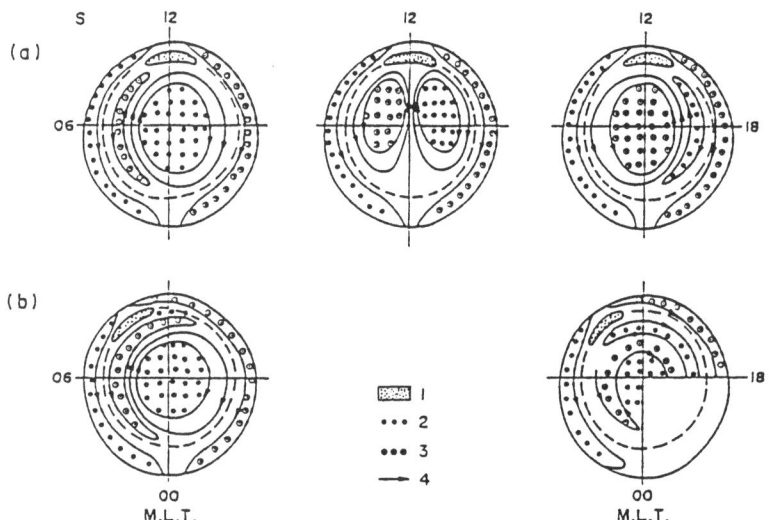

Figure 7. (a) Convection and field-aligned current patterns proposed by Potemra et al. (1984) and Zanetti et al. (1984) for $B_z > 0$, $B_y > 0$ and $B_y < 0$. (b) Convection and field-aligned currents obtained on the basis of meridian chain magnetic data: (1) — Day-time cusp region, (2) — downward field-aligned currents, (3) — upward currents, (4) — convection flows. (After Troshichev et al., 1988)

A dense chain of magnetometers spanning the region poleward of the cleft is necessary to resolve the fine structure of the dayside convection currents. It is also an advantage to have the line of stations located approximately perpendicular to the invariant latitude circles since there may be longitudinal variations which influence the interpretation. Such a chain exists in Greenland in the Northern Hemisphere. A similar chain exists in the Southern Hemisphere in Antarctica. Data from this chain have been used in a study by Troshichev et al. (1988). While the model calculations of Friis-Christensen and Vennerstrøm (1989) are based upon a number of sheet currents of varying width and intensity, Troshichev et al. (1988) use a model of Kotikov et al. (1986) based upon a large number of equidistantly spaced line currents ($n = 50$) to model the observed ground magnetic perturbations . Their results shown in Figure 7 reveal the existence of multiple sheets of oppositely directed currents for $B_z > 0$. The additional current filaments are located between the region 1 currents and the NBZ currents (Iijima et al., 1984) especially near local magnetic noon. Their results further indicate that there is an asymmetry in the current pattern corresponding to the different signs of the B_y component of the IMF. For $B_y > 0$ the observed additional field-aligned current sheets are of low intensity and adjacent to the region 1 current, and the resulting pattern is similar to the system proposed by Potemra et al. (1984) and Zanetti et al. (1984). For $B_y < 0$ the additional field-aligned current systems are found to intensify, widen, and shift to the pole.

4. Ionospheric Traveling Convection Vortices

While the polar cap magnetic perturbations are closely related to the currents associated with the ionospheric plasma convection, a particular type of magnetic pulsations is characteristic of the region just equatorward of the polar cap. In this region irregular pulsations with periods of about 3 to 8 minutes are very often observed. Kleymenova et al. (1982) concluded that the latitude of maximum occurrence of these irregular pulsations corresponds to the equatorward boundary of the cusp, statistically defined by means of particle data. Using today's terminology the pulsations are probably located in the cleft, since they occur in an interval of about 8 to 10 hours centered around magnetic local noon.

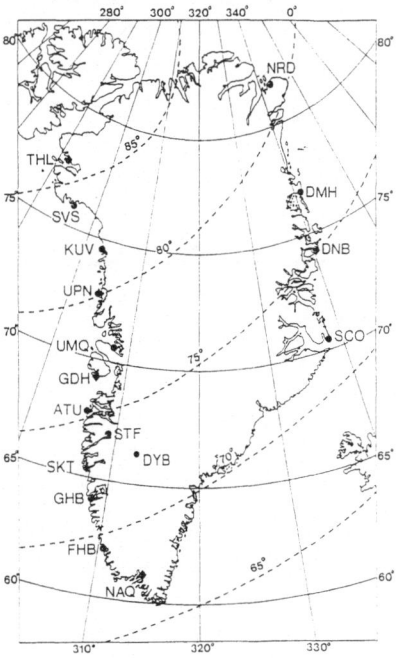

Figure 8. Map showing the location of the Greenland magnetometer chain stations. The dashed lines indicate curves of equal invariant latitude.

Recently these specific variations have attracted increased attention because theoretical considerations by Saunders et al. (1984), Lee (1986) and Southwood (1985; 1987) indicated the possibility of observing an ionospheric signature of sporadic reconnection processes at the dayside magnetopause (flux transfer events) (Russell and Elphic, 1979). Lanzerotti et al. (1986; 1987) used magnetometer stations near the northern and southern cusp to support the idea of the existence of a localized Hall current system consistent with the flux transfer event signature proposed by Saunders et al. (1984) and Lee (1986). Lanzerotti et al. (1987) observed a magnetic impulse event measured at three ground-based stations, Søndre Strømfjord (SS) and Iqaluit (FB) in the Northern Hemisphere and at the conjugate station South Pole (SP) in Antarctica. They modelled the magnetic perturbation with a single Hall current loop moving poleward across the station. From the similarity in the patterns in the Northern and in the Southern Hemisphere and the sign of the vertical disturbance they conclude that the event occurred on closed field-lines.

GREENLAND CHAIN 1 MIN MAGNETIC FIELD—H COMPONENT

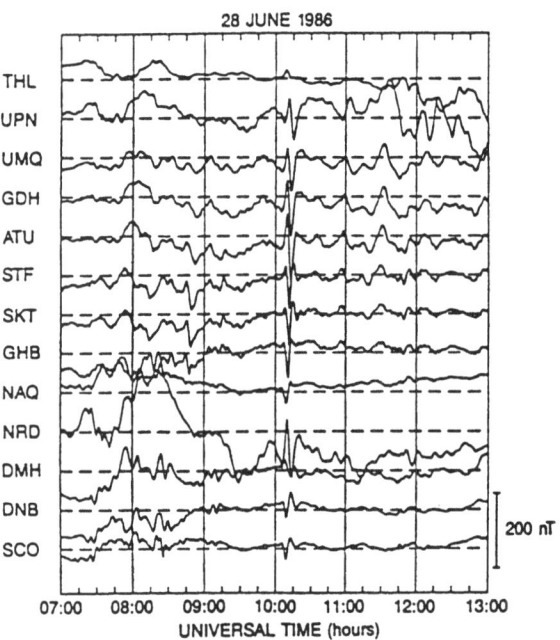

Figure 9. Stacked plot of H-component recordings from the Greenland chain of magnetometers for 0700 to 1300 UT on June 28, 1986. (After Friis-Christensen et al., 1988).

With only one magnetometer station it is not possible to find a unique solution to the current pattern, its velocity, and its direction of motion. With an array of closely spaced stations, however, it is possible to reduce the ambiguity considerably. Such an array was established in Greenland in the Summer of 1986 by supplementing the chain on the West Coast of Greenland with additional temporary stations around Søndre Strømfjord. A map showing the location of these magnetometer stations is shown in Figure 8. The stations to the south of and including GDH recorded 20-second magnetic field measurements. An east-west separation, in particular between the stations SKT and DYB may be used to derive an estimate of possible east-west movements of the current systems. Figure 9 shows a stacked plot of the H-component recordings from the magnetometer stations along the West Coast of Greenland. A magnetic impulse event is taking place around 10:10 UT, approximately four hours prior to magnetic local noon. The event is similar, in all three components, to the event discussed by Lanzerotti et al. (1987), and a phase-shift between the two east-west separated stations indicted a significant westward motion of the source region.

Friis-Christensen et al. (1988) assume that the perturbation is caused by a moving current structure which changes only little in time. Using this assumption the meridian chain of magnetometers is used to "scan" the pattern as it moves across the line of stations. With an estimated velocity of 4 km/s Figure 10 shows the total horizontal magnetic perturbation vectors in a coordinate system moving with the current pattern. The perturbation vectors have been rotated by 90° counterclockwise to indicate the direction of the ionospheric plasma convection. The derived convection pattern consists of a twin-vortex with a scale size of 1000 km moving westward (tailward) roughly along a line connecting the centers of the two vortices. From the hodographs derived for this event Friis-Christensen et al. (1989) find that the trajectory of the current system has a slight equatorward motion corresponding to an antisunward motion approximately along the cleft.

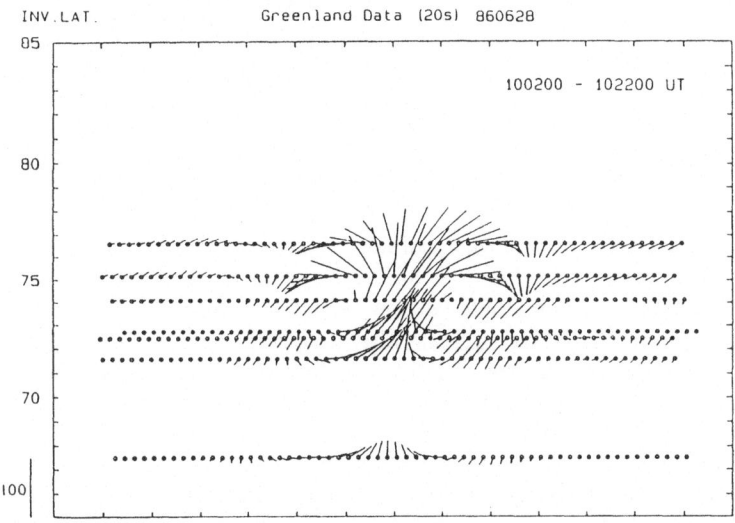

Figure 10. Total horizontal magnetic perturbation vectors have been rotated by 90° counterclockwise and plotted every 20 seconds from 10:02:00 to 10:22:00 UT on June 28, 1986. For each time the position of the vectors have been off-set to the right by a distance corresponding to 80 km to account for an assumed 4 km/s westward motion (to the left) of the pattern.

While the event shown in Figure 10 was moving westward (tailward) in the morning sector, Figure 11 shows an event in the afternoon sector around 16 UT, which is moving eastward and tailward. The equivalent convection vectors in this figure have been plotted from right to left consistent with a direction of motion from left to right, i.e. eastward (tailward) in the postnoon sector. It is seen that an additional vortex is located next to the main vortex. It is a common phenomenon in the magnetometer data that pulsations occur seldom in form of single cycle events, but more often as multiple cycles.

Although the irregular pulsations may correspond to a variety of different current systems, the observations indicate that a basic type of variation consists of a twin-vortex of ionospheric current system, which again is associated with a pair of oppositely directed

162

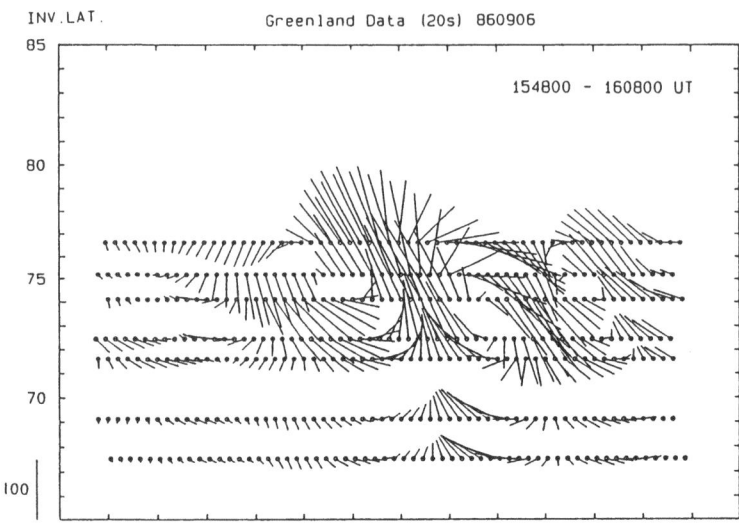

Figure 11. Plot similar to Figure 10 for 15:48:00 to 16:08:00 UT on Sep. 06, 1986. In this case, however, the position of the vectors have been off-set to the left to account for an assumed eastward motion (to the right) of the pattern.

field-aligned current filaments moving tailward in the prenoon and postnoon cleft region. The undistorted twin-vortices are seen during magnetically quiet conditions. When the magnetic activity increases, multiple vortices, apparently consisting of sequences of twin-vortices may appear. Single vortices are also seen, but they seem to be the result of a process which distorts part of the basic twin-vortex system, for instance local conductivity enhancement due to electron precipitation possibly associated with the upward current filament (Friis-Christensen et al., 1989).

None of the examined events correspond to either of the two proposed models of FTE's, the Saunders-Lee type of a single poleward moving Hall current vortex, or the Southwood type of a twin-vortex. Although the convection pattern resembles the twin-vortex proposed by Southwood (1987) the scale size of the vortex is larger than what has been predicted for flux transfer events. Furthermore, the direction of motion perpendicular to the motion predicted by the Southwood model means that the event does not correspond to a net transport of plasma from closed to open field-lines, since this would require an electric field perpendicular to the direction of motion to create the necessary $J \times B$ force. Finally the speed of the motion of the pattern is significantly larger than the typical ionospheric plasma convection velocity.

McHenry et al. (1989) investigated a number of events in the Greenland magnetometer data which showed signatures of multiple vortices. Similar to the impulsive events, the multiple vortices also travel in a westward direction in the prenoon sector and eastward in the postnoon sector. McHenry et al. (1989) do not find a clear IMF B_z dependence, but their results indicate a higher probability of detecting vortices when the solar wind velocity is high. Comparisons with nearly simultaneous particle measurements on board the low-altitude satellite DMSP F7 show that the pulsations probably originate near the inner edge of the magnetopause boundary layer.

Although only few magnetometer chains exist which have a sufficient spatial resolution around the cleft region, it is possible to observe the equatorward part of the twin-vortex current system also at auroral zone latitudes. Glassmeier et al. (1989) found a number of events using the Scandinavian Magnetometer Array. They find that their events are observed predominantly during moderately quiet intervals ($Kp \sim 0$) with a pronounced peak of occurrence frequency near 0830 magnetic local time.

Potemra et al. (1989) compare ground-based signatures of ionospheric traveling vortices with transverse magnetic field oscillations observed by the VIKING and AMPTE/CCE satellites and concluded that the vortices are caused by periodic variations in the solar wind density which create a tailward traveling large-scale magnetosphere wave pulse which excites local field line resonant oscillations.

According to Friis-Christensen et al. (1988) comparisons with IMF and solar wind plasma data indicated that their impulsive event could be generated by a motion of a structure in the magnetopause boundary possibly created by either a sudden change in the solar wind dynamic pressure and/or by a sudden change in the reconnection regions, caused by an abrupt change in the IMF orientation. Such a readjustment might propagate downstream associated with field-aligned currents mapping down to the ionosphere as observed with the chain of magnetometers in the morning as well as in the afternoon sector.

5. Conclusions

In this paper we have briefly reviewed recent studies of polar cap and cleft magnetic perturbations, with emphasis on the fine-structure and dynamics of these variations. Both the polar cap and the cleft magnetic observations show that they are very sensitive to processes at the boundary of the magnetopause. It has been demonstrated that dense arrays of magnetometers provide a unique tool to resolve the ionospheric signatures of these processes, which may be difficult to do with other geophysical instruments because simultaneous good temporal and spatial resolution is essential. One of the main factors controlling high-latitude electric fields and currents is the interplanetary magnetic field (IMF). The polar cap currents respond immediately to changes in the IMF, but it typically takes about 20 minutes to build up a new steady state condition. During this transition time, complex structures may be seen in the ionospheric currents. With sufficient spatial resolution it may be possible to resolve the currents into components which have different relationships to the IMF.

A number of ground-based observations of impulsive nature have been attributed to flux transfer events at the magnetopause. More detailed analysis have shown that some of these observations may have alternative and more likely explanations. Observations indicate that density variation in the solar wind could create magnetopause motions which may give rise to specific pulse-like or oscillatory variations in the ground-based magnetic measurements near the cleft. These variations have features similar to those proposed for the ionospheric signatures of flux transfer events.

References

Banks P. M., Araki, T, Clauer, C. R., Maurice, J. P. St., and Foster, J. C. (1984) 'The interplanetary electric field, cleft currents, and plasma convection in the polar caps', *Planet. Space Sci., 32*, 1551-1557

Clauer, C. R., Banks, P. M., Smith, A. Q., Jørgensen, T. S., Friis-Christensen, E., Vennerstrøm, S., Wickwar, V. B., Kelly, J. D., and Doupnik, J. (1984) 'Observations of interplanetary magnetic field and of ionospheric plasma convection in the vicinity of the dayside polar cleft', *Geophys. Res. Lett., 11*, 891-984.

Clauer C. R. and Friis-Christensen, E. (1988) 'High-latitude dayside electric fields and currents during strong northward IMF: Observations and model simulation', *J. Geophys. Res., 93*, 2749-2757.

Friis-Christensen, E. (1986) 'Solar wind control of the polar cusp', in Y. Kamide and J. A. Slavin, eds., *Solar Wind - Magnetosphere Coupling*, Terra, Tokyo, pp. 423-440.

Friis-Christensen, E. (1989) 'Ground magnetic perturbations in the polar cap and cleft: Relationship with the IMF', this volume.

Friis-Christensen, E., Kamide, Y., Richmond A. D., and Matsushita, S. (1985) 'Interplanetary magnetic field control of high-latitude electric fields and currents determined from Greenland magnetometer data', *J. Geophys. Res., 90*, 1325-1328.

Friis-Christensen, E. and Lassen, K. (1989) 'Large scale distribution of discrete auroras and field-aligned currents', presented at the *International Conference on Auroral Physics, Cambridge July 10-15, 1988.*

Friis-Christensen E. and Vennerstrøm, S. (1989), Manuscript in preparation.

Friis-Christensen, E. and J. Wilhjelm (1975) 'Polar cap currents for different directions of the inter-planetary magnetic field in the Y-Z plane',*J. Geophys. Res., 80*, 1248-1260.

Friis-Christensen, E., McHenry, M. A., Clauer, C. R., and Vennerstrøm, S. (1988) 'Ionospheric traveling convection vortices observed near the polar cleft: A triggered response to sudden changes in the solar wind', *Geophys. Res. Lett., 15*, 253-256.

Friis-Christensen, E., Vennerstrøm, S., Clauer, C. R., and McHenry, M. A. (1989) 'Irregular magnetic pulsations in the polar cleft, caused by traveling ionospheric convection vortices', *Adv. Space Res.*, in press.

Glassmeier K.-H., Hönisch, M. , and Untiedt, J (1989) 'Ground-based and satellite observations of traveling magnetospheric convection twin-vortices', *J. Geophys. Res.*, in press.

Iijima, T., Potemra, T. A., Zanetti, L. J.,and Bythrow, P. F. (1984) 'Large-scale Birkeland currents in the dayside polar region during strongly northward IMF — A new Birkeland current system', *J. Geophys. Res., 89*, 7441-7452.

Jørgensen, T. S., Friis-Christensen, E., Wickwar, V. B., Kelly, J. D., Clauer, C. R., and Banks, P. M. (1984) 'On the reversal from "sunward" to "antisunward" plasma convection in the dayside high-latitude ionosphere', *Geophys. Res. Lett., 11*, 887-890.

Kleymenova, N. G., Bolshakova, O. V., Troitskaya, V. A., and Friis-Christensen, E. (1982) 'Long period geomagnetic fluctuations and the polar chorus at latitudes corresponding to the daytime polar cusp', *Geomagnetism and Aeronomy, 22*, 580-581.

Kotikov, A. L., Latov, Yu. O., and Troshichev, O. A. (1986) 'Structure of auroral electrojet by the data from a meridional chain of magnetic stations', *Proc. International Symp. Polar Geomagnetic Phenomena*, Souzdal, 25-31 May 1986.

Lanzerotti, J. L., Lee, L. C., Maclennan, C. G., Wolfe, A., and Medford, L. V. (1986) 'Possible evidence of flux transfer events in the polar ionosphere', *Geophys. Res. Lett. 13*, 1089-1092.

Lanzerotti, J. L., Hunsucker, R. D., Rice, D., Lee, L. C., Wolfe, A., Maclennan, C. G., and Medford, L. V. (1987) 'Ionosphere and ground-based response to field-aligned currents near the magneto-spheric cusp regions', *J. Geophys. Res., 92*, 7739-7743.

Lee L. C. (1986) 'Magnetic flux transfer at the Earth's magnetopause', in Y. Kamide and J. A. Slavin, eds., *Solar Wind - Magnetosphere Coupling*, Terra, Tokyo, pp. 297-314.

McHenry, M. A., Clauer, C. R., Friis-Christensen, E., and Kelly, J. D. (1989) 'Observations of ionospheric convection vortices signatures of momentum transfer', *Adv. Space Res.*, in press.

Potemra, T. A., Zanetti, L. J., Bythrow, P. F., Lui, A. T. Y., and Iijima, T. (1984) 'B_y-dependent convection patterns during northward interplanetary magnetic field', *J. Geophys. Res.*, *89*, 9753–9760.

Potemra, T. A., Zanetti, L. J., Takahashi, K., Erlandsen, R. E., Lühr, H., Marklund, G. T., Block, L. P., and Lazarus, A. (1989) 'Multi satellite and surface observations of transient ULF waves', *J. Geophys. Res.* , in press.

Reiff, P. H.. and Burch, J. L. (1985) 'IMF B_y dependent plasma flow and Birkeland currents in the dayside magnetosphere. 2. A global model for northward and southward IMF', *J. Geophys. Res.*, *90*, 1595–1609.

Risbeth, H., Smith, P. R., Cowley, S. W. H., Willis, D. M., van Eyken, A. P., Bromage, B. J. I., and Crothers, S. R. (1985) 'Ionospheric response to changes in the interplanetary magnetic field observed by EISCAT and AMPTE-UKS',*Nature*, *318*, 451–452.

Russell, C. T. and Elphic, R. C. (1979) 'ISEE observations of flux transfer events at the dayside magnetopause', *Geophys. Res. Lett.*, *6*, 33–36.

Sandholt, P.E., Jacobsen, B., Lybekk, B., Egeland, A., Meng, C.-I., Newell, P. T., Rich, F. J., and Weber, E. J. (1988) 'Structure and dynamics in the polar cleft: Coordinated satellite and ground-based observations in the prenoon sector', *Report 88-09, Department of Physics, University of Oslo.*

Saunders, M. A., Russell, C. T., and Sckopke, N. (1984) 'Flux transfer events: Scale size and interior structure', *Geophys. Res. Lett.*, *11*, 131–134.

Southwood, D. J., (1985) 'Theoretical aspects of ionosphere - magnetosphere - solar wind coupling', *Adv. Space Res.*, *5*, 7–14.

Southwood, D. J., (1987) 'The ionospheric signature of flux transfer events, *J. Geophys. Res.*, *92*, 3207–3213.

Troshichev, O. A., Bolotninskaya, B. D., and Kotikov, A. L. (1988) 'B_y dependent currents in the southern polar region during positive B_z', *Planet. Space Sci.*, *36*, 523–529.

Zanetti, L. J., Potemra, T. A., Iijima, T., Baumjohann, W. and Bythrow, P. E. (1984) 'Ionospheric and Birkeland current distributions for northward interplanetary magnetic field: inferred polar convection', *J. Geophys. Res.*, *89*, 7453–7458.

ULF PULSATIONS IN THE POLAR CUSP AND CAP

Karl-Heinz Glassmeier
Institut für Geophysik und Meteorologie
Universität zu Köln
Albertus-Magnus-Platz, D-5 Köln 41
Federal Republic of Germany

ABSTRACT. ULF waves are a means for the elctromagnetic coupling and communication between different plasma regimes. With respect to this, ULF pulsation research in recent years has concentrated much on polar cusp and cap regions as it is here where the Earth's magnetic field configuration allows direct access of magnetosheath plasma and waves into the magnetosphere proper. Satellite observations in the cusp region exhibit the existence of strong fluctuations of the magnetic field as well as plasma density and flow, probably generated by e.g. drift wave or Kelvin-Helmholtz instabilities in the nonuniform cusp plasma regimes. At the ground a variety of different ULF pulsations are found - regular pulsation trains, irregular activity, transient perturbations etc. A strong dependence of this pulsation activity on upstream solar wind conditions is found, where the correlation with the IMF cone angle and the IMF B_z component prevails. The often observed spiky or transient magnetic field variations at cusp latitudes suggest a clear correlation with flux transfer events at the dayside magnetosphere or dynamic pressure changes of the magnetosheath flow.

1. Introduction

ULF waves are of importance with respect to two major roles they play in space plasmas. First, ULF waves are generated by a variety of different plasma instabilities such as drift wave, Kelvin-Helmholtz or Rayleigh-Taylor instabilities and other, kinetic types of plasma instabilities [for a review see Southwood and Hughes, 1983]. Their existence thus tells us about the presence of macroscopic or microscopic gradients in space plasmas, and how they can be released.

Second, ULF waves also serve to communicate local changes of the plasma configuration under consideration to the whole system regarded. For example, Pi2 pulsations frequently observed in the nightside of the Earth's magnetosphere during magnetospheric substorms, may be regarded as the transient response of the nightside magnetosphere to sudden disruptions of the neutral sheet current [e.g. Baumjohann and Glassmeier, 1984]. ULF pulsations of the dayside magnetosphere have been interpreted by e.g. Reid and Holzer [1975] as the response of the dayside magnetosphere-ionosphere coupling system due to time-varying field line reconnection at the Earth's magnetopause. ULF waves therefore may be regarded as a means of the electromagnetic coupling between distant space plasma regions and the Earth's ionosphere and surface.

A number of reviews on ULF waves in the polar cusps and cap regions have been published in recent years [D'Angelo, 1977, Troitskaya et al., 1980; Fraser-Smith, 1982;

P. E. Sandholt and A. Egeland (eds.), Electromagnetic Coupling in the Polar Clefts and Caps, 167–186.
© *1989 by Kluwer Academic Publishers.*

Amoldy et al., 1988; Fukunishi and Lanzerotti, 1988], and show the increasing interest in polar latitude ULF pulsations. The present knowledge on these waves is summarized in Figure 1, which will be the guideline for the present paper, too. We shall first give an overview on spacecraft observations of hydromagnetic waves within the exterior cusp and adjacent magnetosheath as well as at high-altitudes. A review on ground-based polar cap and cusp ULF pulsation observations in the Pc3-5 period range is followed by a more detailed discussion on more recent observations of transient or impulsive magnetic field perturbations in the dayside cusp regions. Possible source mechanisms for the observed waves are discussed with special emphasize to flux transfer events [e.g. Southwood et al., 1988] and dynamic pressure changes at the Earth's magnetopause.

2. Spacecraft Observations Within The Dayside Cusp

The general structure of the dayside cusp region as shown in Figure 1 has recently been summarized by Sckopke [1985]. Major regions which have been identified by satellite observations during the last two decades are the low-latitude boundary layer, the entry layer, i.e. a plasma regime on magnetospheric field lines just equatorward of the cusp, the exterior cusp, i.e. the plasma regime bounded by the cusp-like indentation of the magnetopause and the free streaming magnetosheath plasma, and the plasma mantle, a regime of significant plasma flux out of the magnetosphere. Fluctuations of the magnetic field in the cusp region and the adjacent magnetosheath have frequently been observed [Russell et al., 1971; Fairfield and Ness, 1972; D'Angelo, 1973; Fairfield and Hones, 1978] and have been attributed to a filamentary, field-aligned current carrying plasma [Fairfield and Ness, 1972], or turbulence due to Kelvin-Helmholtz and drift wave instabilities.

Figure 2 shows high-resolution data from the IMP 6 satellite [Fairfield and Hones, 1978] when passing through a region of downward plasma flow in the mid-altitude cusp. Immediately after the satellite traversed the boundary between the trapped particle region and the cusp (the boundary is indicated by the peak of the average electron energy at 1622 UT) large irregular plasma density and magnetic field magnitude fluctuations are observed. The period of the magnetic field perturbations is of order 1 min and less, their amplitude clearly increases with increasing plasma density. That both plasma density and magnetic field fluctuations are observed, points towards the wave nature of these perturbations, rather than indicating the presence of field-aligned current structures.

Typical power spectra of ULF magnetic field fluctuations within the cusp and the adjacent magnetosheath are displayed in Figure 3 [Fairfield and Hones, 1978]. Within the cusp region, for frequencies up to a few Hertz the spectrum is well described by a power law with spectral index - 2, while the spectrum within the magnetosheath drops to much lower values near the proton gyro-frequency (f_{gp} in Figure 3). Plasma wave observations can thus be used to trace the cusp-magnetosheath boundary, too. Fairfield and Hones [1978] furthermore note that magnetic field fluctuations at frequencies below the proton gyro-frequency increase at larger radial distances and higher latitude within the cusp.

Similar cusp sprectra as shown in Figure 3 have also been reported about by Russell et al. [1971], who point out that the spectral power of transverse fluctuations is usually larger than that of field-parallel fluctuations with $\delta b_\perp / \delta b_\parallel$ of order 3. This points toward an Alfvénic character of the turbulence, but would also be compatible with an interpretation in terms of moving current structures. The very nature of the observed turbulence can only be resolved from joint plasma and magnetic field observations. Such correlated studies have not (at least to the knowledge of the present author) been performed up

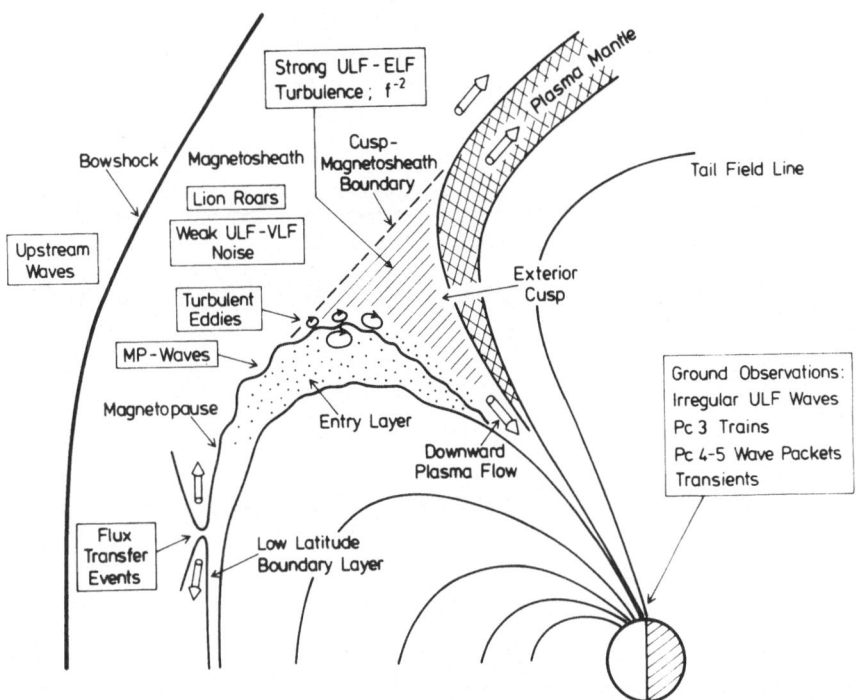

Fig. 1 Schematic view of the polar cusp region illustrating important space plasma physical features of this region.

to now. But Paschmann et al. [1976] report about highly irregular plasma flow and eddy formation within the exterior cusp, which might be interpreted as an indication of the hydromagnetic nature of the turbulence, too. It should be noted that Rezeau et al. [1986] have been able to identify magnetosheath magnetic and electric field fluctuations as Alfvénic type turbulence with spectral slope -2.5. A similar study is missing for the turbulence observed in the high-altitude polar cusp regions.

Explanation and interpretation of the observed spectral index, -2, is not straight forward. It is tempting to compare these spectra with well known Kolmogoroff-Obukow or Kraichnan-type spectra with spectral indices -5/3 and -3/2, respectively. However, these latter spectra represent spatial wave number spectra,

$$P(k) \sim k^{-\alpha}$$

Only if the plasma flow is super-sonic or super-Alfvénic, as in the solar wind, or if the satellite is moving faster than the phase velocity of any of the waves comprising the turbulence, the fluctuations can be treated as "frozen" in the plasma, and the observed temporal variations can be transformed into spatial variations via

$$k = 2\pi f / v_s$$

where v_s is the satellite's velocity in the plasma frame of reference. This "frozen-in" condition is generally not fulfilled in the cusp region, where the flow is usually sub-sonic

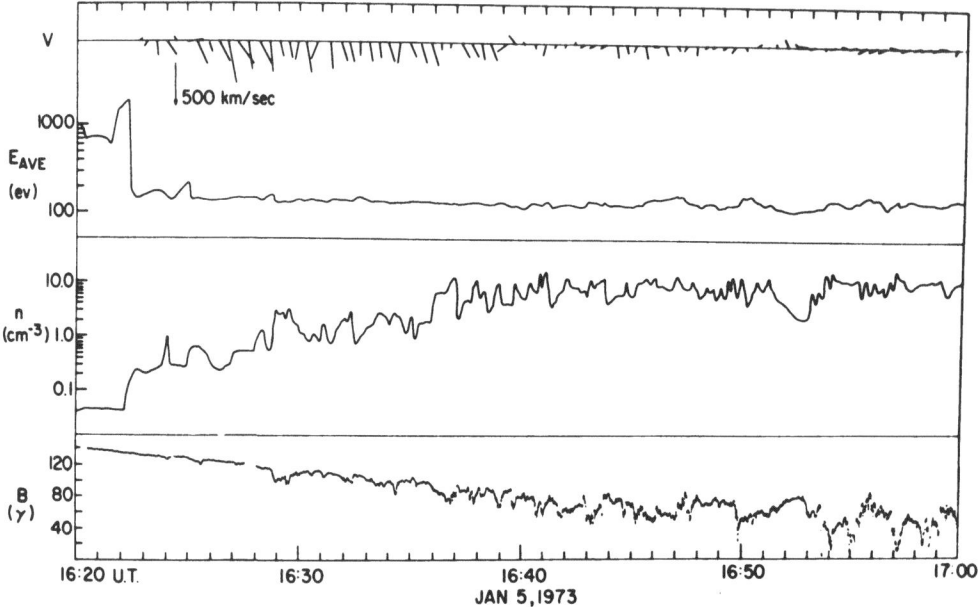

Fig. 2 High-resolution data from the IMP 6 satellite. V represents proton velocity; downward vectors indicate anti-sunward flow down field lines. E_{AVE} represents average electron energy, n and B are plasma density and magnetic field magnitude, respectively [Fairfield and Hones, 1978].

and sub-Alfvénic. A unique interpretation of the observed spectral shape is thus not straightforward.

3. Ground-Magnetic Observations In The Polar Cusp And Cap Regions

Noting that satellite observations exhibit the existence of strong magnetic field, plasma density and flow fluctuations in polar cusp regions leads to the question what signatures there are of these fluctuations at the ground. The observations indicate a variety of different types of ULF pulsations to occur in these regions, such as regular, irregular or transient pulsations occurring over a broad frequency spectrum of 1 mHz - 1 Hz. One of the earliest systematic studies of the question of the nature of Pc 4,5 pulsation activity at polar cusp and cap latitudes is that by Rostoker et al. [1972]. Figure 4 shows their estimated latitude of maximum activity for different frequencies for two magnetic local time intervals, 1000 - 1400 (left panel of Figure 4) and 1800 - 0600 MLT (right panel). It is evident that during day times around local magnetic noon broadband spectral ULF activity occurs north of 75° geomagnetic latitude, while a similar enhancement during night times is absent. This clear enhancement of ULF wave power near local magnetic noon is quite in accord with the early satellite observations [e.g. Russell et al., 1971; Fairfield and Ness, 1972] and also the more recent studies by Olson [1986], and indicates enhanced ULF wave activity at cusp region latitudes.

An average power spectral density function obtained from 237 days of Cape Parry observations is displayed in Figure 5. It shows a clear power law with spectral index

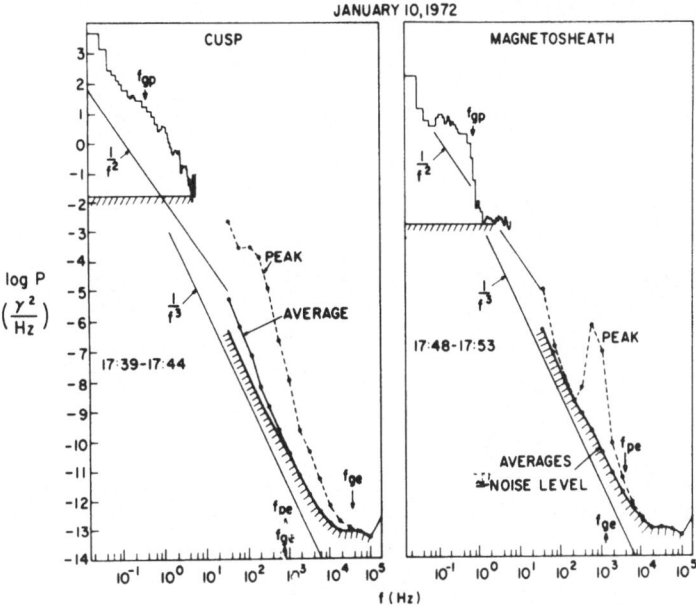

Fig. 3 Magnetic field power spectra on either side of the exterior cusp/magnetosheath boundary as seen by IMP 6 [Fairfield and Ness, 1978].

-2.6, i.e. somewhat steeper than observed by the satellites in the polar cusp (see Figure 3). However, again we like to point out that a physical interpretation of the spectral index is not at all clear. The ionospheric modification [e.g. Prikner and Wagner, 1982] of the magnetospheric perturbations when observed at the ground may partly be able to explain the observed spectral slope [Prikner, pers. communication, 1988].

Another characteristic type of ULF pulsation, seen both within the polar cap at a station such as Vostok and in the polar cusp, are long-period (3-10 min) irregular pulsation events, a typical example of which is shown in Figure 6 [Troitskaya et al., 1980; Bolshakova and Troitskaya, 1982]. This type of pulsation has been called IPCL pulsation by Troitskaya and coworkers. Amplitudes of IPCL pulsations reach up to 20 nT at Vostok and exhibit a strong dependence from upstream solar wind parameters [e.g. Bolshakova and Troitskaya, 1982]. As can be seen from the examples shown in Figure 6, the IPCL pulsation events are clearly switched-on after a southward turning of the IMF. Bolshakova and Troitskaya [1982] furthermore point out a clear correlation of the IPCL amplitude with the solar wind velocity.

Another example of what might be classified as an IPCL pulsation event is given in Figure 7 [Sibeck et al., 1988]. It displays ground-magnetometer observations from South Pole Station together with concurrent magnetic field observations from the AMPTE-IRM satellite. The ULF pulsation event seen on the ground between 1521 -1630 UT on September 10, 1984 is similar to those shown is Figure 6. The IRM satellite was located upstream of the Earth's bow shock at about 13 R_E [Sibeck et al., 1988] during the occurence of the ground-magnetic pulsation and detected a clear southward turning of the IMF direction at about 1518 UT, which might have triggered the pulsation event.

The concidence of a southward turning of the IMF and the switch-on of IPCL pul-

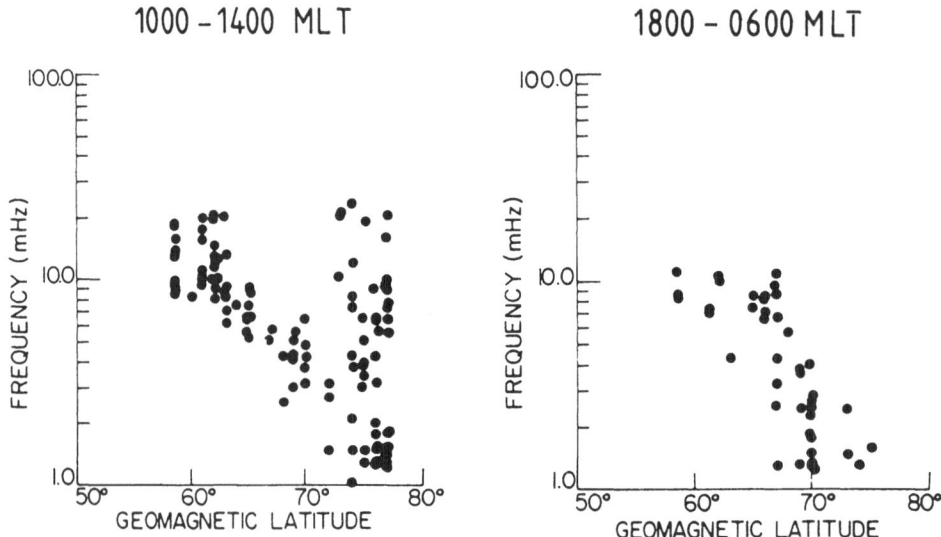

1000 – 1400 MLT

1800 – 0600 MLT

Fig. 4 Estimated latitudes of ULF pulsation intensity maxima versus signal frequency for the time intervals 1000-1400 (left panel) and 1800-0600 UT (right panel) magnetic local time [Rostoker et al., 1972].

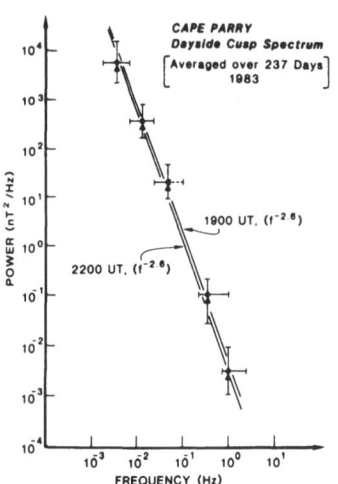

Fig. 5 Average power spectral density function obtained from Cape Parry magnetometer stations for the polar cusp region [Olson, 1986].

sation trains suggests magnetic field line reconnection at the dayside magnetopause as a probable source mechanism [Bolshakova and Troitskaya, 1982]. This suggestion is not in contradiction to Bolshakova and Troitskaya's finding that IPCL pulsation amplitude are often largest at Vostok during northward IMF. This may be explained as a signature of reconnection of northward directed interplanetary magnetic fields and magnetospheric tail field lines.

Sibeck et al. [1988], not mentioning the connection between southward turning of

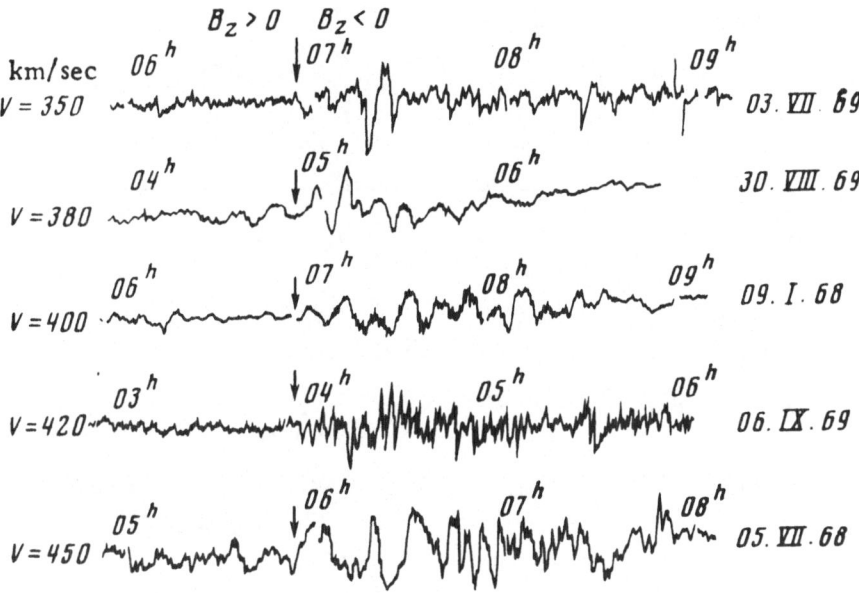

Fig. 6 Ground-magnetic observations from a Soviet station exhibiting a clear correlation between southward turnings of the IMF (B_z is the IMF z-component) and the onset of IPCL pulsations. Amplitude of the pulsations is about 20 nT ; the exact scale value is not given in the paper by Bolshakova and Troitskaya [1982], where this figure is adapted from.

the IMF and their observed pulsation event explicitly, however, give convincing evidence that their pulsation event correlates well with dynamic pressure variations,too, seen in the upstream solar wind and regard the observed pulsation event as the magnetospheric response to these variations.

A number of recent studies [Bolshakova and Troitskaya, 1984; Engebretson et al., 1986; Wolfe et al., 1987; Yumoto et al., 1987] deal in particular with the question whether observed Pc 3 activity is correlated with upstream solar wind conditions. Such a correlation has been found at least for low- and mid-latitude Pc 3 [see Odera, 1986 for a review]. For polar cusp and cap latitudes the above mentioned studies yield different and contradicting results, however. They all agree with Pc 3 activity being strongest at cusp latitudes near local magnetic noon as clearly shown in Figure 8. It displays a statistical summary of South Pole Station wave magnetometer activity in the period band 1-80 s [Engebretson et al., 1986] and shows the percentage days in 1983 during which activity in the chosen period band was above a certain threshold. A broad peak of activity occurs between 0900-1300 MLT, quite in accord with Rostoker et al.'s [1972] earlier work on Pc4-5 pulsations. A secondary maximum between 2100 and 0200 MLT indicates nightside activity, such as magnetic substorms and related pulsations.

Concerning the dependence of Pc 3 activity on upstream solar wind conditions, both Yumoto et al. [1987] and Wolfe et al. [1987] found no clear dependence from the IMF cone angle, but report about a clear correlation between the Pc 3 activity level and the solar wind velocity, as shown in Figure 9 [Wolfe et al., 1987]. However, it should be

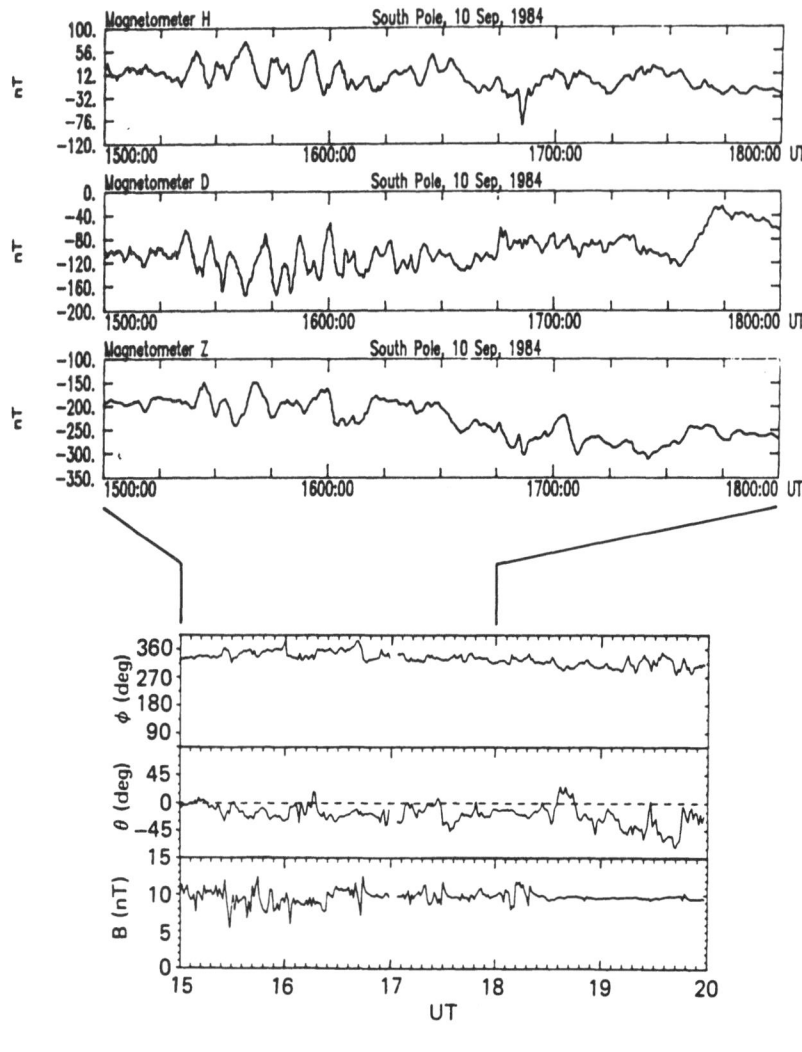

Fig. 7 Ground-magnetic observations made at South Pole Station and concurrent IMF observations made by the AMPTE-IRM satellite on September 10, 1984. The IRM observations are given in GSE coordinates, where $\theta > (<) \, 0^o$ denotes northward (southward) IMF [Sibeck et al., 1988].

noted that though Figure 9 suggests a linear relationship between Pc 3 activity and solar wind speed, the correlation coefficient found, r = 0.42, is not very large. Also, as pointed out by Wolfe et al. [1987], the solar wind speed used in their analysis is that measured by ISEE 3 one hour earlier than the occurrence of Pc 3 at South Pole Station. But Figure 9 also exhibits that for the ensemble chosen the solar wind speed varied significantly, between 250 and 780 km/s. This suggests that at least a variable delay time should be

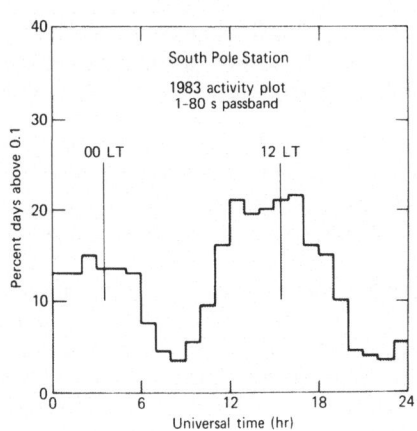

Fig. 8 Occurrence of ULF pulsation activity in the period band 1-80 s as seen at South Pole Station in 1983 [Engebretson et al., 1986].

incorporated in the statistical studies. However, Yumoto et al. [1987] used observations from the IMP 8 satellite, where no significant delay time has to be taken into account, and have been able to confirm Wolfe et al.'s results.

Wolfe et al. [1987] also found a weaker dependence of Pc 3 activity from the IMF B_Z component and derived at the following predictive equation for the Pc 3 power

$$\log \text{power Pc } 3 = (-2.8 \pm 0.13) + (0.0037 \pm 0.0003) \, v_S - (0.027 \pm 0.0067) \, B_Z$$

where the power is given in $(\text{nT})^2$, the solar wind velocity v_S in (km/s), and B_Z in (nT). The strong dependence of the Pc 3 activity on the solar wind velocity is interpreted by Yumoto et al. [1987] and Wolfe et al. [1987] as clear evidence for a Kelvin Helmholtz instability of the high-latitude magnetosphere as a possible source mechanism for the observed wave activity.

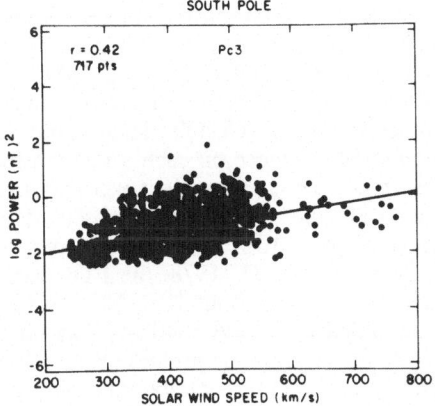

Fig. 9 Scatter plot of Pc3 band activity level versus solar wind speed [Wolfe et al., 1987].

Quite in contradiction to Yumoto et al. [1987], and Wolfe et al.'s [1987] results, Engebretson et al. [1986] report about a strong correlation between Pc 3 activity and the X component of the IMF, i.e. the IMF cone angle. The reason for these contradictory results is the single event approach by Engebretson et al. [1986]. They discriminate between narrowband and broadband activity and are thereby selecting larger-amplitude, monochromatic Pc 3 events, for which the above mentioned IMF cone angle dependence holds. As Yumoto et al. [1987] and Wolfe et al. [1987] do not distinguish between different types of Pc 3 activity, their broadband activity analysed contains contributions from both, Pc 3 pulsation events and broadband Pi activity.

The dependence of Pc 3 pulsation occurrence of the IMF cone angle as found by Engebretson et al. [1986] is in accord with the well-established correlation between low- and mid-latitude Pc 3-4 pulsations and upstream waves [e.g. Odera, 1986], and suggests transmission of upstream waves through the Earth's bow shock and magnetopause into the magnetosphere proper as the source mechanism.

More recently, Slawinski et al. [1988] extended the previous statistical analysis by Wolfe et al. [1987]. In the frequency range 20-50 mHz they found two different branches, exhibiting different correlation with the IMF magnitude B_T. Below 28 mHz no correlation with B_T was found, while the Pc3 activity above 28 mHz shows a linear dependence of the signals frequency from B_T. Thus for this latter branch Slawinski et al. [1988] suggest direct transmission of interplanetary ion cyclotron waves into the magnetosphere too, while for the other branch a variety of sources, such as surface waves or ionospheric processes, are suggested.

The transmission of hydromagnetic waves through the Earth's bow shock has been theoretically investigated by e.g. McKenzie and Westphal [1969], who found that the amplitude of transmitted Alfvén waves is about three times larger than that of the incident wave. However, McKenzie and Westphal [1969] have treated the bow shock as a fast hydromagnetic shock, neglecting all the detailed shock structure and its influence on the transmission process. Further theoretical work seems neccessary to take into account the turbulent nature of the bowshock and magnetosheath to fully understand how upstream waves can enter the magnetosphere proper.

4. Flux Transfer Events And Polar Cusp/Cap Pulsations

The above mentioned studies on polar cusp/cap ULF pulsations suggest a strong dependence of the ULF pulsation activity on upstream solar wind conditions. In particular, studies by e.g. Bolshakova and Troitskaya [1982] or Sibeck et al. [1988] emphazise a dependence on the IMF B_z-component. In the last decade much effort has been devoted to the study of so-called flux transfer events (FTE), i.e. localized, patchy magnetic field reconnection events [for a recent speculation on FTE's see Southwood et al. 1988]. In their introductory study on FTE's Russell and Elphic [1978] already suggested the possibility that FTE's might be a possible source mechanism for Pc5 pulsations within the Earth's magnetosphere. Probably the first observational evidence on such a relationship has been given by Glassmeier et al. [1984]. They report about a transient Pc5 pulsation event which occurred at the same time when an FTE-like signature was observed by the ISEE 1/2 satellite pair near the magnetopause (Figure 10). ISEE 1/2 entered the magnetosphere at about 1025 UT ($\approx$ 1536 MLT) on August 31, 1978, the FTE occurs at about 1540 MLT. The ground magnetic pulsations starts at about 1040 UT ($\approx$ 1310 MLT) in the Scandinavian sector and has been recorded at around 1640 MLT as far to the east of Scandinavia as the Soviet observatory Dixon (DIK). The ground observations of this particular event therefore cover a magnetic local time range of about 6 hours and also that magnetic local time portion where the ground signature of the FTE is expected.

Later studies concentrated on the question 'What is the ionospheric or ground signature of flux transfer events?' [e.g. Goertz et al., 1985; Cowley, 1986]. This question is of eminent importance for the question on characteristics of polar cusp/cap ULF pulsations,

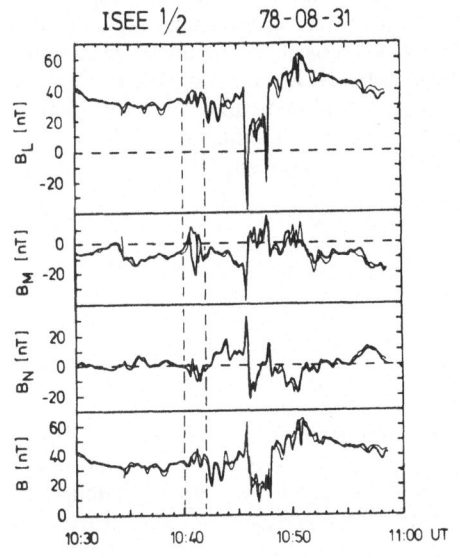

ISEE 1/2   78-08-31

Fig. 10 ISEE 1/2 magnetic field observations (upper panel; in boundary normal coordinates) together with GEOS 2 magnetic and electric field observations, b_r and E_ϕ, respectively, and ground-magnetic observations from a station in Northern Scandinavia (MAT; close to the footpoint of the GEOS 2 field line) and at the Soviet observatory Dixon (DIK) east of Scandinavia . The dashed lines in the upper panel indicate the time interval when the FTE has been observed [after Glassmeier et al., 1984].

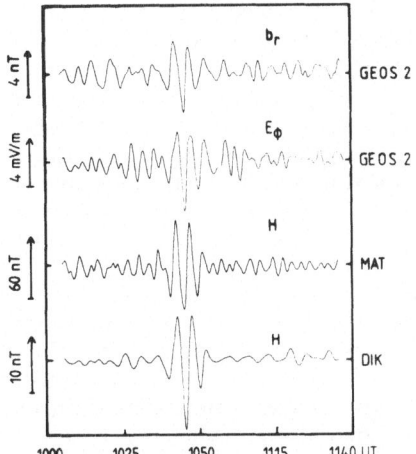

as any newly reconnected flux tube will be convected poleward through the polar cusp or along the polar cap-auroral oval boundary and might give rise to transient perturbations, seen at the ground as wave-like, impulsive fluctuations. It should be noted that already Bolshakova and Troitskaya [1982] associated their IPCL pulsations with patchy reconnection at the magnetopause.

Recently, Lanzerotti and colleagues in a series of papers [see Lanzerotti and MacLennan, 1988 and references therein] analysed a peculiar type of magnetic field variation seen at cusp-latitude stations such as South Pole Station or Sondre Stromfjord (Figure 11). It is characterized by a bi-polar deflection of the H-component and a mono-polar, positive or negative deflection of the D-component. The usually strong deflections of the Z-component point toward a localized ionospheric current system as the cause of these geomagnetic field variations.

Lanzerotti and colleagues associate these transient magnetic field variations with the possible ground signature of flux transfer events. The actual shape of the ionospheric current system of FTEs is still a matter of debate. Two different models have been

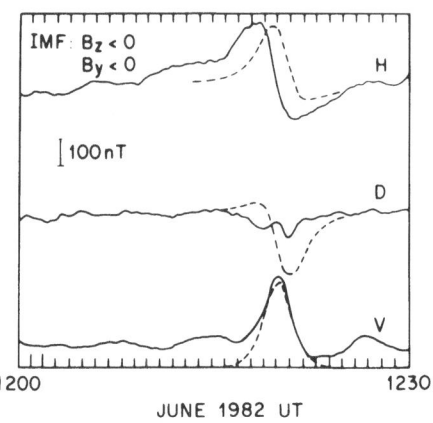

Fig. 11 South Pole Station magnetometer data exhibiting transient magnetic field variations possibly related to flux transfer events. H, D, and Z denote northward, eastward, and vertically downward deflections of the Earth's magnetic field [Lanzerotti and MacLennan, 1988].

suggested sofar, the coaxial model and the perpendicular dipole model (Figure 12), based on work by Lee [1986] and Southwood [1987], respectively. Within the coaxial model it is assumed that the field-aligned currents of the FTE helical field lines [Saunders et al., 1984] are closed in the ionosphere via divergent Pedersen currents and distributed return field-aligned currents (Figure 12), while the perpendicular dipole model assumes an asymmetric current flow with localized field-aligned currents in the north and south which close via meridional Pedersen currents. The associated Hall current system is similar to that of a perpendicular (to the north-south direction) dipole sheet current. Both current systems can explain the observed bi-polar and mono-polar magnetic field variations observed by Lanzerotti and MacLennan [1988], as a comparison of Figures 11 and 12 shows. Only if the location of the ground magnetometer station with respect to the current system is known, a distinction between both models is possible. However, Lanzerotti and coworkers only had available data from one ground station, by which a precise determination of the station location relative to the current system was not possible, and thus prohibited a unique identification of the current system associated with their observed transient magnetic field perturbations.

More recently Hönisch and Glassmeier [1986], Friis-Christensen et al. [1988], and Glassmeier et al. [1988] reported about the observation of transient magnetic variations similar to those reported about by Lanzerotti and MacLennan [1988] and others. For their studies they used data from the IMS Scandinavian Magnetometer Array [Glassmeier et al., 1988] and data from the Greenland Chain of magnetometer stations [Friis-Christensen et al., 1988]. They have been able to show that the transient variations they observed are due to the rapid westward (tailward) overhead propagation of a twin-vortex ionospheric current system as schematically shown in the middle part of Figure 12 (the parallel dipole model). The propagation speed found was of order 2.5-4 km/s. The equivalent current system actually observed by Glassmeier et al. [1988] for one of their transient events is displayed in Figure 13 and shows two current vortices, rotating clockwise in the west and counterclockwise in the east. This extended shapshot of the equivalent current system generating the observed transients has been constructed by properly combining current vector distributions at different times and from the full set of 36 stations of the IMS Scandinavian Magnetometer Array into one huge current pattern taking into account the estimated westward propagation velocity. This merging technique has successfully been used by Kunkel et al. [1986], too, and reference is made to their work for further details.

If the transient magnetic variations studies by e.g. Lanzerotti and MacLennan [1988] and the similar ones analysed by Glassmeier et al. [1988] and Friis- Christensen et al. [1988] are indeed ground signatures of flux transfer events, the associated ionospheric

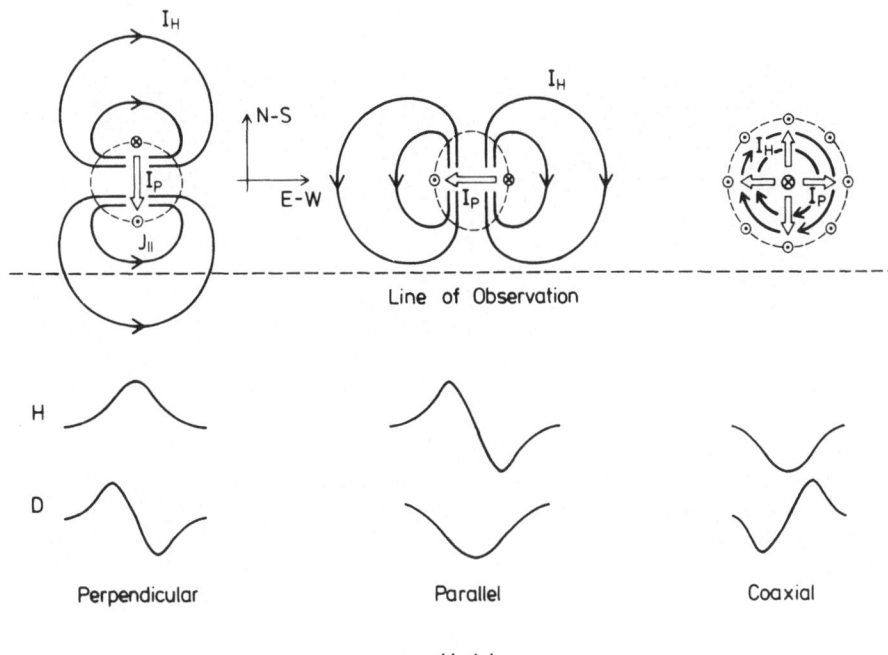

Fig. 12 Three different models of field-aligned current flow into and out of the ionosphere together with the associated Pedersen and Hall current flow in a horizontally uniform ionosphere. The ground magnetic signatures for the H and D components are sketched in the bottom part of the figure.

equivalent current system is certainly different than previously assumed. The current system in Figure 13 and the corresponding one in the study by Friis-Christensen et al. [1988; their Figure 4] may be classified as a parallel dipole model (see Figure 12).

Furthermore, Glassmeier et al. [1988] and Friis-Christensen et al. [1988] pointed out that the scale size of the observed twin-vortex current structure ($\approx$ 2000 km at ionospheric heights) is not compatible with the scale expected for the ionospheric signature of FTE's ($\approx$ 200 km). Friis-Christensen et al. [1988] have also been able to show that at least one of their events studied is closely related to a sudden drop of the solar wind dynamic pressure and interpret their observations as evidence of the triggered response of the polar magnetosphere to sudden changes in the solar wind. Thus it is not at all clear whether the above described transient variations (see Figure 11) are ground signatures of flux transfer events.

However, impulsive, transient magnetic variations of a similar type as those reported above have recently been discussed by Oguti et al. [1988], too, in association with dayside auroral activity. They point out that their observed magnetic impulses are often accompanied by local short-lived enhancements in the E-W movement of auroral structures. For some of the analysed events Oguti et al. [1988] report westward expansion speeds of auroral structures of about 5 km/s, a value in agreement with the westward convection velocity found by Glassmeier et al. [1988] or Friis-Christensen et al. [1988] for their transients. Oguti et al. [1988] discuss their magnetic impulses in connetcion

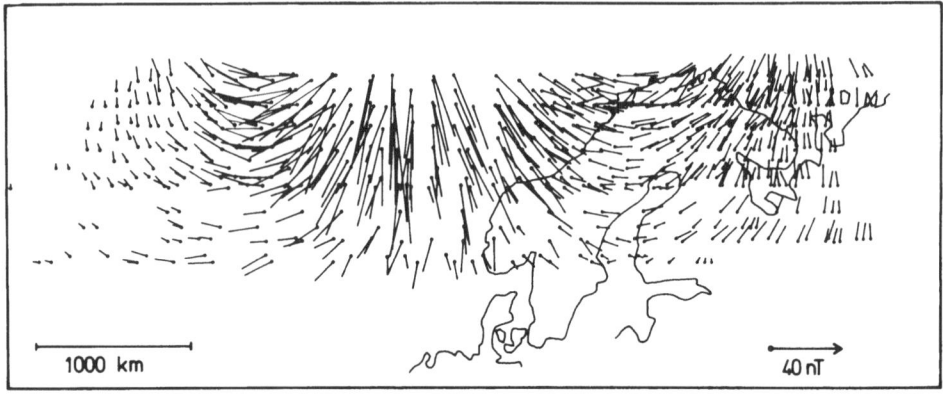

Fig. 13 Distribution of equivalent current vectors of the December 24, 1978 transient magnetic field variation event as constructed from instantaneous equivalent current vector distributions and applying a merging technique (see also text). The solid line is the coastline of Northern Scandinavia. The northern most ground-magnetometer station, data of which are used in this is figure is located at 67° geomagnetic latitude [Glassmeier et al., 1988].

with the ground signature of FTE's, too.

5. Source Mechanisms Of Polar Cusp/Cap ULF Pulsations

A number of different source mechanisms for the generation of ULF pulsations in the polar cusp/cap have been suggested and will be briefly summarized here.

5.1 Upstream Waves

Upstream waves, generated by reflected ion beams in the Earth's foreshock, are thought to be responsible for most of the observed narrowband Pc3 activity [e.g. Engebretson et al., 1986]. Two models are decribed in the literature to explain the occurrence of Pc 3 pulsations at low- and midlatitudes, the surface model and the signal model [Odera, 1986]. While the former assumes Kelvin-Helmholtz instability at the magnetopause, the later suggests upstreams waves convected back towards the magnetopause and transmitted into the magnetosphere as the source of Pc3 Pulsations. The dependence of the Pc3 activity observed in the polar cusp/cap regions on solar wind speed [e.g. Yumoto et al., 1987] and on the IMF cone angle [Engebretson et al., 1986] suggests both models to be likely candidates to explain the observations.

5.2 Kelvin-Helmholtz Instability

As first shown by e.g. Heikkila and Winningham [1971] the polar cusp is a region with a net plasma flow down magnetic field lines into the magnetosphere proper. At the cusp boundaries this net down flow may give rise to a significant shear flow, which can be unstable to the Kelvin-Helmholtz instability. A detailed analysis by D'Angelo

[1973] shows that Kelvin-Helmholtz instability during periods of intense net down flow is indeed a possible generation mechanism. However, D'Angelo [1973] has not been able to give convincing arguments to explain the observed spectra, both seen by satellites as well as on the ground.

5.3 Drift Wave Instability

The polar cusp region is characterized by a number of different plasma regimes and thus exhibits large plasma nonuniformities (cf. Figure 1). Both Rostoker et al. [1972] and D'Angelo [1973] point out that drift waves can be generated by, for example plasma pressure gradients between open and closed field lines. However, as suggested by D'Angelo [1973] drift wave instability produces a turbulence spectrum $P(f) \sim f^{-5}$, i.e. much steeper than those observed (cf. Figures 3 and 5).

5.4 Particle Precipitation and Ionospheric Modification

Evidence for a more indirect source mechanism for Pc3 pulsations has been given by Engebretson et al. [1988]. They report about precipitating magnetosheath-like electrons in polar cusp regions that are modulated with frequencies similar to those of upstream waves. Via modification of the ionospheric conductivity these precipitating electrons can give rise to Pc3 pulsations.

5.5 Patchy Reconnection at the Dayside Magnetosphere

The possibility of a connection between transient magnetic field variations at polar cusp latitudes and flux transfer events has already been discussed above. In a series of paper [see Lee et al., 1988 and references therein] Lee and colleaques have developed the multiple X-line model of magnetic field reconnection (MXR-model). The MXR model not only provides for a convincing explanation for the generation of flux transfer events [LaBelle-Hamer et al., 1988], but is also able to explain the generation of ULF waves at cusp latitudes [Lee et al., 1988]. Two different types of ULF signals are predicted by the MXR-model: an impulsive and a continous fluctuation part. Figure 14 shows a perspective view of the three-dimensional magnetic field configuration asscociated with the MXR-model. Here we like to concentrate on the open field lines A and B. The reconnection process may be looked at as the sudden switch-on of a localized current, ($\vec{J}_R$ in Figure 14) oppositely directed to the magnetopause sheet current. This current also gives rise to the bending of the field lines A and B, and its $\vec{j} \times \vec{B}$ forces drag the newly opened magnetospheric field lines poleward. The reconnection process furthermore gives rise to a dawn-to-dusk electric field, $\vec{E}$, between the newly opened field lines A and B. Note that $\vec{E}$ and $\vec{J}_R$ are anti-parallel and thus the localized reconnection region represents an electromotive force or generator region. At the onset of the MXR-process an Alfvén wave impulse is launched from the reconnection side and travels down to the ionosphere where it is reflected. This bouncing Alfvén wave is the impulsive part of ULF waves generated by the MXR-process. It may be regarded as the dayside equivalent of transient Pi2 pulsations observed in the nightside magnetosphere during substorms [cf. Baumjohann and Glassmeier, 1984]. In fact, J_R and the associated field-aligned currents flowing along field lines A and B (see Figure 14) into the ionosphere may be regarded as the dayside equivalent of the substorm current wedge.

The field-aligned current carried by this Alfvén wave pulse is initially upward (downward) at the dayside (dawnside) of the reconnection region. It gives rise to a Hall current pattern in the ionosphere similar to the parallel dipole model in Figure 14 and in

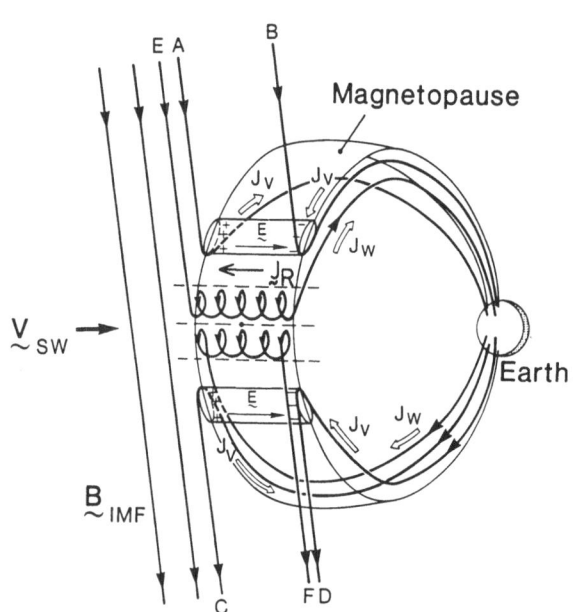

Fig. 14 A perspective view of the three-dimensional magnetic field configuration associated with multiple X line reconncetion at the dayside magnetopause [Lee et al., 1988].

agreement to the current systems deduced for transient magnetic variations studied by Glassmeier et al. [1988] and Friis- Christensen et al. [1988].

It is a basic feature of the MXR-process that the interplanetary and the geomagnetic field lines reconnect along several X lines on the dayside magnetopause. According to Lee et al. [1988] a series of elongated plasma clouds or magnetic islands may thus be generated at the magnetopause. These islands will compress and distort closed magnetospheric field lines adjacent to the reconnection sites, i.e. are pulled along the magnetopause, pressed in on the boundary like a ball moved on the surface of a ballon [Glassmeier et al., 1984] and thus constitute an effective source mechanism of compressible boundary waves, much as they are generated by a Kelvin-Helmholtz instability.

This aspect of the MXR-model, generation of compressible perturbations at the magnetopause, is another appealing point in the MXR-model. However, further theoretical and experimental studies are necessary to firmly establish the MXR-process as a possible generation mechanism for cusp-latitude pulsations.

6. Summary and Conclusions

In recent years the interest in studying ULF pulsations at polar cusp latitudes and within the polar cap significantly increased. The reason is that ULF pulsations play a unique role for the communication between different plasma regimes and provide for a means of the electromagnetic coupling between them.

At polar cusp latitudes and within the polar cap geomagnetic field lines are open field lines and directly connect the Earth's surface with plasma regimes such as the magnetosheath, the entry layer or the exterior cusp. ULF waves propagating down these field lines are recorded at the ground as ULF pulsations.

As reported in this paper a variety of different types of ULF pulsations in the frequency range 1 mHz - 1 Hz has been observed. Regular, longperiod (3-10 min) pulsations trains (termed IPCL pulsations by Troitskaya et al., 1980) have been reported as well as narrowband Pc3 pulsations or more irregular Pi activity. Also often seen spiky, more transient magnetic field variations may be regarded as a member of the ULF pulsation family observed in the polar cusp and cap regions. A typical power spectrum of these different magnetic field fluctuations exhibits a clear power law with spectral index -2.6 (see Figure 5). Statistical studies also show that ULF activity is usually strongest around magnetic local noon.

The high-altitude counterpart of ULF pulsations are strong magnetic field, plasma density, and plasma flow perturbations observed in the different plasma regimes of the dayside cusp (see Figure 2). As at the ground power spectra at high-altitudes may be described by a power law, but with a slightly shallower spectral decay (spectral index -2).

The most exciting developement in the analysis of polar cusp/cap ULF pulsations in the last couple of years is the detection of a clear dependence of the occurrence of such pulsations from solar wind conditions, such as the IMF B_z component or dynamic pressure variations. Especially the IMF B_z-dependence is directly connected to the question 'What is the ground signature of flux transfer events?'. It is this question, where ULF pulsations research in the polar cusp/cap regions can provide important clues to unravel the physical processes of the magnetosheath-magnetosphere interaction at the dayside magnetopause.

There are a couple of scientific problems concerning ULF pulsations in the polar cusp and cap regions, which should be tackled in the future. Some of these are:

o How do upstream waves in the Pc3 period range interact with the bow shock and how do they propagate through the turbulent plasmas of the magnetosheath and the cusp?
o What are the characteristics of the observed cusp turbulence at high- altitudes? Is this turbulence Alfvénic type turbulence as observed in the solar wind ? Correlated observations of the magnetic field and plasma flow as well as plasma density are required. The upcoming CLUSTER mission will certainly help to answer this question.
o Further studies in the polar cap are necessary to distinguish between ULF pulsations generated on cusp field lines and subsequently propagating toward higher geomagnetic latitudes and ULF pulsations generated on polar cap field lines by dynamic processes in the tail of the magnetosphere.
o What is the ground signature of flux transfer events and how are they related to ULF pulsation studies? Further coordinated ground-satellite observations are required. Also, the ionospheric modification of an Alfvén wave generated by FTEs should be studied in more detail. In particular, the influence of horizontally non-uniform ionospheric conductivity distributions has to be investigated [cf. Glassmeier, 1988 and references therein], as they significantly influence the ground signature of ULF pulsations. Furthermore, what is the signature in space and at the ground of time-dependent reconncetion between northward directed IMF field lines and magnetotail fieldlines ?

Acknowledgement. I am grateful to all those colleagues who made available reprints and unpublished preprints to me. Helpful discussions with M. Engebretson, K. Prikner, J.V. Olson, and A. Wolfe are gratefully acknowledged. Thanks are also due to A. Egeland and P.E. Sandholt for organizing a stimulating and fruitful workshop at Lillehammer, where this review has been presented. Special thanks are to F.M. Neubauer for his steady support during the preparation of this review.

7. REFERENCES

Arnoldy, R.L., L.J. Cahill, M.J. Engebretson, L.J. Lanzerotti, A. Wolfe, Review of hydro-magnetic wave studies in the Antarctic, *Rev. Geophys.*, **26**, 181-207, 1988.

Baumjohann, W., K.H. Glassmeier, The transient response mechanism and Pi2 pulsations at substorm onset, *Planet. Space Sci.*, **32**, 1361-1368, 1984.

Bolshakova, O.V., V.A. Troitskaya, Pulsed reconnection as a possible source of IPLC pulsations, *Geomagn. Aeron.*, **22**, 723-725, 1982.

Bolshakova, O.V., V.A. Troitskaya, The relation of the high-latitude maximum of Pc3 intensity to the dayside cusp, *Geomagn. Aeron.*, **24**, 633-635, 1984.

Cowley, S.W.H., The impact of recent observations on theoretical understanding of solar wind - magnetosphere interactions. *J. Geomag. Geoelectr.*, **38**, 1223 - 1256, 1986.

D'Angelo, N., Ultralow frequency fluctuations at the polar cusp boundaries, *J. Geophys. Res.*, **78**, 1206-1209, 1973.

D'Angelo, N., Plasma waves and instabilities in the polar cusp: A review, *Rev. Geophys. Space Phys.*, **15**, 299-307, 1977.

Engebretson, M.J., C.I. Meng, R.L. Arnoldy, L.J. Cahill, Pc3 pulsations observed near the south polar cusp, *J. Geophys. Res.*, **91**, 8909-8917, 1986.

Engebretson, M.J., B.J. Anderson, L.J. Cahill, R.L. Arnoldy, T.J. Rosenberg, D.L. Carpenter, R.H. Eather, L.J. Zanetti, T.A. Potemra, Ionospheric signatures of cusp-latitude Pc3 pulsations, *J.Geophys. Res.*, submitted, 1988.

Fairfield, D.H., N.F. Ness, Imp 5 magnetic field measurements in the high latitude outer magnetosphere near the noon meridian, *J. Geophys. Res.*, **77**, 611-623, 1972.

Fairfield, D.H., E.W. Hones, Imp 6 measurements in the distant polar cusp during substorms, *J. Geophys. Res.*, **83**, 4273-4287, 1978.

Fraser-Smith, A.C., ULF/Lower-ELF electromagnetic field measurements in the polar cap, *Rev. Geophys. Space Phys.*, **20**, 497-512, 1982.

Friis-Christensen, E., M.A. McHenry, C.R. Clauer, and S. Vennerstrom, Ionospheric traveling convection vortices observed near the polar cleft: A triggered response to sudden changes in the solar wind, *Geophys. Res. Lett.*, **15**, 253-256, 1988.

Fukunishi, H., L.J. Lanzerotti, Hydromagnetic waves in the dayside cusp region and ground signatures of flux transfer events, Preprint, 1988.

Glassmeier, K.H., M. Lester, W.A.C. Mier-Jedrzejowicz, C.A. Green, G. Rostoker, D. Orr, U. Wedeken, H. Junginger, E. Amata, Pc5 pulsations and their possible source mechanisms: A case study, *J. Geophys.*, **55**, 108-109, 1984.

Glassmeier, K.H., Reconstruction of the ionospheric influence on ground-based observations of a short duration ULF pulsation event, *Planet. Space Sci.*, **36**, 801-817, 1988.

Glassmeier, K.H., M. Hönisch, J. Untiedt, Ground-based and satellite observations of traveling magnetospheric convection twin-vortices, *J. Geophys. Res.*, in press, 1988.

Goertz,C.K, E. Nielsen, A. Korth, K.H. Glassmeier, C. Haldoupis, P. Hoeg, and D. Hayward, Observation of a possible ground signature of flux transfer events, *J. Geophys. Res.*, **90**, 4069-4081, 1985.

Heikkila, W.J., J.D. Winningham, Penetration of magnetosheath plasma to low altitudes through the dayside magnetospheric cusps, *J. Geophys. Res.*, **76**, 883-891, 1971.

Hönisch, M. and K.H. Glassmeier, Isolated transient magnetic variations in the auroral zone, *EOS, Trans. Amer. Geophys. Union*, **67**, 1163, 1986.

Kunkel, T., W. Baumjohann, J. Untiedt, and R. Greenwald, Electric fields and currents at the Harang discontinuity: A case study, *J. Geophys.*, **59**, 73-86, 1986.

LaBelle-Hamer, A.L., Z.F. Fu, L.C. Lee, A mechanism for patchy reconnection at the dayside magnetopause, *Geophys. Res. Lett.*, **15**, 152-155, 1988.

Lanzerotti, L.J., C.G. MacLennan, Hydromagnetic waves associated with possible flux transfer events, *Astrophys. Space Sci.*, **144**, 279-290, 1988.

Lee, L.C., Magnetic flux transfer at the Earth's magnetopause, in: Y. Kamide, J. Slavin (Eds.), Solar Wind - Magnetosphere Coupling, Terra, Tokyo, 1986.

Lee, L.C., Y. Shi, L.J. Lanzerotti, A mechanism for the generation of cusp region hydromagnetic waves, *J. Geophys. Res.*, **93**, 7578-7585, 1988.

McKenzie, J.F., K.O. Westphal, Transmission of Alfvén waves through the Earth's bow shock, *Planet. Space Sci.*, **17**, 1029-1037, 1969.

Odera, T.J., Solar wind controlled pulsations: A review, *Rev. Geophys.*, **24**, 55, 1986.

Oguti, T., T. Yamamoto, K. Hayashi, S. Kokubun, A. Egeland, J.A. Holtet, Dayside auroral activity and related magnetic impulses in the polar cusp region, *J. Geomagn. Geoelectr.*, **40**, 387-408, 1988.

Olson, J.V., ULF signatures of the polar cusp, *J. Geophys. Res.*, **91** , 10055-10062, 1986.

Paschmann, G., G. Haerendel, N. Sckopke, H. Rosenbauer, P.C. Hegdecock, Plasma and magnetic field characteristics of the distant polar cusp near local noon: The entry layer, *J. Geophys. Res.*, **81**, 28883-2899, 1976.

Prikner, K., V. Wagner, Numerical solution of ionospheric filtration of ULF waves, Part II: Applications, *Travaux Geophysiques*, XXX, No. 577, 231-257, 1982.

Reid, G.C., T.E. Holzer, The response of the dayside magnetosphere-ionosphere system to time-varying field line reconnection at the magnetosphere. 2. Erosion event of March 27, 1968, *J. Geophys. Res.*, **80**, 2050-2056, 1975.

Rezeau, L., S. Perraut, A. Roux, Electromagnetic fluctuations in the vicinity of the magnetopause, *Geophys. Res. Lett.*, **13**, 1093-1096, 1986.

Rostoker, G., J.C. Samson, Y. Higuchi, Occurrence of Pc 4,5 micropulsation activity at the polar cusp, *J. Geophys. Res.*, **77**, 4700-4706, 1972.

Russell, C.T., C.R. Chappel, M.D. Montgomery, M. Neugebauer, F.L. Scarf, Ogo 5 observations of the polar cusp on November 1, 1968, *J. Geophys. Res.*, **76**, 6743-6764, 1971.

Russell, C.T., R.C. Elphic, Initial ISEE magnetometer results: Magnetopause observations, *Space Sci. Rev.*, **22**, 681-692, 1978.

Saunders, M.A., C.T. Russell, N. Sckopke, Flux transfer events: Scale size and interior structure, *Geophys. Res. Lett.*, **11**, 131-133, 1984.

Sckopke, N., Plasma and field observations in the exterior cusp, entry layer, and plasma mantle, in: J.A. Holtet, A. Egeland (Eds.), The polar cusp, pp. 1-7, D. Reidel, Dordrecht, 1985.

Sibeck, D.G., W. Baumjohann, R.C. Elphic, D.H. Fairfield, W.B. Gail, J.F. Fennell, L.J. Lanzerotti, R.E. Lopez, H. Luehr, A.T.Y. Lui, C.G. MacLennan, R.W. McEntire, T.A. Potemra, T.J. Rosenberg, K. Takahashi, The magnetospheric response to 8 minute-period strong-amplitude solar wind dynamic pressure variations, *J. Geophys. Res.*, submitted, 1988.

Slawinski, R., D. Venkatesan, A. Wolfe, L.J. Lanzerotti, C.G. MacLennan, Transmission of solar wind hydromagnetic energy into the terrestrial magnetosphere, *Geophys. Res. Lett.*, in press, 1988.

Southwood, D.J., W.J. Hughes, Theory of hydromagnetic waves in the magnetosphere, *Space Sci. Rev.*, **35**, 301-366, 1983.

Southwood, D.J., The ionospheric signature of flux transfer events, *J. Geophys. Res.*, **92**, 3207-3213, 1987.

Southwood, D.J., C.J. Farrugia, M.A. Saunders, What are flux transfer events ?, *Planet. Space Sci.*, **36**, 503-508, 1988.

Troitskaya, V.A., O.V. Bolshakava, E.T. Matveeva, Geomagnetic pulsations in the polar cap, *J. Geomagn. Geoelectr.*, **32**, 309-324, 1980.

Wolfe, A., E. Kamen, L.J. Lanzerotti, C.G. MacLennan, J.F. Bamber, D. Venkatesan, ULF geomagnetic power at cusp latitudes in response to upstream solar wind conditions, *J.*

Geophys. Res., **92**, 168-174, 1987.

Yumoto, K., A. Wolfe, T. Terasawa, E.L. Kamen, L.J. Lanzerotti, Dependence of Pc3 magnetic energy spectra at South Pole on upstream solar wind parameters, *J. Geophys. Res.*, **92**, 12437-12442, 1987.

STUDIES OF Pc 1 - Pc 3 GEOMAGNETIC PULSATIONS AT HIGH SOUTHERN LATITUDES:
IMPLICATIONS FOR ORIGIN AND TRANSMISSION

M. J. ENGEBRETSON
Augsburg College
Minneapolis, Minnesota 55454

L. J. CAHILL, JR.
University of Minnesota
Minneapolis, Minnesota 55455

R. L. ARNOLDY
University of New Hampshire
Durham, New Hampshire 03824

B. J. ANDERSON
The Johns Hopkins University
Applied Physics Laboratory
Laurel, Maryland 20707

ABSTRACT: Recent multipoint studies of ULF waves in the Pc 1 to Pc 3 period range using data from search coil magnetometers and other instrumentation at South Pole and McMurdo, Antarctica, located at magnetic latitudes typical of the polar cusp/cleft and polar cap, respectively, have identified an apparent high latitude source for two classes of ULF activity: Pc 1-2 wave trains are associated with ~70% of FTE-like magnetic pulsations observed when South Pole Station is near the dayside cusp, and narrowband Pc 3-4 activity, which appears to peak at cusp/cleft latitudes, is related to radial or near-radial IMF conditions through the generation of "upstream waves" in the ion foreshock. On the basis of these observations we suggest a new mechanism for the transmission of upstream wave signals into the magnetosphere and the generation of Pc 3 pulsations in the dayside outer magnetosphere.

1. Introduction

The low-altitude cusp and cleft regions of the earth's magnetosphere have been increasingly recognized as the site of considerable interaction between magnetospheric and ionospheric plasmas. The complex plasma processes known to occur at ionospheric altitudes and above, along field lines connected to the earth's magnetopause and boundary layers, are not yet well understood. There has been ample evidence that magnetosheath plasma can penetrate to ionospheric altitudes in the cusp region [Heikkila and Winningham, 1971, Frank, 1971]. A growing body of evidence, most recently reviewed by Arnoldy et al. [1988b], also suggests that ULF wave activity from the magnetopause

P. E. Sandholt and A. Egeland (eds.), Electromagnetic Coupling in the Polar Clefts and Caps, 187–201.
© *1989 by Kluwer Academic Publishers.*

and boundary layer region is transported, perhaps even focused, toward the ionospheric footpoint of the cusp [Olson, 1986, Morris and Cole, 1987]. Troitskaya et al. [1971] first discussed solar wind control of high latitude pulsations, and many recent studies have confirmed a correlation between waves generated upstream of the earth's bow shock [Fairfield, 1969] and Pc 3-4 period waves observed at various locations in the dayside magnetosphere [Odera, 1986]. More recent evidence [Arnoldy et al., 1988a] indicates that at least some of the higher frequency Pc 1-2 wave activity observed at high dayside latitudes is also related to interactions between the solar wind and the magnetosphere.

The installation of magnetic field sensors and other instruments at South Pole Station, Antarctica ($\Lambda = -74°$), which during certain hours every day is located under the nominal position of the magnetospheric cleft/cusp region, and at McMurdo, Antarctica deep in the polar cap ($\Lambda = -79°$), has provided a large suite of ground-based data with which to study polar cap and cleft/cusp-related ULF wave phenomena. We present in this paper some of the most recent results obtained using the University of New Hampshire/University of Minnesota wave magnetometers installed at these sites as well as at Sondre Stromfjord, Greenland. Instruments at each site consist of two identical search coils mounted orthogonally in magnetically northward and eastward directions [Taylor et al., 1975]. The dB/dt output of each coil is digitized at a rate of 10 samples/s, thus providing data on ULF pulsations of <= 5 Hz. We will also include correlative observations from two other instruments at South Pole Station (the University of Maryland 30 MHz riometer and the Boston College 427.8 nm photometer); from the magnetometer on board the AMPTE/CCE satellite [Potemra et al., 1985], with equatorial apogee located near L = 9 in midmorning local time during the period studied in this paper; and from the IMP 8 satellite, in the solar wind.

2. Short-period ULF Activity: Pc 1 and 2

A wide variety of ULF pulsations has been observed at auroral and subauroral latitudes in both hemispheres. Nagata et al. [1980] and Fukunishi et al. [1981] have suggested 12 different categories for emissions with frequency greater than 0.1 Hz observed at the Syowa-Husafell conjugate pair. While most of these have also been observed at higher latitudes due to presumed ducting from lower latitude sources, events known variously as Pc 1-2 bursts, IPCP, or IPRP appear to be related to a high latitude source [Arnoldy et al., 1988b].

Figure 1 shows an example of this category of Pc 1-2 pulsation event. These are observed frequently at South Pole and at the roughly conjugate northern cusp latitude station at Sondre Stromfjord, Greenland [Arnoldy et al., 1988a]. In this figure we show a plot of ten minutes of data obtained near local magnetic noon (1530 UT) October 15, 1986. The event at 1530 UT is a clear example of a solitary Pc 5 pulsation of the sort often considered to be the ground signature of a flux transfer event, or FTE [Russell and Elphic, 1979]. A Pc 1-2 event is shown superimposed on this solitary pulsation at both sites, although with evident differences in timing between the two stations. Of 155 such solitary Pc 5 events observed during 1985-1986, 70% have superimposed Pc 1-2 pulsations, suggesting a common origin if not a common generation mechanism. Arnoldy et al. [1988a] suggest that this common origin may lie in the passing of a solar wind pressure pulse, which excites the Pc 1-2 pulsations in the hot boundary layer or entry layer plasma in much the same way that SI and SSC events do deep within the magnetosphere.

Figure 2 shows dynamic power spectra for a similar event observed from 1240 to 1255 UT November 6, 1986 centered on the occurrence of an FTE-like signature at 1248 UT [Arnoldy et al., 1988a]. In each panel, the gray scale is proportional to plane-polarized spectral power according

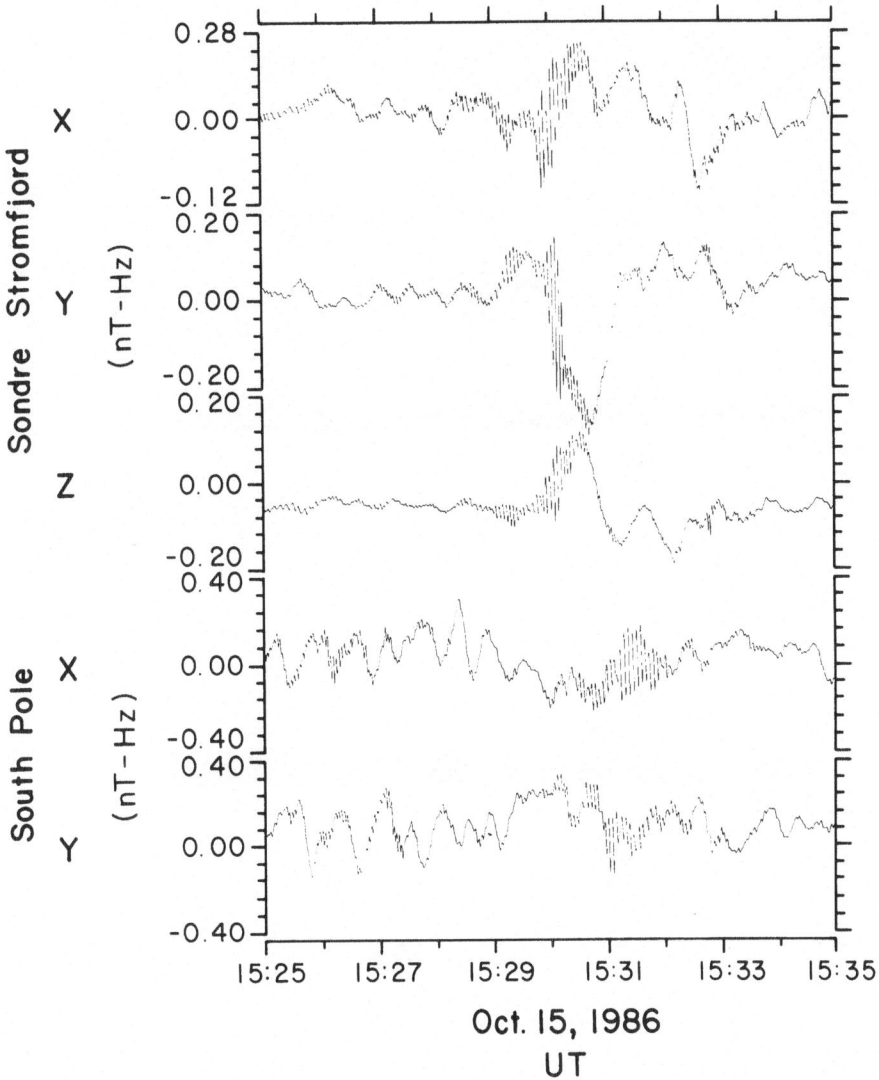

Figure 1. Pc 1-2 event superposed on a solitary pulsation observed from 1525 to 1535 UT October 15, 1985. Traces shown are from search coil magnetometers at Sondre Stromfjord, Greenland (X, Y, and Z, or vertical) and at South Pole Station (X and Y). No vertical component data are available at South Pole.

190

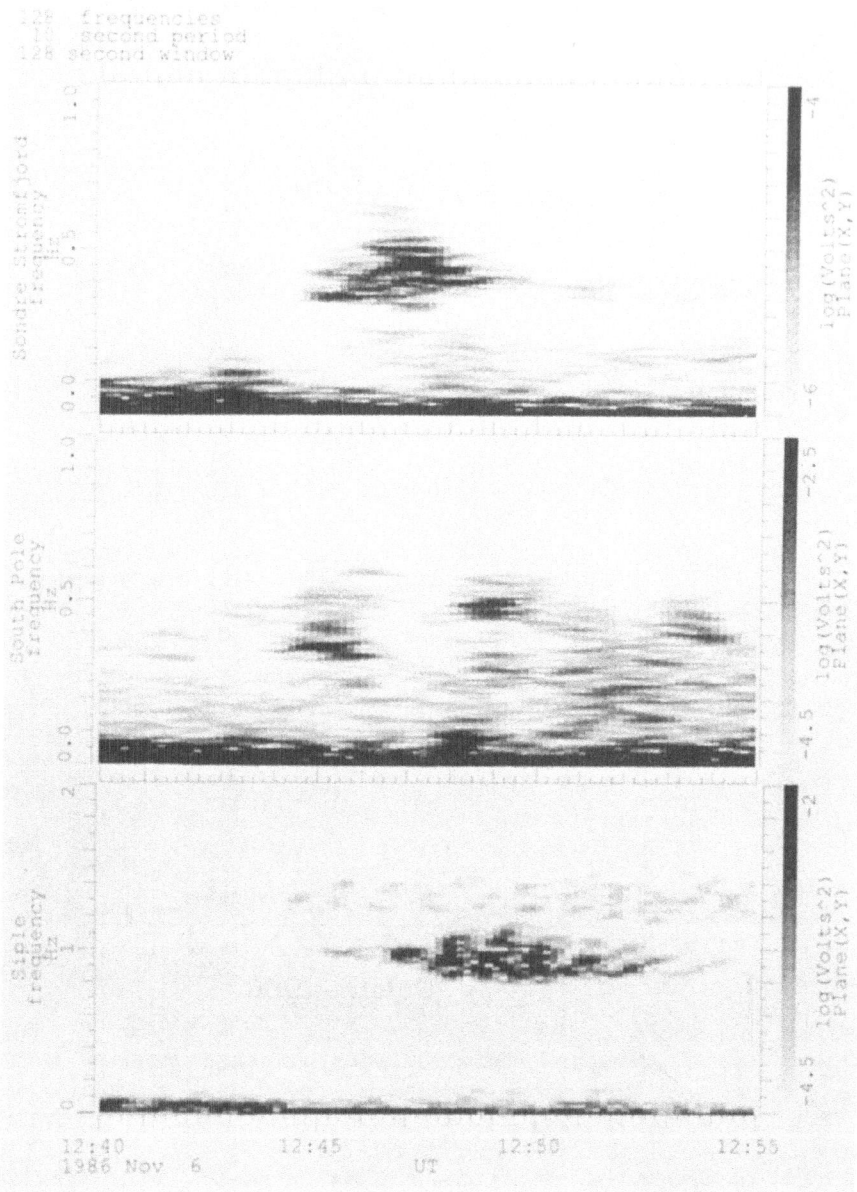

Figure 2. Gray-scale dynamic power spectrogram for the plane-polarized component of data obtained from Sondre Stromfjord, South Pole, and Siple Station, Antarctica for a 15-minute period centered on the occurrence of a solitary pulsation at 1248 UT November 6, 1986.

to the shading shown on the right side of the figure. The three sites represent a magnetic latitudinal span of 14 degrees and a 4-hr spread in local time. The spectrograms show that Pc 1-2 activity was observed at all three sites. Note, however, that the frequency at Siple (L ~ 4.2) was twice that observed at the high-latitude sites. This difference in frequency suggests that the observed Pc 1-2 emissions were generated on or near the field lines at which they were observed. Such a process would be consistent with the well known tendency for periods of positive Dst (sudden storm commencements or sudden impulses) to be associated with the onset of Pc 1 activity at a variety of locations [e.g., Saito and Matsushita, 1967, Kangas et al., 1986].

3. Pc 3-4 and Pi 1 Pulsations

The ion foreshock region of the solar wind, upstream of the earth's bow shock, has been considered a major candidate for the source of dayside Pc 3-4 pulsations for nearly two decades. In this region solar wind ions (mostly protons) reflected upstream from the bow shock interact with the bulk plasma of the solar wind via an ion cyclotron instability to generate Alfven waves in the interplanetary medium [Fairfield, 1969]. The frequency of these waves, as seen in a frame of reference traveling with the earth, is in the Pc 3-4 period range (20 to 120 seconds), with a typical period varying as 160/B sec, for B in nT [Troitskaya et al., 1971]. The ion foreshock develops upstream of the subsolar bow shock when the interplanetary magnetic field (IMF) is oriented in an approximately radial direction, i.e., when the IMF cone angle, $\theta_{xB} = \cos^{-1}$ (Bx/IBI), is closer to 0 or 180° than to 90° [Russell and Hoppe, 1983].

Pc 3 pulsations observed on the ground and in space inside the magnetosphere have been tied to waves generated in this upstream region by several studies, ranging from the pioneering work of Troitskaya et al. [1971] and Greenstadt [1972] to recent studies by Yumoto et al. [1984, 1985] and Engebretson et al. [1986b,c, 1987]. Russell et al. [1983] presented a simple geometrical model explaining the influence of the IMF cone angle in determining the propagation of waves generated in the ion foreshock region to the magnetosheath and magnetopause. Crooker et al. [1981] and Luhmann et al. [1986] provided experimental support for this model with ISEE data by showing a clear correlation between the IMF direction and the presence of energetic ions in the magnetosheath, and levels of magnetosheath turbulence, respectively.

There remains some controversy over the range of latitudes to which upstream Pc 3-4 pulsations can reach. It has been well established that low- and midlatitude Pc 3-4 pulsations are correlated well with upstream waves [Greenstadt et al., 1979, Wolfe et al., 1980, Wolfe and Meloni, 1981, Yumoto et al., 1984, 1985]. Evidence for correlations between higher latitude Pc 3-4 pulsation activity and upstream waves has been mixed, however. Although Soviet observers have consistently reported excellent correlations at high and very high latitudes [Plyasova-Bakounina et al., 1978, 1986, Bolshakova and Troitskaya, 1984] other studies, for example those by Yumoto et al. [1984, 1985] found significantly poorer correlations between the IMF cone angle and wave power in the Pc 3-4 band in the auroral zone than at lower latitudes.

Wolfe et al. [1987] and Yumoto et al. [1987] used fluxgate magnetometer data from South Pole in statistical studies of power levels in the Pc 3 to Pc 5 period ranges, and found that correlations such as those described above did not appear for high latitude (cusp) pulsation activity. They suggested that the dominant energy source of the ULF power they observed was a Kelvin-Helmholtz instability at the magnetopause. On the other hand, Engebretson et al. [1986b] used data from a search coil magnetometer in an attempt to separate types of pulsations in the Pc 3-4 period range. Their study, based on high time resolution data near local noon during a two-month period, separated

SOUTH POLE STATION YRDAY=86116 UT=36128 APR 26, 86
FREQUENCY (mHz)
YBB
RIO2
PHO2
LOG POWER
UT 10:00 11:00 12:00 13:00 14:00 15:00

AMPTE CCE MAGNETOMETER YRDAY=86116 UT=31338 APR 26, 86
FREQUENCY (mHz)
BR
BN
LOG POWER
UT 08:00 10:00 12:00 14:00 16:00 18:00 20:00 22:00 00:00

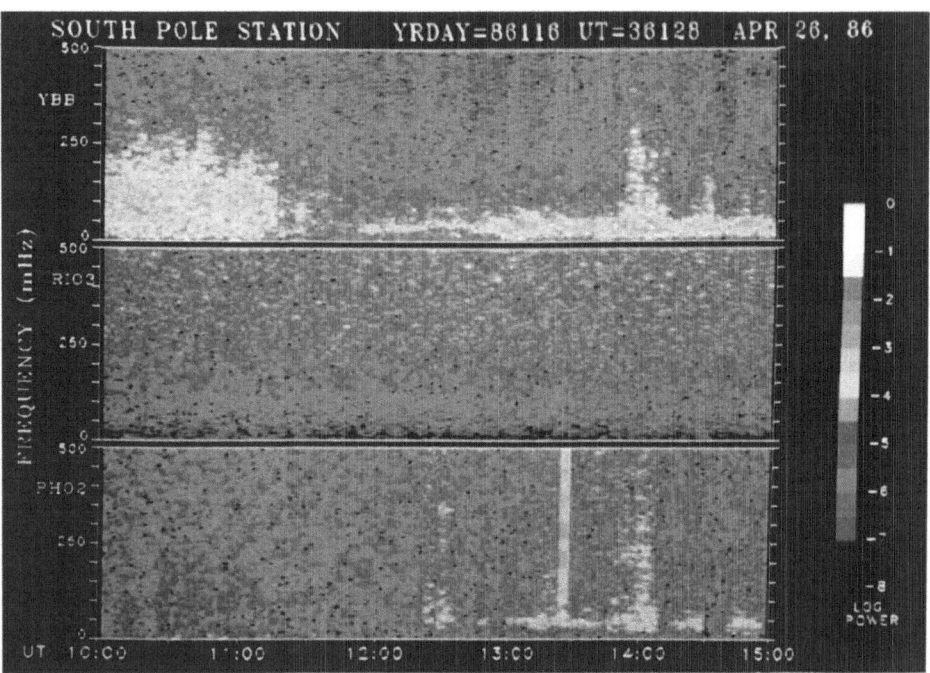

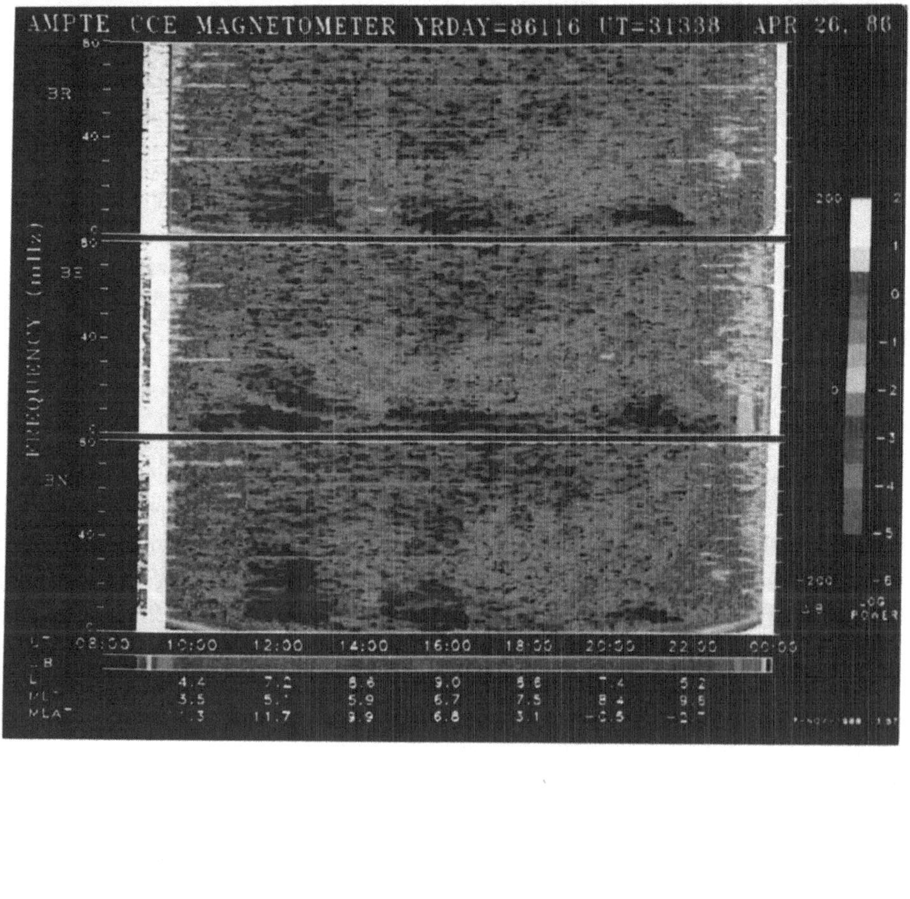

ULF wave events into broadband Pi 1 components and narrowband Pc 3-4 components on the basis of waveform plots. Comparison of the narrowband Pc 3-4 pulsations with hourly averaged data from the IMP 8 satellite showed a clear correlation in both occurrence and frequency, while the statistically dominant Pi 1 activity did not. Because the Pi 1 (broadband) component correlated well with the presence of strong precipitation of energetic electrons, as observed by particle sensors on the DMSP F6 and F7 and DE 2 satellites, Engebretson et al. [1986b] concluded that these pulsations were of local, ionospheric origin.

A Pc 3-4 event with simultaneous ULF modulation of VLF signals reported by Lanzerotti et al. [1986] was also ascribed to an upstream source, and a recent study by Slawinski et al. [1988], using the same instrumentation at South Pole Station as Wolfe et al. [1987] but incorporating an automated procedure to distinguish broadband from narrowband activity, also has found a correlation between the frequency of one class of narrowband pulsations and the IMF magnitude.

Earlier studies of broadband Pi 1-2 activity (~5 to 50 s period) at auroral and subauroral latitudes have shown that its occurrence is governed by precipitation of energetic (>20 keV) auroral electrons [Arnoldy et al., 1982, Engebretson et al., 1986a, Oguti et al., 1984]. It is hence related to substorms, and negative IMF Bz.

3.1 GROUND-LEVEL OBSERVATIONS

As an example of these two categories and of their control by the IMF, we present here observations from South Pole during April 1986. Figure 3 shows a spectrogram of data for a five-hour period from 1000 to 1500 UT April 26, 1986 (0630 to 1130 MLT). Each trace shows relative spectral power based on raw sensor voltages, not geophysical units. From top to bottom the panels show data from the Y component of the wave magnetometer, oriented positive north (equatorward) along the geomagnetic meridian, the 30 MHz riometer, and the 427.8 nm photometer. The riometer and photometer have identical 30° half angle fields of view. The frequency scale, shown to the left of each panel, goes from 0 to 500 mHz, and power levels of each component are displayed at each time and frequency according to the color bar legend at the right edge of the figure.

The most prominent wave features in the magnetometer data are intense broadband signals from 1000 to nearly 1130 UT (and occasionally later, as near 1400 UT), and narrowband power, centered near 30-40 mHz, beginning with moderate power levels from 1150 to 1245 UT and continuing with greater intensity from 1245 to 1500 UT. The frequency range of the more broadband signals is from below 10 mHz to above 250 mHz, while the more narrowband signals appear to be confined to the region between 10 and 70 mHz.

Figure 3 (facing page, top). Dynamic power spectra of ULF wave activity from three instruments at South Pole Station, Antarctica, from 1000 to 1500 UT, April 26, 1986. From top to bottom are color-coded power levels for the Y (north-south) component of the wave magnetometer (YBB), 30-MHz riometer (RIO3), and 427.8 nm photometer (PHO2). Local time at South Pole Station is UT minus 3.5 hours.

Figure 4 (facing page, bottom). Three component dynamic power spectra of magnetic field data from the AMPTE/CCE satellite for a full orbit from 0840 to 2400 UT April 26, 1986.

The riometer, which is sensitive to variations in charge density in the lower ionosphere caused by energetic electrons (>20 keV), shows little or no Pc 3-4 or Pi 1 activity during this period, but the photometer, sensitive to auroral light generated in the upper ionosphere (including that caused by electrons with energy below 1 keV), shows power enhancements similar in frequency to those of the wave magnetometer during much of the interval of narrowband activity.

Magnetic field data obtained simultaneously at McMurdo from 0130 to 0630 local time (not shown) revealed very similar features, but with considerably lower power levels. Weak broadband power appeared from 1000 to 1130 UT, nearly the same times it was observed at South Pole, and only very weak broadband features appeared between 1130 and 1500 UT. Narrowband activity centered near 30-40 mHz appeared from 1200 to 1500 UT, with variations in intensity in both magnetic field components simultaneous with those observed (with much higher amplitude) at South Pole. Only background noise was evident in the McMurdo riometer and photometer data.

3.2 Pc 3-5 OBSERVATIONS IN THE MAGNETOSPHERE

Figure 4 shows 3-dimensional ULF pulsation data obtained during this period by the AMPTE/CCE magnetic fields experiment in the dayside magnetosphere. This plot covers a full 16-hour orbit of the satellite, from perigee to apogee near 1600 UT and back to perigee. The frequency scale, shown on the left, goes from 0 to 80 mHz, and the narrow bottom panel displays variations in field magnitude from the IGRF-80 model value. The three scales at the bottom of the figure denote the L shell, magnetic local time, and magnetic latitude of the satellite. Spectrograms such as this are described in greater detail by Engebretson et al. [1986c, 1987]. Background noise levels drop to their lowest level near 1115 UT, as the satellite magnetometer switches to its most sensitive range. A single narrowband tone is evident in the azimuthal (BE) component, with frequency falling from 15 mHz near 1115 to near 5 mHz near 1400 UT. No such tone is evident in the radial (BR) or magnetically northward (BN) component. A higher frequency azimuthal signal begins near 40 mHz at 1200 UT, and several frequencies appear, again only in this component, near 1300 UT. The simultaneous occurrence of ULF pulsations at several frequencies in the Pc 3-5 range was first reported by Takahashi and McPherron [1982], also using spectrogram displays. We identify the first and lowest frequency in the center panel of Figure 4 as the fundamental field line resonance and the 40 mHz signal beginning at 1200 UT as the third harmonic. The second, fourth, and a faint fifth harmonic appear after 1300 UT. Harmonic structure continues at a roughly constant power level until 2100 UT, but because of the varying magnetic latitude of the AMPTE/CCE spacecraft, the relative power in the various azimuthal harmonics changes. For example, from 1800 to 2000 UT the second and fourth harmonics appear brighter than the first and third harmonics. This is consistent with accepted models of ULF field line resonances [Engebretson et al., 1986c].

General increases in broadband power appear in all three components of Figure 4, for example near 1400 UT and from 1630 to 2030 UT, and coincide with brightenings of the harmonics in the azimuthal direction, but are not of comparable power. The increase at 1400 UT corresponds well with a burst of broadband pulsation activity at South Pole.

Before 1100 UT the satellite's magnetometer was in a less sensitive range and the satellite was located in the pre-dawn local time sector where fundamental mode resonances dominate and harmonic Pc 3 activity is rarely observed [Anderson et al., 1988]. There is no evidence of increased ULF power in the Pc 3/Pi 1 range at AMPTE/CCE during the first and most intense period of broadband activity seen at South Pole from 1000 to nearly 1130 UT, but we cannot rule out the presence of a moderate level of broadband activity before 1100 UT.

3.3 IMF CONTROL OF Pc 3-4 ACTIVITY

IMF data for April 26, 1986 from the IMP 8 magnetometer are shown in Figure 5. During this time the IMP 8 satellite was located more than 20 R_E upstream from the earth. The upper three panels display the field in solar magnetospheric coordinates, and the lower three panels show field magnitude, cone angle (with 0° indicating a field aligned with the X axis and 180° indicating a field aligned antiparallel to the X axis), and elevation angle (with 90° indicating a field aligned with the Z axis and -90° indicating a field aligned antiparallel to the Z axis), respectively. From 1000 to 1100 UT the field is predominantly in the Y direction, with a small southward (-Z) component, and the cone angle is near 90°. The Z component gradually turns positive, and remains positive or near zero for the remainder of the day. The X component becomes strongly positive after 1100, and the Y component gradually decreases until 1300. After a brief interval of almost purely radial field near 1300 UT the IMF is predominantly radial, even during a brief interval of field reorientation near 1450 UT; the cone angle drops below 45° near 1130 UT and remains below 30° from 1300 to 2300 UT.

Comparing the IMF data to the pulsation data shown in Figures 3 and 4, we find that narrowband dayside Pc 3-4 pulsations observed at South Pole in local morning and at McMurdo in the nightside polar cap, and harmonic dayside pulsations observed at AMPTE/CCE, are strongly correlated with times of low IMF cone angle observed by IMP 8. The sustained broadband activity observed at both South Pole and McMurdo from 1000 to 1115 UT occurs during a period of negative IMF Z component, but the less intense broadband activity observed at South Pole and at AMPTE/CCE near 1400 UT corresponds to no identifiable variation of IMF at that time. The presence of narrowband pulsations at McMurdo during this period supports the hypothesis that both narrowband and broadband pulsation activity were present at this time.

Thus this study, and statistical studies by Engebretson et al. [1986b] and Slawinski et al. [1988], provide additional evidence that a significant part of the wave energy in the cusp/cleft regions is related to upstream parameters, but that care must be taken to properly distinguish these two types of pulsations.

3.4 Pc 3-4 PULSATIONS IN AURORAL EMISSIONS

We showed in Figure 3 that pulsations in auroral light appeared during some of the times narrowband magnetic field activity occurred at South Pole. Figure 6 provides more quantitative evidence for a correlation between IMF magnitude and auroral pulsation frequencies. Figure 6a shows stacked logarithmic Fourier spectra of magnetometer, riometer, and photometer signals for a 34-min period beginning at 1425 UT April 26, 1986. The vertical scale indicates the power spectral density as a function of frequency. A vertical dotted line has been drawn at the expected frequency for upstream Pc 3-4 pulsations ($f = .006 \times B_{IMF}$), based on the hourly averaged IMF field magnitude measured by the IMP 8 satellite in the solar wind. The frequency of this line lies near the center frequency of the observed increases in power in magnetometer and photometer signals (compare to Figure 3). Spectra of simultaneous magnetometer data from McMurdo reveal an enhancement in wave activity in the same frequency range observed at South Pole, but no enhancement in either riometer or photometer data.

Figure 6b shows spectra for the interval 1800 to 1834 UT April 25, 1986, characterized by moderately radial IMF ($\theta_{xB} = 145°$) and with modest fluctuations about an hourly average magnitude of only 2.9 nT. Enhanced power is again evident at the expected frequency for upstream pulsations, with nearly identical frequency structure, in three of the four traces shown. Again the riometer signal is devoid of power enhancements.

196

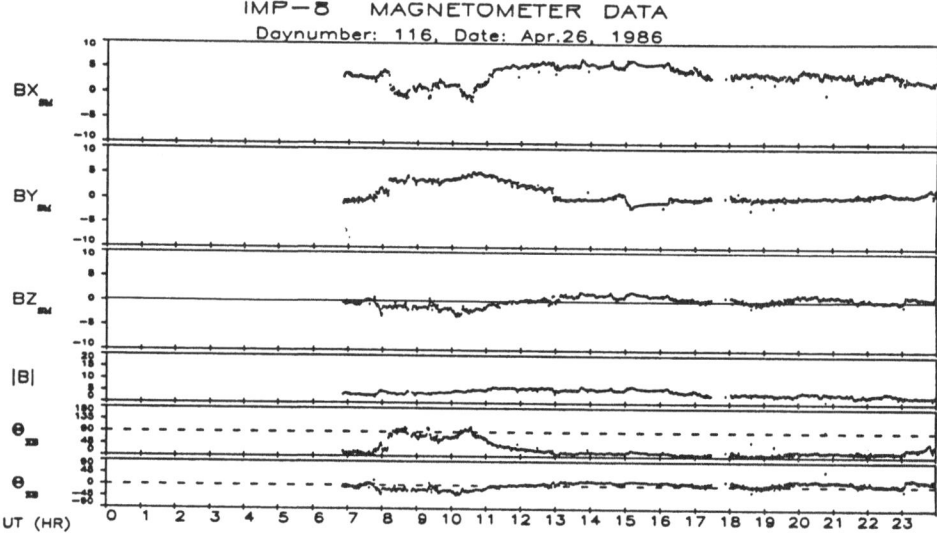

Figure 5. Plot of interplanetary magnetic field (IMF) data obtained by the IMP 8 spacecraft for April 26, 1986. The top three panels show the magnetic field components in a solar magnetospheric coordinate system. The lower panels display field magnitude, cone angle, and latitude, respectively.

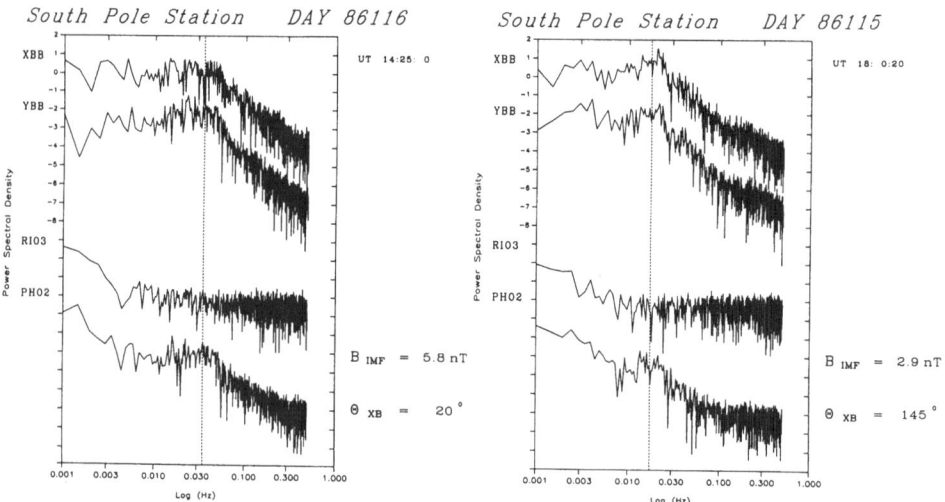

Figure 6. Logarithmic power spectra of data obtained at South Pole Station a) from 1425 to 1459 UT April 26, 1986 and b) from 1800 to 1834 UT April 25, 1986. The traces from top to bottom are from east-west and north-south magnetic field components, 30 MHz riometer, and 427.8 nm photometer. Also given are the hourly average magnitude and cone angle of the interplanetary magnetic field (IMF). The vertical line represents the predicted center frequency of upstream waves, determined by the IMF magnitude.

Figures 3 and 6 are typical examples of our multiple-instrument data set: Pc 3-4 magnetic pulsations observed simultaneously in South Pole and McMurdo magnetometer signals are often observed in South Pole photometer signals with the same IMF-controlled frequency but are not seen in McMurdo photometer data. We found modulations of auroral light at Pc 3-4 frequencies at South Pole to be highly localized in time (on the scale of individual wave packets), with a general trend for greater occurrence frequency as the amplitude of the magnetic pulsations becomes larger. Both of these effects are consistent with the narrower field of view of the photometer and with a varying location and/or latitude for the precipitation responsible for the auroral emissions. However, occurrences of Pc 3-4 pulsations at South Pole and McMurdo are never accompanied by corresponding pulsations in the riometer, even though when Pi 1 pulsations occur overhead the signals are often nearly identical in riometer and photometer data.

By analogy to the ionospheric generation of Pi 1 pulsations by precipitation of plasma sheet electrons [Engebretson et al, 1986a], the simultaneous observations of optical and magnetic Pc 3-4 pulsations at cusp/cleft latitudes shown here suggest a common source in auroral electron precipitation. Moreover, the consistent absence of riometer variations during these periods puts a clear upper bound on the energy of the precipitating electrons causing the auroral light. We suggest that these electrons originate in the cleft/boundary layer region, and that these particles have energies of 1 keV or lower. The local times during which we have observed these auroral pulsations, from before 0830 until after 1500 MLT, are consistent with the precipitating electrons being located within the longitudinally extended cleft region. As defined with low altitude DMSP satellite data by Newell and Meng [1988], this region is characterized by the presence of precipitating electrons with energies slightly above those of the cusp itself, slightly lower intensities, and extending in longitude in both directions from slightly before local noon.

4. Implications for Origin and Transmission of Pc 3-4 Pulsations

Theoretical studies of the transport of solar wind generated wave energy into the magnetosphere have until recently focused on direct entry of wave energy across the magnetopause, usually near the equator, followed by mode conversion to generate resonant transverse pulsations. The earliest mechanism, suggested by Verzariu [1973], was transmission of compressional wave power, but with severe attenuation, directly across tangential discontinuities at the equatorial, subsolar magnetopause. Wolfe and Kaufmann [1975] found support for Verzariu's model in data from Explorer 12 near the subsolar magnetopause, and Greenstadt et al. [1983], using ISEE-1 and -2 data on either side of the magnetopause, confirmed Verzariu's prediction of a roughly 2 orders of magnitude attenuation of power across the subsolar magnetopause.

Observations by Bolshakova and Troitskaya [1984], Plyasova-Bakounina et al. [1986], and Morris and Cole [1987] have all shown the presence of a Pc 3 amplitude maximum at polar cusp/cleft latitudes. These results as well as observations by Lanzerotti et al. [1986] and Engebretson et al. [1986b] at South Pole and Engebretson et al. [1987] at AMPTE/CCE and AMPTE/IRM appear to support models of high latitude entry, although the latter study indicated that both mechanisms are probably operative in the outer magnetosphere near local noon.

Theoretical calculations by Kwok and Lee [1984] suggested considerably greater efficiency for wave transmission through rotational discontinuities, which for a nearly radial IMF field direction might be more dominant at high (and possibly cleft) latitudes. They also showed that, in contrast to the severely attenuated transmission of Verzariu's mechanism, wave power could even be amplified across a rotational (open) magnetopause boundary. The only restriction in their model, of course, is that reconnection must occur in order for such wave transmission to take place. Our observations at South Pole have found the occurrence and amplitude of Pc 3-4 pulsations to have little if any dependence on the sign of IMF Bz or, when the IMF is radial, on the sign of Bx. This suggests little dependence on the IMF directions usually associated with reconnection.

The comparison of bandwidth between South Pole and AMPTE/CCE observations reported here is in agreement with recent multipoint studies by Kato et al. [1985] and Tonegawa et al. [1985], which showed that cleft-associated pulsation activity is more broadband than activity at lower magnetic latitudes. We have found that pulsation periods observed at high southern latitudes appear to be centered on those expected for the ion foreshock upstream of the earth's bow shock, and are more "band-limited" than monochromatic. The (multiple) periods observed at AMPTE/CCE in the dayside magnetosphere lie within this "band-limited" range of frequencies and are consistent with local field line resonances. Kato et al. [1985] found, by comparing ground data at two stations at the same longitude, one in the poleward edge of the cusp region and one at lower latitude in the auroral zone, that while Pc 3 and Pc 4/5 power was often observed during quiet times in near simultaneity at the two stations, the spectral peaks were much sharper at the auroral zone station. They noted that the more broadband character of the cusp region waves (on open field lines) and narrow spectral peaks in the auroral zone (on closed field lines) was consistent with the ability to excite a broad range of resonant harmonics on field lines equatorward of the cusp. The azimuthal propagation pattern noted by Tonegawa et al. [1985], who compared cleft-latitude and auroral zone pulsation patterns, also supported propagation from a cleft-related rather than an equatorial magnetopause source.

Additional evidence for a possible cusp/cleft transmission mechanism comes from a statistical study by Anderson et al. [1988] of the diurnal and radial distribution of azimuthally polarized harmonic pulsations in the Pc 3-5 range observed by the equatorially orbiting AMPTE/CCE satellite. For Pc 3-4 activity there was a broad, nearly uniform local time distribution on the dayside but a rapid cutoff in intensity shortly before local dawn, while for Pc 5 activity there was a broad maximum centered near local dawn. This local time distribution appears to conflict with models of transmission of compressional wave energy across the subsolar equatorial magnetopause and subsequent exponential decay of wave amplitude due to coupling with transverse modes.

Although the Kwok and Lee model is consistent with the large amplitude of Pc 3-4 pulsations observed at cusp/cleft latitudes, the lack of consistency with expected reconnection signatures, and the observed modulations in auroral brightness and the inferred energy of the electrons producing them, suggest a particle precipitation source for the Pc 3-4 activity at South Pole, and this would require a different kind of high latitude entry mechanism. We speculate that magnetosheath fluctuations of the sort associated with upstream waves by Crooker et al. [1981] and Luhmann et al. [1986] might be the source of the observed auroral pulsations, if they are able to cause the periodic injection of magnetosheath plasma into the dayside boundary layer. Any fraction of this injected plasma which precipitates would impinge on the ionosphere at latitudes near or overhead of South Pole during daytime hours. Periodic precipitation of particles would modulate the ionization and conductivity of the cusp/cleft ionosphere. By changing the Pedersen conductivity in the ionospheric foot of the dayside Region 1-2 current system, these modulations in conductivity would cause a time-varying pattern to be imposed on the dayside Region 1-2 current systems.

According to this model no wave mode coupling is required, and the waves need not propagate across field lines at any point in the outer magnetosphere. Rather, modulated Region 2 currents can set up transverse magnetic perturbations, hence launching transverse magnetic pulsations near the foot of closed field lines. Note that a sheet-like perturbation in the Region 2 currents will produce an azimuthal deflection in magnetic field, which is exactly the polarization observed for resonant harmonic pulsations in the outer dayside magnetosphere by Takahashi and McPherron [1982] and Engebretson et al. [1986c, 1987]. In this manner magnetosheath turbulence can be transmitted inward to L shells in the dayside magnetosphere to the extent of the inner edge of the Region 2 current sheet, i.e., in a region covering most of the magnetosphere outside of the plasmapause. Further characterization of this suggested entry process will be presented in a subsequent paper.

In summary, although ULF pulsation activity at high latitudes shares many features with lower latitude observations, observations at cusp/cleft latitudes have revealed an apparent high latitude source for some events in both the Pc 1-2 and Pc 3-4 ranges. Cusp/cleft Pc 3 activity appears to be due to precipitation of electrons with energies characteristic of the magnetosheath or dayside boundary layer. We suggest that these ~<1 kev precipitating electrons may play an important role, via modification of ionospheric conductivities and hence Birkeland currents, in the transmission of upstream wave signals into the magnetosphere and in the generation of Pc 3 pulsations in the dayside outer magnetosphere.

5. Acknowledgements

We thank Ronald P. Lepping of NASA/Goddard Space Flight Center for permission to use IMP 8 magnetometer data, and we thank T. J. Rosenberg and R H. Eather for permission to use unpublished data from South Pole Station and McMurdo. We acknowledge helpful contributions by T. A. Potemra, P. T. Newell and D. G. Sibeck. This research was supported by National Science Foundation Grant DPP-86-13272 to the University of New Hampshire and by subcontracts to Augsburg College and the University of Minnesota, and by NASA under Task I of contract N00024-85-C-5301 to The Johns Hopkins University Applied Physics Laboratory and by subcontract to Augsburg College. Additional support for computational work at Augsburg College was provided by National Science Foundation Grant ATM-86-06388.

6. References

Anderson, B. J., M. J. Engebretson, L. J. Zanetti, T. A. Potemra, and M. H. Acuna (1988) 'A statistical study of Pc 3-5 pulsations observed on the dayside by the AMPTE/CCE magnetic fields experiment', in press, Proceedings of the 1987 Cambridge Workshop in Theoretical Geoplasma Physics, MIT.

Arnoldy, R. L., K. Dragoon, L. J. Cahill, Jr., S. B. Mende, T. J. Rosenberg, and L. J. Lanzerotti (1982) 'Detailed correlations of magnetic field and riometer observations at L = 4.2 with pulsating aurora', J. Geophys. Res. 87, 10449.

Arnoldy, R. L., M. J. Engebretson, and L. J. Cahill, Jr. (1988a) 'Bursts of Pc 1-2 near the ionospheric footprint of the cusp and their relationship to flux transfer events', J. Geophys. Res. 93, 1007.

Arnoldy, R. L., L. J. Cahill, Jr., M. J. Engebretson, L. J. Lanzerotti, and A. Wolfe (1988b) 'Review of hydromagnetic wave studies in the Antarctic', Rev. Geophys. 26, 181.

Bol'shakova, O. V., and V. A. Troitskaya (1984) 'The relation of the high-latitude maximum of Pc 3 intensity to the dayside cusp', Geomagnetism and Aeronomy (English translation) 24, 633.

Crooker, N. U., T. E. Eastman, L. A. Frank, E. J. Smith, and C. T. Russell (1981) 'Energetic magnetosheath ions and the interplanetary magnetic field orientation', J. Geophys. Res. 86, 4455.

Engebretson, M. J., L. J. Cahill, Jr., J. D. Winningham, T. J. Rosenberg, R. L. Arnoldy, N. C. Maynard, M. Sugiura, and J. H. Doolittle (1986a) 'Relations between morning sector Pi 1 pulsation activity and particle and field characteristics observed by the DE 2 satellite', J. Geophys. Res. 91, 1535.

Engebretson, M. J., C. -I. Meng, R. L. Arnoldy, and L. J. Cahill, Jr. (1986b) 'Pc 3 pulsations observed near the south polar cusp', J. Geophys. Res. 91, 8909.

Engebretson, M. J., L. J. Zanetti, T. A. Potemra, and M. H. Acuna (1986c) 'Harmonically structured ULF pulsations observed by the AMPTE/CCE magnetic field experiment', Geophys. Res. Lett. 13, 905.

Engebretson, M. J., L. J. Zanetti, T. A. Potemra, W. Baumjohann, H. Luehr, and M. H. Acuna (1987) 'Simultaneous observation of Pc 3-4 pulsations in the solar wind and in the earth's magnetosphere', J. Geophys. Res. 92, 10053.

Fairfield, D. H. (1969) 'Bow shock associated waves observed in the far upstream interplanetary medium', J. Geophys. Res. 74, 3541.

Frank, L. A. (1971) 'Plasma in the earth's polar magnetosphere', J. Geophys. Res. 76, 5202.

Fukunishi, H., T. Toya, K. Koike, M. Kuwashima, and M. Kawamura (1981)'Classification of hydromagnetic emissions based on frequency-time spectra', J. Geophys. Res. 86, 9092.

Greenstadt, E. W. (1972) 'Field-determined oscillations in the magnetosheath as possible source of medium-period, daytime micropulsations', Solar Terrestrial Relations Symposium, University of Calgary, Calgary, Alberta, Canada.

Greenstadt, E. W., H. J. Singer, C. T. Russell, and J. V. Olson (1979) 'IMF orientation, solar wind velocity, and Pc 3-4 signals: a joint distribution', J. Geophys. Res. 84, 527.

Greenstadt, E. W., M. M. Mellott, R. L. McPherron, C. T. Russell, H. J. Singer, and D. J. Knecht (1983) 'Transfer of pulsation-related wave activity across the magnetopause; observation of corresponding spectra by ISEE-1 and ISEE-2', Geophys. Res., Lett. 10, 659.

Heikkila, W. J., and J. D. Winningham (1971) 'Penetration of magnetosheath plasma to low altitudes through the dayside magnetospheric cusps', J. Geophys. Res., 76, 883.

Kangas, J., A. Aikio, and J. V. Olson (1986) 'Multistation correlation of ULF pulsation spectra associated with sudden impulses', Planet. Space Sci. 34, 543.

Kato, Y., Y. Tonegawa, and K. Tomomura (1985) 'Dynamic spectral study of Pc 3-5 magnetic pulsations observed in the north polar cusp region', Mem. Natl. Inst. Polar Res. 38, 58.

Kwok, Y. C., and L. C. Lee (1984) 'Transmission of magnetohydrodynamic waves through the rotational discontinuity at the earth's magnetopause', J. Geophys. Res. 89, 10697.

Lanzerotti, L. J., C. G. Maclennan, L. V. Medford, and D. L. Carpenter (1986) 'Study of a QP/GP event at very high latitudes', J. Geophys. Res. 91, 375.

Luhmann, J. G., C. T. Russell, and R. C. Elphic (1986) 'Spatial distributions of magnetic field fluctuations in the dayside magnetosheath', J. Geophys. Res. 91, 1711.

Morris, R. J., and K. D. Cole (1987) 'Pc 3 magnetic pulsations at Davis, Antarctica', Planet. Space Sci. 35, 1437.

Nagata, T., T. Hirasawa, H. Fukunishi, M. Ayukawa, N. Sato, and R. Fujii (1980) 'ULF-VLF waves observed at the Syowa Station-Iceland conjugate pair', Mem. Natl. Inst. Polar Res. Spec. Issue, Jpn. 16, 25.

Newell, P. T., and C. -I. Meng (1988) 'The polar cusp and the cleft/boundary layer: Low altitude identification and statistical local time variation' J. Geophys. Res. 93 (in press).

Odera, T. J. (1986) 'Solar wind controlled pulsations, a review', Rev. Geophys. 24, 55.

Oguti, T., J. H. Meek, and K. Hayashi (1984) 'Multiple correlation between auroral and magnetic pulsations', J. Geophys. Res. 89, 2295.

Olson, J. V. (1986) 'ULF signatures of the polar cusp', J. Geophys. Res. 91, 10055.

Plyasova-Bakounina, T. A., Y. Y. Golikov, and V. A. Troitskaya (1978) 'Pulsations in the solar wind and on the ground', Planet. Space Sci. 26, 547.

Plyasova-Bakounina, T. A., Troitskaya, V. A., Muench, J. W., and H. F. Gauler (1986) 'Super-high-latitude maximum of Pc 2-4 intensity', Acta Geodaetica, Geophysica, et Montanistica Hungarica 21.

Potemra, T. A., L. J. Zanetti, and M. H. Acuna (1985) 'The AMPTE CCE magnetic field experiment', IEEE Trans. Geosci. Remote Sensing GE-23, 246.

Russell, C. T., and R. C. Elphic (1979) 'ISEE observations of flux transfer events at the dayside magnetopause', Geophys. Res. Lett. 6, 33.

Russell, C. T., and M. M. Hoppe (1983) 'Upstream waves and particles', Space Sci. Rev. 34, 155.

Russell, C. T., J. G. Luhmann, T. J. Odera, and W. F. Stuart (1983) 'The rate of occurrence of dayside Pc 3,4 pulsations: The L-value dependence of the IMF cone angle effect', Geophys. Res. Lett. 10, 663.

Saito, T., and S. Matsushita (1967) 'Geomagnetic pulsations associated with sudden commencements and sudden impulses', Planet. Space Sci. 15, 573.

Slawinski, R., D. Venkatesan, A. Wolfe, L. J. Lanzerotti, and C. G. Maclennan (1988) 'Transmission of solar wind hydromagnetic energy into the terrestrial magnetosphere', Geophys. Res. Lett. 15, 1275.

Takahashi, K., and R. L. McPherron (1982) 'Harmonic structure of Pc 3-4 pulsations', J. Geophys. Res. 87, 1504.

Taylor, W. W. L., B. K. Parady, P. B. Lewis, R. L. Arnoldy, and L. J. Cahill, Jr. (1975) 'Initial results from the search coil magnetometer at Siple, Antarctica', J. Geophys. Res. 80, 4762.

Tonegawa, Y., H. Fukunishi, L. J. Lanzerotti, C. G. Maclennan, L. V. Medford, and D. L. Carpenter (1985) 'Studies of the energy source for hydromagnetic waves at auroral latitudes', Mem. Natl. Inst. Polar Res. 38, 73.

Troitskaya, V. A., Plyasova-Bakounina, T. A., and A. V. Gul'elmi (1971) 'Relationship between Pc 2-4 pulsations and the interplanetary magnetic field', Dokl. Akad. Nauk SSSR 197, 1312.

Verzariu, P. (1973) 'Reflection and refraction of hydromagnetic waves at the magnetopause', Planet. Space Sci. 21, 2213.

Wolfe, A., and R. L. Kaufmann (1975) 'MHD wave transmission and production near the magnetopause', J. Geophys. Res. 80, 1764.

Wolfe, A., L. J. Lanzerotti, and C. G. Maclennan (1980) 'Dependence of hydromagnetic energy spectra on solar wind velocity and interplanetary magnetic field direction', J. Geophys. Res. 85, 114.

Wolfe, A., and A. Meloni (1981) 'ULF geomagnetic power near L = 4, 6, Relationship to upstream solar wind quantities', J. Geophys. Res. 86, 7507.

Wolfe, A., E. Kamen, L. J. Lanzerotti, C. G. Maclennan, J. F. Bamber, and D. Venkatesan (1987) 'ULF geomagnetic power at cusp latitudes in response to upstream solar wind conditions', J. Geophys. Res. 92, 168.

Yumoto, K., T. Saito, B. T. Tsurutani, E. J. Smith, and S.-I. Akasofu (1984) 'Relationship between the IMF magnitude and Pc 3 magnetic pulsations in the magnetosphere', J. Geophys. Res. 89, 9731.

Yumoto, K., T. Saito, S.-I. Akasofu, B. T. Tsurutani, and E. J. Smith (1985) 'Propagation mechanism of daytime Pc 3-4 pulsations observed at synchronous orbit and multiple ground-based stations', J. Geophys. Res. 90, 6439.

Yumoto, K., A. Wolfe, T. Terasawa, E. L. Kamen, and L. J. Lanzerotti (1987) 'Dependence of Pc 3 magnetic energy spectra at South Pole on upstream solar wind parameters', J. Geophys. Res. 92, 12437.

TRANSMISSION OF SOLAR WIND HYDROMAGNETIC ENERGY INTO THE HIGH LATITUDE MAGNETOSPHERE

A. WOLFE[1]
L. J. LANZEROTTI
C. G. MACLENNAN
R. SLAWINSKI[2]
D. VENKATESAN[2]
AT&T Bell Laboratories
600 Mountain Avenue
Murray Hill, NJ 07974, U.S.A.

ABSTRACT. This article highlights some of our recent research on the contributions of upstream hydromagnetic waves to geomagnetic fluctuations observed in the Pc 3-4 frequency range near magnetospheric cusp latitudes. Results presented show the dependence of Pc 3 occurrence at South Pole on solar wind speed and interplanetary magnetic field magnitude and direction. Although some evidence clearly supports the idea of an upstream wave source for high latitude, dayside Pc 3 energy, we conclude that other source(s) must be more important to account for most of the observed activity in the Pc 3-4 band. Further research is needed to clarify the different controls between broad-banded and nearly monochromatic wave occurrences as well as to relate these high latitude conclusions to observations made deeper within the magnetosphere.

1. Introduction

Energy is transmitted into the Earth's magnetosphere from the solar wind by a variety of mechanisms. One of the most obvious, and yet still not well understood, is the transmission of hydromagnetic energy across the magnetopause under the influence of the solar wind flow around the magnetosphere. Another mechanism arises from the sporadic interconnection of the interplanetary and geomagnetic field lines. Over the last several years our group has concentrated on the study of magnetospheric hydromagnetic energy at high geomagnetic latitudes, using ground-based and satellite-based instrumentation. Some of this work is contained in several papers (Wolfe et al., 1987; Yumoto et al., 1987; Lanzerotti et al., 1986, 1987; Lanzerotti and Maclennan, 1988; Slawinski et al., 1988; Bamber et al., 1988;

---

1. also at New York City Technical College

2. also at University of Calgary

P. E. Sandholt and A. Egeland (eds.), Electromagnetic Coupling in the Polar Clefts and Caps, 203–209.
© *1989 by Kluwer Academic Publishers.*

Lanzerotti, 1989) as well as in a recent review article by Arnoldy et al. (1988). In this paper, we provide a brief summary of several of the more important results from our work, with the expectation that it will be placed in the context of other papers presented in this volume in studies of high latitude geomagnetic phenomena.

The ground-based observations summarized here were made at South Pole station (SP), Antarctica, and Iqaluit (IQ), Northwest Territories, Canada. These locations are shown in Fig. 1, where the Antarctic continent has been mapped along geomagnetic field lines to the northern hemisphere (courtesy M. Rycroft and R. Greenwald). During local daytime, South Pole and Iqaluit are close to the latitude of the

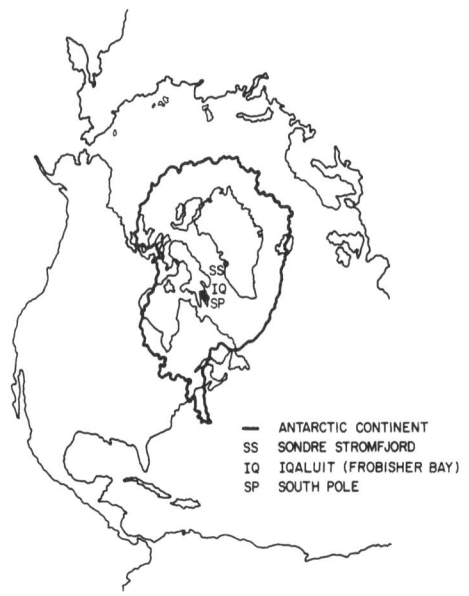

Figure 1. Antarctica mapped in geomagnetic coordinates to the northern hemisphere. The relative geomagnetic locations of Iqaluit (IQ), Sondre Stromfjord (SS) and South Pole (SP) are indicated. Figure courtesy of M. Rycroft and R. Greenwald.

geomagnetic cusp, but of course this location can vary considerably with geomagnetic activity. The instrumentation is described in several of the papers cited in the previous paragraph.

2.   Geomagnetic Power and Interplanetary Conditions

Studies of the influences of solar wind conditions on geomagnetic processes began in a quantitative way shortly after the discovery of the solar wind, with the investigations by Snyder et al. (1963) of the relationship of the geomagnetic activity index to the solar wind velocity. Investigations related specifically to hydromagnetic wave processes are contained in review articles ranging from those of Saito (1969) and Jacobs (1970), to the more recent reviews by Lanzerotti and Southwood (1979) and

Odera (1986). We have investigated the relationships of the power levels of hydromagnetic fluctuations in the classical "Pc" bands (f ∼ 0.001 − 0.1 Hz) as a function of interplanetary conditions. First, we have investigated the relationships between the power levels in two adjacent hydromagnetic bands classified as Pc 3 (period 10-45 sec.) and Pc 4 (period 45-150 sec.). The results in Fig. 2, taken from Wolfe et al. 1987, show that for data acquired during a three month interval at South Pole station the power levels in the two bands are closely related. The correlation coefficient for the number of points is significant to beyond the 99.9% confidence level. Indeed the correlation coefficient of r ∼ 0.9 suggests that at least 80% of the fluctuations in these bands are related to one another.

This observation is different from that reported by Wolfe et al. (1987) for a similar study of hydromagnetic power at lower latitudes (L ∼ 2). In this case, a much larger

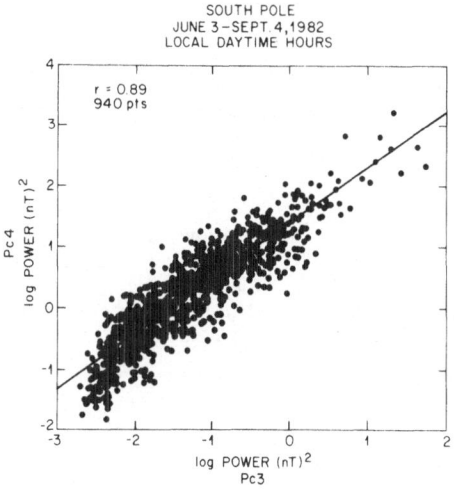

Figure 2. Scatter plot of hourly powers in Pc 4 versus Pc 3. A significant linear relationship is found for these broadband powers at South Pole.

spread was found between the power levels of the two bands; the linear correlation coefficient of ∼0.36 for 48 data points (statistical significance at the 98% confidence level) shows that they are not significantly related. A limitation in the Wolfe et al. study of the low latitude data was that the time interval examined (∼ 4 days) was much shorter than for the South Pole results shown in Fig. 2. This low latitude result obviously needs further investigation and confirmation. Such a study should consist of observations spaced in geomagnetic latitude as well.

Of considerable interest in hydromagnetic wave studies is the relationship of Pc 3 variations to interplanetary conditions. Much of this interest is motivated by the observation that ion cyclotron waves in the interplanetary medium are typically found to occur in this frequency range. The geomagnetic power levels in this frequency band were studied by Yumoto et al. (1987) as a function of solar wind velocity and the direction of the interplanetary magnetic field with respect to the Earth-Sun line (the so called interplanetary magnetic field (IMF) cone angle). These results are shown in Figure 3, where hourly average geomagnetic power levels in the Pc 3 band for hour 15-16 UT (11-12 geomagnetic local time) are plotted as a function of the hourly solar wind speed and IMF cone angle θ_{xB}. Here θ_{xB} is defined as

$$\theta_{xB} = \cos^{-1}(|B_x|/B).$$

Interplanetary conditions were monitored by the ISEE-3 spacecraft for the same three months of data as were analyzed in Fig. 2. A one hour time delay is included for the effects at ISEE-3 near the libration point to be observed on Earth. The linear correlation coefficients between the Pc 3 band power and θ_{xB} for the number of data points shown clearly demonstrate that the logs of the geomagnetic power levels are much more closely related to the solar wind speed in this band at these geomagnetic

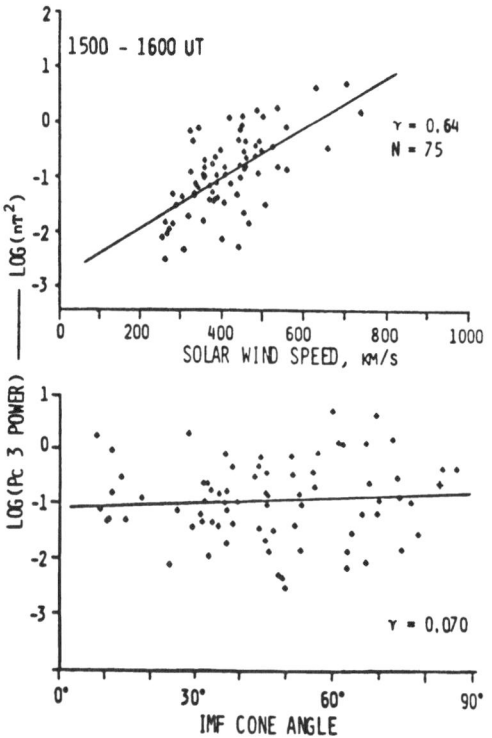

Figure 3. Correlations of V_{SW} and θ_{xB} derived from ISEE 3 data with broadband Pc 3 magnetic power at South Pole in the hourly interval near local noon, i.e., near the cusp.

latitudes than they are to the IMF cone angles. Indeed, the results for the IMF cone angle basically show that the broadband Pc 3 power is unrelated to the interplanetary field direction.

This relationship of the geomagnetic power to the solar wind velocity for this high latitude station is substantially different than is found for investigations at lower latitudes. There, much higher correlation with the interplanetary magnetic field direction is found. (For example, see such work as that contained in Gul'elmi, 1974; Greenstadt and Olson, 1976, 1979; Wolfe, 1980).

3. Hydromagnetic Waves and the Interplanetary Magnetic Field

In addition to studies of geomagnetic power as noted in the previous section, we have also conducted an investigation of more highly monochromatic signals in order to examine distinct waves that may be seen at the higher latitudes. Engebretson et al. (1986) showed that for ten distinct events identified visually from South Pole data (recorded near local noon) there was a good relation between the existence of these Pc 3 waves and the interplanetary magnetic field direction. That is, the visually selected waves were most evident during those times when the interplanetary field cone angle was the smallest; that is, the interplanetary field was directed radially from the Earth's magnetopause to the sun ($\theta_{xB} \sim 0$).

We have calculated the power spectra for South Pole and Iqaluit geomagnetic data at hourly intervals during local daytime hours for the interval July 25 - December 25, 1985. From the power we have subtracted a power law fit to each spectra and examined the peaks in the spectra in the Pc 3 band as a function of their amplitude and their Q-value (where we define Q-value as the ratio of the peak frequency to the frequency bandwidth at half peak amplitude). This procedure allows a quantitative evaluation of the data; for example, the data could be sorted as to amplitude power and as to the Q-value of the signal.

Slawinski et al. (1988) have extensively discussed this analysis procedure and the results of this study. Shown in Fig. 4 are the results, where the closed circles represent the selected data with frequency independent of interplanetary magnetic field magnitude. The open circles represent the selected data points for signals that had a high linear correlation value after eliminating the closed circle points. All of the data on this figure were selected with a power amplitude criteria of 4 $(nT)^2$/Hz. The selected events are found to be nearly monochromatic (high Q) pulsations. Two

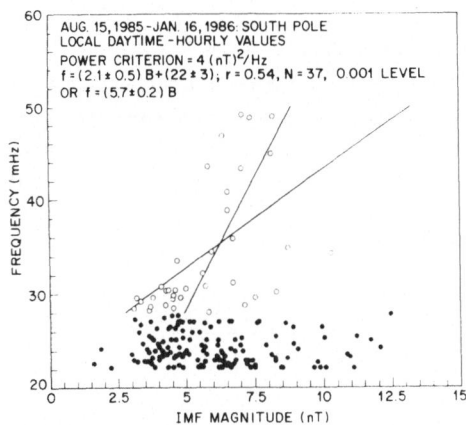

Figure 4. Peak frequency f in Pc 3 band measured at South Pole station vs. magnitude of IMF. Equations are given for best fit lines with and without intercept for events with $f \geq 28$ mHz. The line through the origin agrees reasonably with theory.

straight line fits to the data are also shown in the figure. The line with the lower slope is an unconstrained fit to the open circle points, with the frequency dependence given in the figure. Also shown is the correlation coefficient for 37 data points significant to the 99.9% confidence level.

Constraining the linear fit to the open circle points to pass through the origin of the frequency versus IMF plot yields the line with the steep slope and the functional dependence on the frequency as given in the figure, $f = (5.7 \pm 0.2)B$. This fit, constrained to pass through the origin, is consistent with the theoretical discussions of Gul'elmi (1974) as to the appropriate dependence of the ion cyclotron wave frequency on interplanetary magnetic field magnitude. It is also in agreement with the slope that Gul'elmi (1974) found in an analysis of Pc 3-frequency magnetic variations in chart records at a mid-latitude station. It is interesting to note that the analysis procedure used here provides fewer data points in the frequency-dependent branch than in the frequency-independent branch, a feature which is not apparent in the results of Gul'elmi (1974).

A similar study was done with data from IQ for the same time interval (approximately three weeks of data were missing from IQ because of instrument problems). The results obtained were similar to those shown here for the South Pole data, as would be expected from the near conjugacy of the stations along a magnetopause flux tube.

4. Summary

We have outlined herein several of the recent investigations of high latitude hydromagnetic wave phenomena as measured at ground stations whose flux tubes can pass near the dayside magnetopause region. Such studies are providing new results on the energy transfer process from the interplanetary medium across the magnetopause and into the magnetosphere proper. Our finding in Fig. 2 that Pc 4 and Pc 3 broadband powers are highly correlated at high latitudes implies either a single dominant source for both frequency bands or different wave sources excited by the same dominant control. Even though Fig. 4 shows a significant f vs. B relationship for nearly monochromatic pulsations, most of the events occur in the frequency independent portion with $f \lesssim 28$ mHz. Therefore, we find some support for the idea of an upstream wave source, but only for the high frequency Pc 3 events. No clear evidence has been reported by our group or in the literature which suggests an upstream wave source for Pc 4 events that affects the hydromagnetic wave structure near cusp latitudes. In light of Fig. 2, the dominant source for Pc 3 and Pc 4 must be other than upstream waves. Our work (Fig. 3 and Wolfe et al., 1987) suggests that the solar wind speed plays a significant role for the generation of Pc 3 and Pc 4 power at high latitudes. Our work is continuing and is being extended to correlative studies deeper within the magnetosphere.

ACKNOWLEDGEMENTS. The work at New York City Technical College was supported in part by the National Science Foundation under grant DPP-86-18074.

References

Arnoldy, R. L., L. J. Cahill, Jr., M. J. Engebretson, L. J. Lanzerotti and A. Wolfe, Review of hydromagnetic wave studies in the Antarctic, *Reviews of Geophysics*, **26**, 181, 1988.

Bamber, J. F., D. Venkatesan, L. J. Lanzerotti, A. Wolfe, K. Makita and C. G. Maclennan, Comparison of the modeled region of open field lines with the observed poleward boundary of auroral electron precipitation, *Annales Geophysicae*, **6**, 3, 265, 1988.

Engebretson, M. J., C. I. Meng, R. L. Arnoldy and L. J. Cahill, Jr., Pc 3 pulsations observed near the south polar cusp, *J. Geophys. Res.*, **90**, 8909, 1986.

Greenstadt, E. W., and J. V. Olson, Pc 3, 4 activity and interplanetary field orientation, *J. Geophys. Res.*, **81**, 5911, 1976.

Greenstadt, E. W., and J. V. Olson, Geomagnetic pulsation signals and hourly distributions of IMF orientations, *J. Geophys. Res.*, **84**, 1493, 1979.

Gul'elmi, A. V., Diagnostics of the magnetosphere and interplanetary medium by means of pulsations, *Space Sci. Rev.*, **16**, 331, 1974.

Jacobs, J. A., *Geomagnetic micropulsations*, Springer-Verlag, Berlin, 1970.

Lanzerotti, L. J., Conjugate spacecraft and ground-based studies of hydromagnetic phenomena near the magnetopause, *Advances in Space Research*, in press, 1989.

Lanzerotti, L. J., R. D. Hunsucker, D. Rice, L. C. Lee, A. Wolfe, C. G. Maclennan, and L. V. Medford, Ionosphere and ground-based response to field-aligned currents near the magnetospheric cusp regions, *J. Geophys. Res.*, **92**, 7739, 1987.

Lanzerotti, L. J., L. C. Lee, C. G. Maclennan, A. Wolfe, and L. V. Medford, Possible evidence of flux transfer events in the polar ionosphere, *Geophys. Res. Lett.*, **13**, 1089, 1986.

Lanzerotti, L. J. and C. G. Maclennan, Hydromagnetic waves associated with possible flux transfer events, *Astrophys. Space Sci.*, **44**, 279, 1988.

Lanzerotti, L. J. and D. J. Southwood, Hydromagnetic waves, in *Solar System Plasma Physics*, **3**, 109, North Holland, 1979.

Odera, T. J., Solar wind controlled pulsations: A review, *Rev. Geophys.*, **24**, 55, 1986.

Saito, T., Geomagnetic pulsations, *Space Sci. Rev.*, **10**, 319, 1969.

Slawinski, R., D. Venkatesan, A. Wolfe, L. J. Lanzerotti and C. G. Maclennan, Transmission of solar wind hydromagnetic energy into the magnetosphere, *Geophys. Res. Lett.*, **15**, 11, 1275, 1988.

Snyder, C. W., M. Neugebauer and U. R. Rao, The solar wind velocity and its correlation with cosmic ray variations and with solar and geomagnetic activity, *J. Geophys. Res.*, **68**, 6361, 1963.

Wolfe, A., Dependence of mid-latitude hydromagnetic energy spectra on solar wind speed and interplanetary magnetic field direction, *J. Geophys. Res.*, **85**, 5977, 1980.

Wolfe, A., E. Kamen, L. J. Lanzerotti, C. G. Maclennan, J. F. Bamber and D. Venkatesan, ULF geomagnetic power at cusp latitudes in response to upstream solar wind conditions, *J. Geophys. Res.*, **92**, 168, 1987.

Yumoto, K., A. Wolfe, T. Terasawa, E. L. Kamen and L. J. Lanzerotti, Dependence of Pc 3 magnetic energy spectra at South Pole on upstream solar wind parameters, *J. Geophys. Res.*, **92**, 12, 437, 1987.

ELF AND VLF WAVES IN THE POLAR CLEFTS AND CAPS

Edgar A. Bering, III
Physics Department
University of Houston
Houston, Texas, 77204-5504, U.S.A.

ABSTRACT. This paper reviews recent observations of electrostatic and electromagnetic waves with frequencies between 10 Hz and 30 kHz in the polar cusps, clefts and caps. Primary emphasis is placed on discussion of the observations. However, a brief presentation is made a case study of the convective beam amplification model of auroral hiss production in order to provide a context for the interpretation of the observations. In general, the observed waves were found to be associated with regions of field-aligned electron transport, both upward and downward. Modes and emission types similar to those seen elsewhere in the auroral zone are the dominant features of the wave spectra that were observed.

1. Introduction

This paper will review observations of extremely low frequency (ELF) and very low frequency (VLF) waves in the polar cusps, clefts and caps that have been reported since the previous NATO Advanced Research Workshop on the cusps and clefts in 1983. Papers at the previous session concentrated on ELF waves in the cusp and clefts (Maynard, 1984; Holtet *et al.*, 1984). This paper, therefore, will concentrate on the VLF band and the polar cap. Lengthier reviews covering these topics include papers by Fraser-Smith (1982), Arnoldy *et al.*, (1988), and Temerin (1988).

There are a number of possible organization schemes that can be used in a paper of this nature, such as sorting by region, altitude, source mechanisms or wave band. My choice will be to organize overall by wave band starting with the VLF band. Each section will be subdivided by region.

The study of plasma waves in space is important for many reasons, not the least of which is the intrinsic fascination of the problem. Often, a knowledge of the waves present is crucial to understanding a given precipitation or plasma transport process. In addition, plasma waves can be useful tools for remote sensing the state of the particle population that produced them. This paper is intended primarily to be a review of the data. However, since the production mechanism of some VLF waves provides a context for relating data to our picture of the cusp, a brief presentation of a recent advance in this area will be made. Since I have been charged with covering a large region in frequency and space, this paper will not try to make some central point of physics.

P. E. Sandholt and A. Egeland (eds.), Electromagnetic Coupling in the Polar Clefts and Caps, 211–228.
© *1989 by Kluwer Academic Publishers.*

2. VLF Waves

2.1 SOURCE MECHANISMS

Mechanisms that may produce VLF waves in the magnetosphere include the cyclotron instability driven by loss-cone anisotropy, coherently amplified Cherenkov emission and the cyclotron maser instability. The first instability occurs primarily on closed field lines at lower latitudes than we are concerned with here. The third instability is believed to be responsible for the production of the Z-mode waves that are seen at high altitude over the polar cap (Gurnett *et al.*, 1983) that will be discussed below. Since these Z-mode waves can propagate over large distances perpendicular to B in the polar cap, they have not yet been used as remote sensing tools and I will not, therefore, examine the instability further. The second instability, often called the convective beam amplification (CBA) model (Swift and Kan, 1975; Maggs, 1976), produces nearly electrostatic whistler mode waves propagating near the resonance cone. For down going waves produced on auroral field lines, this cone is quite narrow. Even for upgoing waves produced by the lower energy electrons in downward current regions, the resonance cone is defined well enough to allow ray-tracing to locate the source region for most events. Therefore, this second case is one where VLF waves can be used as a remote sensing tool provided we understand the production mechanism.

2.1.1 *The CBA Model.* One of the more significant advances in VLF waves studies during the past four years was the successful case study validation of the CBA model (Bering *et al.*, 1987). The model can be thought of as a two step process. A field-aligned beam of electrons such as the primary auroral beam or an upward moving return flux can easily have a parallel velocity greater than the phase velocity of whistler mode waves near the resonance cone. Under these circumstances, the beam will emit whistler mode waves via the Landau resonance, i.e. Cherenkov radiation. These emissions are then convectively amplified. Like most other plasma instabilities, the waves draw energy from a region of positive slope in the distribution function, in this case on the low energy side of the beam itself. Thus, the waves are produced with phase velocities parallel to B, moving in the same direction as and with a speed comparable to the parallel velocity of the emitting electrons. Downgoing auroral beam electrons produce the downward propagating "auroral" VLF hiss that is often observed at low altitudes near auroral arcs (Bering *et al.*, 1987 and references therein). The lower energy upgoing electrons that carry downward currents produce the upward propagating waves at larger resonance angles that are observed as VLF saucers (James, 1976).

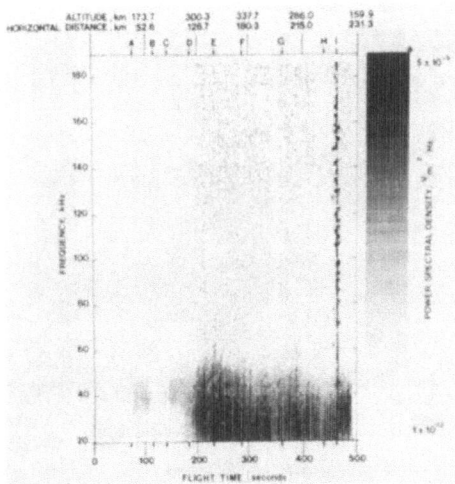

Figure 1. Spectrogram of the electric field from 20 to 185 kHz, plotted as a function of time from the launch of sounding rocket 29.007. Spectra are 4-s (10 spins of the rocket) average spectra of VLF swept frequency analyzer data (Bering *et al.*, 1987).

The case study reported by Bering *et al.* (1987) used data from a sounding rocket flight that passed over a quiet, evening sector auroral arc and descended into the evening sector polar cap (Robinson *et al.*, 1981). Figure 1 shows the electric field spectrum from 20 to 180 kHz observed over the entirety of the rocket flight. The arc was encountered at ~259 s and exited at ~365 s. The figure show an extended region of broad-band auroral hiss that was most intense just equatorward of the arc and extends for some distance poleward of the arc. Figure 2 shows the electric field spectrum observed from 0 to 8 kHz during the portion of the flight when the payload was apparently in the polar cap (Robinson *et al.*, 1981). In this figure, the lower edge of the VLF hiss band can be seen terminating at the local lower hybrid frequency (~7.5 kHz). Several spikes of sub-lower hybrid (sub-LHR) emission can be seen in the figure. Sub-LHR emissions such as these are relatively common in the polar cap (see below). However, the sources of these emissions remain to be explained. The intense ELF emissions seen at the bottom of the figure are presumably indicative of Doppler shifted plasma irregularities (Temerin, 1978, 1979).

In validating the CBA model, Bering *et al.*, considered the measured velocity distribution functions of electrons from 30 eV to 25 keV, such as the example shown in Figure 3 from the equatorial edge of the arc. An analytic model of the auroral electron distribution function (Maggs and Lotko, 1981) was successively least squares fit to the data from the entire flight, binned in four second intervals. These analytic models were then adiabatically convected back to the source altitude implied by the pitch angle distribution parameter used in the fit. At this altitude, the distribution functions found in a region a few kilometers thick on the equatorial edge of the arc appeared to have been unstable to VLF wave growth. The VLF power flux spectrum was then obtained by integrating the wave kinetic equation corresponding to the lowest order WKB approximation along

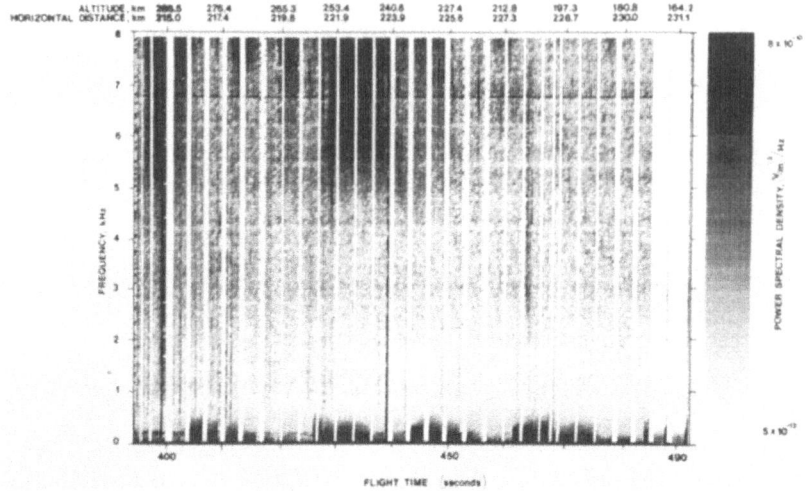

Figure 2. Detail of the last 90 s of Fig. 1, showing 1 spin average spectra.

ray paths in an inhomogenous auroral arc model. Figure 4 shows a comparison of the model spectra with the spectrum plotted at arrow "F" in Figure 1. Case 4, which used very reasonable arc parameters, shows excellent quantitative agreement with the data. Successful validations of this type are not common, as yet, in space plasma physics. This success gives us confidence in our ability to use this model to draw credible inferences about source region conditions from VLF waves observations.

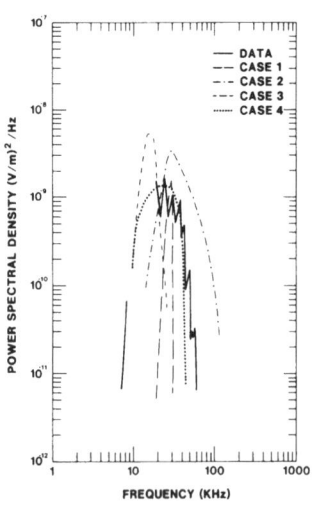

Figure 4. Comparison of the measured and predicted VLF power for the model discussed in the text (Bering et al., 1988).

215

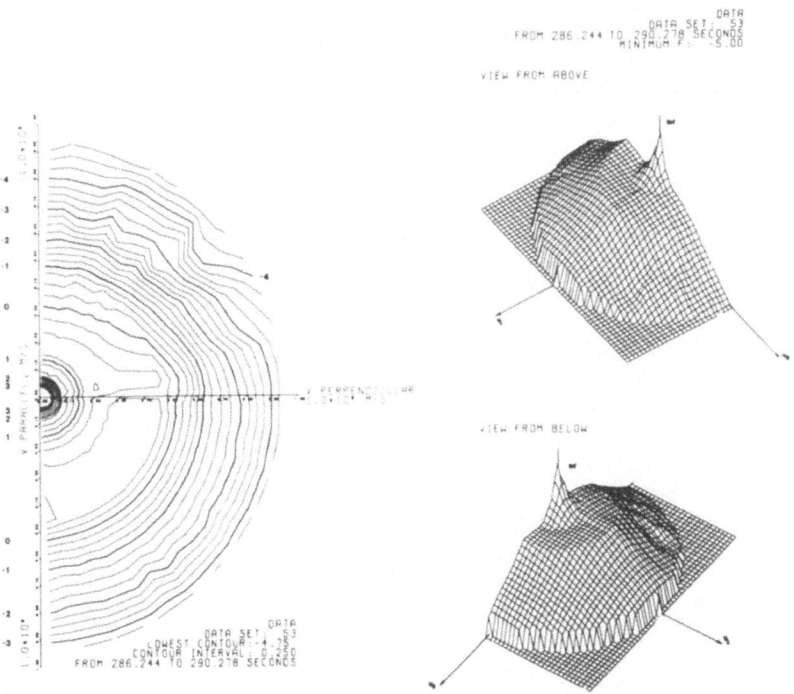

Figure 3. Contour map and two three dimensional views of the electron velocity distribution function seen on the same rocket flight as Figs. 1 and 2. The logarithm to the base 10 of the distribution function is plotted in units of s^3km^{-6}. The heavy contours labeled with the bold-faced numbers are an order of magnitude apart. The velocity axes on the contour plot extend to $\pm 1. \times 10^8$ m/s in $v_\parallel$ and from 0 to $1. \times 10^8$ m/s in $v_\perp$. The base of the 3-D view is at the integer logarithmic level given in the upper right hand corner and the peak is twelve decades above the base level (Bering *et al.*, 1987).

2.2 CUSP AND CLEFT VLF OBSERVATIONS

An example of the use of the CBA model in the interpretation of polar cusp data can be found in a recent statistical study of ground-based VLF data obtained at South Pole station (Tian *et al.*, 1988). This study used several years of ELF-VLF receiver data obtained at the cusp latitude ($\Lambda=74.5°$) South Pole station. The output from the South Pole ELF-VLF receiver is connected to 5 bandpass filters covering 0.5-1, 1-2, 2-4, 11-13, and 31-38 kHz. Power levels in these frequency bands were sampled at 10 second intervals and digitally recorded. Tian *et al.* used one minute averages converted to equivalent electric field strength. An event was said to have occurred in a given wave band in an hour if a single 1-minute sample exceeded 10 μV/m.

South Pole, 1986

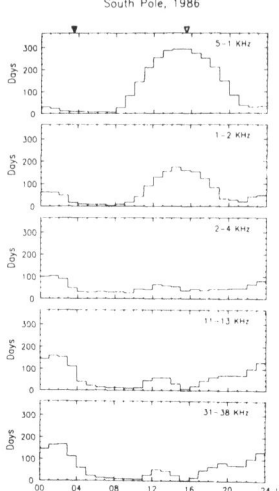

Figure 5. Hourly occurence statistics for five frequencies for intensities in excess of 10μV/m during 1986 at South Pole Station. Magnetic noon (midight) are denoted by the open (solid) triangle (Tian *et al.*, 1988).

South Pole, Winter, 1986

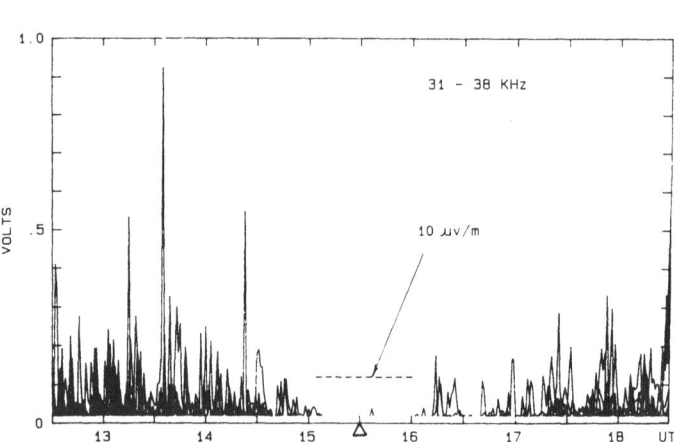

Figure 6. Superposition of VLF emission events in the 31-38 kHz band occurring in winter, 1986 at South Pole. Only the time interval around magnetic noon is shown (Tian *et al.*, 1988).

Figure 5 shows the number of days during an entire year when events above this threshold were found at South Pole for each one UT hour bin in each of these 5 frequency bands. Magnetic noon at South Pole occurs at 1530 UT. Figure 5 shows clearly that there is a peak in the occurrence of ELF activity near magnetic noon and gap in the occurrence of VLF activity at the same time. Figure 6 shows a superposed epoch plot of the data observed near noon in the 31-38 kHz channel for all 90 days of austral winter, 1986. (The authors presented all four seasons. I selected to reproduce the winter figure because of the relative absence of ionospheric absorption and consequent fading during winter.) The noon gap can seen in Figure 6 as a complete void in auroral hiss activity near local noon. Tian *et al.* suggest that the absence of VLF hiss is indicative of the absence of an accelerated auroral electron beam capable of producing the hiss. The authors go further and suggest that the absence of the auroral beam and the presence of elevated ELF activity are jointly indicative of the presence of the cusp over South Pole near noon.

It should be noted that this interpretation of the gap is not the only explanation that is possible. The VLF hiss spectra presented above have relatively steep upper bandedges that appear in the models to be related to the plasma frequency at production altitude. A decrease in this frequency below 31 kHz or any other downshift in the hiss band near magnetic noon (James, 1973) could also account for the gap. It should also be noted that the absence of VLF hiss in the vicinity of the cusp is a result that appears to be somewhat in contrast to the satellite results that will be discussed in the next few paragraphs.

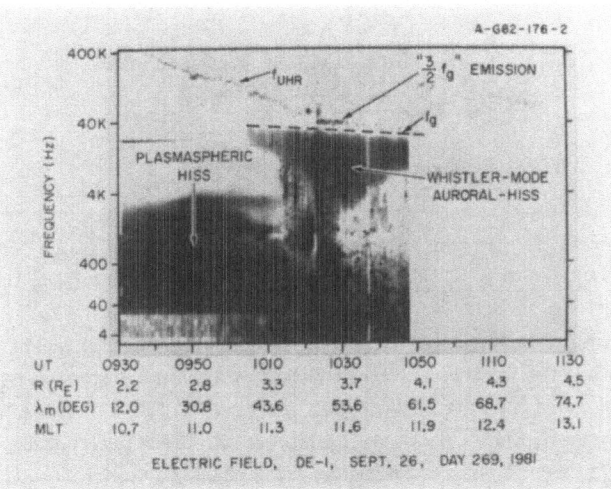

Figure 7. A dayside pass of the DE-1 spacecraft through the auroral zone showing a funnel-shaped auroral hiss event and $3f_g/2$ electrostatic emissions associated with the polar cusp (Gurnett *et al.*, 1983).

A typical example of the VLF spectrum of the cusp as seen by a high-altitude spacecraft is shown in Figure 7. This figure shows a spectrogram of the electric field measured by the DE-1 spacecraft as it passed over the noon sector auroral oval (Gurnett *et al.*, 1983). In the center of the figure we see the characteristic funnel shape signature produced by whistler mode auroral hiss propagating upward from a low altitude source. Ray tracing studies of similar events show that these funnel events are produced by lower altitude sources that are narrow width upgoing beams of low energy electrons (Lin *et al.*, 1984).

218

The magnetic local time of the observations in Figure 7 was ~1130 MLT, suggesting that these data were obtained in the vicinity of the cusp or cleft. The presence of upgoing VLF waves in the cusp does not disagree with the ground-based results of Tian *et al.*, (1988) which dealt only with down-going waves. The contrast is nonetheless quite interesting.

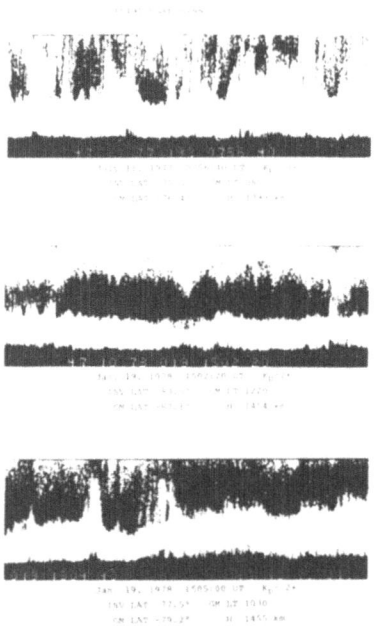

Figure 8. Typical spectra of polar cusp and cleft hisses observed in the dayside polar cap on July 11, 1977 ($K_p=3$) and on January 19, 1978 ($K_p=2-$) by the ISIS-2 (Ondoh *et al.*, 1981).

Another interesting data base that has been the source of several recent papers is the collection of ISIS I and II VLF receiver data recorded by a team of Japanese investigators at Syowa Station, Antarctica (Nagata *et al.*, 1980; Ondoh *et al.*, 1981). An example of Ondoh *et al.*'s results is shown in Figure 8. This figure presents three spectrograms of wideband data obtained during dayside passes under conditions of low to moderate activity. Judging from position alone (since no particle data were shown), the top and bottom panels were most likely from dawnside cleft passes. The middle panel is more likely to have been obtained during a cusp pass. Poynting vectors were not given in the paper. However, the relatively low altitude of the observations (~1400 km) and the absence of the "saucer" signature of upgoing waves (James, 1976) suggests that ISIS was seeing downgoing waves during these passes.

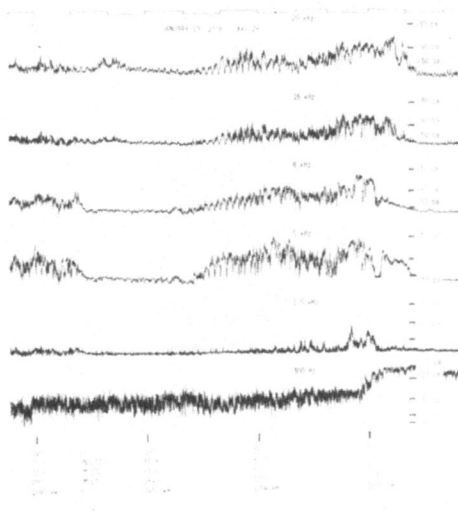

Figure 9. Narrow-band intensity data for the same ISIS pass as the middle and bottom panels of Fig. 8 (Ondoh *et al.*, 1981).

The observation by ISIS of downgoing VLF hiss in the cusp appears to be in direct contradiction to the results of Tian *et al.* (1988) discussed above. However, if one examines Figure 8, it appears that the hiss band in the middle panel has a lower frequency upper bandedge that does either of the two "cleft" hiss bands shown. This suggestion is confirmed by the data shown in Figure 9, which presents power levels in several bandpass filter channels for the same orbit as the bottom two panels in Figure 8. The power in the 20 kHz channel can be seen to have increased by more than 10 db in the 3 minutes between the middle panel at 1502 UT and the bottom panel at 1505 UT. The observation that the frequency range of the hiss band is lower in the cusp than elsewhere is consistent with previous results (James, 1973). Thus, the apparant contradiction between the results of Tian *et al.* and Ondoh *et al.* can be resolved if one attributes the midday gap found in the ground-based observations to a narrowing or lowering of the hiss band rather than to its absence.

2.3 POLAR CAP VLF OBSERVATIONS

I have found the literature on ground level observations of VLF emissions in the polar cap to be remarkably sparse. Apart from early work such as the studies by Ungstrup and Jackerott (1963) and Jørgensen (1966), very little has been done in this area. Recently, Fraser-Smith *et al.* (1987) have installed ELF-VLF radiometers at Thule, Greenland and Arrival Heights, Antarctica which should produce interesting data in the near future.

220

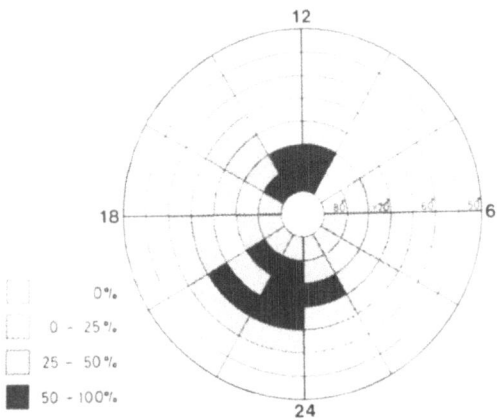

Figure 10. Polar plot of VLF hiss occurence percent observed by ISIS 2 during austral winter plotted in geomagnetic latitude – magnetic local time coordinates (Nagata *et al.*, 1980).

There certainly are VLF emissions to be seen in the polar cap, albeit not as many as at lower latitude. Figure 10 shows the percentage of the time that VLF hiss is seen over the austral polar cap during winter by a survey of spectrograms similar to those shown in Figure 8 (Nagata *et al.*, 1980). Clearly, hiss can be found at nominally polar cap latitudes some of the time.

Before inquiring as to the sources of this hiss, one must consider the apparently contradictory results obtained from a different survey of the same data base by Ondoh (1988). Rather than examining spectrograms, Ondoh used an automated technique to survey narrow-band data similar to that shown in Figure 9. He defined his events in terms of relative power increase in dB, and defined broadband hiss as the simulttaneous appearance of an event in all channels from 5 to 20 kHz. He found hiss defined in this fashion showed an occurence maximum in the low latitude polar cap at around $\Lambda=80°$, not in the auroral zone, a result that is in disagreement with essentially all the rest of the literature. I believe that the source of this discrepancy can be found in Ondoh's definition of hiss, one that includes a broad and inflexible bandwidth. Neither the data of Figures 1 and 2, nor the data of the middle panel of Figure 8 would meet Ondoh's criteria. In the first case, the LHR is above 5 kHz, a situation that is common, and in the second case, the upper bandedge is below 20 kHz. Nonetheless, both events can clearly be seen in the spectrograms to be examples of hiss observations.

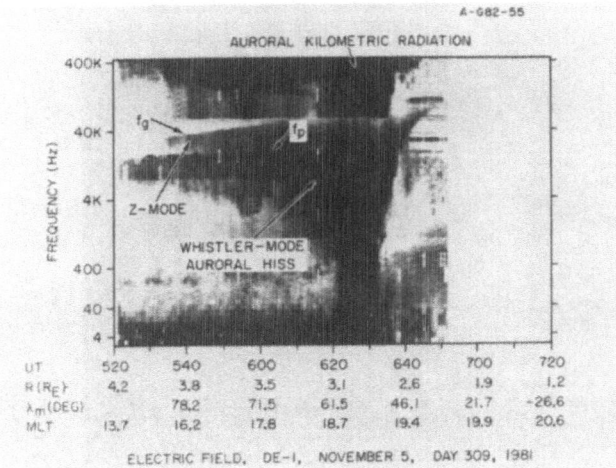

Figure 11. A representative spectrogram of electric field intensities for a nightside crossing of the auroral field lines by DE-1. The plasma density over the polar region is relatively low, with $f_p \ll f_g$ (Gurnett et al. 1983).

One of the more interesting questions in the study of VLF waves in the polar cap asks if there are any new emissions, waves modes, source mechanisms and source regions to be found in the polar cap. The general answer to these questions is a disappointing, and rather surprising "no". One major exception is a wave mode, the Z-mode, that can be seen at high altitude over the polar cap and that is not often identified elsewhere (Gurnett et al., 1983). The Z-mode is an electromagnetic mode that propagates in the band between the upper hybrid resonance and the L=0 cutoff. The polarization of the mode is R-X in the upper portion of the band and L-X in the lower portion. A typical Z-mode observation is shown in Figure 11 (Gurnett et al., 1983). This figure presents a spectrogram of the electric field measured by the DE-1 spacecraft as it crossed the polar cap. The various features of interest are labeled on the figure. A band of Z-mode radiation is noted over the polar cap. As the spacecraft moves toward the evening sector auroral zone, the Z-mode band blends into a region with the characteristic funnel shape of whistler mode auroral hiss. Ray tracing studies have shown that the Z mode emissions found in the polar cap have not been locally produced, but have instead been produced on auroral field lines at high altitude (Menietti and Lin, 1985, 1987). These waves are not, therefore, true polar cap emissions.

222

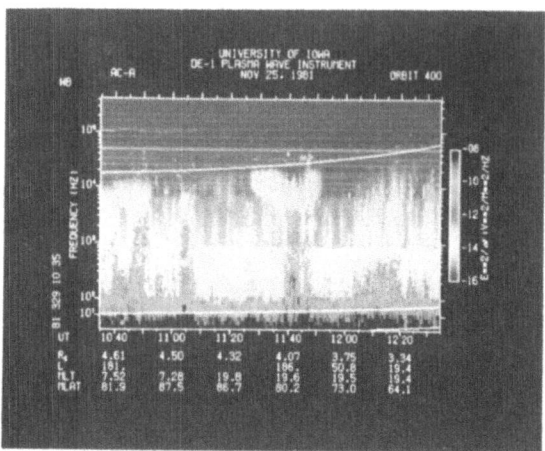

Figure 12. Spectrogram of electric field intensities during a traversal of the DE-1 spacecraft over the northern auroral zones and polar cap on November 8, 1981. The spacecraft traverses magnetic field lines threading the transpolar arc during 1546-1600 UT. Broadband electrostatic noise and a funnel-shaped auroral hiss event are observed in this time period (Frank *et al.*, 1986).

In fact, all of the polar cap VLF emissions that I have found in the literature where the particle data have also been presented appear to have had an auroral source of some sort, either by propagation in from the auroral zone, or by production by familiar auroral mechanisms taking place in polar cap arcs (e.g. Persoon *et al.*, 1983). Figures 12 and 13 show examples of polar cap arc emissions. The fountain feature in the middle of Figure 12, another electric field spectrogram obtained by DE-1, is indicative of whistler mode hiss being produced by a theta aurora (Frank *et al.*, 1986). Figure 13 shows the electric field wave spectrum and energy flux due to precipitating electrons measured by a sounding rocket that was traversing a polar cap arc (Weber *et al.*, 1988) in a situation somewhat akin to the event shown in Figures 1 and 2. The presence of VLF hiss near the arc is clear in this Figure. According to the model picture discussed above, this arc is producing VLF hiss at higher altitude that is refracting into the area around the arc. Cyclotron damping below the production region is thought to be responsible for the gap in the hiss within the arc itself. Since some authors do not consider polar cap arcs to be a genuine part of the polar cap (Chiu and Gorney, 1983), it is not clear that waves associated with these arcs should be considered to be polar cap waves in the strict sense of the term. There are some examples in the literature of what appears to be isolated bursts of hiss in the polar cap (e.g. Nagata *et al.*, 1980). However, the particle data needed to interpret the wave data were not presented. My own view is that it is most likely that all of these isolated polar cap hiss observations can ultimately be associated with the presence of polar cap precipitation.

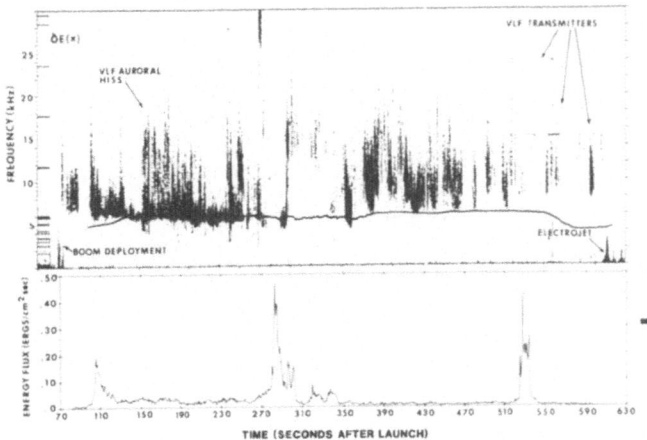

Figure 13. 0-30 kHz digital sonagram of the electric field waves data obtained by a Black Brant IX payload flown from Sondrestrom, Greenland across a polar cap arc. Also shown (solid line) is the local plasma lower hybrid frequency computed from the ion mass spectrometer measurements. The total electron energy flux is shown for reference (Weber *et al.*, 1988).

3. ELF Waves

3.1 CUSP AND CLEFT ELF OBSERVATIONS

The ELF band, as noted above, has been discussed more extensively at prior meetings than was the VLF. Furthermore, Temerin (1988) did an excellent job of covering this topic. Therefore, this section of this paper will be limited to the presentation of a single example, one that has been selected because it ties together an event (Erlandson *et al.*, 1988; Potemra, 1988) and a wave mode (Lysak and Temerin, 1987; Temerin, 1988) that were presented by others earlier in the meeting.

The event that I am refering to was a VIKING spacecraft encounter with the cusp that was noteworthy because of the pronounced diamagnetic depression of the total magnetic field strength that was observed (Potemra, 1988). Figure 14 presents spectra of components of B obtained near the time of local minimum in field strength (Erlandson *et al.*, 1988). The spectra show a peak between 18 and 27 Hz in the perpendicular components. These waves were left hand polarized and are quite similar to previously reporterd observations (Fredericks and Russell, 1973). They may have been an example of the Alfvén-ion cyclotron mode (Lysak and Temerin, 1987; Temerin, 1988; Boehm *et al.*, 1988). These waves are believed to be generated by the precipitating electrons in the cusp (Temerin and Lysak, 1984).

3.2 POLAR CAP ELF OBSERVATIONS

Relative to the auroral zone, the polar cap is a quiet region in the ELF (Gurnett *et al.*, 1984). However, Fraser-Smith (1982) has pointed out that "quiet" in this context usu-

224

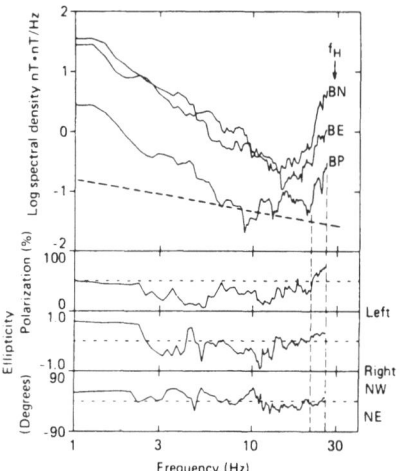

Figure 14. The magnetic field power spectral density from 2233:40 to 2233:45 UT, 24 July 1986 measured by the VIKING spacecraft during a high-altitude pass over the dayside auroral zone. The dashed line is an approximate background level determined equatorward of the dayside oval. The degree of polarization, ellipticity, and angle from north of the major axis of the polarization ellipse are also shown (Erlandson *et al.*, 1988).

ally implies the absence of discrete structured emissions that stand out noticeably above background and that essentially no attention has been given to quantifying and understanding this background. In polar cap electric field measurements, part of the background is due to electrostatic turbulence, which is ubiquitous in the polar cap (Kintner, 1988). Fraser-Smith *et al.* (1987) have begun a global program of magnetic background noise measurements that includes stations in the polar cap. A sample of preliminary results from this work is shown in Figure 15, which shows the diurnal variation in ELF noise observed on the ground from Arrival Heights, Antarctica on 3 February, 1987. Identifying the sources of these signals remains a task for the future.

Despite the relative quiet of the polar cap, I have found two examples in the literature of what appear to be a polar cap emissions in the ELF, one of which was shown above in Figure 13 and the other of which is presented in Figure 16 (Gallagher *et al.*, 1986). This figure shows an electric field spectrogram obtained at 3 R_E over the polar cap. The emissions that are shown in the Figure near 1 kHz were identified by Gallagher *et al.* as ion acoustic and lower hybrid emissions. Similar lower hybrid emissions can be seen in Figure 13. The source of the emission shown in Figure 16 is not well understood at present.

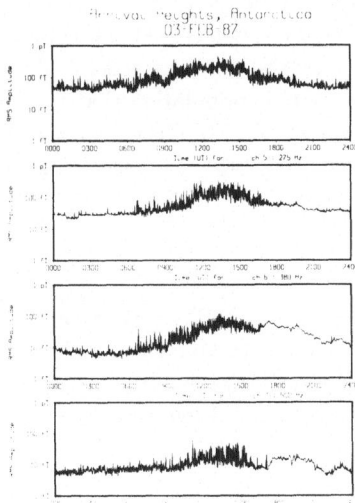

Figure 15. Variation of the one-minute rms average values of electromagnetic noise intensity in four narrow-band channels of the Arrival Heights, Antarctica ELF-VLF radiometer over the UT day 3 February 1987 (Fraser-Smith *et al.*, 1987).

Figure 16. Spectrogram of the electric field intensities observed by DE-1 during a polar cap pass on 14 October 1981 (Gallagher *et al.*, 1986).

4. Summary

I have presented several examples of what I consider to be some of the more interesting recent VLF and ELF wave observations from the polar cusps, clefts and caps. In general, the wave modes and emission mechanisms that have been discussed are similar if not identical to VLF and ELF modes and mechanisms that are found in the nightside auroral zone. The understanding of these waves that has been gained elsewhere is being used to interpret cusp, cleft and cap wave observations and make VLF waves into a useful tool for remote sensing the state of the particle population that produced them. There is no evidence yet available that the unique physical conditions of the cusps and polar caps give rise to any wave emission mechanisms that have not been found elsewhere.

Acknowledgments

I wish to thank Drs. R. Arnoldy, L. Cahill, R. Erlandson, A. Fraser-Smith, D. Gallagher, D. Gurnett, R. Helliwell, T. Hirasawa, M. Kelley, P. Kintner, J. Olson, T. Ondoh, T. Rosenberg, M. Temerin and E. Weber for providing figures and for useful discussions. This work was supported in part by National Science Foundation Grant DPP-8614091.

References

Arnoldy, R. L., L. J. Cahill, Jr., M. J. Engebretsen, L. J. Lanzerotti, and A. Wolfe (1988) 'Review of Hydromagnetic Wave Studies in the Antarctic', *Revs. Geophys.*, **26**, 181–207.

Bering, E. A., J. E. Maggs, and H. R. Anderson (1987) 'The plasma wave environment of an auroral arc. 3. VLF Hiss', *J. Geophys. Res.*, **92**, 7581–7605.

Boehm, M. H., C. W. Carlson, J. P. McFadden, J. H. Clemmons, and F. S. Mozer (1988) 'High resolution sounding rocket observations of large amplitude Alfvén waves', submitted to *J. Geophys. Res.*, **93**.

Chiu, Y. T., and D. J. Gorney (1983) 'Eddy intrusion of hot plasma into the polar cap and formation of polar-cap arcs', *Geophys. Res. Lett.*, **10**, 463.

Erlandson, R. E., L. J. Zanetti, T. A. Potemra, M. André, and L. Matson (1988) 'Observation of electromagnetic ion cyclotron waves and hot plasma in the polar cusp', *Geophys. Res. Lett.*, **15**, 421–424.

Frank, L. A., J. D. Craven, D. A. Gurnett, S. D. Shawhan, D. R. Weimer, J. L. Burch, J. D. Winningham, C. R. Chappell, J. H. Waite, R. A. Heelis, N. C. Maynard, M. Sugiura, W. K. Peterson, and E. G. Shelley (1986) 'The theta aurora', *J. Geophys. Res.*, **91**, 3177–3224.

Fraser-Smith, A. C. (1982) 'ULF/Lower ELF electromagnetic field measurements in the polar caps', *Revs. Geophys.*, **20**, 497.

Fraser-Smith, A. C., R. A. Helliwell, B. R. Fortnam, P. R. McGill, and C. C. Teague (1987) 'A new global survey of ELF/VLF radio noise', presented at the *NATO/AGARD Conference on Performance of Military Radio Communication Systems*, Lisbon, Portugal.

Fredericks, R. W., and C. T. Russell (1973) 'Ion cyclotron waves observed in the polar cusp', *J. Geophys. Res.*, **78**, 2917.

Gallagher, D. L., J. D. Menietti, J. L. Burch, A. M. Persoon, J. H. Waite, Jr., and C. R. Chappell (1986) 'Evidence of high densities and ion outflows in the polar cap during the recovery phase', *J. Geophys. Res.*, **91**, 3321.

Gurnett, D. A., R. L. Huff, J. D. Menietti, J. L. Burch, J. D. Winningham, and S. D. Shawhan (1984) 'Correlated low-frequency electric and magnetic noiose along the auroral field lines', *J. Geophys. Res.,* **89**, 8971–8985.

Gurnett, D. A., S. D. Shawhan, and R. R. Shaw (1983) 'Auroral hiss, Z mode radiation, and auroral kilometric radiation in the polar magnetosphere: DE-1 observations', *J. Geophys. Res.,* **88**, 329–340.

Holtet, J. A., S. Aasheim, A. Egeland, and P. E. Sandholt (1985) 'Low frequency waves at the dayside auroral oval', in J. A. Holtet and A. Egeland, (eds.), *The Polar Cusp,* D. Reidel Publishing, Dordrecht, pp., 323–335.

James, H. G. (1973) 'Whistler mode hiss at low and medium frequencies in the dayside cusp ionosphere', *J. Geophys. Res.,* **78**, 4578.

James, H. G. (1976) 'VLF saucers', *J. Geophys. Res.,* **81**, 501.

Jørgensen, T. S. (1966) 'Morphology of VLF hiss zones and their correlation with particle precipitation events', *J.Geophys.Res.,* **71**, 1367.

Kintner, P. M. (1988) 'A technique for fully specifying plasma waves', in press *Proceedings of the Yosemite Conference.*

Lin, C. S., J. L. Burch, S. D. Shawhan, and D. A. Gurnett (1984) 'Correlation of auroral hiss and upward electron beams near the polar cusp', *J. Geophys. Res.,* **89**, 925.

Lysak, R. L., and M. A. Temerin (1983) 'Generation of Alfvén-ion cyclotron waves on auroral field lines in the presence of heavy ions', *Geophys. Res. Lett.,* **10**, 643–646.

Maggs, J. E. (1976) 'Coherent generation of VLF hiss', *J. Geophys. Res.,* **81**, 1707.

Maggs, J. E., and W. Lotko (1981) 'Altitude dependent model of the auroral beam and beam-generated electrostatic noise', *J. Geophys. Res.,* **86**, 3439.

Maynard, N. C. (1985) 'Structure in the dc and ac electric fields associated with the dayside cusp region', in J. A. Holtet and A. Egeland, (eds.), *The Polar Cusp,* D. Reidel Publishing, Dordrecht, pp., 305–322.

Menietti, J. D., and C. S. Lin (1985) 'Ray tracing of Z-mode emissions from source regions in the high altitude auroral zone', *Geophys. Res. Lett.,* **12**, 385.

Menietti, J. D., and C. S. Lin (1987) 'Ray tracing survey of Z mode emissions from source regions in the high alitude auroral zone', *J. Geophys. Res.,* **91**, 13559.

Nagata, T., T. Hirasawa, H. Fukunishi, and N. Sato (1980) 'Selected results obtained at Syowa Station, Antarctica, by Reception of Kyokko and Isis', *Mem. Natl. Inst. Polar Res. (Jpn),* **16**, 84–94.

Ondoh, T. (1988) 'Polar occurence map of broad-band auroral hiss observed by ISIS satellites', *J. Radio Res. Lab.,,* **35**, 1–14.

Ondoh, T.,Y. Nakamura, and T. Murakami (1981) 'Characteristics of VLF saucers and auroral hisses from ISIS satellites received at Syowa', *Mem. Natl. Inst. Polar Res., (Jpn),* **18**, 54–71.

Persoon, A. M., D. A. Gurnett and S. D. Shawhan (1983) 'Polar cap electron densities from DE 1 plasma wave observations', *J. Geophys. Res.,* **88**, 10123.

Potemra, T. A. (1988) 'Large scale polar current systems', paper presented at the *NATO Advanced Research Workshop on Electromagnetic Coupling in the Polar Clefts and Caps,* Lillehammer, Norway.

Robinson, R. M., E. A. Bering, R. R. Vondrak, H. R. Anderson, and P. H. Cloutier (1981) 'Simultaneous rocket and radar measurements of currents in an auroral arc', *J. Geophys. Res.*, **86**, 7703.

Swift, D. W., and J. R. Kan (1975) 'A theory of auroral hiss and implications on the origin of auroral electrons', *J. Geophys. Res.*, **80**, 985.

Temerin, M. (1978) 'The polarization, frequency and wavelengths of high latitude turbulence', *J. Geophys. Res.*, **83**, 2609.

Temerin, M. (1979) 'Polarization of high latitude ionospheric turbulence as determined by analysis of data from the OV1-17 satellite', *J. Geophys. Res.*, **84**, 5935.

Temerin, M. (1988) 'Small scale, low frequency electric field structures in the polar cleft and cap and the acceleration of electrons and ions', paper presented at the *NATO Advanced Research Workshop on Electromagnetic Coupling in the Polar Clefts and Caps*, Lillehammer, Norway.

Temerin, M., M. Boehm, J. McFadden, C. W. Carlson and W. Lotko (1986) 'Production of flickering aurora and field-aligned electron flux by electromagnetic ion cyclotron waves', *J. Geophys. Res.*, **91**, 5769.

Temerin, M., and R. L. Lysak (1984) 'Electromagnetic ion cyclotron mode (ELF) waves generated by auroral electron precipitation', *J. Geophys. Res.*, **89**, 2849–2859.

Tian, B.-N., T. J. Rosenberg, U. S. Inan, D. L. Carpenter, and F. T. Berkey (1988) 'Observations of ELF/VLF emissions at the geographic South Pole', *EOS, Transactions, A.G.U.*, **69**, 444.

Unstrup, E., and I. M. Jackerott (1963) 'Observations of chorus below 1500 cycles per second at Godhavn, Greenland, from July 1957 to December, 1961', *J. Geophys. Res.*, **68**, 2141.

Weber, E. J., M. C. Kelley, J. O. Ballenthin, S. Basu, H. C. Carlson, J. R. Fleischman, D. A. Hardy, N. C. Maynard, R. F. Pfaff, P. Rodriguez, R. E. Sheehan, and M. Smiddy (1988) 'Rocket measurements within a polar cap arc: Plasma, particle and electric circuit parameters', in press, *J. Geophys. Res.*.

Pc 1-5 GEOMAGNETIC PULSATIONS AND 750 HZ ELF ACTIVITY AT GROUND LEVEL
IN THE NORTHERN AND SOUTHERN HEMISPHERES

Natsuo Sato
National Institute of Polar Research, 9-10, Kaga 1-chome,
Itabashi-ku, Tokyo 173, Japan

Thorsteinn Saemundsson
Science Institute, University of Iceland, Dunhaga 3, Reykjavik
107, Iceland

ABSTRACT. The statistical characteristics of the seasonal and diurnal
dependence of the occurrence of Pc 1-5 magnetic pulsations and 750 Hz
ELF emission are examined, using simultaneous observations at the
geomagnetically conjugate stations Syowa in Antarctica and Husafell and
Tjörnes in Iceland which lie in the auroral zones. The evidence of a
strong seasonal and diurnal dependence in the occurrence of emission
suggests that sunlight in the topside ionosphere, by causing an increase
in electron density, affects the propagation characteristics of waves
from the magnetosphere to the ground through the ionosphere. That is,
the wave intensity of Pc 3-5 magnetic pulsations is damped when the
waves propagate through the daylight ionosphere where the electron
density is high. On the other hand, the intensity of unstructured
Pc 1-2 band pulsations and 750 Hz ELF emission is damped drastically
when the waves propagate through the dark ionosphere where the density
is low.

1. INTRODUCTION

Unstructured Pc 1-2 band magnetic pulsations and 750 Hz ELF emissions
are believed to be generated by the mechanism of ion-cyclotron and
electron-cyclotron resonance near the equatorial plane on the higher
latitude side of the auroral zone and the polar cusp regions (Heacock,
1974; Troitskaya et al., 1980; Ungstrup and Jackerott, 1963; Egeland et
al., 1965). The generation of Pc 3-5 magnetic pulsations is thought to
be due to broad band noise in the magnetosphere which excites resonant
oscillations at specific frequencies, appropriate to particular field
line lengths and plasma densities (Chen and Hasegawa, 1974; Southwood,
1974).
    Simultaneous measurements at geomagnetically conjugate points
provide an important diagnostic tool to probe the separation of
magnetospheric and ionospheric phenomena. In this paper, we report on
the statistical characteristics of the seasonal and diurnal dependence

229

P. E. Sandholt and A. Egeland (eds.), Electromagnetic Coupling in the Polar Clefts and Caps, 229–238.
© *1989 by Kluwer Academic Publishers.*

of Pc 1-5 magnetic pulsations and 750 Hz ELF emissions, using simultaneous observations at Syowa in Antarctica (-69.0°, 39.6°; -66.2°, 71.4°, L=6.2 in geographic and corrected geomagnetic coordinates), and Husafell (64.7°, -21.0°; 65.9°, 69.4°. L=6.0) and Tjörnes (66.2°, -17.1°; 66.8°, 73.8°, L=6.5) in Iceland. The conjugate point of Syowa is located midway between Husafell and Tjörnes. The geomagnetic local time (MLT) is almost equal to Universal Time (within 20 min.) at the three stations (Ono, 1987). The characteristics of the seasonal and diurnal dependence of the waves observed at these conjugate stations may give an indication of the generation and propagation mechanisms of the waves in the magnetosphere and ionosphere.

2. Unstructured Pc 1-2 Band Emissions

A classification of Pc 1 magnetic pulsations in the frequency-local time domain has been reported by Fukunishi et al (1981), using data obtained at Syowa. We have now examined the seasonal and diurnal dependence of Pc 1 magnetic pulsations, using one year of simultaneous observations at the conjugate pair of stations Syowa and Husafell. The results reveal that unstructured Pc 1-2 band emissions show obvious seasonal and diurnal variations. For other types of pulsations, clear conjugacy is obtained. We shall therefore limit our discussion to the statistical characteristics of unstructured Pc 1-2 band emissions.

Figure 1 shows the frequency-time spectra of the H component of

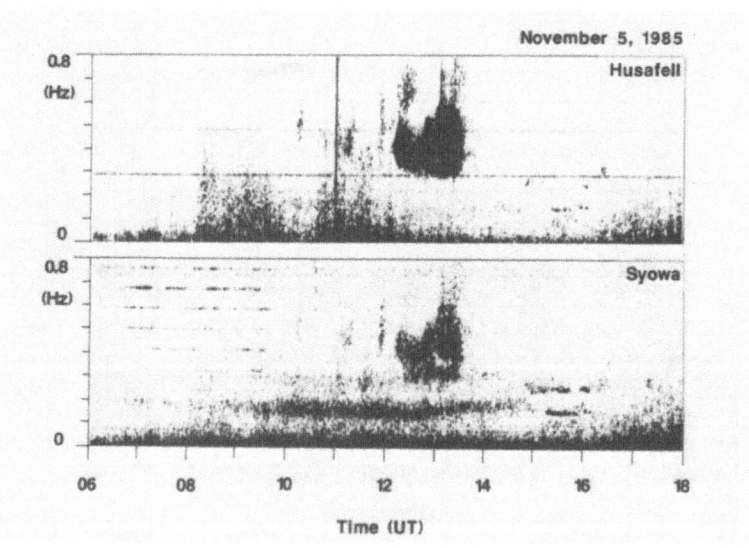

Figure 1. Frequency-time spectra of the H component of Pc 1 magnetic pulsations observed simultaneously at the geomagnetically conjugate stations of Husafell in Iceland and Syowa Station in Antarctica.

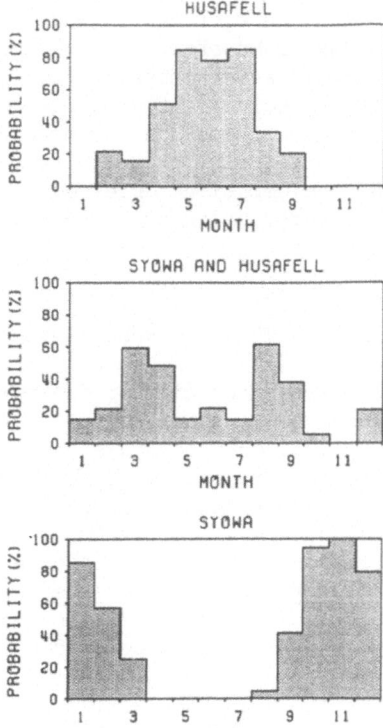

Figure 2. Seasonal dependence of the probability of occurrence of unstructured Pc 1-2 band emissions for nonconjugate events where emissions were received only at Husafell (Upper panel), conjugate events (middle) and non-conjugate events where emissions were received only at Syowa (bottom).

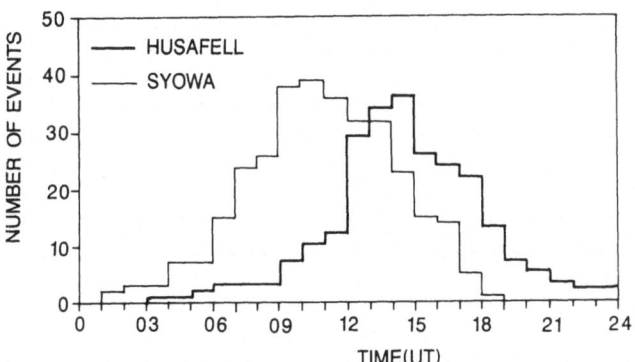

Figure 3. Diurnal dependence of the occurrence of unstructured Pc 1-2 band emission for non-conjugate events.

Pc 1 magnetic pulsations observed simultaneously at Husafell and Syowa on November 5, 1985. According to the classification by Fukunishi et al (1981), hydromagnetic (HM) chorus type emissions are observed simultaneously at the conjugate pair of stations at ~1210-1330 UT in the frequency range of ~0.3-0.8 Hz. On the other hand, unstructured Pc 1-2 band emissions are observed only at Syowa at ~08-15 UT in the frequency range of ~0.1-0.2 Hz. Figure 2 shows the seasonal dependence of

occurrence of unstructured Pc 1-2 band emissions. It is obvious that
the occurrence of non-conjugate cases (when emissions are received only
at Syowa or Husafell) is consentrated in the summer hemisphere, and that
conjugacy cases (emissions are received simultaneously at conjugate
stations) occur mostly in the equinox season. The diurnal variation of
occurrence of non-conjugate cases is shown in Figure 3. Emissions which
are observed only at Syowa are at a maximum around 10 UT. On the other
hand, the emissions observed only at Husafell have a maximum around
14 UT, accompanied by a reduction in morning occurrences in comparison
to Syowa.

3. PC 3-5 MAGNETIC PULSATIONS

The seasonal and diurnal dependence of Pc 3-5 magnetic pulsation power
are examined, using two years of simultaneous observations beginning on
November 5, 1985 at Syowa and Tjörnes. Figure 4 shows the ratio of the
dynamic power spectra of the H component at Syowa to those at Tjörnes,
averaged over four seasons. In this analysis, power spectra and
coherency are successively calculated with a time window of 20 minutes.
Power spectra with good coherency (>0.75) between the stations were
selected first, and then the power density, averaged over three months,
was derived for every 20 minutes in Universal Time. The color scale
indicates a ratio of power density at Syowa to that at Tjörnes, i.e.,
the red, yellow and blue colors show that wave power is relatively
higher at Syowa than at Tjörnes (red), almost the same at both stations
(yellow), and higher at Tjörnes than at Syowa (blue). From this figure,
it is found that there are two predominant spectral bands, the Pc 3
range with a frequency of ~20-60 mHz and the Pc 4-5 range with a
frequency of ~3-15 mHz. The power ratio of Pc 3-5 magnetic pulsations
between the stations shows obvious seasonal and diurnal dependence. The
power density is higher at Tjörnes than at Syowa during the northern
winter season. Conversely, it is higher at Syowa than at Tjörnes during
the northern summer season. It is noticeable during northern winter
that the power density is slightly higher at Syowa than at Tjörnes in
the longer period range of 5-10 mHz at ~03-05 UT and also in the range
of ~2-7 mHz at ~16-22 UT. Furthermore, during the morning in northern
winter, Pc 3 range magnetic pulsations show a power density which is
higher at Tjörnes than at Syowa by more than 0.4 dB. During northern
summer, the power density at Syowa is higher than at Tjörnes by less
than 0.2 dB. In the equinox season, the diurnal variation of the ratio
of wave power at the conjugate stations is approximately the same in
northern spring as in northern autumn. Pc 3 and Pc 5 pulsations which
have a higher power at Tjörnes than at Syowa appear in the morning at
~05-11 UT in the frequency range of ~20-60 mHz, and at ~05-11 UT in the
range of ~3-7 mHz, respectively. On the other hand, in the cases where
power density is higher at Syowa than at Tjörnes, Pc 3 pulsations appear
during the afternoon-evening (~13-21 UT) in the frequency range of
~20-40 mHz, and Pc 5 range pulsations appear during the early morning
(~00-04 UT) in the frequency range of ~6-10 mHz and during the
afternoon-evening (~12-20 UT) in the frequency range of ~4-7 mHz.

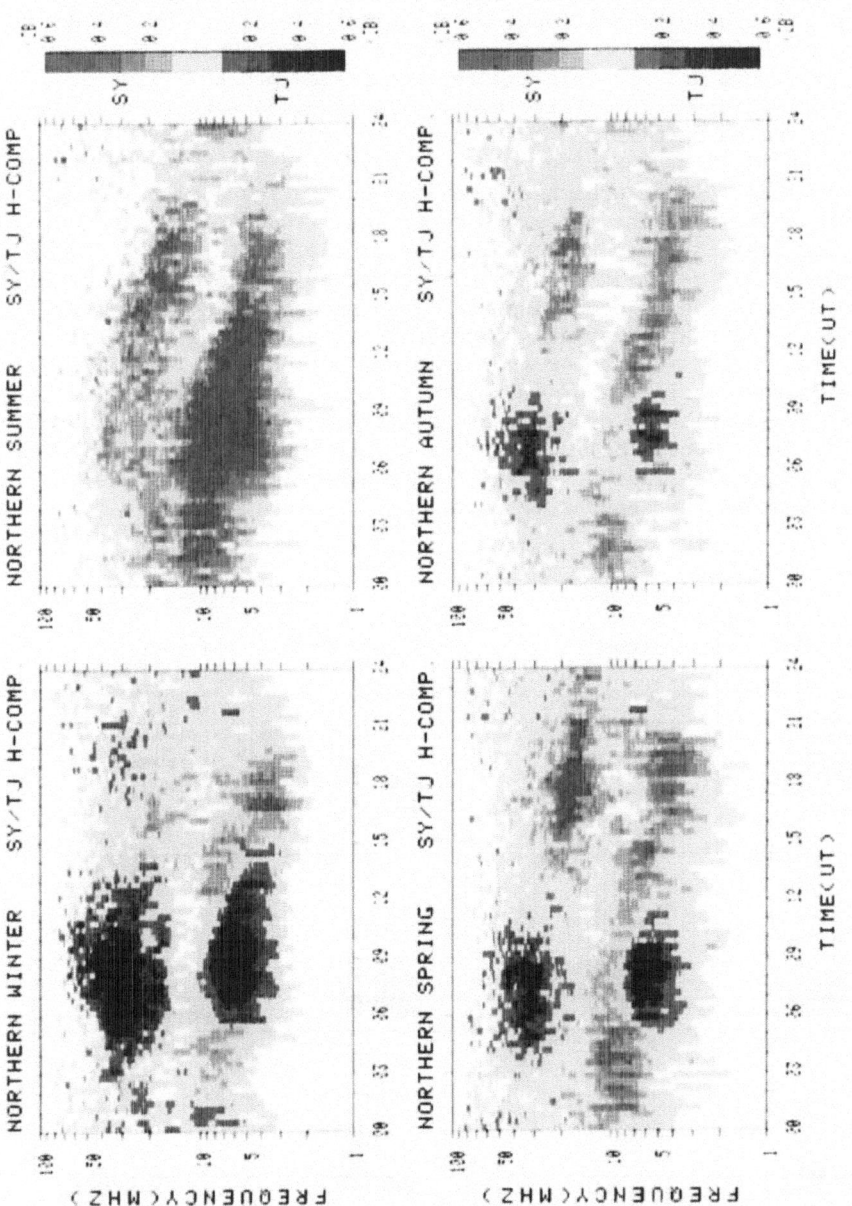

Figure 4. Ratio of the dynamic power spectra of the H component at Syowa to those at Tjörnes, averaged over three months, using 2 years of simultaneous observations. The color scale indicates the ratio of power density (dB) between the two stations.

234

4. 750 HZ BAND AND ELF EMISSION ACTIVITIES

Statistical characteristics of ELF emission occurrence have been
examined, using 2 years of digital recordings of 750 Hz intensity
obtained at two conjugate stations. The diurnal variations of 750 Hz
emissions and their relation to season, as observed at Husafell, are
displayed in Figure 5a. In this figure, occurrence probabilities are
displayed for different intensity levels. It is evident that emissions
occur mostly in the daytime and that the probability of occurrence is
highest around 12-13 UT. Furthermore, it is obvious that the occurrence
of higher intensity levels reaches a maximum during the summer season.
It is worth noting that the emissions occur less frequently in winter
than in other seasons. Figure 5b shows the diurnal variation of 750 Hz
emissions and their relation to season as observed at Syowa Station.
The seasonal variation of occurrence at Syowa shows almost the same
characteristics as that observed at Husafell. That is, the emissions
show a maximum during the summer and a minimum during the winter season.
As for the diurnal variation of emissions, the maximum is around
11-12 UT at Syowa, with an emphasis on the morning period when compared
to Husafell. It is worth noting that the occurrence probability reaches
a maximum 1-2 hours earlier at Syowa than at Husafell.

Figure 6 shows the seasonal variation of relative emission

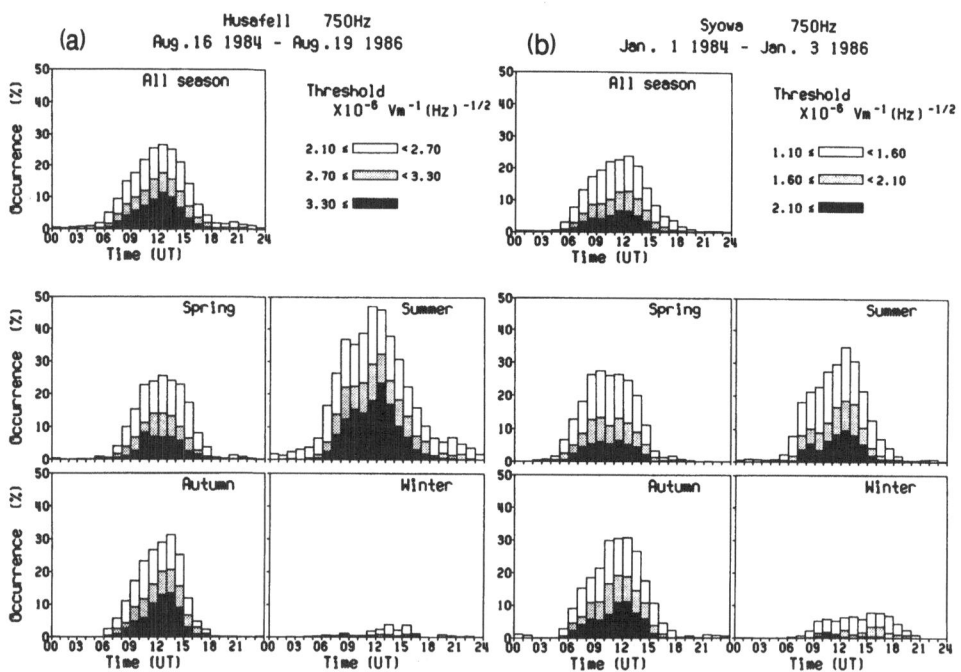

Figure 5. Seasonal characteristics of the diurnal variation of 750 Hz
emission observed at (a) Husafell and (b) Syowa Station.

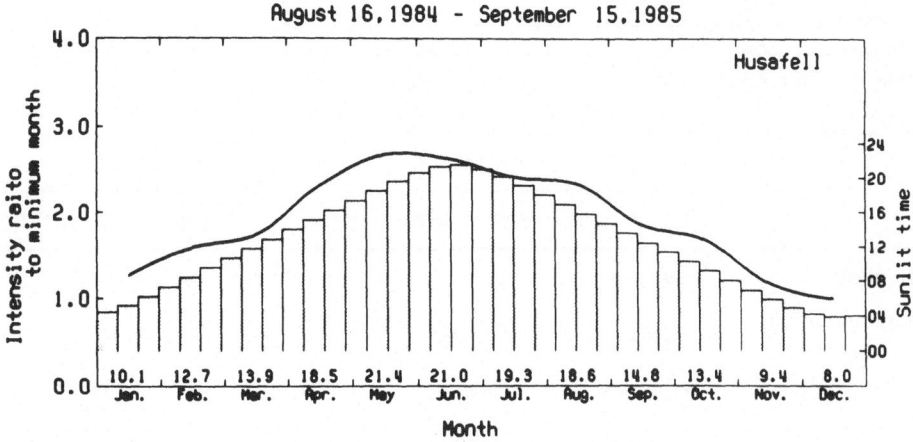

Figure 6. Seasonal variations of emission intensity at Husafell, taking
as unity the emission intensity of the minimum month and the hours of
sunlight over Husafell.

intensity, i.e., intensity relative to the minimum monthly average for
observed during 10-14 MLT at Husafell also shown on the hours of
sunlight in the ionosphere over Husafell. It is clear that the emission
intensity is closely related to the time of sunlight.

5. Summary and Discussion

Seasonal and diurnal dependence of different types of electromagnetic
waves observed at the Syowa-Iceland conjugate stations are summarized in
Table 1. Some kinds of emission shown in the table are not discussed in
this paper.
        A knowledge of the seasonal and diurnal dependence of
electromagnetic wave occurrences and their geomagnetic conjugacy is
important for the study of the general conditions for the generation and
propagation of these waves. The strong seasonal and diurnal dependence
of the occurrence of emission described above suggests that the sunlight
in the topside ionosphere, by causing an increase in electron density,
affects the propagation of waves from the magnetosphere to the ground
through the ionosphere. That is, the wave intensity of Pc 3-5 magnetic
pulsations is damped when the waves propagate through the daylight
ionosphere where the electron density is high (summer hemisphere). On
the other hand, the emission intensity of unstructured Pc 1-2 band
emissions and 750 Hz ELF emissions is depressed drastically when the
waves propagate through the dark ionosphere where the density is low
(winter hemisphere).
        The diurnal variation of electromagnetic wave intensity at
conjugate stations during equinox is also explained by the effects of
sunlight in the ionosphere. One of the important points which should be
mentioned here is the difference in magnetic local time of maximum
occurrence in the conjugate regions. The geographic local time (LT) at

Table 1: Summary of Seasonal and Diurnal (during equinox season) Dependence of Natural Electromagnetic Wave Occurrence Observed at Geomagnetic Conjugate Stations in the Auroral Zones, in Iceland and at Syowa in the Antarctic.

Type	Seasonal Dependence		Diurnal Dependence	
	Max.season	Min.season	Morning(MLT)	Afternoon(MLT)
HM chorus	equinox	?	IC$\geq$SY	IC$\leq$SY
Unstructured Pc 1-2 band	summer	winter	IC<<SY	IC>>SY
Pc 3 (H-Comp)	equinox	summer	IC>>SY	IC$\leq$SY
Pc 3 (D-Comp)	equinox	summer	IC$\geq$SY	IC$\leq$SY
Pc 5 (H-Comp)	equinox	summer	IC>>SY	IC<<SY
Pc 5 (D-Comp)	equinox	summer	IC$\geq$SY	IC$\leq$SY
750 Hz emission	summer	winter	IC<<SY	IC>>SY
2 kHz emission (daytime)	?	?	IC$\geq$SY	IC$\leq$SY
4 kHz emission (daytime)	winter	summer	IC>>SY	IC>SY

Syowa is approximately MLT (~UT) plus 3 hours and that at Iceland is approximately MLT (~UT) minus 1 hour. Thus the diurnal dependence of the difference in Pc 3-5 pulsation power at the two stations may be ascribed to the difference in geographic local time (solar zenith angle). Pc 3-5 pulsation at stations in Iceland tend to be strongest relative to that at Syowa around 09 UT, at the time of geographical noon (~12 LT) at Syowa, On the other hand, the relative power maximum at Syowa is around 16 UT when the sun has set at Syowa (~19 LT) but Iceland is still sunlit (~15 LT). It is suggested that sunlight effects in the ionosphere may control the Pc 3-5 magnetic pulsation intensity observed at ground level even if MLT and season are the same at the conjugate stations.

Previous studies of the diurnal dependence of Pc 3-5 magnetic pulsation power at conjugate stations near L=4 have been examined by Lanzerotti and Robbins (1973) using 3 days of data near the December solstice. They demonstrated that the power density was largest at the daylight station of Siple in Antarctica, and suggested that magnetic pulsation power may be associated with the S_q current system through field-aligned currents because field-aligned current density is larger in the daylight hemisphere as reported by Fujii et al (1982). Our results obtained in the auroral zone (L~6) are completely opposite to the results obtained in the sub-auroral zone (~4) by Lanzerotti and Robbins. That is, our results cannot be explained by the enhancement of field-aligned currents. Instead, we must postulate a strong damping or shielding mechanism for Pc 3-5 pulsations when the waves propagate from the magnetosphere to the ground through the ionosphere in the daylight hemisphere.

The characteristics of the seasonal and diurnal dependence of the occurrence of unstructured Pc 1-2 band and 750 Hz ELF emissions is opposite to that of Pc 3-5 magnetic pulsations. Unstructured Pc 1-2 band and ELF emission intensities show a minimum in winter. The diurnal variation shows that the emission of Pc 1-2 band and 750 Hz ELF reaches maximum about 1-4 hours earlier at Syowa than at Husafell. Why the emission shows such a strong seasonal variation is still unknown, but some possibilities will be discussed here. The plasma produced by sunlight in the summer hemisphere may propagate to the other hemisphere along the geomagnetic field lines as a polar wind. Thus the asymmetry of ambient plasma density between the northern and southern hemispheres near the equatorial plane in the magnetosphere may be cancelled quickly. According to a previous study (Ito et al., 1986), the occurrence of emission is also related to geomagnetic activity. However, we could distinguish the effect of geomagnetic activity from the seasonal variation of emission using simultaneous data observed at conjugate stations (not shown in this paper).

Wave absorption effects in the ionosphere are difficult to identify from data observed on the ground. However, satellite data indicate a strong seasonal dependence of ELF emission intensities between the winter and summer hemispheres in the topside ionosphere (Kelly et al., 1975). That is, the intensities observed in the summer hemisphere are three times as strong as those in the winter hemisphere. Measurements by satellites give almost the same results, statistically, as those obtained on the ground. This suggests that the effect of the seasonal variation of wave absorption may be weak relative to the seasonal asymmetry of occurrence or propagation effects. The effects of asymmetry of wave duct enhancement and wave propagation from the magnetosphere to the ionosphere provide a more reasonable explanation of the seasonal variation of emission intensities because a density irregularity and a density gradient along the geomagnetic field line may be produced in the sunlit hemisphere.

We have now examined and discussed the seasonal and diurnal dependence of wave occurrences observed at geomagnetically conjugate stations in the auroral zones. Our conclusion is that electromagnetic wave phenomena observed on the ground are strongly controlled by sunlight effects. Sunlight may affect the asymmetry of wave duct enhancement and wave propagation from the magnetosphere to the ionosphere in both hemispheres. On the basis of this conclusion, we suggest that electromagnetic wave phenomena observed in the polar cusp and cleft regions on the ground may exhibit even stronger sunlight effects because most of these regions are located at a higher geographic latitude and have longer periods of daylight and darkness than the auroral zones. The Universal Time dependence in this region is also important because the difference of location between the geographic pole and geomagnetic pole causes an asymmetry of interaction between the solar wind and the magnetosphere.

ACKNOWLEDGEMENTS

We thank all members of the 24-27th Japanese Antarctic Expedition for the data acquisition at Syowa Station. We are also grateful to S. Johannesson, A. Egilson, J. Marvinson and J. Sveinsson for their careful support in maintaining the system in Iceland. We thank all members of the Upper Atmosphere Physics and Data Analysis Division, NIPR, for their kind support. The project in Iceland is supported by a Grant-in-Aid for Overseas Scientific Survey 60041085 and 63041130 and for Science Research B (61460051) from the Ministry of Education, Science and Culture, Japan.

REFERENCES

Chen, L., and Hasegawa, 'A theory of long-period magnetic pulsations, 1. Steady state excitation of field line resonance', J. Geophys. Res., 79, 1924, 1974.

Egeland, A., G. Gustafsson, S. Olsen, J. Arons, and W. Barron, 'Auroral emissions centered at 700 cycles per second',J. Geophys. Res., 70, 1079, 1965.

Fujii, R., T. Iijima, T. A. Potemra, and M. Sugiura, 'Seasonal dependence of large-scale Birkeland currents, Geophys. Res. Lett., 8, 1103, 1981.

Fukunishi, H., T. Toya, K. Koike, M. Kuwashima, and K. Kawamura, 'Classification of hydromagnetic emissions based on frequency-time spectra', J. Geophys. Res., 86, 9092, 1981.

Heacock, R. R., 'Midday Pc 1-2 pulsations observed at subcleft location', J. Geophys. Res., 79, 4239, 1974.

Ito, K., S. Shibuya, K. Maezawa, and N. Sato, 'Statistical characteristics of 750-Hz band ELF emissions observed at Syowa Station', Mem. Natl Inst. Polar Res., Spec. Issue, 10, 1986.

Kelly, M. C., B. T. Tsurutani, and F. S. Mozer, 'Properties of ELF electromagnetic waves in and above the earth's ionosphere deduced from plasma wave experiment on the OVI-17 and OGO 6 satellite, J. Geophys. Res., 80, 1975.

Lanzerotti, L. J., and M. F. Robbins, 'ULF geomagnetic power near L=4, 1. Quiet day power spectra at conjugate points during December solstice, J. Geophys. Res., 78, 3816, 1973.

Ono, T., 'Temporal variation of geomagnetic conjugacy in Syowa-Iceland pair', Mem. Natl Inst. Polar Res., Spec. Issue, 48, 46, 1987.

Southwood, D. J., 'Some features of field line resonance in the magnetosphere', Planet. Space Sci., 22, 483, 1974.

Troitskaya, V. A., O. V. Bolshakova, and E. T. Matveeva, 'Geomagnetic pulsations in the polar cap', J. Geomag. Geoelectr., 32, 309, 1980.

Ungstrup, E., and L. M. Jackerott, 'Observation of chorus below 1500 cycles per second at Godhaven, Greenland, from July 1957 to December 1961', J. Geophys. Res., 68, 2141, 1963.

GROUND MAGNETIC PERTURBATIONS
IN THE POLAR CAP AND CLEFT:
RELATIONSHIP WITH THE IMF

E. FRIIS-CHRISTENSEN
Danish Meteorological Institute
Division of Geophysics
100 Lyngbyvej
DK-2100, Copenhagen
Denmark

ABSTRACT.   Average empirical models of the magnetospheric and ionospheric electrodynamic parameters play an important role as a frame of reference and also as boundary conditions for the more advanced dynamical models. Furthermore the empirical models may be used to test our physical models and concepts. With the vast amount of geophysical and solar data which is now available, a number of empirical models have appeared, but the different empirical models are not all comparable due to a difference in experimental methods, averaging techniques, and observing conditions. In this review we will describe various average models of electric fields and currents and compare them with models obtained using ground-based magnetometer measurements. It is our conclusion that the major differences between the models are not primarily due to the different measuring techniques, but can be explained taking the merits and demerits of the different averaging and normalization techniques into consideration. This is important to take into account when trying to understand the physics, which the empirical models represent.

1. Introduction

Magnetospheric physics today is in a state where there is a need for an instantaneous electrodynamic model of the solar wind-magnetosphere coupling. In spite of that, the quality of our models and the amount of synoptic data available is still too limited to provide us with a tool which can be used to predict the time-development of geophysical parameters, particularly in the high-latitude magnetopause and its projection along the field-lines to the ionosphere. A minimum requirement regarding proposed instantaneous models is that they should incorporate, as a special case, the average patterns of high-latitude electric fields and currents and their relationship with the IMF. There is therefore a need for average empirical models which can be used as a frame of reference for the more advanced dynamical models. Furthermore, average models of the magnetospheric and ionospheric electrodynamic parameters still contain features which cannot be immediately understood in terms of basic physical processes. Some of these features are undoubtedly due to deficiencies in our physical models, but some of the features may be related to limitations inherent in the various measurements and techniques used to provide an average model.

In a companion paper (Friis-Christensen, 1989) the structure and dynamics of the ionospheric current systems have been described with particular emphasis on those aspects which are difficult to resolve in average models. In the present paper we will deal with the

P. E. Sandholt and A. Egeland (eds.), Electromagnetic Coupling in the Polar Clefts and Caps, 239–251.
© 1989 by Kluwer Academic Publishers.

average models of electric fields and currents, since, in spite of the fact that many data are now available, the interpretation of the various results is still not unambiguous. One of the problems with the different empirical and theoretical models is that they are not all comparable due to different experimental methods, averaging techniques and observing conditions. It is our conclusion that the major differences in the models are not primarily due differences in the measuring techniques. Different averaging and normalization techniques have a significant effect on the the empirical models, and this aspect should not be neglected in the geophysical applications of the models.

The average electric field and current models have been based primarily on measurements of magnetic fields, electric fields and ion drifts, on satellites as well as from the ground using radar probing systems and magnetometer chains. This review will focus on the results obtained using ground-based magnetometers, but comparisons with results obtained using different measurements will be a major part of the review. In section 2 average magnetic perturbation patters will be described in terms of equivalent current patterns. In section 3 electric potential patterns derived from the equivalent current systems will be compared with more direct measurements of electric fields. The equivalent current patterns together with average conductivity models may be used to derive field-aligned or Birkeland current patterns which in section 4 are compared with patterns derived using magnetic field measurements on board low-altitude satellites. Various merits and demerits which characterize the different averaging techniques will be discussed in section 5. It is concluded, that no single averaging technique is sufficiently satisfactory to be used in the derivation of an empirical electrodynamic model. There is therefore a need to develop more sophisticated ways of averaging the large amount of geophysical and solar data which is now available.

2. Average Equivalent Current Systems for Various IMF Orientations

For a long time ground-based magnetometer measurements provided the primary source of information which could be used to derive global models of magnetospheric electrodynamics. It was, however, not until the first measurements from a satellite outside the magnetosphere, in the solar wind, that it was possible to relate the variations in the geomagnetic perturbation pattern directly to external sources. Traditionally ground-magnetic measurements are displayed in form of equivalent currents although it is well-known that the real current system is three-dimensional, with a significant contribution from Birkeland currents flowing from the magnetosphere into the ionosphere along the magnetic field-lines. Equivalent current systems are still used as the first step in the derivation of the real current system using advanced computer codes developed during the International Magnetosphere Study (IMS) when a number of meridian chains in the polar regions were established.

Most of the published results using ground-based magnetometer data are based upon data from the Northern Hemisphere, where the number of observatories is larger, and the distribution of stations is more favorable for statistical analysis. In 1975, however, a meridian chain of magnetometer stations from $-75.6°$ to $-88.4°$ corrected geomagnetic latitude was established in Antarctica (Mansurov et al., 1981). Using four months of data from this meridian chain Papitashvili et al. (1983) performed a regression analysis of the coupling between the average hourly values of the B_x, B_y, and B_z components of the IMF and the horizontal components of the magnetic perturbations. The results of their

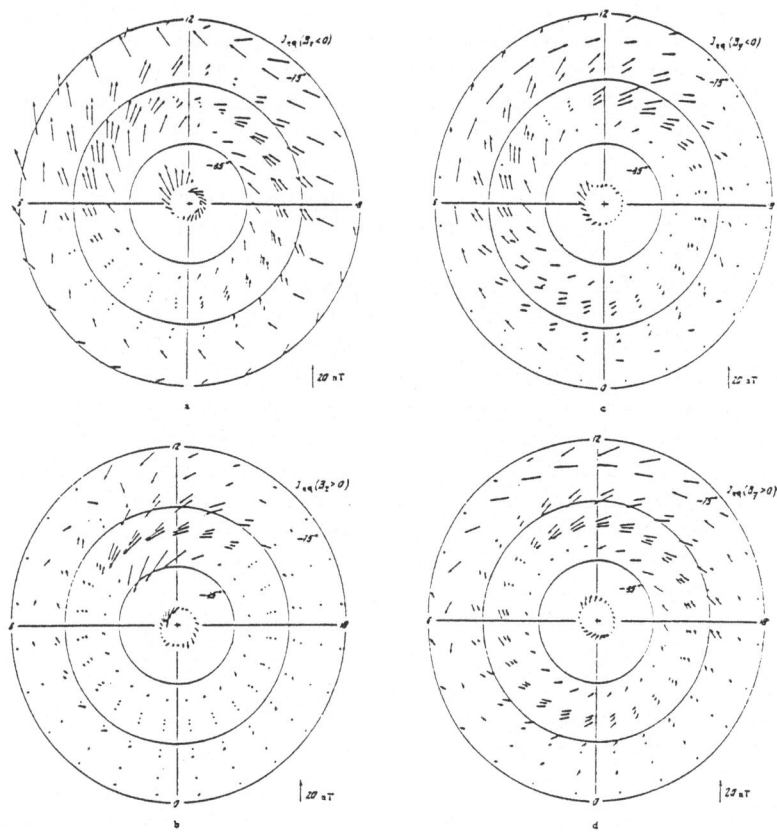

Figure 1. Equivalent current vectors derived from a chain of magnetometers in the Southern Hemisphere. The vectors correspond to a change of 1 nT in the IMF B_z and B_y component, respectively, and have been calculated by a linear regression analysis. (After Papitashvili et al., 1983).

analysis is reproduced in Figure 1 in form of equivalent current vectors in polar diagrams corresponding to different conditions of the IMF. The results may be compared with the results from a similar analysis by Friis-Christensen (1979) who used 20 minute average data from the meridian magnetometer chain in Greenland in the Northern Hemisphere. The meridian chain in Antarctica is located at latitudes closer to the invariant pole than the Greenland chain so only the high-latitude part of the equivalent current system poleward of $-75°$ is shown in Figure 1. Papitashvili et al. (1983) conclude that the gross features of the current systems are similar to those obtained from the Northern Hemisphere, but there are differences. For the B_y controlled parts of the equivalent current systems they report on large vectors in the nightside compared with the Northern Hemisphere. This difference could be a result of a larger time difference between magnetic local noon and geographic local noon in the data from the Antarctic chain (typically around 5 hours) compared to the Greenland chain (only about two hours). Papitashvili et al. (1983) find relatively large differences between the systems for $B_y > 0$ and $B_y < 0$. An asymmetry between the systems for different signs of B_y was also found and discussed by Friis-Christensen (1986) based on Northern Hemisphere data.

An interesting observation by Papitashvili et al. (1983) is the existence of regions limited in latitude and magnetic local time around the cusp where the geomagnetic response to IMF variations is strikingly decreased. This reduction in the regression coefficients around magnetic local noon was in fact already observed by Friis-Christensen and Wilhjelm (1975) using Northern Hemisphere magnetic observatory data. None of the papers, however, provide an explanation for this feature. Friis-Christensen and Wilhjelm(1975) propose that the DPY-current might be associated with a local return current at lower latitudes. The magnetic effect from this return current opposes the effect from the primary DPY-current so that the magnetic effects cancel in a small region near noon. Papitashvili et al. (1983) speculate that the effect could be due to an additional system of field-aligned currents flowing into the dayside cusp and out from the ionosphere on the nightside.

The recent findings of a possible distinction between the cleft and the "cusp proper" (Newell and Meng, 1989) and the association between the cleft particle precipitation observed on low-altitude satellites and the region 1a currents (Friis-Christensen and Lassen, 1989), could provide an explanation for the peculiar reduction near the cusp of the geomagnetic response to solar wind changes. The existence of the "cusp proper" means that there is a limited region topologically separated from the cleft and its associated low-latitude boundary layer. The boundary layer is the possible origin of the region 1 Birkeland currents, which are the driving currents for the large-scale ionospheric currents seen in ground magnetic observations. A discontinuity in the particle precipitation regions indicates a separation of different magnetopause regimes which would be associated with an apparent discontinuity in the currents flowing in the ionospheric projection of these regions. Further investigations using combined sets of ground- and satellite data will be necessary to clarify the possible ground-based geomagnetic signature of the cusp proper.

3. Empirical Electric Potential Patterns

The driving mechanism for the electrical coupling between the solar wind and the magnetosphere is reflected most directly in the convection pattern. The large-scale distribution of the ionospheric electric field is mapped into the magnetosphere along the magnetic field-lines which are assumed to be equipotential lines at least as long as the steady state convection is concerned. Knowing the ionospheric height-integrated conductivity it is possible to invert the equivalent current pattern by solving the differential equation resulting from using Ohm's law together with a separation of the ionospheric currents into a toroidal and a polodial part as described by Kamide et al. (1981). The instantaneous height-integrated conductivity is difficult to obtain on a global scale, but several attempts have been made to use a combination of UV imagery, field-aligned current observations and satellite and radar observations of electric fields to derive an electric potential distribution in optimal agreement with the actually measured parameters (Richmond and Kamide, 1987; Marklund et al., 1987).

For the average convection patterns, however, which we deal with here, the situation is quite different. Local conductivity enhancements which are characteristic of disturbed conditions, average out and empirical height-integrated conductivity models may be used. Feldstein et al. (1984) used such an empirical conductivity model together with Northern Hemisphere magnetic observatory data to derive an empirical model of the high-latitude ionospheric electric field. Friis-Christensen et al. (1985) used an average conductivity model by Kamide and Matsushita (1979) and data from a dense chain of magnetometers in Greenland to obtain an empirical model of electric fields and currents for 9 different directions of the IMF vector in the $B_y - B_z$ plane. In Figure 2 is shown the results corresponding to $B_z = 0$. The predominant east-west directed convective flow near magnetic local noon is seen to be nearly totally controlled by the B_y component of the IMF, consistent with the earlier results concerning the ionospheric Hall currents (DPY currents) in this region.

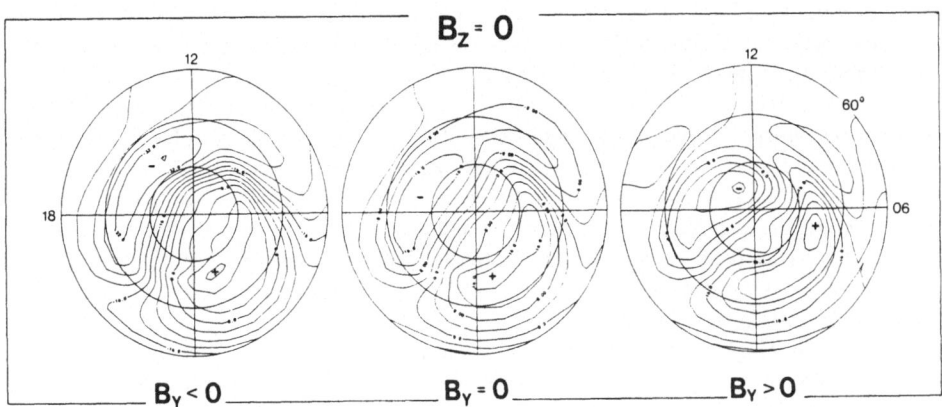

Figure 2. Empirical electric potential patterns for IMF $B_z = 0$ and various B_y conditions obtained from ground-magnetic measurements and an average conductivity model. (Reproduced from Friis-Christensen et al., 1985).

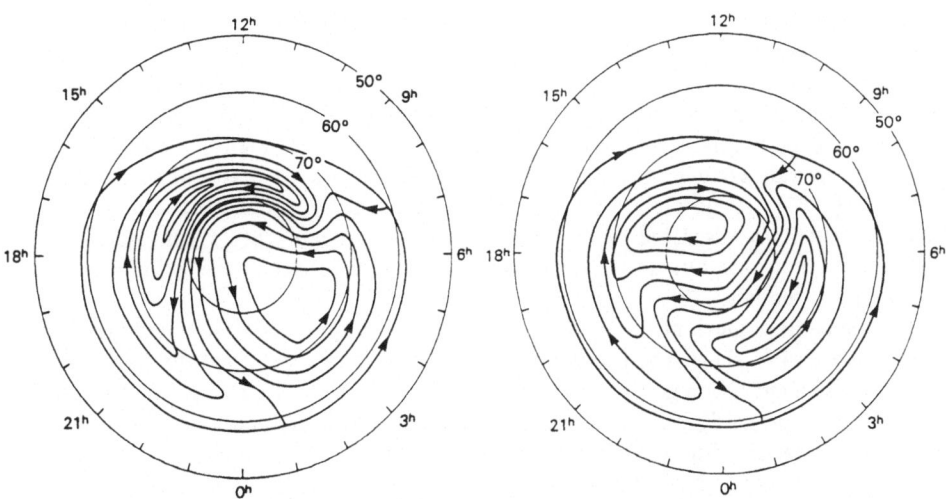

Figure 3. Empirical electric potential pattern for IMF weakly northward and $B_y < 0$ and $B_y > 0$. The patterns have been obtained from satellite electric field measurements by Heppner and Maynard(1987).

Foster et al. (1986) presented empirical convection patterns derived from electric field measurements with the incoherent scatter radar facility at Millstone Hill. Data from a large number of experiments were sorted into magnetic latitude-magnetic local time bins according to Kp and IMF B_z and B_y components. The convection pattern in the polar cap has been extrapolated since the Millstone Hill radar did not provide any experimental data poleward of 75° invariant latitude. The patterns for $B_y < 0$ and $B_y > 0$ are therefore much less detailed in the high-latitude part, and especially the east-west component of the flow in the noon region is underestimated compared to the empirical model derived from the ground-based magnetometer measurements.

The averaging techniques used to derive the empirical models of electric fields and currents have certain demerits which are evident in the resulting statistical convection patterns. The solar wind-magnetosphere coupling is a complicated process which can probably not be fully described solely by the value of the IMF components. Therefore the resulting magnetosheric and ionospheric electrodynamic parameters will be accompanied by a considerable uncertainty, and any pure averaging technique will tend to smooth out sharp boundaries. In particular, if some of the features in the convection pattern are of a bi-modal nature a simple averaging technique may create average patterns which do not represent possible and realistic states at all. In spite of this potential problem, the results from ground-based magnetometer observations display features which are relatively detailed, especially in the high-latitude day-sector.

Heppner and Maynard (1987) used a quite different "averaging" technique when constructing their empirical high-latitude electric field models based upon a large number of electric field measurements from the Dynamics Explorer 2 satellite. The technique they used is based upon a "pattern recognition" scheme, where each satellite pass is classified using a limited number of possible patterns. Within each class of electric field signatures typical values of primary parameters like the location of the convection reversal, electric field maximum, and low-latitude boundaries were determined and used in the construction of a potential pattern. The procedure has the advantage compared to the "true" averaging that primary boundaries, in particular those used to "normalize" the data, are conserved and maybe even enhanced. So we would expect to see convection patterns constructed in this way to be more conspicuous than those resulting from average models for the same solar wind conditions. Comparing the result of Heppner and Maynard (1987) for $B_y < 0$ and $B_y > 0$ in Figure 3 with the corresponding patterns in Figure 2 we will notice a very good overall agreement, but it is obvious that the two-cell patterns of Heppner and Maynard (1987) are more distorted than those based on the ground-based magnetometer measurements. As discussed above this difference is probably not a consequence of the different observations, but rather a consequence of the difference in the technique to derive the empirical models.

The nightside is much more detailed in the model of Heppner and Maynard (1987). This model very clearly shows the Harang discontinuity as a persistent feature independent on the B_y component although the flow poleward of the Harang discontinuity is very different for different signs of B_y. The Harang discontinuity does not stand out very clearly, if at all, in the model based on the ground magnetic perturbations. Comparing with the electric field model by Foster et al. (1986) it is seen that this model is also missing a persistent Harang discontinuity. We interpret this to be a natural consequence of the averaging technique used, since the latter two models were obtained using similar averaging methods. Considering the magnetometer observations, the Harang discontinuity is characterized by a region with oppositely directed electrojets, with the westward electrojet located poleward of the eastward electrojet in the pre-midnight sector. Since the latitude of the electrojets is not fixed, and does not vary systematically with the IMF components, the averaged magnetic effect from these currents will be small and therefore not give rise to large average electric fields.

The patterns from all the empirical models appear to be distorted two-cell patterns in contrast to the schematic and qualitative models proposed by Reiff and Burch (1985) based upon plasma, magnetic field, and electric field observations from Dynamics Explorers 1 and 2 satellites. For weakly northward IMF their models is characterized by three convection cells, the familiar symmetric two-cell system and a third and independent "lobe-cell" located entirely in the polar cap on open magnetic field lines. Although the difference between a distorted two-cell system and a three-cell system is of principal nature and theoretically important with regard to the understanding of the magnetospheric processes which generate the electric field, it is probably not possible to distinguish between the models just on the basis of average empirical models. A three-cell system would tend to be averaged into a distorted two-cell system, because there is no way to distinguish between electric field components from different sources in the resultant average data. The complexity of the patterns, particularly in the dayside high-latitude sector indicates a multitude of independent solar wind-magnetosphere coupling processes, all having their origin in different parts of the magnetopause.

4. Birkeland Current Patterns

The Birkeland currents connect the ionosphere with the distant magnetospheric regions. The pattern they form is a mapping of the magnetosphere into the ionosphere along the magnetic field-lines. Knowing the height-integrated ionospheric Hall- and Pedersen conductivities and the electric potential distribution it is possible to calculate the distribution of Birkeland currents, since the three-dimensional divergence of the total current systems is zero everywhere. In the previous section it was shown that the electric potential distribution could be derived from a sufficiently large number of ground-based magnetometer observations. It is therefore possible to derive the Birkeland current pattern from ground-based magnetometer observations provided that the height-integrated conductivity distribution is known.

Using the electric potential distribution and the average conductivity model described in the previous section, a distribution of Birkeland currents calculated for different IMF B_y and B_z components was presented by Friis-Christensen et al. (1985) and reproduced here in Figure 4. Although the zero contour line has been deleted for simplicity, it is evident that their results are completely missing the field-aligned currents in the region of the Harang discontinuity around midnight. As discussed in the previous section this is a natural consequence of the averaging technique used in the derivation and is not due to the fact that they dealt with ground-based observations. The averaging technique reduces the current intensity in regions of muliple oppositely directed field-aligned current sheets, which is the case in the Harang discontinuity where the westward and the eastward electrojets meet and overlap. This might be a concern also on the dayside, where the currents are also known to be very variable. Here, however, the variability seems to be more systematically related to the IMF components and a consistent set of field-aligned currents is, in fact,

246

FIELD-ALIGNED CURRENTS

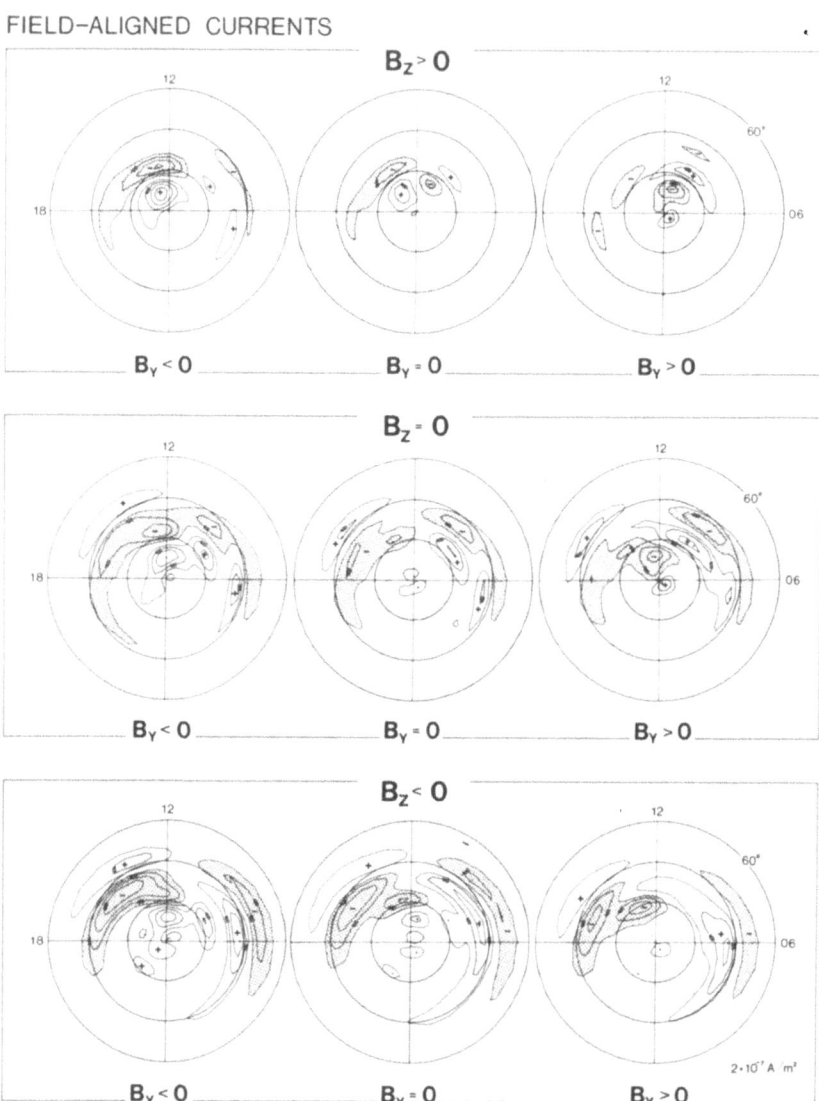

Figure 4. Distribution of Birkeland currents for different interplanetary magnetic field orientations obtained using ground-based magnetometer measurements from Greenland. The shaded regions indicate upward current. (After Friis-Christensen et al., 1985).

observed.

We may compare the distribution of field-aligned currents of Figure 4 with the "traditional" Iijima and Potemra (1976) model shown in Figure 5. This model is based upon a large number of satellite passes across the auroral oval. The width, location and direction

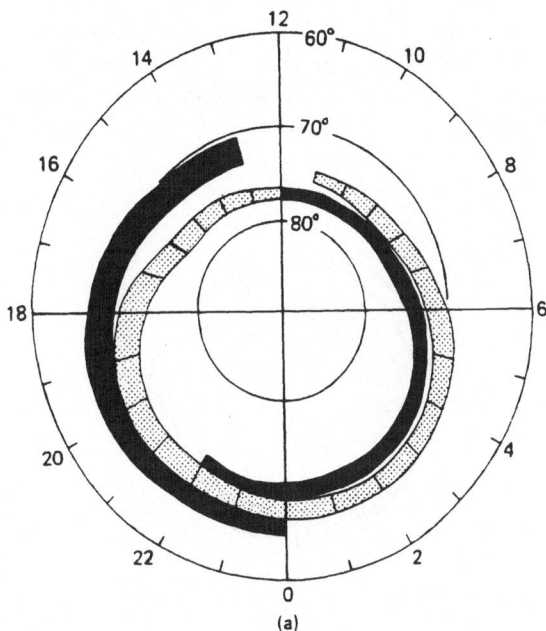

Figure 5. Statistical distribution of field-aligned currents. The filled areas indicate downward current, the blank areas indicate upward current. (After Iijima and Potemra, 1978).

of the Birkeland current sheets has been estimated for all magnetic local times for different levels of magnetic activity determined by the AL index. The resultant system contains the qualitative features of the current systems, but is a schematic rather than a quantitative model. The empirical model of Friis-Christensen et al. (1985) is a quantitative average model in the sense that the pattern as well as the intensity of the Birkeland currents was calculated to be consistent with the average ground magnetic perturbations and an assumed average conductivity model. A necessary consequence of the different averaging techniques used in these two studies is that the current distribution in the "average" model in Figure 4 seems to be drastically smoothed compared to the more idealized distribution in Figure 5.

The average models for $B_y \neq 0$ in Figure 4 show a large-scale Birkeland current region in the polar cap. The direction of this current is downward for $B_y < 0$ and upward for $B_y > 0$. This current, located poleward of the region 1 current was originally referred to as the "cusp" currents by Iijima and Potemra (1976). Recently simultaneous measurements of particles and magnetic fields on low-altitude satellites have provided direct evidence that these currents are associated with the plasma mantle (Bythrow et al., 1988; Erlandsen et al., 1988). For $B_y = 0$ and $B_z > 0$ Figure 4 shows two localized field-aligned current regions in the polar cap. This system was also observed with the MAGSAT satellite by Iijima et al. (1984) and referred to as the "NBZ" Birkeland current.

A difference between the two models shown in Figure 4 and Figure 5 is a systematic separation near dawn and dusk of the region 1 currents seen in the average data in Figure 4.

This separation is not present in the Iijima-Potemra model. Friis-Christensen et al. (1985) concluded that the distribution of the Birkeland currents indicated a separate source of the dayside and nightside region 1 currents and that one should be careful in generalizing the term "region 1 current" to describe at all local times the poleward part of the double current sheet along the oval.

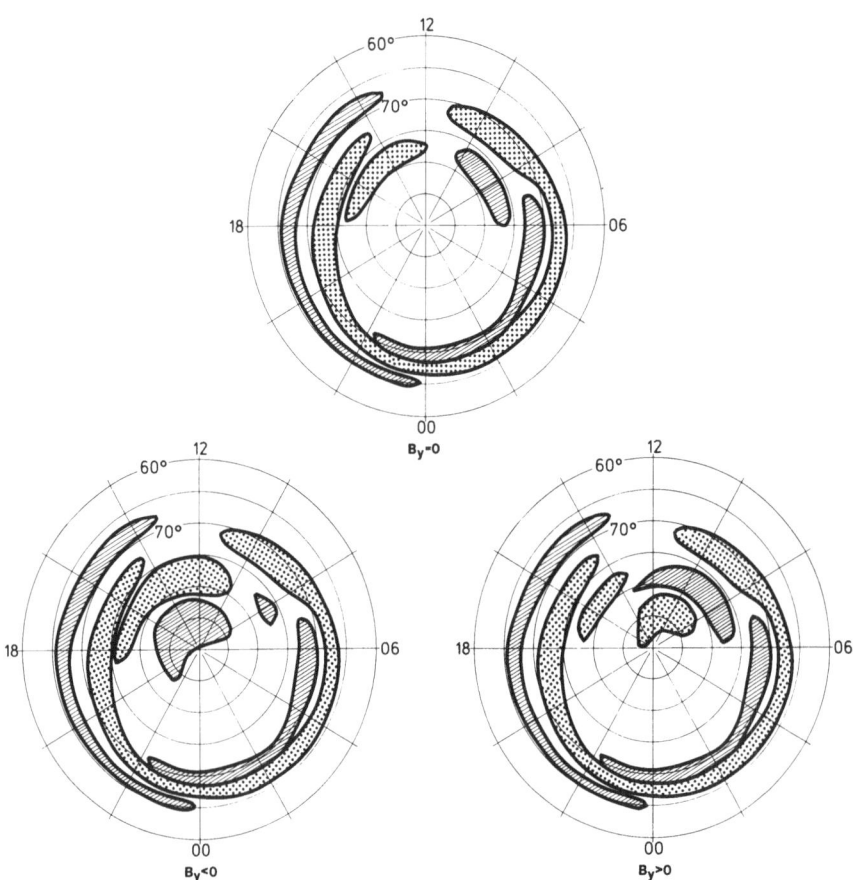

Figure 6. Model distribution of Birkeland currents for different B_y conditions based upon the statistical distribution obtained by satellites (Iijima and Potemra, 1976; 1978) as well as by ground-based measurements (Friis-Christensen et al., 1985). Hatched areas indicate downward currents, dotted areas are upward currents. The dayside region 1 currents (region 1a) together with the mantle Birkeland currents on the poleward side are supposed to be strongest during positive B_z while the nightside region 1 currents (region 1b) together with the region 2 currents are supposed to be strongest for negative B_z and become very weak during positive B_z (After Friis-Christensen and Lassen, 1989).

Friis-Christensen and Lassen (1989) compared the average distribution of Birkeland currents with the statistical distribution of discrete auroras. They find that the distribution of discrete auroras forms separate dayside and nightside maxima, which correspond in latitude and magnetic local time to the two parts of the region 1 currents, the region 1a and the region 1b system, and they conclude that these two systems are independently controlled by the IMF B_y and B_z components. In Figure 6 is shown a schematic diagram of the Birkeland current distribution. In this figure the actually calculated and IMF-dependent average currents in the dayside (including the region 1a currents have been combined with the "traditional" Iijima-Potemra model. The region 1a system is a mainly dayside controlled system which maps to the low-latitude boundary layer (LLBL). The region 1b system is a mainly nightside controlled system connected by magnetic field-lines to the plasma sheet.

5. Averaging techniques: Merits and demerits

In comparing different empirical models concerning high-latitude electrodynamic parameters we have noted significant differences which could not exclusively be attributed to the use of different data sets. Some of the differences are caused by the differences in the techniques used to derive the empirical models. Three principally different methods have been used in the models described in the previous sections. Two of them are based primarily on a quantitative averaging technique, namely the bin-averaging technique used by Foster et al. (1986) and the linear prediction method, where assumed statistical linear relations between the data and key parameters have been derived and used to estimate the electrodynamic parameters for selected values of the key parameters. This method was used by Friis-Christensen (1979), Papitashvili et al. (1983), and Friis-Christensen et al. (1985). The third method is based on pattern recognition supplemented with more or less normalization. This technique was used by Iijima and Potemra (1976) to derive the Birkeland current patterns and by Heppner and Maynard (1987) to create the electric field models.

The bin-averaging and the linear prediction technique allows a quantitative estimate of the average values for a certain parameter for given conditions of the input parameters which may be solar wind parameters or geomagnetic activity indices. Therefore models derived using these techniques can be used to quantitatively estimate changes in the models due to given changes in the input conditions. The demerits of these models are, as we have discussed in the previous sections, that boundaries are drastically smoothed, or even lost, and that average models may predict conditions, which will never occur in reality, because we may be dealing with a bi-modal magnetospheric response to the input parameters.

The techniques based on pattern recognition conserve boundaries, and may even enhance them, since pattern recognition is based on recognition of boundaries. On the other hand, information about relative intensities are often lost. Furthermore, pattern recognition is complicated, and heavily depends on which basic classification scheme has been applied. These models are therefore more apt to reflect preconceived ideas.

6. Conclusions

Several empirical models of high-latitude electrodynamic parameters have been reviewed and compared with each other. Different models based on different techniques and data generally agree on a global scale. But differences appear when comparing quantified parameters derived using different techniques. To a large extent these differences are explainable taking the merits and demerits of the different techniques into consideration. No single technique is capable of providing an adequate description of the whole range of processes affecting the polar ionosphere and a multiplicity of empirical models is necessary to interpret and understand the average state of the high-latitude ionospheric electrodynamics and the relation to the solar wind conditions.

References

Bythrow, P. F., Potemra, T. A.,Erlandsen R. E., Zanetti, L. J., and Klumpar, D. M. (1988) 'Birkeland currents and charged particles in the high-latitude prenoon region: A new interpretation', *J. Geophys. Res., 93*, 9791–9803.

Erlandsen R. E., Zanetti, L.J., Potemra, T. A., Bythrow, P. F.,and Lundin, R. (1988) 'IMF B_y dependence of region 1 Birkeland currents near noon', *J. Geophys. Res., 93*, 9804–9814.

Feldstein, Y. I., Levitin, A. E., Faermark, D. S., Afonina, R. G., and Belov, B. A. (1984) 'Electric fields and potential patterns in the high-latitude ionosphere for different situations in interplanetary space', *Planet. Space Sci., 32*, 907–923.

Foster, J. C., Holt, J. M., Musgrove, R. G., and Evans, D. S. (1986) 'Solar wind dependencies of high-latitude convection and precipitation', in Y. Kamide and J. A. Slavin, eds., *Solar Wind -Magnetosphere Coupling*, Terra, Tokyo, pp. 477–494.

Friis-Christensen, E. (1979) 'The effect of the IMF on convection patterns and equivalent currents in the polar cap and cusp', *Magnetospheric Study 1979*, Japanese IMS Committee, 290–293.

Friis-Christensen, E. (1986) 'Solar wind control of the polar cusp', in Y. Kamide and J. A. Slavin, eds., *Solar Wind-Magnetosphere Coupling*, Terra, Tokyo, pp. 423–440.

Friis-Christensen, E. (1989) 'Ground magnetic perturbations in the polar cap and cleft: Structure and dynamics of ionospheric currents', this volume.

Friis-Christensen, E., Kamide, Y., Richmond A. D., and Matsushita, S. (1985) 'Interplanetary magnetic field control of high-latitude electric fields and currents determined from Greenland magnetometer data', *J. Geophys. Res., 90*, 1325–1328.

Friis-Christensen, E. and Lassen, K. (1989) 'Large scale distribution of discrete auroras and field-aligned currents', presented at the *International Conference on Auroral Physics, Cambridge July 10–15, 1988*.

Friis-Christensen, E. and J. Wilhjelm (1975) 'Polar cap currents for different directions of the interplanetary magnetic field in the Y–Z plane',*J. Geophys. Res., 80*, 1248–1260.

Heppner, J. P. and Maynard, N. C. (1987) 'Empirical high-latitude electric field models', *J. Geophys. Res., 92*, 4467–4489.

Iijima, T. and Potemra, T. A. (1976) 'Field-aligned currents in the dayside cusp observed by TRIAD', *J. Geophys. Res., 81*, 5971–5979.

Iijima, T., and Potemra, T. A. (1978) 'Large-scale characteristics of field-aligned currents associated with substorms', *J. Geophys. Res., 83*, 599–615.

Iijima, T., Potemra, T. A., Zanetti, L. J.,and Bythrow, P. F. (1984) 'Large-scale Birkeland currents in the dayside polar region during strongly northward IMF — A new Birkeland current system', *J. Geophys. Res., 89*, 7441–7452.

Kamide, Y., and Matsushita, S (1979) 'Simulation studies of electric fields and currents in relation to field-aligned currents, 1, Quiet periods', *J. Geophys. Res., 84*, 4083–4098.

Kamide, Y., Richmond, A. D., and Matsushita, S. (1981) 'Estimation of ionospheric electric fields, ionospheric currents, and field-aligned currents from ground magnetic records', *J. Geophys. Res., 86*, 801–813.

Mansurov, S. M., Troshichev, O. A., Zaytzev, A. N., Papitashvili,V. O., Timofeyev, G. A., and Kandibolotskaya, M. A. (1981) 'Characteristics of magnetic disturbances produced by the IMF in the southern polar cap', *Geomagn. Aerom., 21*, 428–430.

Marklund, G. T., Blomberg, L. G., Potemra, T. A., Murphree, J. S., Rich, F. J., and Stasiewicz, K. (1987) 'A new method to derive "instantaneous" high-latitude potential distributions from satellite measurements including auroral imager data, Geophys. Res. Lett., 14, 439–442.

Newell, P.T. and Meng, C.-I. (1989) 'On quantifying the distinction between the low altitude cusp and the cleft/LLBL', this volume.

Papitashvili, V. O., Zaytzev, A. N., and Feldstein, Y. I. (1983) 'Magnetic disturbances generated by the interplanetary magnetic field in the southern polar cap during summer', *Geomagn. Aeron.*, *23*, 506–509.

Reiff, P. H.. and Burch, J. L. (1985) 'IMF B_y dependent plasma flow and Birkeland currents in the dayside magnetosphere. 2. A global model for northward and southward IMF', *J. Geophys. Res.*, *90*, 1595–1609.

Richmond, A. D., and Kamide, Y. (1987) 'Mapping electrodynamic features of the high-latitude ionosphere from localized observations: Technique, *J. Geophys. Res.*, *93*, 5741–5759.

RELATIONSHIPS BETWEEN AURORAL AND MAGNETIC ACTIVITY IN THE POLAR CUSP/CLEFT

Takasi OGUTI
Geophysics Research Laboratory,
University of Tokyo,
Tokyo 113, Japan

ABSTRACT.   Both large- and small-scale dynamics of the cusp aurora and cusp electrodynamics are discussed in connection with electric currents, electric field and ground magnetic perturbations. Sources of the field-aligned electric currents in the dayside polar cusp are most likely located within the boundary layers, both the frontside low latitude and the magnetotail, where induction currents due to relative motion between the sheath plasma and the magnetosphere produces space   charges. The importance of the magnetospheric magnetic field model in the study of the cusp phenomena, especially their magnetospheric source regions, is noted.

1. Introduction

The dayside ionospheric cusp region is connected with the outermost magnetospheric region where the solar wind effects are most directly transferred. The auroral and magnetic activities in the dayside cusp, therefore, could be direct reflections of the processes of particle penetration as well as charge separation in the boundary layer which results in the particle precipitations, electric fields and currents in the dayside cusp.

Four problems of the dayside cusp are discussed in this   paper; i.e., 1) basic electrodynamics of the cusp, involving electric fields (ionospheric drifts),   ionospheric electric currents, field-aligned currents and ground magnetic field perturbations, 2) characteristic radial structures of the dayside discrete auroras, 3)temporal change in the cusp latitude, and 4) localized, intermittent, transient auroral activities associated with magnetic impulses. These phenomena could be various appearances of interaction between the solar wind (sheath plasma) and the magnetosphere. Many studies have already been carried out of these problems. However, there are various controversial ideas, and we are still far away from real understanding them.

The purpose of this paper is to examine the observations of these phenomena, to search the possibility of reasonable explanations of them, and to discuss the necessary points of further observations in reaching the full understanding of the interaction between the solar wind and the

P. E. Sandholt and A. Egeland (eds.), Electromagnetic Coupling in the Polar Clefts and Caps, 253–267.
© 1989 by Kluwer Academic Publishers.

magnetosphere.

2. Basic Electrodynamics of the Dayside Cusp

The findings of IMF(towards or away from the sun) control of high
latitude magnetic field variations by Svalgaard(1973) stimulated broad
interests on the interaction between solar wind and the magnetosphere,
especially on the magnetic field connection. This is closely connected
with the cusp electrodynamics. Friis-Christensen et al.(1972) found that
the crucial factor was the IMF By-component. Friis-Christensen and
Wilhjelm (1975) demonstrated evident control of equivalent current
around the dayside cusp by the IMF By-component. It is shown that a
small ionospheric jet current(DPY current), eastward or westward, occurs
at 75°-80° MLAT in the dayside for the time of positive or negative IMF
By, respectively.

Electric fields(ion drifts) measured by satellites(e.g., Heppner,
1972; Heelis et al., 1976; Heelis, 1984; Heppner, 1977; Heppner and
Maynard, 1987) and on the ground by radars (e.g., Jørgensen et al, 1984;

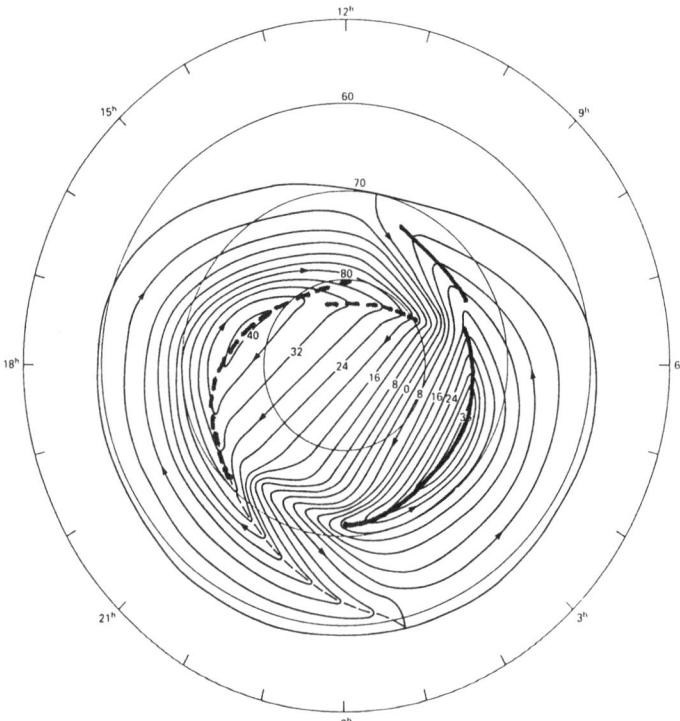

Figure 1. Distribution of ionospheric convection flow in high
latitudes after Heppner and Maynard(1987), along with schematic illustration of field-aligned currents. The flow pattern and the field-aligned
currents are basically consistent with each other, and they are also
consistent with ionospheric currents shown by Friis-Christensen and
Wilhjelm(1975).

Clauer et al., 1984) also showed IMF By-dependence of the field distribution in the dayside high latitudes. Large-scale distribution of field-aligned currents (e.g., Iijima et al., 1978; McDiarmid et al., 1979; Erlandson et al., 1988) changes as the IMF By changes. Figure 1 shows an example of electric potential(drift flow lines) distribution obtained by satellite measurements(Heppner and Maynard, 1987) along with a schematical illustration of field-aligned currents.

If we neglect the IMF By effect, the basic pattern is two-cell convection, poleward in the dayside cusp, associated with downward region-1 field-aligned current in the dawn side and upward field-aligned current in the duskside. These downward and upward region-1 field-aligned currents are collocated with counterclockwise and clockwise flow reversals, and, clockwise and counterclockwise ionospheric Hall currents reversals, respectively. In a little higher latitudes, there are the so-called "cusp" or "NBZ" field-aligned currents with polarity reverse to

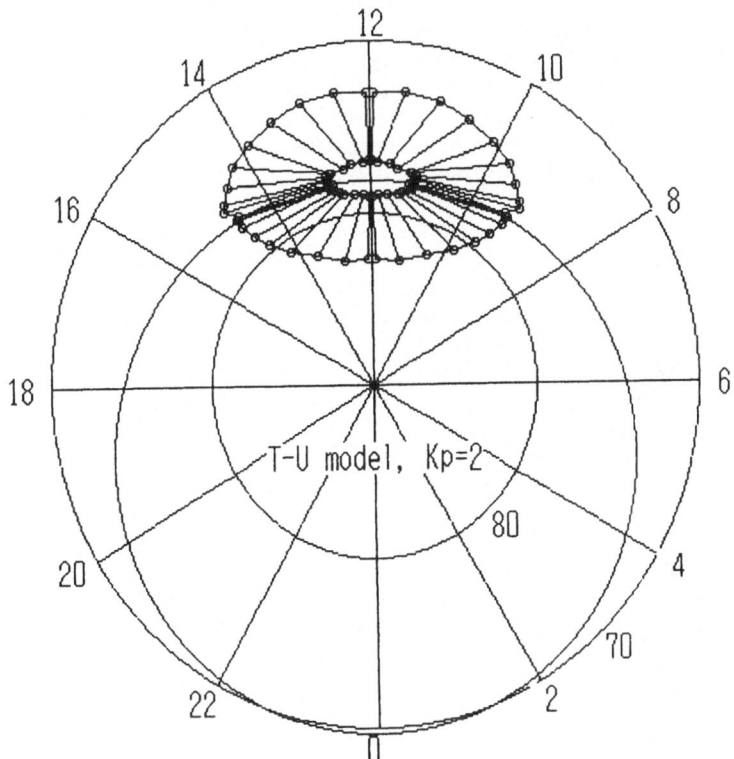

Figure 2. Projection of the outermost magnetospheric region(0.3-1 RE inside) along field lines onto the dayside ionosphere. An external field model proposed by Tsyganenko and Usmanov(1982) is used. The sunward half (or a little larger) of each elliptic loci are traced from the equatorial plane(Z=0) and the rest half from the X=0 plane. The radial lines correspond to those near the magnetopause from 0.3 to 1 RE inside the magnetosphere.

that of the region 1 field-aligned currents. The IMF By effects mostly occur in the "cusp" currents. This basic structure of the dayside cusp persists even when the IMF Bz is positive. However, magnitude of the region 1 field-aligned current decreases a little while that of the "cusp" currents largely increases when the IMF Bz is positive.

This figure, together with the equivalent current patterns given by Friis-Christensen and Wilhjelm (1975), indicates that the distribution of all the field-aligned currents, ionospheric electric field(drifts), ionospheric currents and ground magnetic field variations are basically consistent with each other. The ground magnetic field deflections are mostly due to ionospheric Hall currents. The observations appear now well established, although this has not yet been confirmed with simultaneous data set.

The problem is the implication of these observations, especially the field-aligned currents. How could the dayside region-1 currents in the lower latitudes and the "cusp" currents or "NBZ" currents in higher latitudes be understood? As Oguti(1989) pointed out, the boundary layer with thickness of 1 R_E on the entire magnetopause is likely linked with a spatially limited dayside cusp region as shown in Figure 2, based upon a semi-empirical magnetic field model proposed by Tsyganenko and Usmanov (1982). If this is the case, the most probable guess is that the region 1 currents within this region is connected with the inner edge of the low latitude boundary layer, and the "cusp" currents are linked with the plasma mantle region, or more generally with the circum-tail boundary layer(Figure 3).

These field-aligned currents in the dayside cusp must be consequences of the tangential stress (i.e., motional induction currents) operating at the entire magnetopause both in the frontside low latitude boundary layers and circum-tail boundary layers(Oguti, 1989). A general feature of the induction currents in the tail boundary layer and the resulting field-aligned currents inside the tail magnetosphere, that must be connected with the dayside "cusp" field-aligned currents, are schematically illustrated in Figure 3.

In connection with this, the east-west extent of the dayside cusp, which could be connected with the entire outermost magnetosphere, is estimated to be about 1000-1500 km. If the "dayside reconnection" makes the polar cap field lines open, the effect of "reconnection"(flows of the connected field lines or the electric fields) must occur within this 1000-1500 km extent. Thus, the east-west electric potential across the dayside cusp at 1000-1500 km extent must be equal to the total cross-polar cap potential (60-80 kV), resulting in a strong eastward electric field in the dayside cusp. However, the observation of electric field (Figure 2) does not show such a large east-west electric field in the dayside cusp although a large modification of the north-south electric field is seen due to the IMF By effect. The electric potential of 15-20 kV, expected from the anti-sunward flow within the existing low latitude boundary layer, is almost enough to explain the east-west electric field in the dayside cusp. We do not need any additional potential contribution from the "dayside reconnection".

These facts strongly suggest that the stationary "dayside reconnection" is not operative, or, the "dayside reconnection" plays little

Currents to or from
inner edge of low latitude mixing(boundary) layer

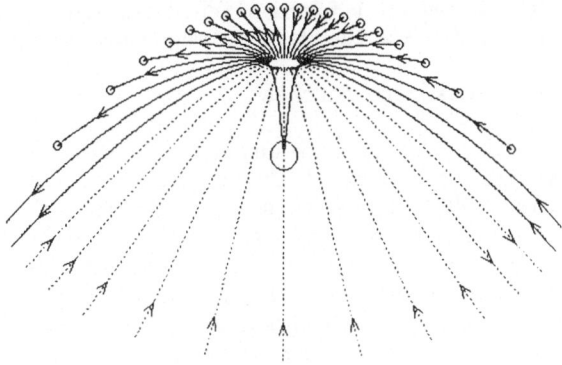

Currents to or from
high latitude(plasma mantle) mixing(boundary) layer

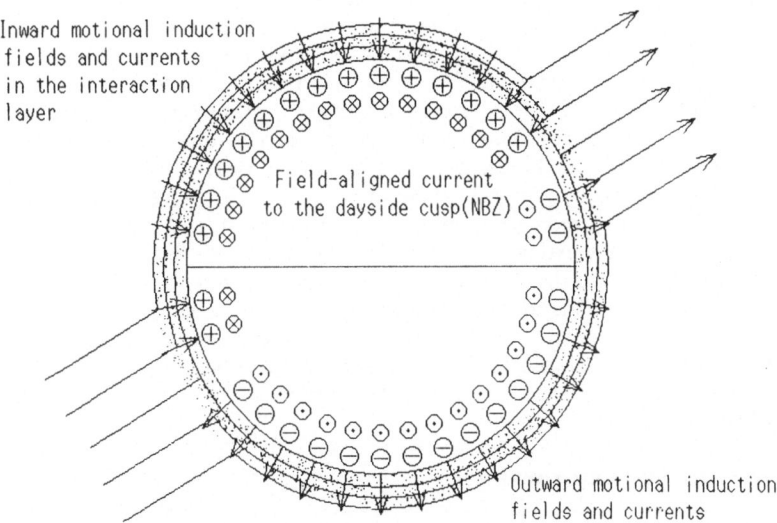

Figure 3. Field-aligned currents expected from the motional induction currents within the boundary layers. The top panel shows the distribution of field-aligned currents both from the inner edge of frontside low latitude boundary layer(solid curves) and tail boundary layer(dashed curves), projected onto X-Y plane. The bottom panel indicates a cross-sectional illustration of the tail where the motional induction currents within the tail boundary layer with negative IMF By and small positive IMF Bz are shown, referring the field-aligned currents in the top panel.

role in driving the magnetospheric convection even if it is operative. It is almost certain that the most part(outside the dayside elliptic region in Figure 2) of the polar cap potential(total minus LLBL contribution), as well as the region 1 field-aligned current is supplied from the tail plasma sheet, not from the dayside magnetopause(Oguti, 1989).

3. Radial Structures of Discrete Auroras

A radial structure of discrete auroras, focusing to the dayside cusp, is one of important features of the dayside auroras(e.g., Akasofu, 1976; Meng and Lundin, 1986). Since this structure is limited around the dayside cusp, the radial auroras could be connected with some outermost structures of the magnetosphere. Each auroral arc of this radial structures may have some connection with the aurora which intermittently develops associated with magnetic impulsive variation as discussed later, but direct evidence of this connection is not yet known.

In connection with the magnetic field linkage, the magnetospheric source regions of this radially-structured auroras are an interesting matter of examination. Meng and Lundin(1986) associate these auroras with plasmoids which penetrate the dayside magnetopause into the magnetosphere. As seen in Figure 3, the radial extent with length of 1 R_E at the magnetopause is projected to the radial lines around the dayside cusp, shorter in the north-south direction and longer in the east-west direction. For the dayside half(a little larger), the radial lines of this figure are traced from the equatorial plane($10°$ separation in azimuthal angle), and for the rest half from the dawn-dusk plane(X=0, $10°$ separation in latitude). In comparing these radial lines with observations of the dayside auroral structures, it is suggested that the radial auroral structures are most likely connected with the dawn and dusk(to the tail) flank rather than with the dayside magnetopause. The tail flank appears to be the most favourable region of interaction between the sheath plasma and magnetospheric plasma in terms of the magnetic flux conservation. However, the field linkage alone does not answer the question about where the sheath plasma gets inside. The plasma penetration is also expected to occur at the dayside magnetospheric entry layer

4. Variation in the Cusp Latitude

Distribution of the magnetically active zone and its latitudinal shift with general magnetic activity were shown by Harang(1946). Distribution of auroral belt (auroral oval) and its dependence on geomagnetic activity(Q-index) was shown by Feldstein and Starkov(1967) using IGY-IGC all-sky auroral films. Akasofu(1972), Vorobjev et al.(1975), Eather et al. (1979) and Sandholt et al.(1980) all showed decrease in the dayside cusp latitude with increase in magnetic activity in the nightside. Eather (1984, 1985) claims that the dayside oval latitude is more closely correlated with the nightside magnetic activity than IMF Bz.

On the other hand Sandholt et al.(1983) and Meng(1983) claim closer correlation of the cusp latitude with IMF Bz. A variation of this idea is the relationship between the time-delayed IMF Bz and the cusp latitude. For example, Horwitz and Akasofu(1977), and Carbary and Meng (1986a, b) obtained characteristic delay time of the dayside auroral shift behind the southward turning of the IMF of 17 minutes and 30 minutes, respectively. Carbary and Meng(1986b) claim that the cusp latitude has higher correlation with time-delayed IMF-Bz than AE-index.

This contention is mainly due to different ideas on the cause of the oval latitude shifts. The former stresses importance of the mag-netospheric electric currents(enhancements in the region 1 field-aligned currents and DP-1 current) at the time of a magnetospheric substorm and the latter is based upon the idea of magnetic flux erosion from dayside magnetosphere to the tail lobe for the time of negative IMF Bz, due to the dayside magnetic field merging.

Since the magnetic field configuration is determined by electric currents, the cusp latitude must depend on all the large-scale currents such as equatorial ring currents, C-F currents, cross-tail currents and field-aligned currents. The "erosion" effects should also be replaced by these currents. The time-delayed correlation between IMF Bz and cusp latitude really suggests the response time of the magnetospheric electric currents to a change in the IMF. All of these currents vary during the course of a substorm. It may be rather difficult to distin-guish the specific effect, e.g., of "erosion" or AE-index, from the entire current effects.

The magnetic field configuration and its change at the time of a substorm is not yet established. We need further examination of the magnetic field changes in the magnetosphere to reach a decisive conclusion about what is the principal cause of the change in cusp latitude. The two contentions may not be as much different from each other as the authors from both sides insist.

5. Transient, Intermittent, Small-scale Dynamics

In the early phase of research, not many studies have been carried out on the local, transient activity in the dayside high latitude. Oguti(1969) examined the time sequence of a substorm, and found that a spike-like positive variation often occurs in the dayside cusp occurring about 20-30 minutes ahead of a substorm in the nightside. It was also noted that magnetic disturbances tended to occur in the dayside 30-40 minutes after the onset of nighttime substorms. Kaneda(1973) showed a precursory auroral activity in the dayside before a nightside auroral breakup. Vorobjev et al.(1975) noted that brightening of the dayside aurora tended to occur about 40 minutes after the initiation of the substorm in the nightside. It is not yet clear whether the "precursor" is the real precursor, as well as the 30-40 minute delay really indicates the delayed effect, of the nightside activity. The intermit-tent, transient auroral activity, associated with magnetic impulses has recently been focused again in connection with the so-called "flux transfer events" (Russell and Elphic, 1979).

Vorobjev et al.(1975), and Horwitz and Akasofu (1977) noted that typical intermittent poleward movements of dayside auroral structures often occur for the time interval when substorm is in progress in the nightside. Sandholt et al.(1986) and Kokubun et al.(1988), on the other hand, associated the intermittent poleward movements of auroras with the concurrent, slow equatorward shifts of the dayside oval, for the time of growth phase. Eather(1984) exemplified the poleward drifting auroras by using meridian displays. He showed that fast poleward movements of auroral structures occur in the dayside, and equatorward movements occur in the late afternoon. We have so many varieties. A general tendency of the time sequence is not yet established.

Lanzerotti et al.(1986) analyzed magnetic impulses at southern polar cusp region, and suggested a traveling single vortex current for explaining these impulses. They noted these magnetic impulses as possible ground signatures of the "flux transfer events" model proposed by Lee and Fu (1985). Sandholt et al.(1986) and Kokubun et al.(1988) show that the magnetic impulse is associated with poleward expansion of the cusp aurora. Oguti et al.(1988) emphasize the enhancements in east-west drifts of auroral structures during the auroral expansion in association with the magnetic impulses. Lanzerotti et al.(1987) examined the conjugate relationship between magnetic field variations at South Pole and those at Frobisher Bay.

Spatial distribution of the equivalent currents of similar magnetic impulses were studied by Friis-Christensen et al.(1988), using magnetic field measurements at Greenland chain stations. They obtained westward moving twin-vortex currents in contrast to the result by Lanzerotti et al. (1986) and connected this with a sudden change in solar wind para-meters, especially IMF By and Bz, and/or sudden decrease in the solar wind dynamic pressure. The ionospheric flow bursts observed by radars

GREENLAND CHAIN MAGNETIC PERTURBATIONS
PLOTTED AS EQUIVALENT CONVECTION

28 JUNE 1986
10:06-10:21 UT

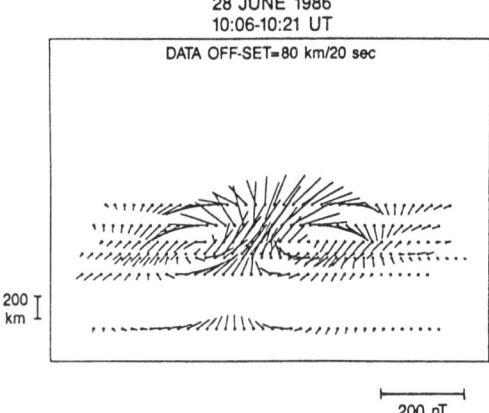

DATA OFF-SET=80 km/20 sec

200 km

200 nT

Figure 4. Distribution of ionospheric drifts estimated from an impulsive magnetic field variations along chain stations in Greenland (after Friis-Christensen et al., 1988). Twin-vortex system is evident.

(e.g., Goertz et al., 1985; Willis et al., 1986; Todd et al., 1986) are connected with the concurrent magnetic field impulses. Three such flow bursts from 06:00 to 07:50 UT on 27 October, 1984 obtained by the EISCAT radar(Willis et al., 1986) along the line of sight(approximately north-south component), were associated with concurrent magnetic impulses at Svalbard by Kokubun et al.(1988).

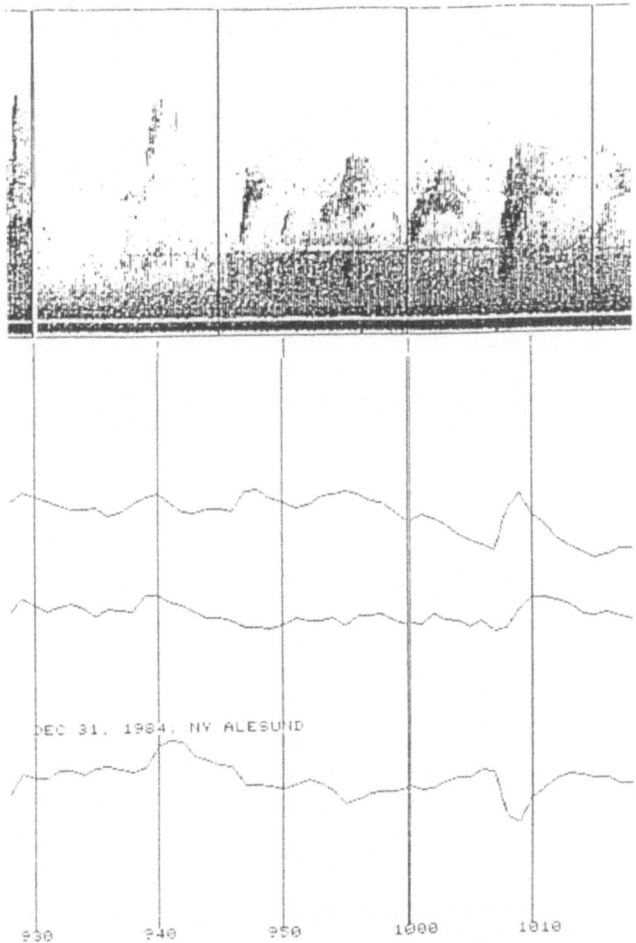

Figure 5. Meridian display of five sequential, localized, dayside auroral expansions(top panel) associated with magnetic impulses(bottom panel) at Ny Alesund. The last expansion as well as the magnetic field deflection at 1009 UT, December 31, 1984, is most outstanding. The Xm-component(upper plot) deflects northward for all the five events. Note that the Z-component deflects upward(downward on the plot) for the last 4 events which occur south of the station, while downward for the first event which takes place in the north of the station.

Based upon HILAT data, Baker et al.(1986) reported relationships between field-aligned currents and ionospheric drifts. The strong drift was eastward in the cusp region in the northern high latitudes. On the

other hand, Sandholt and Egeland(1988), using also HILAT data, showed
that the poleward expansion aurora in the dayside cusp was associated
with strong westward drifts(northward electric fields). Their result
also shows that the spatial structure of the electric field could be as
small as 50 km in latitudinal extents. The discrepancy in the flow
direction between them could be due to probable difference in the IMF-
By, negative in the case of Baker et al.(1986) whereas positive in the
case of Sandholt and Egeland(1988). These results of ionospheric drifts
are consistent with fast east-west movements (drifts) of auroral
structures studied by Oguti et al.(1988), using a TV camera at Svalbard.

6. Sources of the Dayside Transient Activity

The essential point of the dayside auroral activity is an intermittent, short-lived, spatially limited enhancement in luminosity associated with a north-south electric field and often associated with a
magnetic impulsive variation. The dayside aurora frequently expands
polewards as seen in Figure 5. The movement of aurora in the east-west

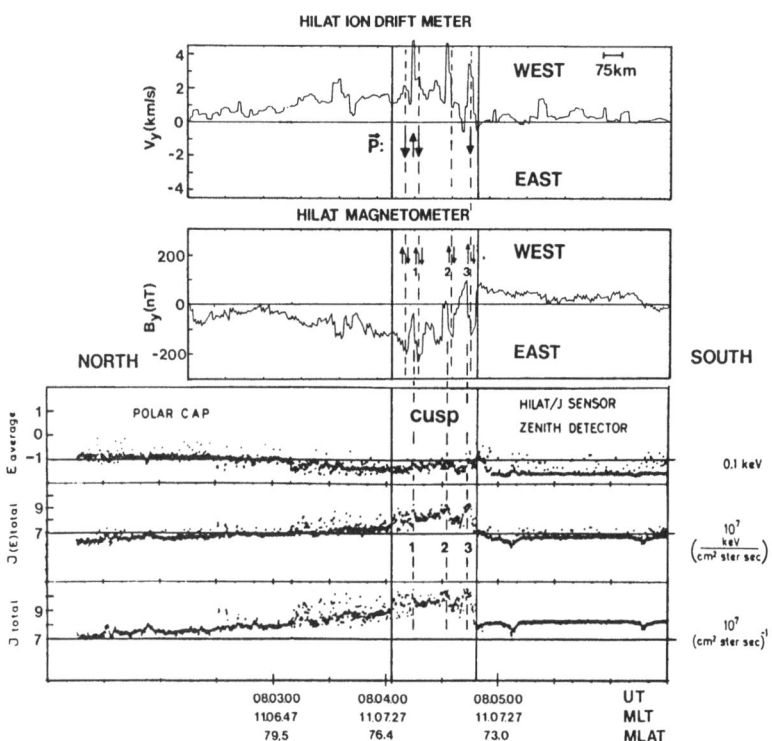

Figure 6. Electric fields(ion drifts), magnetic field deflections,
and electron precipitations around three transient dayside auroral
activities measured by HILAT satellite above Svalbard(after Sandholt and
Egeland, 1988). Strong electric fields are limited within small latitudinal extents.

direction is much faster than northward expansion. The electric field fast grows, probably associated with the enhancement in luminosity. The latitude range of the strong electric field could be as small as 50 km with an east-west extent of several hundreds to 1000 km. Multiple structures of such strong electric field zone occur(Figure 6).

Sudden, local enhancements in auroral luminosity, electric fields (ionospheric ion drifts), and associated magnetic impulsive variations appear almost consistently, in terms of conductivity increase and strong, latitudinally localized electric field(Oguti et al., 1988; Sandholt et al., 1989). The two dimensional distribution of the current system is sometimes estimated from magnetic field observations assuming that the stationary pattern holds during an event. However, evidently this is not the case. The magnetic impulsive variation should be understood taking the temporal change in intensity and distribution pattern of the current into consideration. Observation of the temporal-spatial change in the current is one of the crucial requirements for understanding the basic structure of this event. Another way is theoretical examination of the current system around a local enhancement in conductivity based upon the observation of electric fields. In contrast to the currents around pulsating auroral patches(e.g., Oguti and Hayashi, 1984, 1985), non-linear conductivity enhancement, as well as the transient effect, is essential in this case. A study of this point will be presented elsewhere.

The magnetospheric origin of the local, transient, strong electric field, as well as related electron precipitations, is not yet known. Many authors(e.g., Goertz et al., 1985; Todd et al., 1986; Willis et al., 1986; Lanzerotti et al., 1986, 1987; Sandholt et al., 1986) suggest the possible connection with dayside transient phenomena such as the sporadic, localized magnetic field reconnection(or "flux transfer" event, FTE). However, as Kokubun et al.(1988) and Friis-Christensen et al.(1988) discuss, not all the dayside transient events are consistent with the flux transfer models(e.g., Lee and Fu, 1985; Southwood, 1987).

We have some problems in relating the characteristic dayside transient phenomena with the "flux transfer" events. First, phenomenologically, the "flux transfer" event merely indicate a characteristic change in magnetic field around the magnetopause. It is not yet confirmed at all that this indicates the real flux transfer. Further, we do not yet know the spatial structure of the "flux transfer" event which could be connected with that of the ionospheric phenomena. These facts mean that the predictions of the ground signature of the "flux transfer" event, such as those of the models by Lee and Fu(1985) and Southwood (1987), are not yet fully reliable to relate the ground observations with the "flux transfer" events. Secondly, we have not yet enough information about temporal and spatial changes in the current system associated with the local dayside auroral activity, which can give a crucial constraint on the structure of the current in the magnetosphere. Although Friis-Christensen et al.(1988) show two dimensional distribution of the ionospheric drifts(currents), their result is obtained on an assumption of the stationarity of the drift pattern which is not suitable for the study of this essentially transient phenomenon.

The transient, local enhancement in electric field in the dayside cusp may suggest the transient change in magnetic field on the magneto-pause. However, it appears too early to relate the transient phenomena with local "reconnection" since the major energy here in the magneto-sheath is the plasma kinetic energy not the magnetic energy, and the sheath plasma is highly turbulent. The temporal change in the magnetic field must be associated with an inductive electric field which could contribute to a charge separation and a subsequent static electric field. This is not necessarily called "reconnection".

The plasmoid injection idea may also be promising. If there are small, nearly non-magnetized plasmoids or o-type plasmoids in the turbu-lent magnetosheath, they could penetrate the magnetopause into the magnetosphere. These plasmoids more likely occur in the magnetosheath when the sheath plasma is more turbulent. Note that the plasmoid injection here, is quite different from that proposed by Lemaire (1977) and Lundin and Dubinin (1984).

It is worth noting that the direction of the short-lived magnetic impulse is not always the same as that of the background large-scale magnetic field deflection examined with hourly values by Friis-Christensen and Wilhjelm (1975). Sometimes, the magnetic impulse is directed opposite to the background deflection. Although we need temporal and spatial information of the current system in order to reach a decisive conclusion, a possibility exists that the small-scale current direction is quite different from the large-scale, ambient current. A study of the sheath plasma dynamics is crucial in this connection.

7. Conclusion

The basic cusp electrodynamics are likely understood in terms of the macroscopic motional induction effect between the solar wind and the magnetosphere within the boundary layers. However, crucial information to relate the sporadic, small-scale enhancements in electric fields and magnetic field impulses with the "flux transfer" events is still lacking. We may guess the possible connections between them but the evidence is still very poor. This is mostly due to the fact that we do not yet know the spatial-temporal structures of the "flux transfer" event around the magnetopause as well as those of the electric currents of the transient phenomena in the dayside cusp. We also do not yet know their basic relationships with transient changes in the solar wind pressure and magnetic field. Simultaneous data sets of the solar wind parameters, the magnetospheric and the ionospheric signatures, with sufficient temporal and spatial resolution are necessary in order to reach a decisive conclusion.

Another basic problem is uncertainty of the field linkage between the ionosphere and the outer magnetosphere. This is especially important in the study of dayside cusp phenomena in connection with physical processes in the boundary layers. An accurate knowledge of the magnetic field distribution will provide us a great advancement in the study of the solar wind-magnetosphere interaction.

References
Akasofu, S.-I., Midday auroras at the south pole during magnetic sub-

storms, J. Geophys. Res., **77**, 2303-2308, 1972.

Akasofu, S.-I., Recent progress in studies of DMSP auroral photographs, Space Sci. Rev., **19**, 169-215, 1976.

Baker, K. B., R. A. Greenwald, A. D. M. Walker, P. F. Bythrow, L. J. Zanetti, T. A. Potemra, D. A. Hardy, F. J. Rich and C. L. Rino, A case study of plasma processes in the dayside cleft, J. Geophys. Res., **91**, 3130-3144, 1986.

Carbary, J. F. and C.-I. Meng, Relations between the interplanetary magnetic field Bz, AE index, and cusp latitude, J. Geophys. Res., **91**, 1549-1556, 1986a.

Carbary, J. F. and C.-I. Meng, Correlation of cusp latitude with Bz and AE(12) using nearly one year's data, J. Geophys. Res., **91**, 10047-10054, 1986b.

Clauer, C. T., P. M. Banks, A. Q. Smith, T. S. Jørgensen, E. Friis-Christensen, S. Vennerstrøm, V. B. Wickwar, J. D. Kelly and J. Doupnik, Observation of interplanetary magnetic field and of iono-spheric plasma convection in the vicinity of the dayside polar cleft, Geophys. Res. Lett., **11**, 891-894, 1984.

Eather, R. H., S. B. Mende and E. J. Weber, Dayside aurora and relevance to substorm current systems and dayside merging, J. Geophys. Res., **84**, 3339-3359, 1979.

Eather, R. H., Dayside auroral dynamics, J. Geophys. Res., **89**, 1695-1700, 1984.

Eather, R. H., Polar cusp dynamics, J. Geophys. Res., **90**, 1569-1576, 1985.

Erlandson, R. E., L. J. Zanetti, T. A. Potemra, P. F. Bythrow and R. Lundin, IMF By dependence of region 1 Birkeland current near noon, J. Geophys. Res., **93**, 9804-9814, 1988.

Feldstein, Y. I. and G. V. Starkov, Dynamics of auroral belt and polar geomagnetic disturbances, Planet. Space Sci., **15**, 209-229, 1967.

Friis-Christensen, E., K. Lassen, J. Wilhjelm, J. M. Wilcox, W. Gonzalez and D. S. Colburn, Critical component of the interplanetary magnetic field responsible for large geomagnetic effects in the polar cap, J. Geophys. Res., **77**, 3371-3376, 1972.

Friis-Christensen, E. and J. Wilhjelm, Polar cap currents for different directions of the interplanetary magnetic field in the Y-Z plane, J. Geophys. Res., **80**, 1248-1260, 1975.

Friis-Christensen, E., M. A. McHenry, C. R. Clauer and S. Vennerstrøm, Ionospheric traveling convection vortices observed near the polar cleft: A triggered response to sudden changes in the solar wind, Geophys. Res. Lett., **15**, 253-256, 1988.

Goertz, C. K., E. Nielsen, A. Korth, K. H. Glassmeier, C. Haldoupis, P. Hoeg and D. Hayward, Observations of a possible ground signature of flux transfer events, J. Geophys. Res., **90**, 4069-4078, 1985.

Harang, L., The mean field of disturbance of polar geomagnetic storms, Terr. Magn. Atmosph. Elec., **51**, 353-380, 1946.

Heelis, R. A., W. B. Hanson and J. L. Burch, Ion convection velocity reversals in the dayside cleft, J. Geophys. Res., **81**, 3803-3809, 1976.

Heelis, R. A., The effects of interplanetary magnetic field orientation on dayside high-latitude ionospheric convection, J. Geophys. Res., **89**, 2873-2880, 1984.

Heppner, J. P., Polar-cap electric field distributions related to the interplanetary magnetic field direction, J. Geophys. Res., 77, 4877-4887, 1972.

Heppner, J. P., Empirical models of high-latitude electric fields, J. Geophys. Res., 82, 1115-1125, 1977.

Heppner, J. P. and N. C. Maynard, Empirical high-latitude electric field models, J. Geophys. Res., 92, 4467-4489, 1987.

Horwitz, J. L. and S.-I. Akasofu, The response of the dayside aurora to sharp northward and southward transitions of the interplanetary magnetic field and to magnetospheric substorms, J. Geophys. Res., 82, 2723-2734, 1977.

Iijima, T., R. Fujii, T. A. Potemra and N. A. Saflekos, Field-aligned currents in the south polar cap and their relationship to the interplanetary magnetic field, J. Geophys. Res., 83, 5595-5603, 1978.

Jørgensen, T. S., E. Friis-Christensen, V. B. Wickwar, J. D. Kelly, C. R. Clauer and P. M. Banks, On the reversal from "sunward" to "antisunward" plasma convection in the dayside high latitude ionosphere, Geophys. Res. Lett., 11, 887-890, 1984.

Kaneda, E., Dayside auroral activity and its relation to substorm, Rept. Ionosph. Space Res. Japan, 27, 207-212, 1973.

Kokubun, S., T. Yamamoto, K. Hayashi, T. Oguti and A. Egeland, Impulsive Pi bursts associated with poleward moving auroras near the polar cusp, J. Geomagn. Geoelectr., 40, 537-551, 1988.

Lanzerotti, L. J., L. C. Lee, C. G. Maclennan, A. Wolfe and L. V. Medford, Possible evidence of flux transfer events in the polar ionosphere, Geophys. Res. Lett., 13, 1089-1092, 1986.

Lanzerotti, L. J., R. D. Hunsucker, D. Rice, L. C. Lee, A. Wolfe, C. G. Maclennan and L. V. Medford, Ionosphere and ground-based response to field-aligned currents near the magnetospheric cusp regions, J. Geophys. Res., 92, 7739-7743, 1987.

Lee, L. C. and Z. F. Fu, A theory of magnetic flux transfer at the Earth's magnetopause, Geophys. Res. Lett., 12, 105-108, 1985.

Lemaire, J., Impulsive penetration of filamentary plasma elements into the magnetosphere of the Earth and Jupiter, Planet. Space Sci., 25, 887-890, 1977.

Lundin, R. and E. Dubinin, Solar wind energy transfer regions inside the dayside magnetopause -I. Evidence for magnetosheath plasma penetration, Planet. Space Sci., 32, 745-755, 1984.

McDiarmid, I. B., J. R. Burrows and M. D. Wilson, Large scale magnetic field perturbations and particle measurements at 1400 km on the dayside, J. Geophys. Res., 84, 1431-1441, 1979.

Meng, C.-I., Case studies of the storm time variation of the polar cusp, J. Geophys. Res., 88, 137-149, 1983.

Meng, C.-I. and R. Lundin, Auroral Morphology of the midday oval, J. Geophys. Res., 91, 1572-1584, 1986.

Oguti, T., Poleward travel of electric current filament in the polar cap region, Rept. Ionosph. Space Res. Japan, 23, 175-184, 1969.

Oguti, T., Questions on the dayside reconnection in connection with magnetospheric convection and open-close boundary, submitted to J. Geomagn. Geoelectr., 1989.

Oguti, T. and K. Hayashi, Multiple correlation between auroral and magnetic pulsations, 2. Determination of electric currents and electric fields around a pulsating auroral patch, J. Geophys. Res., **89**, 7467-7481, 1984.

Oguti, T. and K. Hayashi, Polarization and wave form of magnetic pulsations below pulsating auroras: Magnetic effects of electric currents induced in an ionization tail of a moving auroral patch, J. Geomagn. Geoelectr., **37**, 65-91, 1985.

Oguti, T., T. Yamamoto, K. Hayashi, S. Kokubun, A. Egeland and J. A. Holtet, Dayside auroral activity and related magnetic impulses in the polar cusp region, J. Geomagn. Geoelectr., **40**, 387-408, 1988.

Russell, C. T. and R. C. Elphic, ISEE observations of flux transfer events at the dayside magnetopause, Geophys. Res. Lett., **6**, 33-36, 1979.

Sandholt, P. E., K. Henriksen, C. D. Deehr, G. G. Sivjee, G. J. Romick and A. Egeland, Dayside cusp auroral morphology related to nightside magnetic activity, J. Geophys. Res., **85**, 4132-4138, 1980.

Sandholt, P. E., A. Egeland, B. Lybekk, C. S. Deehr, G. G. Sivjee and G. J. Romick, Effects of interplanetary magnetic field and magnetospheric substorm variations on the dayside aurora, Planet. Space Sci., **31**, 1345-1362, 1983.

Sandholt, P. E., C. S. Deehr, A. Egeland and B. Lybekk, Signatures in the dayside aurora of plasma transfer from the magnetosheath, J. Geophys. Res., **91**, 10063-10079, 1986.

Sandholt, P. E. and A. Egeland, Auroral and magnetic variations in the polar cusp and cleft - signatures of magnetopause boundary layer dynamics, Astrophys. Space Sci., **144**, 171-199, 1988.

Sandholt, P. E., B. Lybekk, A. Egeland, R. Nakamura and T. Oguti, Midday auroral breakup, J. Geomagn. Geoelectr., **41**, in press, 1989.

Southwood, D. J., The ionospheric signature of flux transfer events, J. Geophys. Res., **92**, 3207-3213, 1987.

Svalgaard, L., Polar cap magnetic variations and their relationship with the interplanetary sector structure, J. Geophys. Res., **78**, 2064-2078, 1973.

Todd, H., B. J. I. Bromage, S. W. H. Cowley, M. Lockwood, A. P. van Eyken and D. M. Willis, EISCAT observations of bursts of rapid flow in the high latitude dayside ionosphere, Geophys. Res. Lett., **13**, 909-912, 1986.

Tsyganenko, N. A. and A. V. Usmanov, Determination of the magnetospheric current system parameters and development of experimental geomagnetic field models based on data from IMP and HEOS satellites, Planet. Space Sci., **30**, 985-998, 1982.

Vorobjev, V. G., G. Gustafsson, G. V. Starkov, Y. I. Feldstein and N. F. Shevnina, Dynamics of day and night aurora during substorms, Planet. Space Sci., **23**, 269-278, 1975

Willis, D. M., M. Lockwood, S. W. H. Cowley, A. P. van Eyken, B. J. I. Bromage, H. Rishbeth, P. R. Smith and S. R. Crothers, A survey of simultaneous observations of the high-latitude ionosphere and interplanetary magnetic field with EISCAT and AMPTE-UKS, J. Atmos. Terr. Phys., **48**, 987-1008, 1986.

ELECTRODYNAMICS OF AURORAL AND POLAR CAP ARCS AT VERY HIGH LATITUDES

R. R. Vondrak and R. M. Robinson
Lockheed Palo Alto Research Laboratory
3251 Hanover Street, Palo Alto, CA 94304

ABSTRACT

The electrodynamics of auroral and polar cap arcs at very high latitudes has been studied using data from the Sondrestrom incoherent scatter radar. These experiments have been performed in coordination with ground-based and space-based auroral imagers so that the radar observations can be described in the context of visible auroral features. The radar measures the spatial distribution of electron density and plasma drift, which are used to calculate the ionospheric conductivity, electric field, electric current, and energy characteristics of the precipitating particles. Many types of very high latitude aurora have been studied by this technique. We describe the electrodynamic configuration associated with discrete arcs in the early afternoon (1400 MLT), in the polar cap, and in the late morning. Each of these arcs has a distinct relationship with respect to the large-scale electric fields and field-aligned currents at very high latitudes. This information is combined with similar results obtained while the radar was at Chatanika, Alaska. Together these data can be used as a framework for a phenomenological understanding of auroral precipitation.

INTRODUCTION

Recently, there has been much interest in electrodynamic processes at very high latitudes, stimulated by observations made by ground-based and space-based instruments. These observations have raised many new questions about the nature of solar-terrestrial coupling and have changed substantially our view of the polar cap ionosphere and upper atmosphere. Previously, understanding of these processes has been limited by the lack of coordinated measurements of key electrodynamic parameters. The relocation of the Chatanika incoherent scatter radar to Sondre Stromfjord, Greenland (Kelly, 1983), has presented a new opportunity to study electrodynamic processes at very high latitudes (i.e. invariant latitudes greater than approximately 70^{0}).

P. E. Sandholt and A. Egeland (eds.), Electromagnetic Coupling in the Polar Clefts and Caps, 269–284.
© 1989 by Kluwer Academic Publishers.

In this review we first summarize the understanding of the high latitude electrodynamics of discrete auroral arcs that was obtained with the Chatanika radar. We then describe the results that have been obtained at Sondre Stromfjord. These experiments have yielded information about three distinct types of discrete aurora persistently observed at very high latitudes. The first is an arc-like precipitation feature in the afternoon local time sector. The second is an array of arcs in the morning sector auroral zone. The third is a transpolar arc observed in the late evening sector. We then identify some other dynamic types of aurora that are seen at Sondrestrom. Finally, we show how the Sondrestrom measurements can be combined with the Chatanika results to produce a comprehensive picture of high latitude precipitation and its relationship to the large-scale patterns of convection and field-aligned currents.

ELECTRODYNAMIC CONFIGURATION OF HIGH-LATITUDE AURORA

Numerous studies of high latitude discrete auroral electrodynamics have been based on coordinated measurements made by incoherent scatter radar and polar-orbiting spacecraft (Robinson et al., 1982; Senior et al., 1982; Vondrak and Rich, 1982; Robinson et al., 1984, 1985; Vondrak et al., 1985). These studies pertain primarily to electrodynamics of late evening and early morning sector aurora because they generally relied upon data obtained by the Chatanika incoherent scatter radar while it was located in Alaska. At an invariant latitude of 64.8°, Chatanika was an ideal site for studying the auroral ionosphere at these local times.

The experiments performed at Chatanika were coordinated with overpasses of satellites such as Triad, S3-2, AE-C, and Dynamics Explorer. During these overpasses the radar was usually scanned in the magnetic meridian plane. Measurements made in the magnetic meridian yield important information about the cross-sectional variation of electric fields, conductivities and currents in high latitude arcs. Such observations have been used to construct the diagram shown in Figure 1 (Vondrak, 1981). The bottom panel shows the typical pattern of electron density in the magnetic meridian in the evening sector auroral zone. The principal ionization features in the E-region are the relatively uniform diffuse aurora and the discrete arcs that are isolated by dark bands. Each of these features can be associated with precipitation from a distinct magnetospheric plasma regime. The meridional electric field profile that is usually observed in association with these aurora is shown in the middle panel. The top panel shows the corresponding field-aligned current structure. There are two large-scale field-aligned current sheets corresponding to Region 1 and Region 2 (Iijima and Potemra, 1976). The boundary between the two field-aligned current sheets is generally associated with a localized ionization enhancement within the diffuse aurora referred to as the interface arc. The Region 2 downward field-aligned current coincides with the equatorward portion of the diffuse aurora (Robinson et al., 1982). The dashed lines within

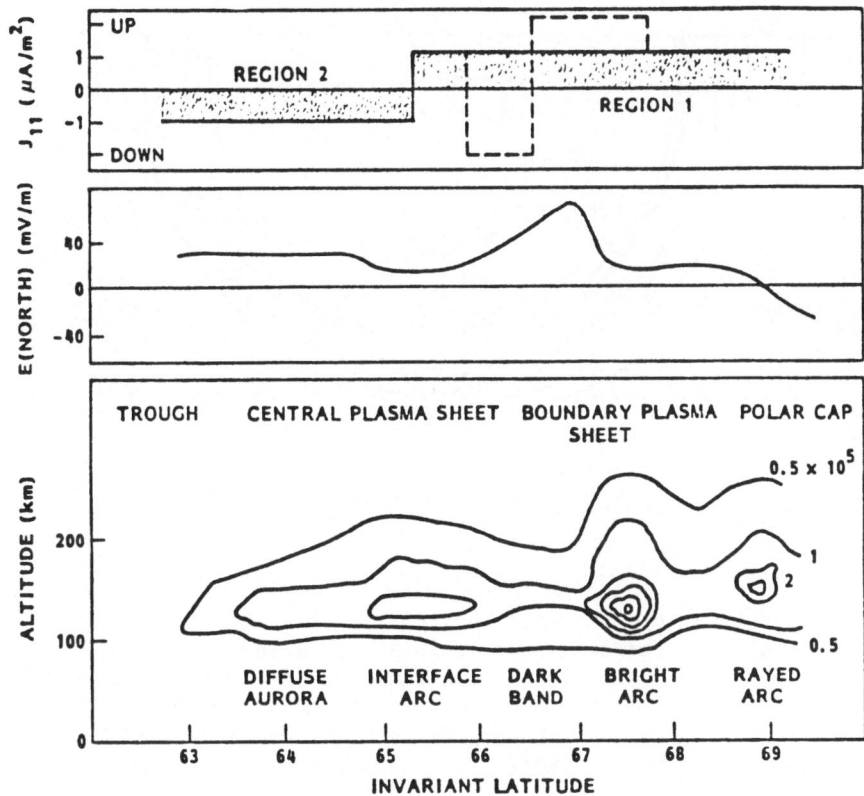

Figure 1. Schematic representation of relationship between electric
field, field-aligned currents, and precipitation-produced
ionization across the evening sector auroral zone (from
Vondrak, 1981).

region 1 indicate the type of perturbation in the large scale-field-
aligned current configuration that is sometimes, but not always,
observed when an auroral arc is present. At higher latitudes there is
sometimes a weak arc near the ion-convection reversal that corresponds
to the polar cap boundary in the early evening and the Harang discontinuity in the midnight sector. This arc seems to represent a
boundary where the ionospheric electric field reverses direction from
northward to southward.

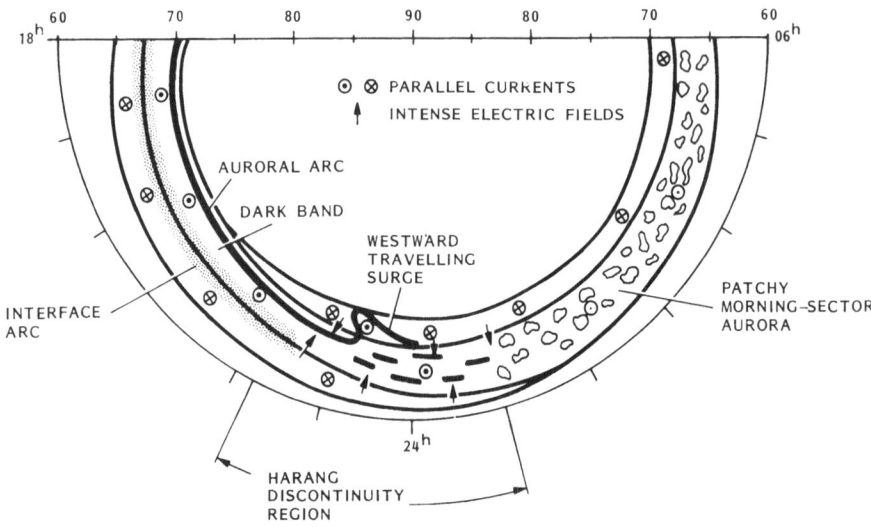

Figure 2. Schematic diagram of the relationship between field-
         aligned currents, and visible aurora in the nightside
         auroral zone. The midnight sector is a transition region
         between the evening sector pattern deduced by Robinson et
         al. (1982) and the morning sector pattern deduced by
         Senior et al. (1982).

    The evening-sector observations described above have been com-
bined with the midnight sector results of Robinson et al. (1985) and
the morning sector results of Senior et al. (1982) to produce the
sketch shown in Figure 2. This diagram shows the spatial relationship
between visual aurora and field-aligned currents over the entire
nightside auroral zone. On the morningside the region 2 current sheet
coincides with patchy morning sector diffuse aurora (Senior et al.,
1982). The region 1 current sheet lies poleward of this aurora. In
the midnight sector near the Harang discontinuity the dominant field-
aligned current is upward. Equatorward of this current a downward
sheet is observed which is the eastern end of the evening sector
region 2 sheet. Poleward of the upward current are several sheets of
varying directions and magnitudes. These sheets are associated with
active aurora during substorms. The most intense electric fields are
located at the poleward and equatorward boundaries of the upward
current sheet. The midnight sector is a transition region in which
the evening sector arc breaks up into a number of isolated arcs.
Westward traveling surges, when they occur, are situated in the
poleward portions of the auroral zone.

VERY HIGH LATITUDE AURORA OBSERVED AT SONDRE STROMFJORD

The incoherent scatter radar at Chatanika was moved to Sondre Stromfjord, Greenland in 1982 and began operations in February, 1983. At an invariant latitude of 74° the radar is at an ideal location to study electrodynamic properties of very high latitude aurora, such as quiet auroral arcs and polar cap arcs. Figure 3 shows the four main types of discrete auroras observed at very high latitudes in DMSP satellite photographs Gussenhoven, 1982. The percentages show the occurrence frequency for each auroral configuration. The E-region field of view at Sondre Stromfjord is shown by the shaded circle. During the last five years Sondrestrom radar observations have been coordinated with other ground-based and space-based measurements to

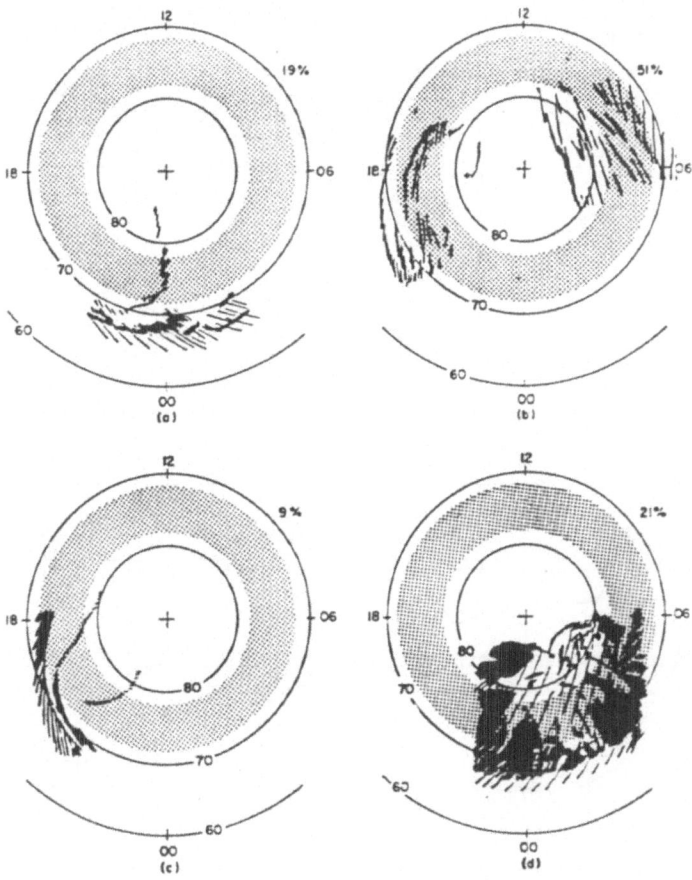

Figure 3. Distribution of very high latitude auroras as determined from DMSP satellite imagery at visible wavelengths (from Gussenhoven, 1982).

study the electrodynamic configuration of these different types of discrete aurora.

The Early-Afternoon (1400MLT) Arc

Ionization produced by electron precipitation in the early after-noon sector is one of the most distinctive features that have been observed in the Sondrestrom radar data (Clauer et al., 1984; Jorgensen et al., 1984). Electric fields and currents associated with this type of ionization enhancement were studied by Robinson et al. (1984). The results of that experiment are shown in Figure 4. Triad magnetometer data were used to show that the enhancements are within the region 1 upward field-aligned current. Electric fields in the enhancement

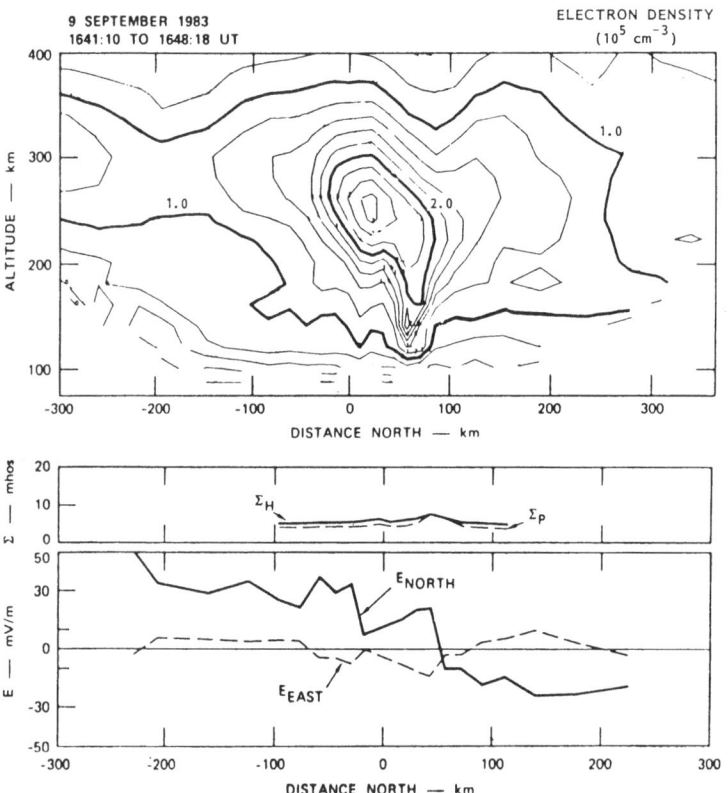

Figure 4.   Electron density, conductivities and electric fields associated with an auroral arc in the early afternoon time sector (Robinson et al., 1984).

satisfy the condition $\nabla \cdot E < 0$.   Thus, electrodynamically this discrete feature resembles evening sector auroral arcs.   However, the
high altitude of the ionization produced by the precipitation indicates lower energy electron precipitation.   Simultaneous electron
spectrometer data from the NOAA satellite were used to show that the
parallel accelerating potential associated with the precipitation was
approximately one kV.   Statistically the field-aligned currents at
this local time are more intense than in the evening sector.   This by
itself would result in higher parallel potentials.   However, Robinson
et al. (1984) suggested that field lines from the afternoon arc thread
the low latitude boundary layer where the plasma density is
several/cm^3.   The much higher source density relative to that in the
tail plasma sheet relieves the requirement for large potential drops
along the field line.   A subsequent study by Evans (1985) confirmed
the lower potential drops and larger source densities associated with
these discrete features.   Lundin and Evans (1985) presented a model in
which the afternoon arc is the ionospheric signature of the direct
penetration of solar wind plasma into the low latitude boundary layer.

The close association of these afternoon arcs with the convection
reversal boundary suggests that they are strongly tied to the configuration of the interplanetary magnetic field (IMF).   In particular, the
reversal in convection in the afternoon sector is associated with
negative values of the y component of the IMF.   This association was
confirmed by Robinson et al. (1986) who correlated the occurrence of
afternoon arcs in radar data with interplanetary magnetic field data.
Thus, observations of this arc may give a direct indication of the y
component of the IMF.

Very High Latitude Arcs in the Morning Sector Auroral Zone

A case study of discrete arcs in the morning sector auroral zone
was presented by Robinson et al., (1987).   For several hours around
local dawn on November 17, 1985, a system of auroral arcs was observed
over Sondrestrom.   These arcs varied in intensity and width, but were
all extended generally in the magnetic east-west direction.   A DMSP
photograph obtained during the observations showed that two types of
arcs were present.   In the south the arcs were localized enhancements
or striations within a broad diffuse aurora.   Poleward of the diffuse
aurora was a system of arc segments that extended over about five
degrees of latitude.   The general configuration was very similar to
that shown in Figure 3b.

The relationship among ionization, electric fields, and field-
aligned currents for these morning-sector arcs was deduced from
simultaneous observations by the radar and by the HILAT satellite and
is shown schematically in Figure 5.   Throughout the region of upward
field-aligned current there is a relatively uniform flux of electrons
with energies near 1 keV.   This produces a background diffuse auroral
glow that is apparent in the ground-based and space-based imaging.
Latitudinally localized ionization enhancements are embedded within

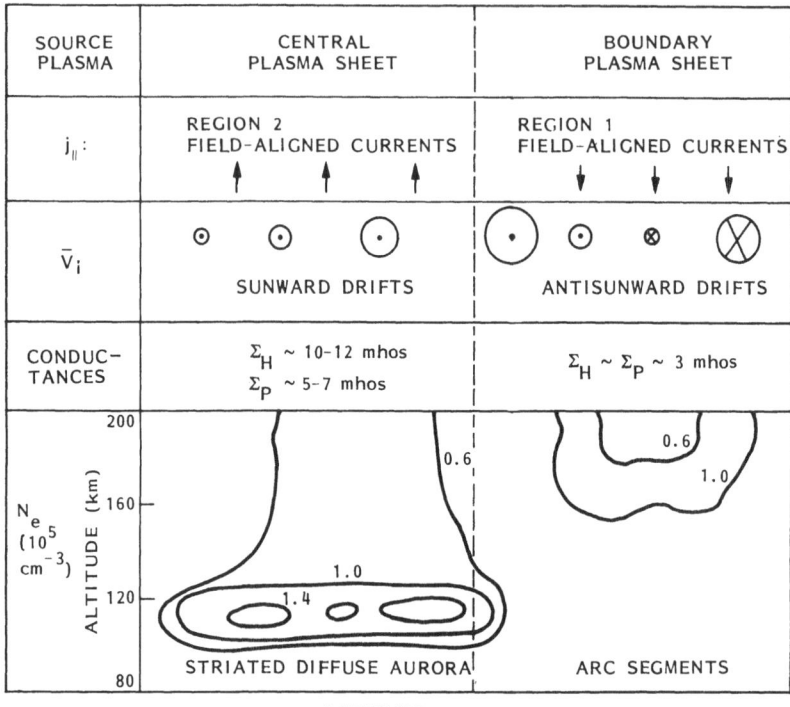

Figure 5. Relationships between electric fields, conductivities and field-aligned currents for morning sector auroral arcs (from Robinson et al., 1987a).

the diffuse auroral region. These structures are east-west extended and are produced by enhanced fluxes of electrons from several keV to 10 keV with no significant changes in the fluxes at lower energies. The electric field in the diffuse auroral region is predominantly southward with a magnitude that increases with latitude to the pole-ward extent of the upward field-aligned current sheet. The poleward boundary of the diffuse auroral precipitation occurs just poleward of the interface between the upward and downward field-aligned currents. The height-integrated conductances in the diffuse aurora are also given in the figure. The Hall conductance varies between 10 and 12 mhos, while the Pedersen conductance is between 5 and 7 mhos. The Hall to Pedersen ratio is about 2 which reflects the 5-keV average energy of the precipitating electrons.

Poleward of the diffuse aurora the fluxes of electrons with energies above 1 keV are lower by several orders of magnitude. The ionization in this region occurs at altitudes above 150 km and is produced by enhanced electron fluxes below 1 keV. These enhanced fluxes are highly localized and produce a system of disconnected arc segments. Although the precipitating fluxes are localized, the ionization produced is spread out considerably, so that individual arcs are not apparent in the cross sections of electron density. This is probably due to the limited spatial and temporal resolution of the radar in the scan mode and indicates that the source of the ionization is transient or moving. Although these arcs are barely visible in the ground-based all-sky camera photographs, careful examination of the photos indicates the presence of auroral rays embedded within them. Because of the high altitude of the peak electron density the height integrated Hall and Pedersen conductances are approximately equal. Electric fields in this region were not measured by the radar, but since the southward fields maximized at the equatorward edge of this region, fields must also have been southward within some distance from this boundary. The region 1 field-aligned currents map to the low-latitude boundary layer within which a convection reversal is typically observed (Bythrow et al., 1981). Thus the electric field may have changed sign farther to the north, and some of the downward field-aligned current observed here may have been on field lines containing antisunward drifting plasma. Field-aligned currents inferred from the radar measurements of electric fields and conductivities are consistent with the field-aligned currents inferred from the HILAT magnetometer data. In the region of the diffuse aurora the southward electric field increases with latitude. Thus the southward Pedersen current driven by this electric field diverges to produce a broad upward field-aligned current approximately coincident with the diffuse aurora.

Sun-Aligned Arcs Connected to the Auroral Oval

A complex auroral form that has received much attention since the availability of satellite-based imagery is the theta aurora. This phenomenon has been extensively studied using data from the Dynamics Explorer satellites (Frank et al., 1986). The theta arcs are polar cap arcs that attach to the oval at both ends. From the point of view of both ionospheric electrodynamics and magnetosphere-ionosphere coupling the location at which the theta arc intersects the oval is an extremely interesting area of investigation

On the evening of 14 November, 1985, the Sondrestrom incoherent scatter radar made measurements of ionization and drifts associated with a sun-aligned polar cap arc near local midnight. All-sky photographs showed that the arc, which appeared to move westward across the Sondrestrom meridian, was connected to the auroral oval. Figure 6 shows the location of auroral forms over Sondrestrom as determined from four all-sky camera photographs. The photographs from which the sketches were derived were taken with a 630.0 nm filtered and light-intensified film camera (Mende et al., 1988). The 630.0 nm filter admits $O(^1D)$ emissions from atomic oxygen which, because of quenching, originate primarily from altitudes above 200 km. The sun-aligned arc was first observed on the horizon to the east at about 2300 UT (2130 MLT). The arc appeared to drift westward crossing the Sondrestrom meridian at 2330 UT. The velocity of the apparent motion was about 300 m/s in a reference frame corotating with the earth. However, the uncertainty in this measurement is large enough so that it is possible that the arc was stationary in local time with the apparent velocity simply due to the rotation of the earth which produces an apparent drift of about 200 m/s at this latitude.

At about 2340 UT, 630.0 nm luminosity from the poleward portion of the auroral oval, which was out of the field of view of the all-sky camera prior to this time, appears on the southern horizon. The all-sky image taken at 2345 UT (Figure 6) shows that the sun-aligned arc

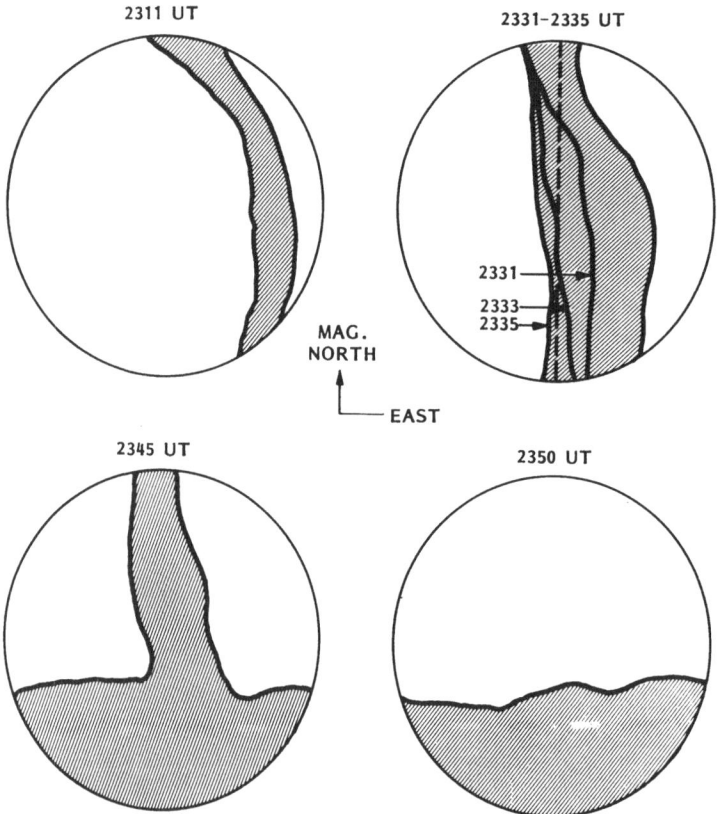

Figure 6. Location of auroral forms over Sondre Stromfjord as determined from all sky camera photographs taken with a 630.0 nm filter on the evening of 14 November 1985. In the second frame the location of the western edge of the sun-aligned arc is shown at three different times. The dashed line shows the magnetic meridian along which the radar was scanning during the 5 minute sequence.

was connected to the auroral oval producing the distinctive pattern of luminosity shown. At 2350 UT, shortly after the northward motion of the poleward edge of the auroral oval, the sun-aligned arc faded, leaving only an east-west band of luminosity to the south.

For the observations discussed here the Sondrestrom incoherent scatter radar was operating in a combination fixed position and scan mode. Although the radar does not measure electric fields and conductivities simultaneously over a two dimensional area, it is possible to construct a simplified model that is consistent with the observations. Such a model is shown in Figure 7. The radar electric field measurements suggest that the antisunward moving plasma in the sun-aligned

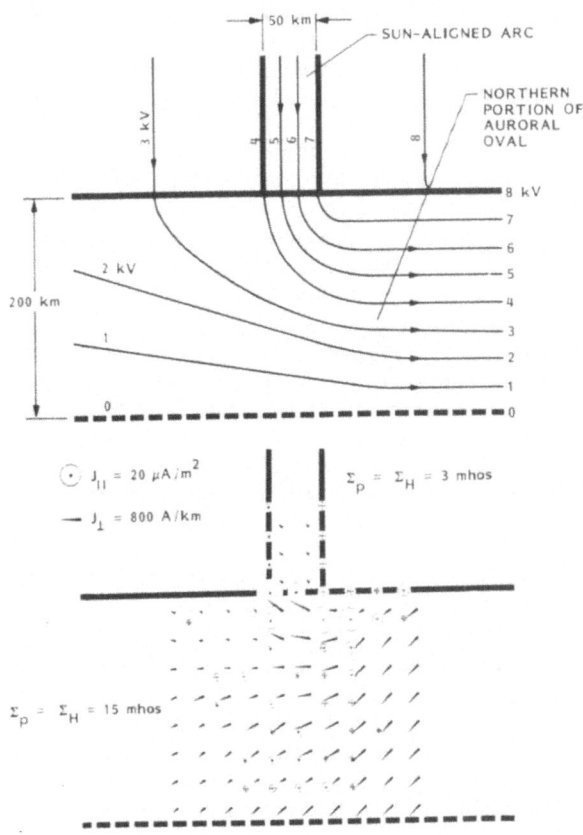

Figure 7. Model convective flow and currents for a sun-aligned arc that connects to the auroral oval (from Robinson et al., 1987.)

arc merges with the eastward flowing plasma in the auroral oval. The convection streamlines in the top panel of Figure 7 have been drawn accordingly. The equipotentials are spaced so that there is a westward electric field in the polar cap arc of 60 mV/m and a southward electric field in the auroral oval that varies from 15 to 40 mV/m from west to east. The modeling was done only for the sun-aligned arc and the northern portion of the auroral oval. The average westward electric field outside of the sun-aligned arc on either side is approximately 10 mV/m. The Hall and Pedersen conductances are assumed to be equal to 3 mhos in the polar cap arc, 15 mhos in the auroral oval, and zero everywhere else. These electric fields and conductances produce the horizontal and vertical currents shown in the bottom portion of the figure. The horizontal currents are indicated by the arrows. In the auroral oval the currents are predominantly southwestward except near the point where the polar cap arc connects. In this region the southward electric field rotates toward the west producing a northwestward current. In the polar cap arc the current is also northwestward, but the magnitude is reduced considerably owing to the lower conductances.

The field-aligned currents computed from the divergence of the horizontal currents are shown by the circles with the radius of the circles proportional to the current density. On the edges of the polar cap arc there are field-aligned currents connected to the westward Pedersen currents in the arc. These field-aligned currents are upward on the western edge and downward on the eastern edge. To the east of the point where the arc connects to the auroral oval there is a zone of downward field-aligned current that connects to the southward Pedersen currents and are thus part of the region 1 field-aligned currents (Iijima and Potemra, 1976). The most intense field-aligned currents are at the junction between the two auroral forms. On the west side of the junction the currents are upward, while on the east side of the junction they are downward. The magnitudes of the currents exceed 20 $\mu A/m^2$.

An implication of the model is that strong upward and downward field-aligned current filaments are present where the polar cap arc meets the auroral oval. The currents are intense enough that the magnetospheric electrons cannot carry the current without a parallel potential along the field lines. This parallel electric field accelerates electrons into the ionosphere producing enhanced auroral emissions. The enhanced auroral luminosity at this location has been reported by Murphree et al. (1987) who noted this phenomenon in images obtained by Viking. The presence of this bright spot at precisely the location predicted in the model lends credence to the derived pattern of electric fields and currents.

Other Types of Discrete Aurora at Very High Latitudes

Isolated sun-aligned arcs drifting across the polar cap have been observed by both optical and radar methods. A simultaneous study was analyzed by Mende et al. (1988). In that study, all-sky, light-intensified images were used to place individual radar drift measurements in the proper location relative to the arcs. Although there was considerable variation in the magnitudes and directions of the velocity, the general trend is for anti-sunward drifts within the arcs. The puzzling aspect of these observations is that theta arcs have been observed to be associated with sunward drifts. Thus, it remains unclear how the arcs observed by the radar relate to the theta aurora studied by Frank et al. (1986).

Another type of auroral form that is frequently observed at Sondrestrom is the westward traveling surges. Recent theoretical studies have attempted to explain the surge phenomenon (e.g. Rothwell et al., 1984; Kan et al., 1984). Imaging data from the Viking spacecraft reported by Shepherd et al. (1987) have called into question the concept of a westward traveling surge. In some cases, there appears to be no westward motion associated with auroral intensifications that accompany substorms. Regardless of whether there is a westward motion associated with these surges, it is known that auroral luminosity brightens and expands poleward of the quiet time oval when the westward electro-jet intensifies. Ahead of the surge the electric fields are northward except at the poleward edge of the auroral oval where strong southward fields are observed. Within the surge the fields are generally small and southward, in anticorrelation with the conductivity.

SUMMARY

The observations at Sondrestrom and at Chatanika can be used to identify the relationships between aurora, electric fields and field-aligned currents. The results of these studies are summarized in Figure 8. In the center of the figure is the large-scale pattern of field-aligned currents at high latitudes as determined by Iijima and Potemra (1976) using data from the Triad satellite. Because many polar-orbiting satellites carry magnetometers to measure these field-aligned currents, this pattern has been useful in ordering data from other instruments. The panels surrounding this pattern show the observed relationships among aurora, electric fields and field-aligned currents at five different local times that have been described in this paper.

The relationships depicted in this figure were compiled from the case studies of auroral precipitation and its associated electrodynamic properties. Although there are many previous studies whose results can be integrated into a comprehensive description of auroral processes, more experiments and analysis are necessary to determine (1) the dependence of auroral precipitation on interplanetary magnetic field direction, (2) the persistence or characteristic lifetime of the precipitation, and (3) the location of the precipitation relative to the large-scale convection pattern. Such a model has profound importance to the study of ionospheric and thermospheric dynamics.

282

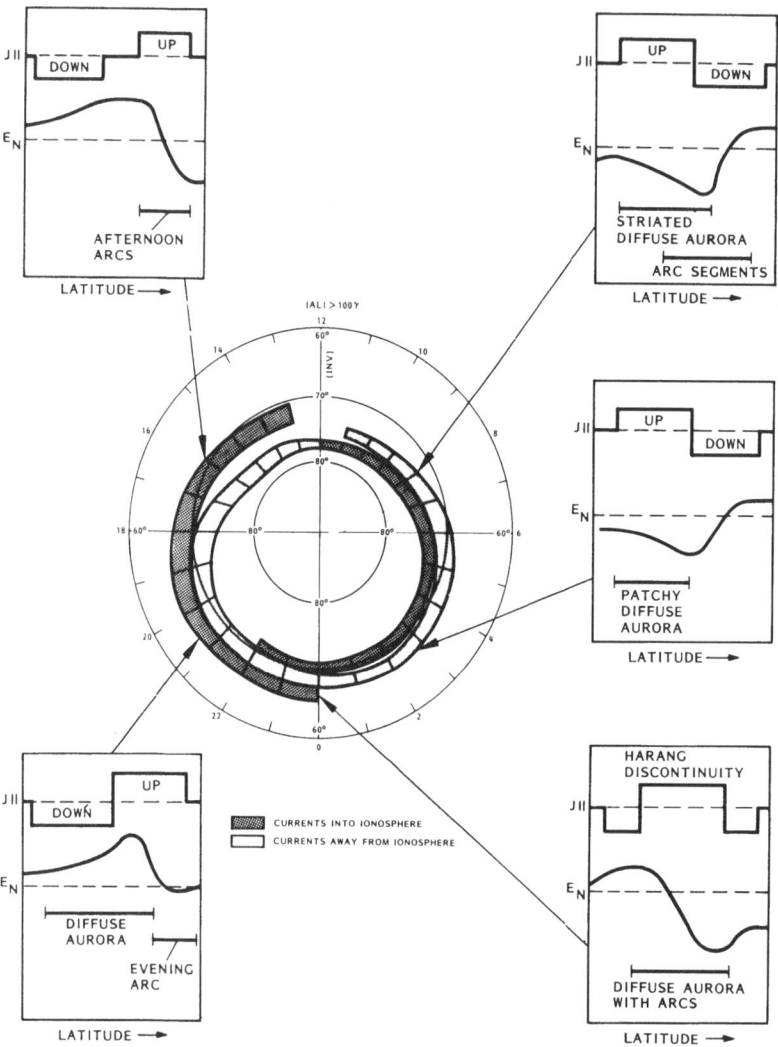

Figure 8. Field-aligned currents, meridional electric fields and
aurora observed at five local times. In the center of
the figure is the statistical pattern of field-aligned
currents observed by Iijima and Potemra (1976).

The primary objective in establishing the relationships discussed above is to achieve a phenomenological understanding of auroral electrodynamics. This means that once a particular configuration of aurora is identified, the associated electrodynamic parameters can be estimated. These relationships can be used to study the behavior of electric fields and currents from ground-based or space-based images of the instantaneous auroral distribution.

ACKNOWLEDGEMENTS

This work was supported by NSF Grant ATM-8717840 and the Lockheed Independent Research Program. The authors are indebted to the SRI International personnel at the Sondre Stromfjord radar site for their help in acquiring the data.

REFERENCES

Bythrow, P. F., R. A. Heelis, W. B. Hanson and R. A. Power, Observational evidence for a bundary layer source of dayside region 1 field-aligned currents, J. Geophys. Res., 86,5577, 1981.

Clauer, C. R., P. M. Banks, A. Q. Smith, T. S. Jorgensen, E. Friis-Christensen, S. Vennerstrom, V. B. Wickwar, J. D. Kelly, and J. Doupnik, Observations of interplanetary magnetic field and ionosphere plasma convection in the vicinity of the dayside polar cleft, Geophys. Res. Lett., 11, 891, 1984.

Evans, D. S., The characteristics of a persistent auroral arc at high latitude in the 1400 MLT sector, in The Polar Cusp, edited by J. A. Holtet and A. Egeland, p. 99, D. Reidel Pub. Co., Dordrecht, Holland, 1985.

Frank, L. A., J. D. Craven, D. A. Gurnett, S. D. Shawhan, D. R. Weimar, J. L. Burch, J. D. Winningham, C. R. Chappell, J. H. Waite, R. A. Heelis, N. C. Maynard, M. Suguira, W. K. Peterson, and E. G. Shelley, The theta aurora, J. Geophys. Res., 91, 3177, 1986.

Gussenhoven, M. S., Extremely high latitude auroras, J. Geophys. Res., 87, 2401, 1982.

Iijima, T. and T. A. Potemra, The amplitude distribution of field-aligned currents at northern high latitudes observed by Triad, J. Geophys. Res., 81, 2165, 1976.

Jorgensen, T. S., E. Friis-Christensen, V. B. Wickwar, J. D. Kelly, C. R. Clauer, and P. M. Banks, On the reversal from "Sunward" to "Antisunward" plasma convection in the dayside high latitude ionosphere, Geophys. Res. Lett., 11, 887, 1984.

Kan, J. R., R. L. Williams, and S.-I. Akasofu, A mechanism for the westward traveling surge during substorm, J. Geophys. Res., 89, 2211, 1984.

Kelly, J. D., Sondrestrom radar--initial results, Geophys. Res. Lett., 10, 1112, 1983.

Lundin, R. and D. S. Evans, Boundary layer plasmas as a source for high-latitude, early afternoon, auroral arcs, Planet, Space Sci., 33, 1389, 1985.

Mende, S. B., J. H. Doolittle, R. M. Robinson and R. R. Vondrak, Plasma drifts associated with a system of sun-aligned arcs in the polar cap, J. Geophys. Res., 93, 256, 1988.

Murphree, J. S., L. L. Cogger, C. D. Anger, D. D. Wallis and G. G. Shepherd, Oval intensifications associated with polar arcs, Geophys. Res. Lett., 14, 403, 1987.

Robinson, R. M., R. R. Vondrak, and T. A. Potemra, Electrodynamic properties of the evening sector ionosphere within the region 2 field-aligned current sheet, J. Geophys. Res., 87, 731, 1982.

Robinson, R. M., D. S. Evans, T. A. Potemra, and J. D. Kelly, Radar and satellite measurements of an F-region ionization enhancement in the post-noon sector, Geophys. Res. Lett., 11, 899, 1984.

Robinson, R. M., F. Rich and R. R. Vondrak, Chatanika radar and S3-2 measurements of auroral zone electrodynamics in the midnight sector, J. Geophys. Res., 90, 8487, 1985.

Robinson, R. M., C. R. Clauer, O. de la Beaujardiere, J. D. Kelly and D. S. Evans, IMF B$_y$ control of ionization and electric fields measured by the Sondrestrom radar, in Solar Wind-Magnetosphere coupling, Y. Kamide Ed., 1986.

Robinson, R. M., R. R. Vondrak, D. Hardy, M. S. Gussenhoven, T. A. Potemra and P. F. Bythrow, Electrodynamics of very high latitude arcs in the morning sector auroral zone, J.Geophys. Res., 93, 913, 1987.

Rothwell, P. L., M. B. Silevitch and L. P. Block, A model for the propagation of the westward traveling surge, J. Geophys. Res., 89, 8941, 1984.

Senior, C., R. M. Robinson, and T. A. Potemra, Relationship between field-aligned currents, diffuse auroral precipitation and the westward electrojet in the early morning sector, J. Geophys. Res., 87, 469, 1982.

Shepherd, G. G., C. D. Anger, J. S. Murphree and A. Vallance-Jones, Auroral intensifications in the evening sector observed by the Viking ultraviolet imager, Geophys. Res. Lett., 14, 395, 1987.

Vondrak, R. R., Chatanika radar measurements of the electrical properties of auroral arcs, in Physics of Auroral Arc Formation, Geophysical Monograph 25, AGU, Washington, D. C., 1981.

Vondrak, R. R. and F. J. Rich, Simultaneous Chatanika radar and S3-2 satellite measurements of ionospheric electrodynamics in the diffuse aurora, J. Geophys. Res., 87, 6173, 1982.

Vondrak, R. R., J. S. Murphree, and C. D. Anger, Remote sensing of high latitude ionization with the ISIS-2 auroral scanning photometer, Radio Science, 20, 439, 1985.

SIMULTANEOUS RADAR AND SATELLITE OBSERVATIONS OF THE POLAR CUSP/CLEFT
AT SONDRE STROMFJORD

C.E. Valladares[1], Su. Basu[1], R.J. Niciejewski[2], and
R.E. Sheehan[3]

[1]Emmanuel College, 400 The Fenway, Boston, MA 02115, USA
[2]University of Michigan, Ann Arbor, MI 48109, USA
[3]Boston College, Chestnut Hill, MA 02167, USA

ABSTRACT. On February 9, 1988 a multi-instrument observation of the
dayside polar ionosphere was made at Sondre Stromfjord, Greenland as
part of a CEDAR high latitude campaign. Our objective was to observe
the low altitude signature of the cusp during southward IMF but under
fairly quiet ionospheric conditions (Kp=2). The HiLat satellite
measured in-situ particle precipitation, field-aligned currents and
velocity shears in the dayside cusp/cleft, while the VHF beacon on board
showed large phase scintillation and saturated intensity scintillation.
The phase spectral index was fairly shallow indicating that large power
spectral density was present up to scales on the order of 100m. During
the spacecraft overflight, the radar was scanning close to the F-region
projection of the satellite track. It is the object of this paper to
compare the radar measured thermal density, Ti and Te features in the
F-region with topside density and velocity structures and particle
precipitation characteristics measured in-situ by the satellite at
830 km. The precipitating electron flux measured by HiLat was also
input to an ionospheric chemistry model that calculates electron density
and temperature profiles for different exospheric temperatures. The
modeling results show that most of the features of the radar data may be
explained in terms of cleft precipitation, consequent higher exospheric
temperatures and enhanced convection.

1. INTRODUCTION

The polar cusp has been defined as the region of very intense and
structured particle precipitation with soft magnetosheath-like energy
spectra (Heikkila, 1985; Vasyliunas, 1985). Typically the average
energy of the electrons deposited in the cusp region is below 100 eV
(Frank, 1971), and the number flux is on the order of
10^9-10^{10} el/cm^2-s-ster (Gussenhoven et al., 1985). The location of the
low altitude cusp is closely related to the convection pattern which
depends on the orientation of the IMF. Potemra et al. (1985) suggested
that the velocity reversal and the location of the region 1 current
could be used as indicators of the equatorward edge of the cusp. Burch

P. E. Sandholt and A. Egeland (eds.), Electromagnetic Coupling in the Polar Clefts and Caps, 285–298.
© 1989 by Kluwer Academic Publishers.

et al. (1985) pointed out that during southward IMF with B_y negative, plasma in the pre-noon region convects mostly eastward except for a narrow zone that convects west. This feature was interpreted in terms of a viscous cell. Recently, Bythrow et al. (1988) and Erlandson et al. (1988) have presented evidence that the cusp precipitation region coincides with the region 1 current system, and that the traditional cusp current flows along lines that map to the plasma mantle.

Roble and Rees (1977) developed a numerical model of the high latitude ionosphere to study the response of the plasma parameters when cusp-type or auroral-type electrons are precipitating. They concluded that for soft precipitation and quiet background conditions the F-region electron temperature rises in a few seconds, but plasma density requires a few minutes to build up. Stamnes et al. (1985) pointed out that the solar EUV input may in general contribute substantially to the electron density and temperature in cusp auroras in stations such as Sondrestrom which has a relatively low geographic latitude.

Several efforts have been made earlier to study the polar cusp region at Sondrestrom. Kelly (1985) presented two types of daytime density enhancements observed with the radar. One type has high Te and is associated with region 1 Birkeland currents, and it is located equatorward of the shear reversal (Foster et al., 1985). The other type of Ne increase does not show Te enhancement, instead is related to large plasma flows which transport plasma from lower latitudes.

This paper presents the results of one multi-technique experiment performed at Sondrestrom with the purpose of investigating the following topics: 1) identify the radar signature of the cusp/cleft region in terms of the geophysical parameters measured by the radar, 2) compare the satellite observations at 830 km with the radar measured Ne and Te at F-region altitudes, and 3) determine background conditions for small-scale structuring in the cusp/cleft region. Further, the energy spectra of the precipitating electrons measured in-situ were used as inputs to a model of production rate and ionospheric chemistry, to determine time scales involved in the ionospheric response to the cusp precipitation.

2. OBSERVATIONS

On February 9, 1988 the Sondre Stromfjord radar (66.99°N, 50.95°W, 74° invariant latitude) was operated in a slow scan mode from 12-15 UT. During the experiment the radar antenna performed elevation scans in the magnetic meridian plane, probing 120 degrees of the sky in 15 minutes. The low scanning speed was selected to provide temperature measurements with small statistical uncertainty and at the same time give adequate latitudinal resolution, 20 km in the F-region. The HiLat satellite crossed the Sondrestrom latitude at 12:55:20 UT. Figure 1 shows the projection of the HiLat track at 350-km altitude, together with the radar coverage at the same altitude. Both, satellite overpass and radar scan, were coordinated to start simultaneously and progress as a north to south motion.

The IMP-8 satellite was in the solar wind on this day. This spacecraft measures the 3 components and the total value of the

FEB. 9, 1988

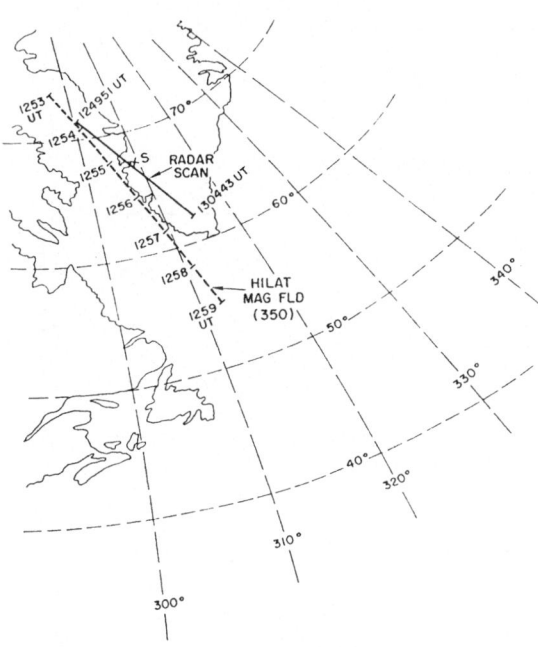

Figure 1. Locations of radar scan and HiLat magnetic field line trace at 350-km altitude for Sondrestrom.

interplanetary magnetic field in the Solar-Magnetospheric coordinate system. During the experiment B_x was positive, the average value was equal to +3 nanoteslas (nT), B_y was negative with a mean value of -2 nT, B_z was mainly negative except for a 15-minute positive excursion from -3 to +1 nT which occurred between 13:10 and 13:25 UT. During the time of the HiLat overpass B_z pointed south. The 3-hour Kp index was 2, the solar sunspot number was 47, and the solar zenith angle (SZA) at Sondrestrom for the time of the measurement was 87°.

2.1. HiLat In-Situ Observations

The HiLat satellite carried both in-situ instrumentation and coherent radio beacons when it was launched in 1983 (Fremouw et al., 1985). The in-situ data will be discussed in this section while the beacon data will be presented in the next section.

Figure 2 shows relevant parameters measured by HiLat during the descending pass of 12:49 UT. Data from 4 instruments on-board HiLat are plotted versus invariant latitude in order to facilitate the comparison with radar data. Panel a shows the number flux measured by the electron flux J sensor (Hardy et al., 1984). The number flux is enhanced almost 2 orders of magnitude between 75.8° and 79°Λ, with respect to the flux level in the polar cap. Panel b is the average energy of the precipitating electrons, as measured by the zenith detector. The mean

288

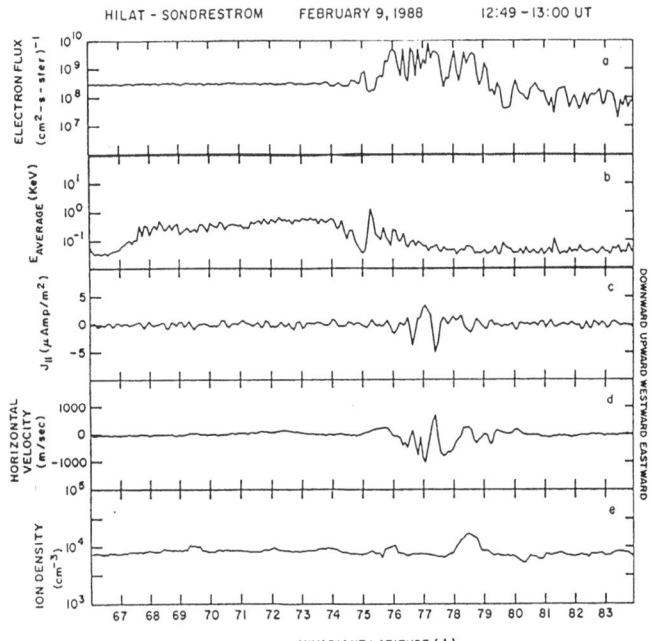

HILAT - SONDRESTROM   FEBRUARY 9, 1988   12:49 – 13:00 UT

INVARIANT LATITUDE (Λ)

Figure 2. HiLat in-situ data of precipitating electron flux (panel a), average energy of electrons (b), field-aligned current intensity (c), horizontal E-W velocity (d), and ion-density (e) at 830-km altitude.

energy of the electrons is below 100 eV at Λ latitudes greater than 76.5°. The high intensity and softness of the electron precipitation, between 76.5° and 79°Λ shows that this region represents a classical cusp precipitation (Potemra et al., 1977; Gussenhoven et al., 1985).

The field-aligned current, labeled $J_{\parallel}$ in panel c, was calculated from the eastward component of the magnetic disturbance (Potemra et al., 1984), positive values refer to downward currents. Following the nomenclature used by Bythrow et al. (1988) and very similar to their morning HiLat pass shown in Plate 3, we find an upward directed current which can be identified with the region 2 current with its peak at 76.5°, followed by a downward region 1 current and then an upward traditional cusp current poleward of 77° invariant. The maximum intensity is almost 5 μAmp m^{-2} in the cusp.

The crosstrack velocity measured by the IDM on-board HiLat (Rich et al., 1984) is shown in panel d. This velocity is very structured and it is directed primarily eastward between 76°-79° except for the period between 77.2° and 77.5°Λ, when it turns westward. This small portion of westward velocities seems to be co-located with the cusp current system.

The ion number density measured by the RPA is shown in panel e of Figure 2. Ni is uniform except for the factor of 2 enhancement in the poleward side of the cleft/cusp region, between 78.1° and 78.9°Λ. This density was measured at the satellite altitude (830 km). In a later section, we comment on the implications of this measurement and the Ni values obtained with the radar.

The RAM velocity from the RPA and the crosstrack velocity measured by the IDM were combined to resolve the plasma velocity vector. The RAM velocity was corrected to compensate for the possible charging of the spacecraft and the uncertainty in the derivation of the RAM velocity $(300-400 \text{ ms}^{-1})$ to provide an agreement with the radar measured drift velocity. Figure 3 shows the flow vector along the satellite trajectory. At invariant latitudes below 75° the plasma velocity is very small as is typical for subauroral latitudes. The invariant latitude range between 76° and 78° is a region with a velocity magnitude almost 1 km and directed northeast. At 77.2° the velocity reverses to a westward convection for about 30 km. This convection pattern is common for B_z southward and B_y negative IMF conditions. Burch et al. (1985) have indicated that limited reversals from east to west then back to east are indicative of the presence of a viscous cell. North of the cusp/cleft region the plasma velocity is eastward and less than 150 ms^{-1}.

HILAT ION DRIFT VELOCITIES

FEBRUARY 9, 1988   1249-1300 UT   1 km / sec

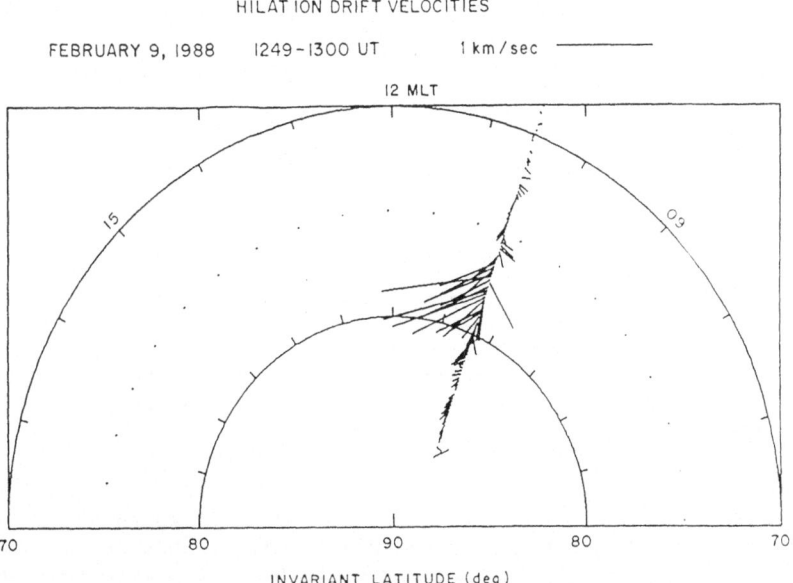

Figure 3. Vector velocities derived by combining the HiLat N-S velocity from the retarding potential analyzer and the E-W velocity measured by the ion-driftmeter.

2.2. Scintillations Using HiLat

Phase and amplitude scintillations are available from the HiLat satellite at 138 and 413 MHz. Total electron content measurements are also available along the slant path between the satellite and the ground station. The F-region of the ionosphere generally introduces the largest perturbation in the signal propagating from the satellite. We thus show in Figure 4, the 350-km intersection of the ray path from the satellite·together with the 350-km HiLat field line track (a part of which was shown in Figure 1) for comparison of the scintillation/TEC behavior in respect to the in-situ data.

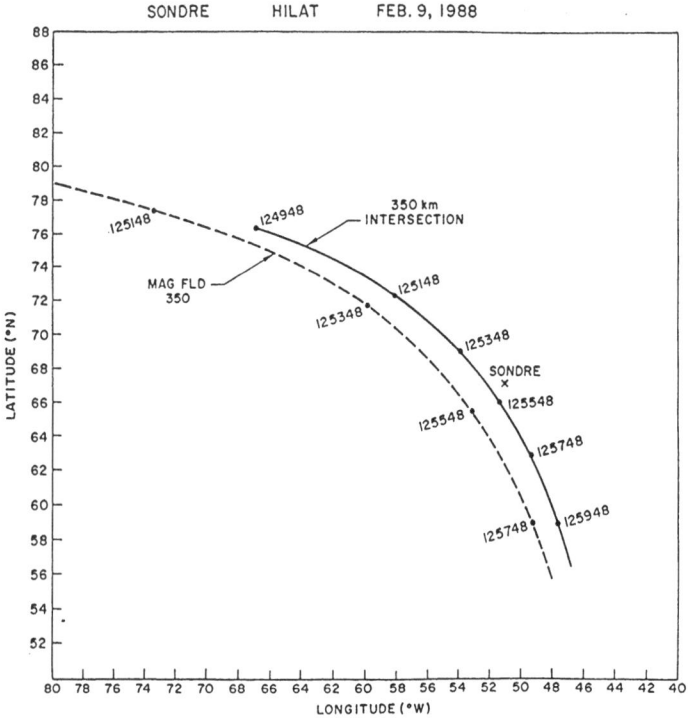

Figure 4. Locations of the ionospheric intersection of the ray path from HiLat and its magnetic field line trace, both referred to 350-km altitude for Sondrestrom.

Figure 5 shows the phase and amplitude scintillations obtained at 137 and 413 MHz from the HiLat satellite. A prominent increase in these parameters is observed between 1251:48-1254:18 UT. In particular, the large amplitude scintillation is noteworthy. The dotted line shows the 137-MHz amplitude scintillation corrected for the effect of the slant path. (The data at elevation angles less than 20° are contaminated by possible multipath effects and hence not considered.) The above time interval corresponds to an invariant latitude interval of 80.2°-75.8° which encompasses the region where both the largest number flux of precipitated electrons are seen (Panel a of Figure 2) and the largest structured velocities are observed (Figure 3). In addition, the spectral analysis of the phase scintillation data (not shown) yields spectral indices of the order of -2.5. This together with the large magnitude of the phase scintillation index indicates larger power spectral densities at short scales (<1 km) which is consistent with the existence of a relatively high level of amplitude scintillation. The

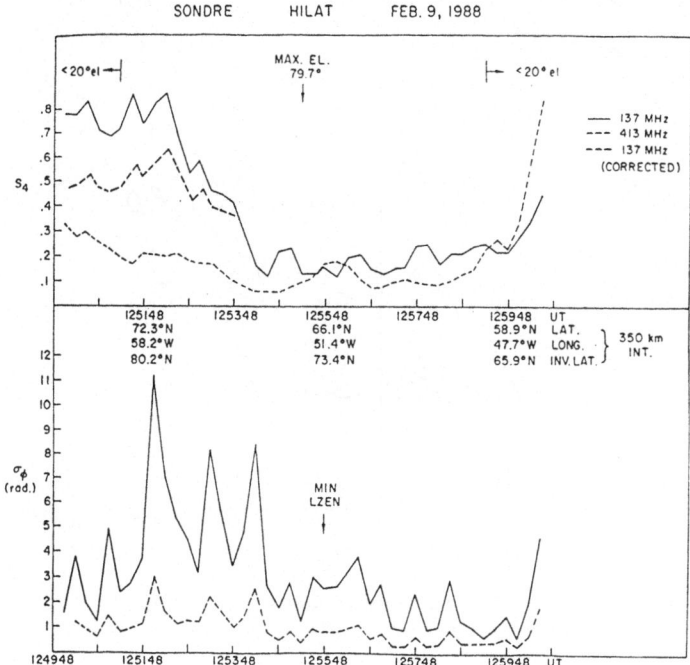

Figure 5. Phase and amplitude scintillations observed by HiLat at 137 and 413 MHz.

position along the orbit of this near-overhead pass where the raypath was aligned with the local magnetic L-shell is indicated as MIN LZEN. Virtually no geometrical enhancement was seen, further confirming the hypothesis that the irregularity region was confined to the cusp/cleft region and did not extend to the overhead location of the station. The equivalent vertical total electron content in the cusp (also not shown) exhibited no significant variations with a value of approximately 3.5×10^{12} el cm^{-2} which is consistent with the integration of the radar profiles in that region.

2.3. Radar Observations

Figure 6 presents the temperature corrected densities for the elevation scan executed simultaneously with the HiLat pass. The maximum number density is 3×10^5 el cm^{-3}. This region of high Ni is located in the southern boundary of the radar field of view with the plasma density decreasing linearly for higher values of SZA at positive northern distances. At 150 km north of the radar Ni reaches its lowest value, 1.8×10^5 el cm^{-3}. However, at a northern distance of 350 km the density increases to a value of 2.6×10^5 el cm^{-3}. Figure 6 shows only a hint of a region of large density while the previous scan, not shown here, and obtained a few minutes earlier, displays better the small region of higher density.

The electron temperature Te and the Te error bars are plotted in

292

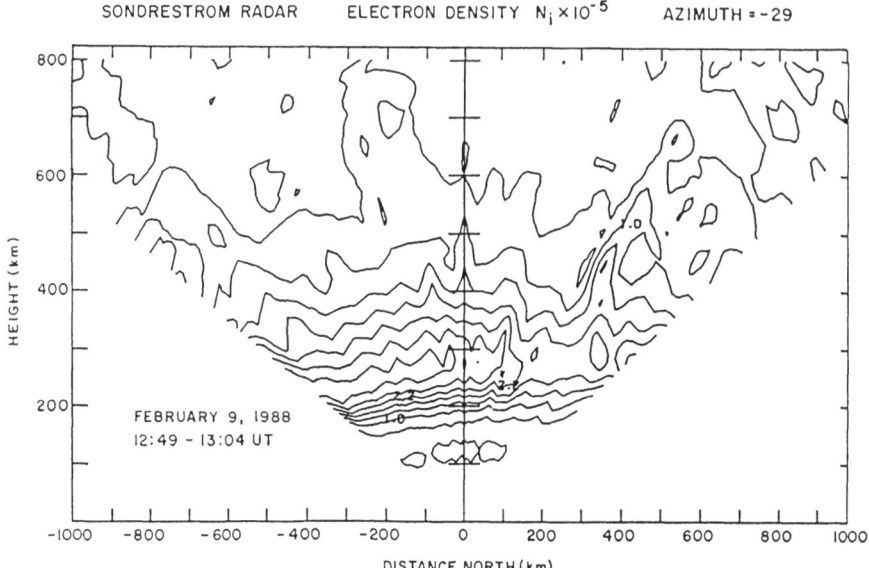

Figure 6. Elevation scan of ionospheric densities made by the
Sondrestrom radar between 12:49-13:04 UT on February 9, 1988.

Figure 7. Te slowly increases for larger southern distances from
Sondrestrom. This is due to more intense solar illumination at slightly
lower SZA. At about 200 km north of the radar site, F-region Te
increases from 2200° to 3300°K, probably in response to the
precipitation measured by HiLat. The F-region Ti, though not shown
here, is also enhanced at 300 km north of the radar. Ti rises
significantly from 1000° to 1500°K in that region. The radar data
implies that for this day the cusp/cleft region consisted of two
well-defined segments: the equatorward side with low Ni and high Te,
and the poleward part with high Ni and high Te.

3. MODEL RESULTS

To study the ionospheric response to the flux of soft electrons, we have
used a numerical model which starts with the energy spectra of the
precipitating electrons measured in-situ by HiLat and then calculates
the electron temperature and the density profile of 14 ionospheric
species (Strickland et al., 1976; Weber et al., 1985, 1988).
    Our 1-dimensional ionospheric chemistry code has 2 distinct parts:
part 1 assumes that the incident flux is isotropic at an upper altitude
boundary and generates production rate and electron plasma heating
profiles; part 2 uses the results from part 1, includes local chemistry,
O^+ diffusion, and electron heat conduction. The MSIS 86 model was used
to provide the density of the atmospheric neutral constituents.
    Figure 8 shows two electron spectra measured by the J-sensor during
the HiLat traversal of the cusp/cleft region, both spectra correspond to
4.5 second averages. Model 1 (panel a) was obtained at 1254:30 UT,

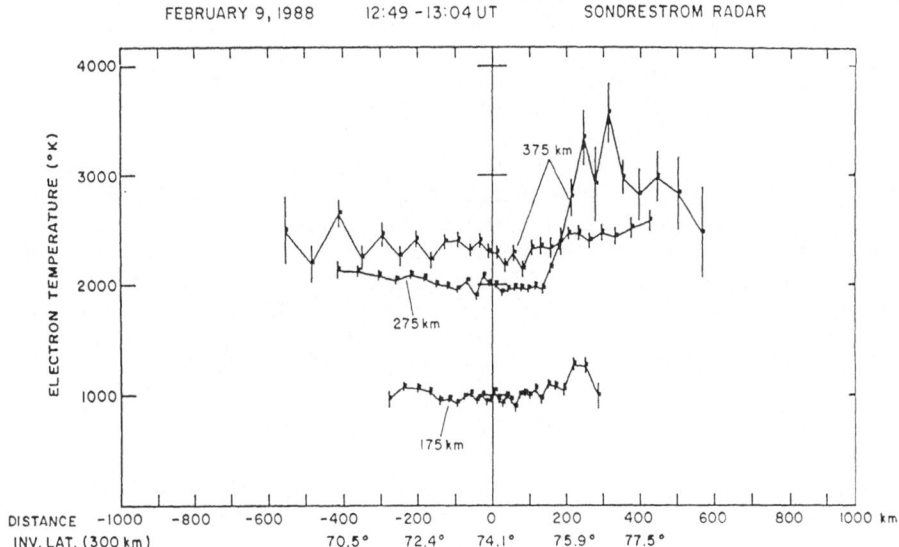

Figure 7. Electron temperatures at 3 altitudes measured by the Sondrestrom radar together with the density measurements shown in Figure 6.

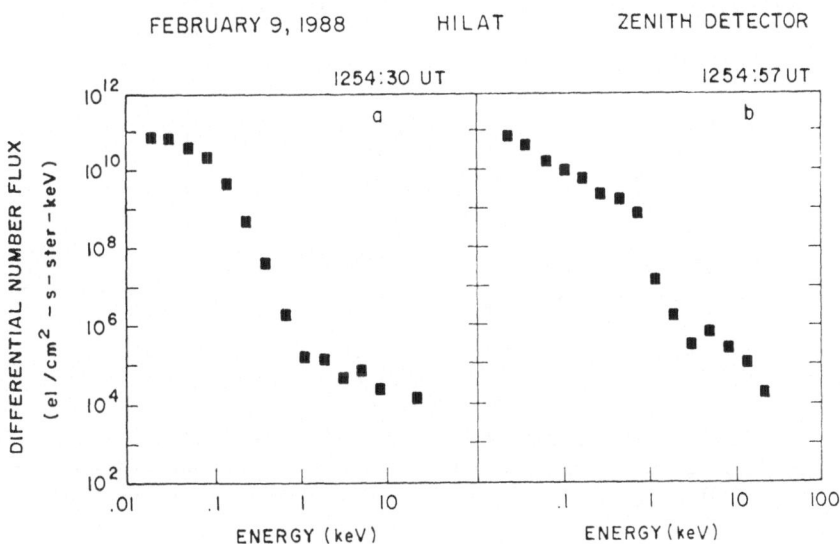

Figure 8. Two spectra of precipitating electrons measured by HiLat. The spectra in panel a was obtained in the cusp proper, while that in panel b was from the cleft region.

294

corresponding to 77.6 invariant latitude, when the spacecraft was in the cusp region. The energy flux was .32 erg/cm^2-s-ster. The spectra of panel a does not show significant acceleration. The spectra of panel b (1254:57 UT, 75.8°Λ, .76 erg/cm^2-s-ster), on the contrary, indicates electrons that have been accelerated to 1 keV. This is considered a typical sub-cusp or cleft electron flux. The cleft spectrum was found to persist for 20s during which the satellite traversed 160 km, whereas the cusp precipitation was observed over a broader region of 240 km.

Figure 9 shows model results using the electron spectra of panel b and two different exospheric temperatures. Each situation starts with the indicated initial electron density profile and then runs as if the flux were precipitating unchanged for 2 minutes. The results indicate that higher exospheric temperature produces increased densities at altitudes greater than 300 km and reduced densities lower down.

FEBRUARY 9, 1988   HILAT   1254:57 UT

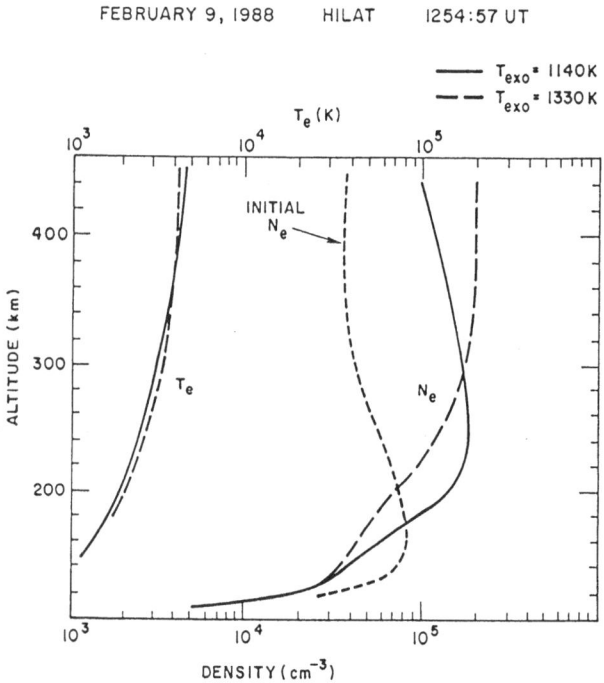

Figure 9. Model results of electron density for electron spectrum shown in Figure 8b. Two different exospheric temperatures are used.

Calculated Te is less affected by the exospheric temperature variation but is somewhat higher than indicated by the radar measurements. The best agreement with corresponding radar densities is with the model profile for T_{exo} = 1140°K which produces a peak density of 2x10^5 at 240 km. Outflowing O$^+$ ions can also affect the number density profile. We tested this hypothesis employing a range of up- and down-going flows in our model and found that this factor had a relatively small effect on the density, particularly at altitudes below 400 km. While model results for the more typical cusp spectrum of Figure 8a are not shown, it is important to note that the model had to be run for 5 minutes in order to obtain a peak density of 2×10^5 cm^{-3}. It should also be considered that another source of ionization in the cusp is precipitating ions (Bythrow et al., 1988). However, since HiLat does not measure the ions, we considered electrons only, which will underestimate the total ionization input.

4. DISCUSSION

One of the most significant results of this joint satellite and radar study shows that it is possible to observe a depletion in the pre-noon F-region ionosphere in conjunction with intense particle precipitation (which causes enhanced Te) and rapid eastward convective flows. It seems probable that these rapid eastward flows convect less dense plasma from the dawnside polar region which is in darkness at this time of the year. The model studies show that steady precipitation over approximately 2 minutes is necessary for the desired density buildup in the ionosphere. Within this time the poleward flow of approximately 1 km s^{-1} will convect plasma more than 1° poleward. Since the cleft-type of precipitation occurred over a region of 160-km latitudinal extent, the convecting plasma was within the precipitation region for the required length of time. The density buildup is seen approximately 1° poleward of the equatorward edge of the cleft precipitation consistent with the model results. The F-region depletion is probably not a result of an exospheric temperature increase alone as the model results indicate that this would probably have caused enhanced densities at the satellite altitude which were not observed in this region. The region poleward of the depletion where an enhancement is observed both in the F-region by the radar and in the topside by the satellite seems to be a consequence of the local particle precipitation, poleward convection of the plasma subjected to more intense precipitation, and the increased Te and Ti which causes an increased scale height of the plasma in the topside.

   The small-scale irregularities as diagnosed by the scintillation technique are found in the region of the largest particle fluxes and structured velocities. The same finding was reported by Baker et al. (1986) in a case study at Sondrestrom utilizing data from HiLat and the Goose Bay HF radar. Unfortunately, they did not have support from the incoherent scatter radar there so that they hypothesized a density enhancement in the F-region and the existence of the generalized gradient drift instability based on the satellite precipitation results alone. However, we have shown here that intense precipitation can actually be co-located with a depletion in density in the F-region so that one has to carefully investigate appropriate conditions for the small-scale irregularity generation. With the observed poleward drift, it is only the poleward gradient seen by the radar north of 77° invariant which should be unstable to the gradient drift instability. However, irregularities are observed starting from 75.8° in a region which should be stable to this mechanism. Other case studies done by Basu et al. (1986, 1988) have shown the importance of velocity shears

and field-aligned currents in generating scintillation producing
irregularities. We would thus like to suggest one of the
transverse-shear driven (Keskinen et al., 1988) or
shear-cum-field-aligned current driven modes (Nishikawa et al., 1988) as
plausible candidates for the small-scale irregularity generation.
Further studies are being planned in the summer and winter polar cusp
for more stringent tests of these mechanisms.

Finally, we would like to point out that it has been suggested in
the literature that the spectra of the incoherent scatter (IS) signal
can be distorted by velocity shears if they are unresolved within the
probed volume (Swartz et al., 1988). Field-aligned currents may also
alter the returned signal producing asymmetries in the spectra (Foster
et al., 1988). In order to determine the level of contamination of our
radar measurements we performed a careful examination of the incoherent
scatter spectra, looking for deviations from the typical double-humped
shape. We found that the anomalies were always smaller than the
statistical uncertainties.

ACKNOWLEDGMENTS

We wish to thank F.J. Rich, R.A. Heelis, and D.A. Hardy for making the
HiLat in-situ data available to us and for many useful discussions.
R.A. Livingston and R.P. Lepping provided, respectively, the HiLat
beacon data and the IMP-8 IMF data. The work at Emmanuel College was
supported by NSF Grant ATM-8715445 and AFGL Contract F19628-86-K-0038.

REFERENCES

Baker, K.B., Greenwald, R.A., Walker, A.D.M., Bythrow, P.F.,
   Zanetti, L.J., Potemra, T.A., Hardy, D.A., Rich, F.J., and Rino, C.L.
   (1986) 'A cast study of plasma processes in the dayside cleft', J.
   Geophys. Res. 91, 3130.
Basu, Su., Basu, S., Senior, C., Weimer, D., Nielsen, E., and
   Fougere, P.F. (1986) 'Velocity shears and sub-km scale irregularities
   in the nighttime auroral F-region', Geophys. Res. Lett. 13, 101.
Basu, Su., Basu, S., MacKenzie, E., Fougere, P.F., Coley, W.R.,
   Maynard, N.C., Winningham, J.D., Sugiura, M., Hanson, W.B., and
   Hoegy, W.R. (1988) 'Simultaneous density and electric field
   fluctuation spectra associated with velocity shears in the auroral
   oval', J. Geophys. Res. 93, 115.
Burch, J.L., Reiff, P.H., Menietti, J.D., Heelis, R.A., Hanson, W.B.,
   Shawhan, S.D., Shelley, E.G., Sugiura, M., Weimer, D.R., and
   Winningham, J.D. (1985) 'IMF B_y-dependent plasma flow and Birkeland
   currents in the dayside magnetosphere, 1. Dynamics Explorer
   observations', J. Geophys. Res. 90, 1577.
Bythrow, P.F., Potemra, T.A., Erlandson, R.E., Zanetti, L.J., and
   Klumpar, D.M. (1988) 'Birkeland currents and charged particles in the
   high-latitude prenoon region: A new interpretation', J. Geophys. Res.
   93, 9791.

Erlandson, R.E., Zanetti, L.J., Potemra, T.A., Bythrow, P.F., and
  Lundin, R. (1988) 'IMF B_y-dependence of region 1 Birkeland current
  near noon', J. Geophys. Res. 93, 9804.
Foster, J.F., Holt, J.M., Kelly, J.D., and Wickwar, V.B. (1985)
  'High-resolution observations of electric fields and F-region plasma
  parameters in the cleft ionosphere', in J.A. Holtet and A. Egeland
  (eds.), The Polar Cusp, NATO ASI Ser. C; Vol. 145, D. Reidel,
  Dordrecht, p. 349.
Foster, J.C., del Pozo, C., Groves, K., and St. Maurice, J.-P. (1988)
  'Radar observations of the onset of current driven instabilities in
  the topside ionosphere', Geophys. Res. Lett. 15, 160.
Frank, L.A. (1971) 'Plasma in the Earth's polar magnetosphere', J.
  Geophys. Res. 76, 5202.
Fremouw, E.J., Carlson, H.C., Potemra, T.A., Bythrow, P.F.,
  Rino, C.L., Vickrey, J.F., Livingston, R.L., Huffman, R.E.,
  Meng, C.I., Hardy, D.A., Rich, F.J., Heelis, R.A., Hanson, W.B., and
  Wittwer, L.A. (1985) 'The HiLat satellite mission', Radio Sci. 20,
  416.
Gussenhoven, M.S., Hardy, D.A., and Carovillano, R.L. (1985) 'Average
  electron precipitation in the polar cusps, cleft and cap', in
  J.A. Holtet and A. Egeland (eds.), The Polar Cusp, NATO ASI Ser. C;
  Vol. 145, D. Reidel, Dordrecht, p. 85.
Hardy, D.A., Huber, A., and Pantazis, J.A. (1984) 'The electron flux J
  sensor for HiLat', The Johns Hopkins/APL Tech. Digest 5, 125.
Heikkila, W.J. (1985) 'Definition of the cusp', in J.A. Holtet and
  A. Egeland (eds.), The Polar Cusp, NATO ASI Ser. C; Vol. 145,
  D. Reidel, Dordrecht, p. 387.
Kelly, J.D. (1985) 'Incoherent-scatter radar observations of the cusp',
  in J.A. Holtet and A. Egeland (eds.), The Polar Cusp, NATO ASI Ser. C;
  Vol. 145, D. Reidel, Dordrecht, p. 337.
Keskinen, M.J., Mitchell, H.G., Fedder, J.A., Satyanarayana, P.,
  Zalesak, S.T., and Huba, J.D. (1988) 'Nonlinear evolution of the
  Kelvin-Helmholtz instability in the high-latitude ionosphere', J.
  Geophys. Res. 93, 137.
Nishikawa, K.-I., Ganguli, G., Lee, Y.C., and Palmadesso, P.J. (1988)
  'Simulation of electrostatic ion instabilities in the presence of
  parallel currents and transverse electric fields', Proceedings of MIT
  Workshop on Polar Cap Dynamics and High Latitude Ionospheric
  Turbulence.
Potemra, T.A. and Zanetti, L.J. (1985) 'Characteristics of large-scale
  Birkeland currents in the cusp and polar regions', in J.A. Holtet and
  A. Egeland (eds.), The Polar Cusp, NATO ASI Ser. C; Vol. 145,
  D. Reidel, Dordrecht, p. 203.
Potemra, T.A., Peterson, W.K., Doering, J.P., Bostrom, C.O.,
  McEntire, R.W., and Hoffman, R.A. (1977) 'Low-energy particle
  observations in the quiet dayside cusp from AE-C and AE-D', J.
  Geophys. Res. 82, 4765.
Potemra, T.A., Bythrow, P.F., Zanetti, L.J., Mobley, F.F., and
  Scheer, L. (1984) 'The HiLat magnetic field experiment', The Johns
  Hopkins/APL Tech. Digest 5, 120.

Rich, F.J., Heelis, R.A., Hanson, W.B., Anderson, P.B., Holt, B.J., Harmon, L.L., Zuccaro, D.R., Lippincott, C.R., Girouard, D., and Sullivan, W.P. (1984) 'Cold plasma measurement on HiLat', The Johns Hopkins/APL Tech. Digest 5, 114.

Roble, R.G. and Rees, M.H. (1977) 'Time-dependent studies of the aurora: Effects of particle precipitation on the dynamic morphology of ionospheric and atmospheric properties', Planet. Space Sci. 25, 991.

Stamnes, K., Rees, M.H., Emery, B.A., and Roble, R.G. (1985) 'Modelling of cusp auroras: The relative impact of solar EUV radiation and soft electron precipitation', in J.A. Holtet and A. Egeland (eds.), The Polar Cusp, NATO ASI Ser. C; Vol. 145, D. Reidel, Dordrecht, p. 137.

Strickland, D.J., Book, D.L., Coffey, T.P., and Fedder, J.A. (1976) 'Transport equation techniques for the deposition of auroral electrons', J. Geophys. Res. 81, 2755.

Swartz, W.E., Providakes, J.F., Kelley, M.C., and Vickrey, J.F. (1988) 'The effect of strong velocity shears on incoherent scatter spectra: A new interpretation of unusual high latitude spectra', Geophys. Res. Lett. 15, 1341.

Vasyliunas, V.M. (1985) 'Summary', in J.A. Holtet and A. Egeland (eds.), The Polar Cusp, NATO ASI Ser. C; Vol. 145, D. Reidel, Dordrecht, p. 411.

Weber, E.J., Tsunoda, R.T., Buchau, J., Sheehan, R.E., Strickland, D.J., Whiting, W., and Moore, J.G. (1985) 'Coordinated measurements of auroral zone plasma enhancements', J. Geophys. Res. 90, 6497.

Weber, E.J., Kelley, M.C., Ballenthin, J.O., Basu, S., Carlson, H.C., Fleischman, J.R., Hardy, D.A., Maynard, N.C., Pfaff, R.F., Rodriguez, P.A., Sheehan, R.A., and Smiddy, M. (1988) 'Rocket measurements within a polar cap arc: plasma, particle, and electric circuit parameters', J. Geophys. Res., in press.

THE ELECTRODYNAMIC SIGNATURE OF SHORT SCALE FIELD ALIGNED CURRENTS, AND ASSOCIATED TURBULENCE IN THE CUSP AND DAYSIDE AURORAL ZONE.

A. BERTHELIER, J.-C. CERISIER, J.-J. BERTHELIER,
CRPE-CNRS, 4 Avenue de Neptune, 94107 Saint-Maur CEDEX, France
J.-M. BOSQUED,
CESR-CNRS, 9 Avenue Colonel Roche, 31400 Toulouse, France
R.A. KOVRAZKHIN
IKI, Academy of Sciences, Moscow, USSR

ABSTRACT. The analysis of the electrodynamic characteristics of a crossing of the southern cusp and dayside auroral zone at an altitude of about 1000km by the AUREOL-3 satellite is conducted from a data set including the DC magnetic and electric fields, particles precipitation, and AC electric and magnetic fields. The convection is analysed both from the differential drift of energetic ions injected at high altitude in the polar cusp, and from direct electric field measurements. The ionospheric conductivity calculated from **E** and **B** variations is shown to be slightly different in the regions of downward and upward currents, and the influence of particle precipitations is discussed. The analysis of short scale structures shows that the fluctuations of the upward current intensity are well correlated with the low energy electron flux variations. Turbulence on AC electric and magnetic signals is observed in the cusp and in the region of upward currents, with intense events close to the maxima of these currents. The variation of the B/E ratio with frequency is delineated and the origin of this turbulence is discussed.

1. Introduction

The dayside high latitude regions of the ionosphere and magnetosphere play a key role in the electrodynamics of the entire system since the physical processes which are acting there govern for a large part the interaction between the solar wind and the Earth's magnetosphere. Particle and field measurements at low altitude have been extensively used to link the large scale structures in the ionosphere to the various boundaries which have been observed at the dayside magnetopause (Vasyliunas, 1979). However, the strong influence of the solar wind parameters and their variability together with the fuzzy and diffusive nature of the dayside interaction region make these studies extremely complex (e.g. Hardy *et al*, 1985, Meng and Lundin, 1986, Zanetti and Potemra, 1986). Moreover several recent results have exemplified the importance of small scale or transient phenomena; in particular it has been shown that the electromagnetic interaction between the solar wind and the earth magnetic field rely essentially on unsteady and small scale processes, as the flux transfer events (Russell and Elphic, 1978, Farrugia *et al*, 1988) and that, at lower altitudes, an intense turbulence is the signature of cusp and cleft regions (Berthelier and Machard, 1985, Erlandson *et al*, 1987, Gurnett and Inan, 1988).

In this paper we report a case study based on the data acquired on board the French-Soviet satellite AUREOL-3 during a crossing of the southern cusp and cleft regions (Orbit 740 South, Nov. 16th, 1981) in the vicinity of Terre Adélie (geographic longitude 140°E, latitude 66.7°S).

P. E. Sandholt and A. Egeland (eds.), Electromagnetic Coupling in the Polar Clefts and Caps, 299–310.
© *1989 by Kluwer Academic Publishers.*

Analysis of particle precipitations, electric and magnetic fields enables us to delineate the large scale characteristics of the cusp and cleft related to conditions prevailing in the solar wind. A detailed study of the electrodynamics of the ionosphere is conducted to emphasize relationships between particle precipitations, conductivities and currents. The second part of the paper deals with the analysis of small scale structures and electromagnetic turbulence observed in the ULF, ELF range; the most intense events are observed to occur close to the maxima of intensity of field aligned currents and their origin in terms of small scale high intensity field aligned currents or electromagnetic Alfven waves is discussed.

2. Data Presentation

2.1. INSTRUMENTATION

AUREOL-3 is a three-axis stabilised spacecraft with the Z axis within $\approx 15°$ of the local vertical and the Y axis within $\approx 15°$ of the perpendicular to the orbit plane, equiped with a rather complete set of instruments aimed at measuring energetic particles, thermal plasma parameters and electric and magnetic fields from DC up to 16 kHz. A detailed description of the experiments may be found in a special issue of Annales de Géophysique (Vol. 38, 1982), and we give in this paragraph only a brief description of the data relevant to this study.

Electron and ion energy spectra from 0.1 to 22 keV are obtained from the SPECTRO experiment, every 1.6 s at 9 pitch angles, while the KUKUSHKA detectors provide electron and ion fluxes in 2 energy channels (here 0.1 and 1.8 keV) at a higher sampling rate (3 up to 100 points/s, depending upon the telemetry mode), along an upward looking direction $\approx 15°$ from the Z axis. Quasi DC electric E and magnetic B field data are acquired at the same sampling rate ; due to a telemetry failure only the Y component of the DC electric field vector is available for the orbit under consideration.

The field aligned currents (FAC) have been deduced from the DC magnetic field components after substraction of the background field, namely ΔB_X along the orbital direction and ΔB_Y perpendicular to it (TRAC experiment). As in the region of interest ΔB_X, ΔB_Y remain quasi proportional (see Figure 2) we can assume that $\partial \Delta B_X / \partial X$ and $\partial \Delta B_Y / \partial X$ are the projections of the variation of ΔB accross a FAC sheet, and calculate its main direction and intensity. Practically the slopes of the ΔB_X, ΔB_Y variations are calculated by means of running averages over a period of time, or an equivalent distance, choosen to correspond to the scale under study ; namely from $\Delta T \approx 10$ s, i.e. $\Delta L \approx 70$ km for medium scale, to $\Delta T \approx 1.6$ s, i.e. $\Delta L \approx 10$ km for small scale in the case here of a 3 points/s sampling rate.

The wave experiment (TBF experiment) provides 2 AC components of the electric field (E_H close to Y direction, E_Z close to vertical), and 3 AC components of the magnetic field. In the telemetry mode used during Orbit 740, all five components are measured in the frequency range from ≈ 10 Hz to 1500 Hz.

Figure 1 (opposite) - This figure shows from top to bottom : ion and electron color coded spectrograms (a complete energy spectrum every 1.6 s), variations of the ion and electron fluxes in the 100 eV and 1.8 keV channels reported versus time upon a time sampling rate of ≈ 3 points/s, running averaged values of field aligned currents over ≈ 70 km, and the indications of the presence of enhanced electric turbulence (green hachures) and of large bursts in electromagnetic turbulence (green dots). A to E refer to the typical regions within the dayside auroral zone, as described in the text.

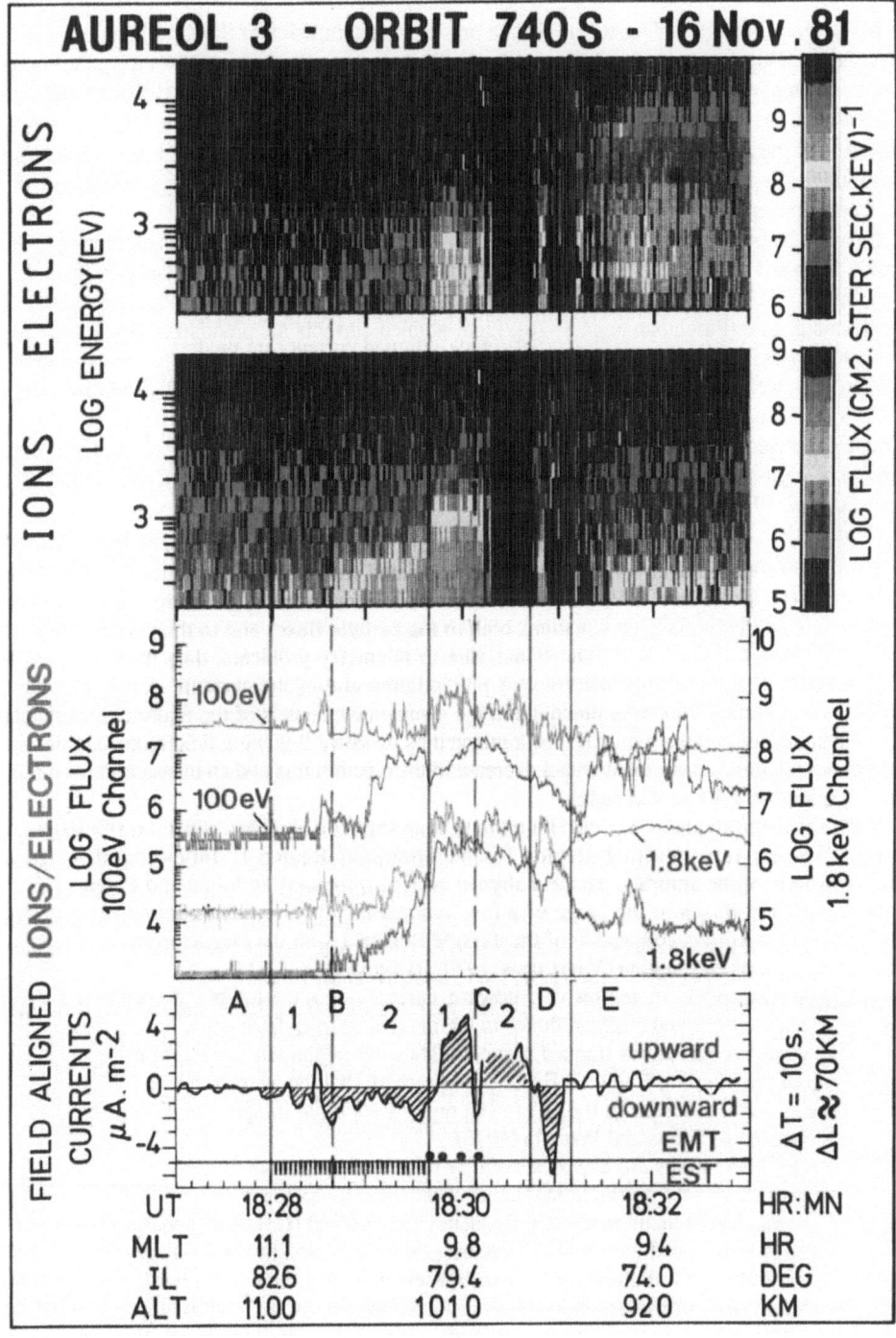

AUREOL 3 - ORBIT 740 S - 16 Nov. 81

2.2. DATA DESCRIPTION

In the composite plot of AUREOL-3 data presented as a function of time in Figure 1 we have disposed from top to bottom : electron and ion color coded spectrograms (SPECTRO), the energy fluxes in 2 separate energy channels, namely 100 eV and 1.8 keV, for both electrons and ions (KUKUSHKA), intensity of the field aligned currents, averaged over ≈ 70 km (TRAC), and an indication of the presence of electromagnetic or electrostatic turbulence (TBF). From this data set of simultaneous measurements, the pass through typical regions within the dayside auroral zone can be recognized as follows.

Leaving the polar cap (A), the satellite enters the region (B), identified as corresponding to the interior cusp and plasma mantle projection, where the particle population is mainly composed of low energy (≈ 100 eV) magnetosheath electrons and ions, part of which will populate ultimately the plasma mantle (Paschman *et al*, 1976). Field-aligned currents are predominantly downward, with an average intensity of about 1-2 $\mu A/m^2$. One observes first, (B1), discrete increases in the electron flux, well associated with fluctuations in the currents, and slightly enhanced low energy ion fluxes. Then, (B2), an increase in the ion flux is mainly seen at lower energies, and the latitude dispersed ion structure appears to be typical of the differential drift of energetic ions injected at high altitudes under the influence of an antisunward convection (Reiff *et al*, 1977, Bosqued *et al*, 1984).

Next to (B), in (C) region, upward currents are observed with intensity values as high as ≈ 5 $\mu A/m^2$ (averaged over ≈ 70 km); they are associated with electron precipitations in the range of about 100eV to a few keV, and with ion precipitations in the same energy range. One can observe in this region large fluctuations both in the particle fluxes and in the currents, with two main maxima (C1, C2). Let us note that, due to telemetry problems, data gaps occur in the measurements of the energy spectrum of precipitation and of the currents in the (C2) region; however the particle flux measurements in the 4 separate channels of the Kukushka experiment remain continuous. Next, region (D) is dominated by downward flowing field aligned currents, as large as ≈ 5 $\mu A/m^2$, associated with a decrease in the electron flux and an increase in the ion flux clearly seen in the 1.8 keV channels.

We have identified the (C) and (D) regions at ionospheric altitudes as the dayside extensions of the respectively morning Region 1, and afternoon Region 1, following the classical denomination of the auroral zone field aligned currents proposed by Iijima and Potemra (1976). The intense ion fluxes in the range of a few hundred eV observed here around noon gives an indication of the direct connection of the dayside Region 1 with the magnetospheric low latitude boundary layer or cleft region (Vasyliunas, 1979, Bythrow *et al*, 1981).

At lower latitudes, in region (E), upward currents of ≈ 0.5 $\mu A/m^2$ are observed on the average, and the observed electron fluxes in the range of energy from some hundred eV to about 10-20 keV, are recognized as trapped particles. Thus the region (E) is thought to be the dayside extension of the morning Region 2 FAC. Let us remark that the average upward and downward directions of FAC observed in the (A) to (E) regions are quite in agreement with the statistical patterns obtained by Iijima and Potemra (1976).

Magnetic fluctuations are also observed by means of the search coil AC magnetometer. In particular strong events of electromagnetic turbulence appearing as bursts of enhanced ULF electric and magnetic fluctuations in the frequency range 10-20 Hz, lasting about 100-500 ms, are seen in the (C) region, and their times of occurrence are indicated by dots in Figure 1. In addition, low frequency electric turbulence is, as usual, observed during the whole crossing of the dayside auroral region. However its intensity is clearly enhanced from 18:28:00 to 18:29:40 UT that is precisely in the (B) region, previously identified as the cusp.

3. Results and Discussion.

We will focus our study on the cusp (B) and on the part of cleft closest to it (C1), first considering the E and B variations, and the ionospheric conductivity, then comparing short-scale FAC to particle precipitations, at last studying the variation as a function of frequency of the E/B ratio in the region of enhanced electromagnetic turbulence.

3.1. ELECTRODYNAMICS OF THE IONOSPHERE

As shown in Figure 2 the variations of ΔB_X, ΔB_Y and ΔE_Y are well correlated in the cusp

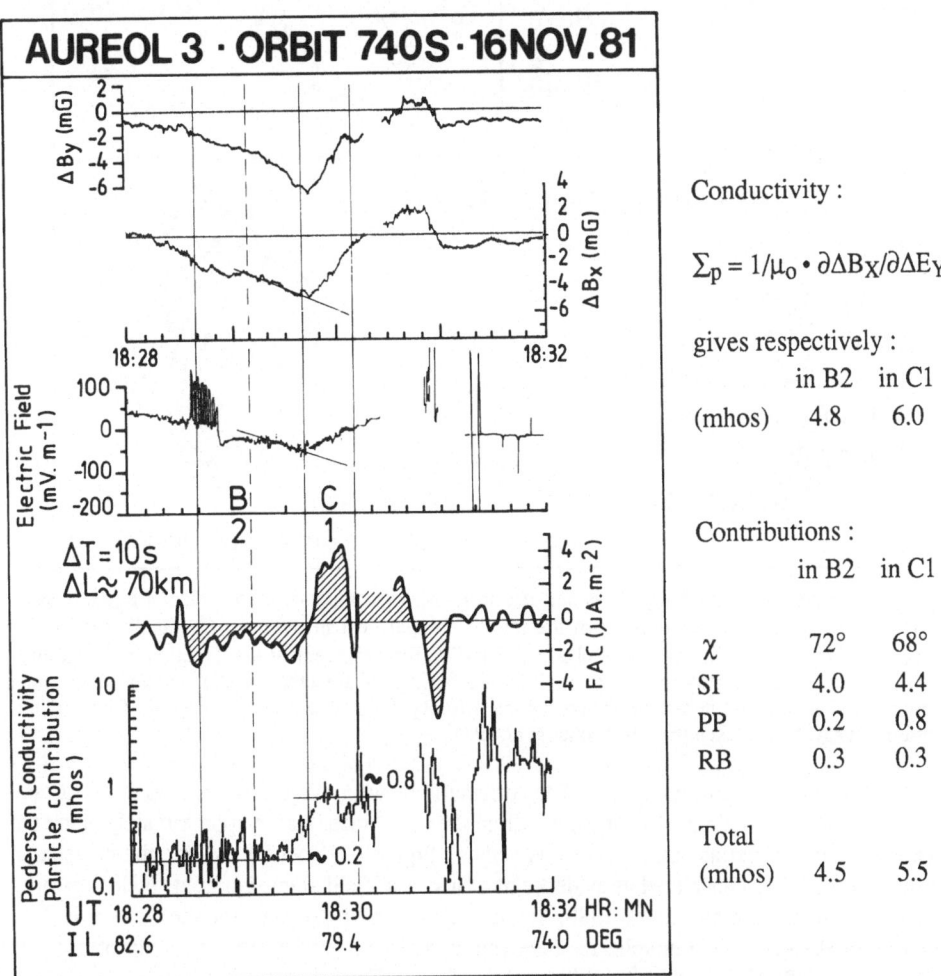

Figure 2 - Variations of ΔB_X, ΔB_Y transverse components of DC magnetic field are compared to variations of ΔE_Y DC electric field in order to calculate ionospheric Pedersen conductivity in the cusp downward FAC (B2), and cleft upward FAC (C1). Estimations of the contributions of particle precipitation (bottom curve, PP), solar ionisation (SI), and radiation background (RB) are given at right for comparison with the upper determination of Σ_p.

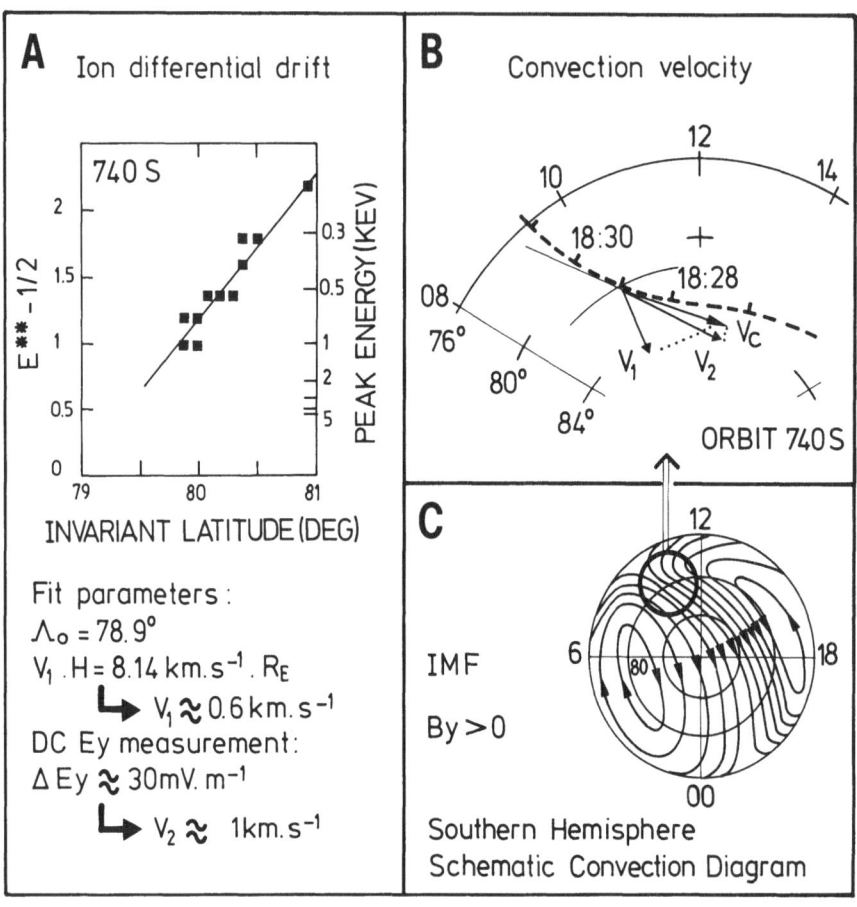

Figure 3 - The convection velocity V_c is obtained from the V_1 component perpendicular to the lines of constant invariant latitude, deduced from the ion differential drift in the cusp (A panel), and the V_2 component along the satellite trajectory obtained from DC Ey measurement. Its direction in the morning cusp (B panel) is approximatly from dawn to dusk, in a fairly good agreement with the dominant direction of convection in the southern hemisphere under IMF By>0, as deduced for example from ground measurements (C panel, taken from Berthelier *et al*, 1974).

and in the neighbour cleft region (C1). The correlation between ΔB_X, ΔB_Y shows that FAC are mainly current sheets, directed at about 45° from the orbit trace, that is approximately along the lines of constant invariant latitude in this region (Figure 3, Panel B). As for the correlation between ΔB, ΔE, it is considered as evidence of a closure of field aligned currents by ionospheric Pedersen currents (Smiddy *et al*, 1980, Sugiura, 1984, and references therein), and it is thus possible to deduce the ionospheric Pedersen conductivity from the ratio of orthogonal components of ΔB, ΔE, namely : $\Sigma_p = 1/\mu_o \cdot \partial \Delta B_X / \partial \Delta E_Y$. The calculated value of Σ_p is slightly lower in (B2) (downward FAC, low energy precipitations), than in (C1) (upward FAC, higher energy fluxes), respectively 4.8 and 6 mhos. These values are consistent with a determination made by taking into account the contributions of the particle precipitations (Reiff *et al*, 1984), of the solar ionisation varying with the solar zenith angle χ (Mehta, 1978), and background radiations (right panel, figure 2).

The convection velocity is calculated from the ion differential drift observed in the cusp, which gives the component in the meridian direction, and from the DC E_Y measurement (figure 3). The main direction from dawn to dusk is consistent with the dissymetric convection cells observed in the southern hemisphere when the IMF B_Y component is positive (as indicated by ISEE 3 data at the time corresponding to Orbit 740), as well as from ground based magneto-meter (Berthelier *et al*, 1974) or satellite measurements (e.g. Heppner, 1972, Bythrow *et al*, 1982, Heelis, 1984).

Region (C1) is also caracterized by intense highly structured upward currents as shown in the extended view of Figure 4 (central curve). The well known correlation between field aligned currents and particle precipitations (Hoffman *et al*, 1984) is shown to apply even at scale as short as a few kilometers, and a regression analysis between the intensity of FAC and electron fluxes

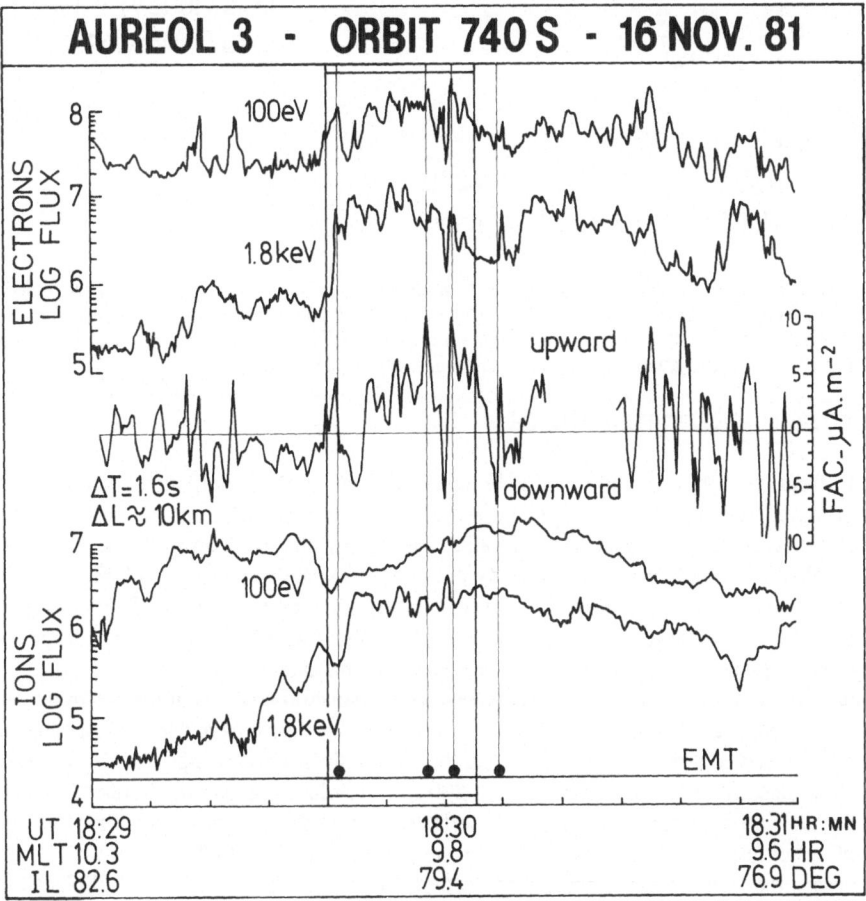

Figure 4 - A good correlation between small scale structures of field aligned currents (central curve) and electron fluxes (upper curves), reported versus time, is clearly seen in the region of the cleft upward currents (C1).

306

gives a correlation coefficient of 0.7 for the 100 eV channel, compared to 0.4 in the case of the 1.8 keV channel, and no correlation is found with the ion precipitation (Figure 5). Thus one can conclude that low energy ($\approx$ 100 eV) electrons play a key role in carrying the small scale field aligned currents in the cleft region.

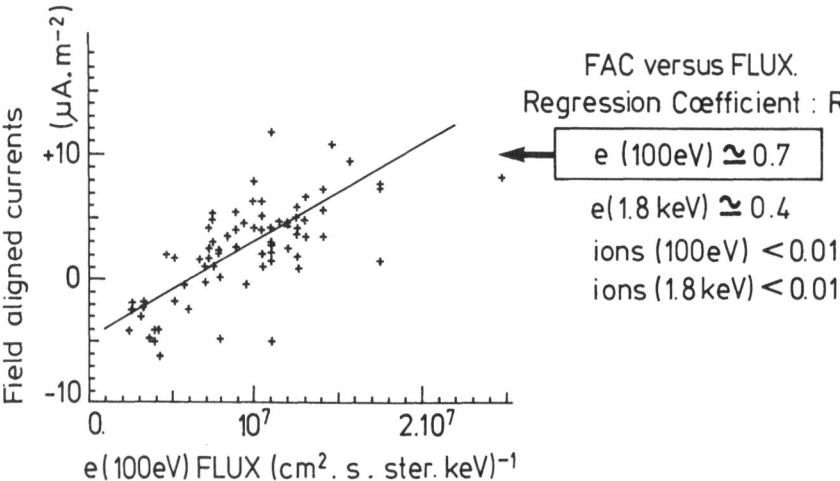

Figure 5 - The intensity of FAC, calculated from DC magnetic variations (3 points/s, running averages over 1.6 s, i.e. $\approx$ 10 km), is reported as a function of electron flux in the 100 eV channel. The value of the regression coefficient R is 0.7, largely higher than in the case of FAC versus 1.8 keV electrons (R = 0.4), and the regression line is drawn. No correlation are seen with ions (R < 0.01).

3.2. ELECTROMAGNETIC TURBULENCE

Apart from the turbulence observed only on the two AC electric components which peaks in the cusp region and displays the typical spectral and temporal variations of the high latitude electrostatic turbulence, several short duration bursts are observed on both the electric and magnetic channels at positions indicated by dots in the expanded time plot of Figure 4. They occur in region (C) which was shown in section 2 to correspond to the cleft or ionospheric projection of the low latitude boundary layer. All of them appear to coincide with peaks of upward field aligned currents and of the 100 eV electron precipitations.

Displayed on Figure 6 (right part) are the results of a spectral analysis of the electromagnetic turbulence : more precisely the B_Y and E_H components of the signal have been Fourier analyzed for the burst observed around 18:29:46 UT and the ratio of the Fourier components of B_Y and E_H plotted as a function of frequency from 10 to 200 Hz. In order to extend the frequency range of the analysis a similar computation has been performed using the B_X and E_Y components of the quasi DC electric and magnetic fields during the period from 18:29:45 to 18:30:05 UT which encompasses the time of occurrence of the selected burst of electromagnetic turbulence. The sampling rate of measurement (3 points/s) just enables to cover a useful frequency range of 0.1 to 1 Hz (left part of Figure 6). Although the analysis was performed on two different sets of data, with components not in the same direction, the two curves do show a striking continuity, with the ratio of the magnetic to the electric components decreasing regularly when the frequency increases.

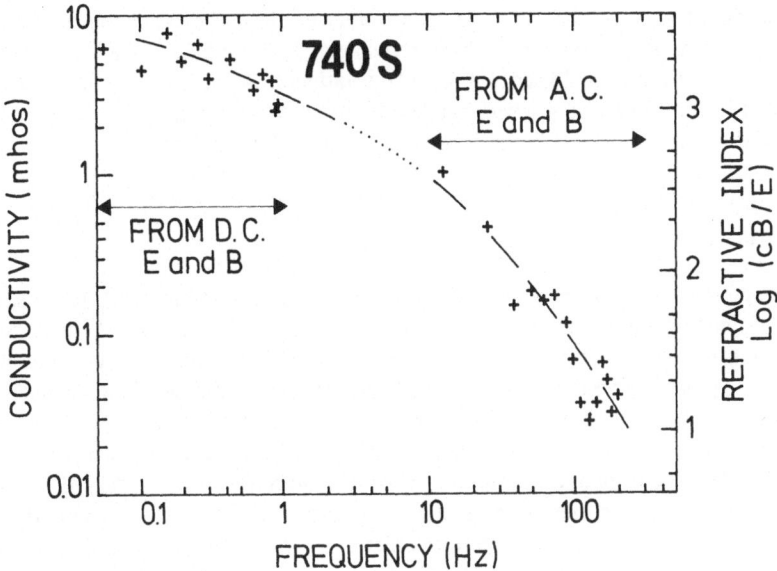

Figure 6 - The B/E ratio is reported versus the frequency, expressed as a conductivity value : B/μ_0E, mhos, left scale, and as a wave refractive index : cB/E, right scale.

Two different ways of interpretation of this result can be proposed. First, if the electric and magnetic fluctuations are due to the motion of the satellite through small scale field aligned currents, the B/μ_0E ratio will measure the integrated Pedersen conductivity of the underlying ionosphere (left scale in figure 6), implicitly assumed to be constant. The second and alternate interpretation considers that the electric and magnetic fluctuations results from the detection of Alfven waves, in which case the cB/E ratio measures the refractive index of the wave, R_A (right scale in figure 6), and should be equal to c/V_A, where c is the light velocity and V_A the Alfven velocity. However in both interpretations the observed variation of the B/E ratio with frequency is in apparent contradiction with the above simple theoretical rules.

Under the first (static) interpretation two arguments can be put forward to account for the decrease of the conductivity with the frequency. First the effect of the conductivity gradients in the divergence of the Pedersen ionospheric curents has not been taken into account in the simple calculation of the B/E ratio, and second it can be supposed that for short transverse scales the closure of the parallel currents does not occur in the total height range of the dynamo region, but only affects a limited height range, thus enhancing the apparent resistivity of the ionosphere.

On the other hand, if the fluctuations are assumed to be the signature of Alfven waves the cB/E ratio must be interpreted as Alfven refractive index, which for the low frequency values leads to an Alfven velocity ≈ 150 km.s^{-1} (f ≈ 0.1 to 1 Hz, $R_A \approx 2.10^3$, Figure 6). Such a velocity is much smaller than the value of 1000 km.s^{-1} ($R_A \approx 3.10^2$) deduced from simultaneous electron density measurements provided by the on board ISOPROBE experiment (D. Lagoutte, private communication). This suggests that at least below ≈ 1 Hz, the observed turbulence is due to small scale field aligned structures through which the satellite is moving. This interpretation is consistent with the observed coincidence between electromagnetic bursts and intense upward currents as mentioned above.

At higher frequencies, the observed variation of the B/E ratio, remains difficult to explain in terms of classical Alfven wave theory. However a similar result has been previously reported in the litterature (Gurnett et al, 1984), and it has been shown by Temerin and Kintner (1988) that oblique kinetic Alfven waves could account for such a variation. A similar wave interpretation of the magnetic turbulence for frequencies up to a few Hz has also been proposed by Chmyrev et al (1985), for observations made with the Intercosmos-Bulgaria-1300 satellite.

4. Conclusions.

A set of complementary measurements obtained during a pass of the AUREOL-3 satellite through the dayside high latitudes has been used to study the electrodynamics of the ionosphere in the cusp and cleft regions. Their position and latitudinal extent have been delineated through the charateristics of particle precipitations and field aligned currents which enable to distinguish the region of direct entry of magnetosheath low energy plasma (cusp) from the ionospheric projection of the low latitude boundary layer (cleft). The extremely good agreement between the values of the integrated Pedersen conductivity deduced respectively from large scale variations of the DC electric and magnetic fields and from a direct calculation taking into account the solar illumination and particle precipitations confirms that at a large scale ($\geq$ 100 km) field aligned currents close in the ionosphere through Pedersen currents. Superimposed on the large scale upward field aligned currents in the cleft we have detected small scale ($\approx$ 10 km) variations which are correlated with enhancements of the low energy ($\approx$ 0.1 keV) precipitating electrons.

Bursts of electromagnetic turbulence lasting from a few tenths to $\approx$ 1 second occur in the vicinity of these enhanced current sheets. From the variations of the electric and magnetic components measured in two frequency ranges 0.1 to 1 Hz and 10 to 200 Hz, a spectral analysis of the E/B ratio has been performed which shows that its value is a decreasing function of frequency over the whole interval. Although the observations above $\approx$ 10 Hz may be explainable in terms of kinetic Alfven waves, this is not the case below 1 Hz where the data most probably correspond to the traversal of small scale striated current sheets.

References

Berthelier, A., J.-J. Berthelier, and C. Guérin, The effect of the east-west component of the interplanetary magnetic field on magnetospheric convection as deduced from magnetic perturbations at high latitudes, *J. Geophys. Res., 79(22)*, 3187-3192, 1974.

Berthelier, A., and C. Machard, Small scale intense field aligned current sheets in the northern polar cusp, in *The Polar Cusp,* ed. by J.A. Holtet and A. Egeland, 235-242, Reidel, 1985.

Berthelier, J.-J., C. Machard, J.-C. Cerisier, A. Berthelier, and J.-M. Bosqued, ULF magnetic turbulence in the high latitude topside ionosphere, *J. Geophys. Res., 93(A6)*, 5701-5712, 1988.

Bosqued, J.-M., J.-A. Sauvaud, H. Rème, J. Crasnier, D. Roux, Yu.I. Galperin, T.M. Moularchik, and V.A. Gladyshev, Evidence for ion energy dispersion in the polar cusp related to a northward directed IMF, in *Results of the ARCAD-3 Project and of the Recent Programs in Magnetospheric and Ionospheric Physics,* 331-341, CNES, France, 1984.

Bythrow, P.F., R.A. Heelis, W.B. Hanson, R.A. Power, and R.A. Hoffman, Observational evidence for a boundary layer source of dayside Region 1 field aligned currents, *J. Geophys. Res., 86(A7)*, 5577-5589, 1981.

Bythrow, P.F., T.A. Potemra, and R.A. Hoffman, Observations of field-aligned currents, particles, and plasma drift in the polar cusps near solstice, *J. Geophys. Res., 87(A7)*, 5131-5139, 1982.

Chmyrev, V.M., V.N. Oraevsky, S.V. Bilichenko, N.V. Isaev, G.A. Stanev, D.K. Teodosiev, and S.I. Shkolnikova, The fine structure of intensive small scale electric and magnetic fields in the high latitude ionosphere as observed by Intercosmos-Bulgaria-1300 satellite, *Planet. Space Sci., 33(12)*, 1383-1388, 1985.

Erlandson, R.E., R. Pottelette, T.A. Potemra, L.J. Zanetti, A.G. Bahnsen, R. Lundin, and M. Hamelin, Impulsive electrostatic waves and field-aligned currents observed in the entry layer, *Geophys. Res. Lett., 14(4)*, 431-434, 1987.

Farrugia, C.J., D.J. Southwood, S.W.H. Cowley, R.P.Rijnbeek, and P.W. Daly, Two-regime flux transfer events, *Planet. Space Sci., 35(6)*, 737-744, 1987.

Gurnett, D.A., R.L. Huff, J.D. Menietti, J.L. Burch, J.D. Winningham, and S.D. Shawhan, Correlated low frequency electric and magnetic noise along auroral field lines, *J. Geophys. Res., 89(A10)*, 8971-8985, 1984.

Gurnett, D.A., and U.S. Inan, Plasma wave observations with the Dynamics Explorer 1 spacecraft, *Rev. Geophys., 26(2)*, 285-316, 1988.

Hardy, D.A., M.S. Gussenhoven, and E. Holeman, A statistical model of auroral electron precipitation, *J. Geophys. Res., 90(A5)*, 4229-4248, 1985.

Heelis, R.A., The effects of interplanetary magnetic field orientation on dayside high-latitude ionospheric convection, *J. Geophys. Res., 89(A5)*, 2873-2880, 1984.

Heppner, J.P., Polar cap electric field distribution related to the interplanetary magnetic field direction, *J. Geophys. Res., 77(25)*, 4877-4887, 1972.

Hoffman, R.A., M. Sugiura, and N.C. Maynard, Current carriers for the field-aligned current system, COSPAR XXV Plenary Meeting, Graz, Austria, 1984.

Iijima,T., and T.A. Potemra, Field aligned currents in the dayside cusp observed by Triad, *J. Geophys. Res, 81(34)*, 5971-5979,1976.

Mehta, N.C., Ionospheric electrodynamics and its coupling to the magnetosphere, Ph.D. thesis, Univ. of Calif., San Diego, 1978.

Meng, C.-I., and R. Lundin, Auroral morphology of the midday oval, *J. Geophys. Res., 91(A2)*, 1572-1584, 1986.

Paschmann, G., N. Sckopke, and H. Grünwaldt, Plasma in the polar cusp and plasma mantle, in *Magnetospheric Particles and Fields*, ed. by B.M. McCormac, 37-46, Reidel, 1976.

Reiff, P.H., T.W. Hill, and J.L. Burch, Solar wind plasma injection at the dayside magnetospheric cusp, *J. Geophys. Res, 82 (4)*, 479-491,1977.

Reiff, P.H., Models of auroral zone conductances, in *Magnetospheric Currents*, ed. by T.A. Potemra, 180-191, AGU Geophys. Monogr., vol. 28, 1984.

Russell, C.T., and R.C. Elphic, Initial ISEE magnetometer results : magnetopause observations, *Space Sci. Rev., 22 (6)*, 681-715, 1978.

Smiddy, M. W.J. Burke, M.C. Kelly, N.A Saflekos, M.S. Gussenhoven, D.A. Hardy, and F.J.Rich, Effects of high-latitude conductivity on observed convection electric fields and Birkeland currents, *J. Geophys. Res., 85(A12)*, 6811-6818, 1980.

Sugiura, M., T. Iyemori, R.A. Hoffman, N.C. Maynard, J.L. Burch, and J.D. Winningham, Relationships between field-aligned currents, electric fields, and particle precipitation as observed by Dynamics Explorer 2, in *Magnetospheric Currents,* ed. by T.A. Potemra, 96-103, AGU Geophys. Monogr., vol. 28, 1984.

Temerin, M., and P.M. Kintner, Review of ionospheric turbulence, Paper submitted to publication, 1988.

Vasyliunas, V.M., Interaction between the magnetospheric boundary layers and the ionosphere,in *Magnetospheric Boundary Layers,* 387-393, ESA SP-148, 1979.

Zanetti, L.J., and T.A. Potemra, The relationship of Birkeland and ionospheric current systems to the interplanetary magnetic field, in *Solar Wind-Magnetosphere Coupling,* ed. by Y. Kamide and J.A. Slavin, 547-562, TERRAPUB, 1986.

POLAR CLEFT STRUCTURE AT 09 MLT: COORDINATED SATELLITE- AND GROUND-BASED OBSERVATIONS

P.E. Sandholt, B. Jacobsen, B. Lybekk and A. Egeland*
C.-I. Meng, and P.T. Newell**,
F.J. Rich, and E.J. Weber***.

*Department of Physics, University of Oslo, P.O.Box 1048 Blindern, 0316 Oslo 3, Norway, **Applied Physics Laboratory, The Johns Hopkins University, Johns Hopkins Road, Maryland 20707, ***Air Force Geophysics Laboratory, Hanscom Air Force Base, Bedford, Massachusetts 01731.

ABSTRACT. Observations of electron and proton precipitation fluxes and Birkeland currents from spacecraft DMSP F7 and simultaneous ground-based optical measurements from Svalbard (~75° geom.lat.) of dayside aurorae (~ 09 MLT) are presented. In these cases, the auroral emissions within the field of view can be separated in different latitudinal zones with corresponding structures in the particle precipitation. The temporal evolution of these structures is monitored by the ground-based instruments, indicating the presence of both stationary cleft emissions produced by soft electron fluxes (energy few hundred eV) and multiple, discrete arcs and arc-fragments corresponding to the precipitation of keV electrons. The latter are associated with narrow sheets of upward flowing Birkeland currents and reduced proton precipitation, suggesting the presence of field-aligned potential drops above the spacecraft. Field-aligned current systems separated in latitude and associated with plasma-sheet-like and LLBL-like particles, respectively, are observed during the satellite pass presented here, which occurred after a sharp IMF B_z polarity change.

1. INTRODUCTION

Combined satellite and ground-based observations in the cusp have revealed discrete, transient auroral forms with large internal northward electric field (~ 100-200 mV/m) and rather complicated Birkeland current signatures /1,2/.

311

Electron energy spectra from above such aurorae, obtained
at ~ 11 MLT, showed peaks on the high-energy side, located
between ~ 200 and 600 eV. Changes of the electron spectral
characteristics with latitude and local time around noon
are now well documented /3,6/. A zone of magnetosheath-like
electrons (energy < 200 eV) is often observed near noon (~
11-13 MLT), corresponding to the midday gap of discrete
aurorae (cusp proper), indicating minor particle energization in this region. It should be noted, however, that
sporadic, short-lived intensifications of discrete aurorae
(dayside "breakups") are observed from the ground, also
within this region /4,5/. Away from noon and at lower
latitudes near noon, within the cleft, particle energization is more common /3,6/. The occurrence rate of ~ 1 keV
electron precipitation and associated discrete aurorae
increases with the distance from magnetic noon.

Torbert and Carlson /7/ presented rocket observations of
electron and proton precipitation at cleft latitudes near
09 MLT that indicated acceleration by parallel electric
fields.

Lundin and Evans /8/ presented a boundary layer model of
observed structures in dayside aurorae. According to this
model, multiple arcs in the post-noon sector are the result
of several plasma injections in various parts off the
"center" of the cusp, or an injected plasma cloud that
breaks up into several filaments.

The large-scale morphology of the dayside section of the
auroral oval has been characterized by Meng and Lundin /9/,
based on DMSP satellite images. Certain differences and
similarities of the auroral structures in the pre- and
postnoon sectors were pointed out.

In this study coordinated satellite and ground-based observations of dayside aurorae (~ 09 MLT) are presented. From
the ground-site on Svalbard, Norway (75° MLAT) optical
emissions were recorded by an image intensified all-sky
camera and a 4-channel system of meridian scanning photometers, covering the latitudinal range ~ 70-80° MLAT. By
these techniques the spatial structure and the time variations of the different individual auroral forms were observed.

Satellite DMSP-F7 carries a high resolution magnetic field
experiment (SSM), electron and ion spectrometers (J/4),
covering the energy range 30 eV - 30 keV, and a number of
other experiments. For a complete description of the DMSP-
F7 instrumentation, see Rich et al. /10/. DMSP F7 is close
to the Svalbard meridian at ~ 06 UT (~ 09 MLT). One pass

which was very close to the scanning plane of the ground photometers was selected for presentation in this report. A more comprehensive material is discussed in Sandholt <u>et al.</u> /11/.

JAN. 23, 1985

SVALBARD PHOTOMETERS

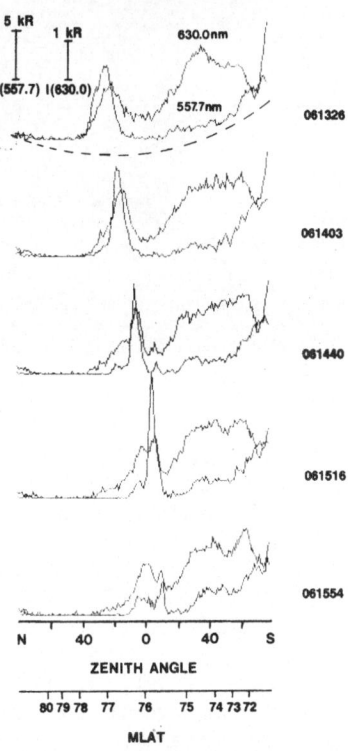

Fig. 1: North-south meridian photometer scans (8-168 degr. from north horizon) at wavelengths 630.0 and 557.7 nm, obtained at Ny Alesund, Svalbard (75.4° MLAT). Start time of each pair of scans in UT is marked on the right side. Intensity scales are given in the upper left corner. MLAT scale corresponds to 250 km emission altitude.

2. THE OBSERVATIONS

Figure 1A shows photometer scans along the magnetic meridian through Ny Alesund, Svalbard, at wavelengths 630.0 and 557.7 nm, during the time period 061326 - 061554 UT (˜ 0920 MLT) on 23 Jan. 1985. Spacecraft DMSP F7 passed above (˜ 800 km) the ground site at Ny Alesund at ˜ 061435 UT. The satellite trajectory was close to the scanning plane of the

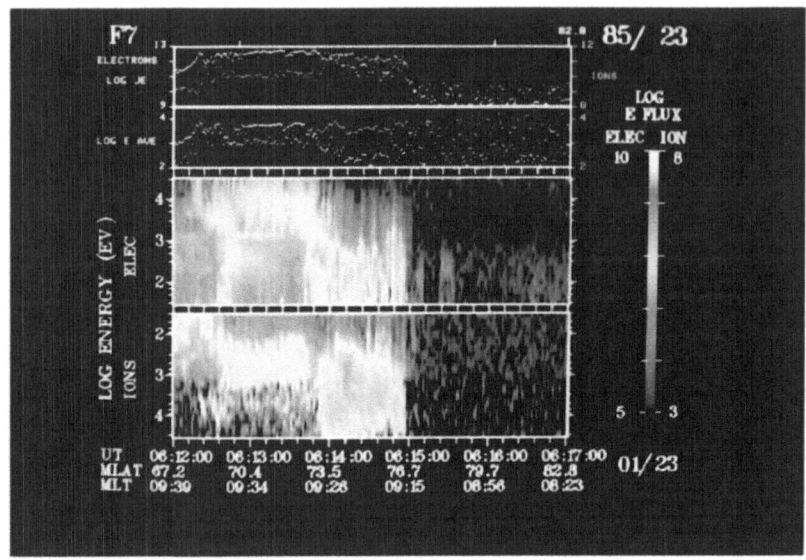

Fig. 2A: Spectrogram of 32 eV to 30 keV electrons and
ions obtained during a DMSP F-7 pass above Ny Alesund
(061435 UT) on 23 Jan., 1985. The main panels show
electron and ion differential energy flux in units of
eV/cm²s sr versus energy in eV.

photometer system on the ground, in this case. Energy
spectra of the electron and proton precipitation fluxes
during this pass are shown in Figure 2A. Figure 2B shows
the electron data (particle flux, energy flux and average
energy) and the east-west component of the magnetic deflection recorded from the spacecraft.

The observations from the ground and the satellite suggest
a separation into different latitudinal zones of different
auroral emission and particle precipitation characteristics.

a) Cleft-like emissions between 15° and 55° south of the
zenith with strong red oxygen emission at 630.0 nm. The
DMSP measurements show a region of soft electron precipitation (few hundred eV) between 73.9 and 75.2° MLAT (close to
Ny Alesund).
b) South of this cusp/cleft zone the green line is enhan
ced more than the red line, corresponding to the somewhat
higher energy of precipitating electrons between 72.7 and
73.9° MLAT (cf. Figures 2A,B).
c) North of the cleft-like emission a slowly equatorward
moving discrete auroral arc is observed (7° north of the
zenith at 061448 UT). The spectral ratio I630.0 nm/I557.7

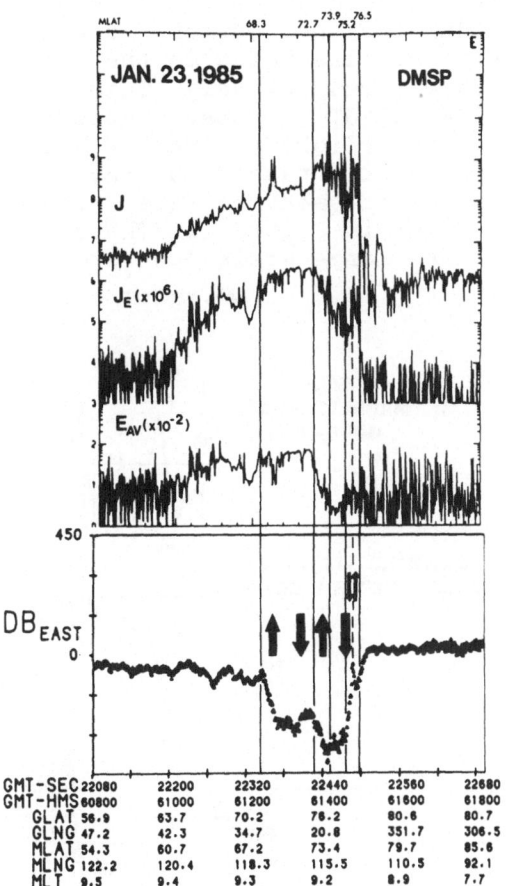

Fig. 2B: Electron precipitation (particle flux-
el./cm² s sr, energy flux - eV/cm² s sr, average energy
- eV) and eastward magnetic deflection (lower panel)
obtained from DMSP F-7. Vertical lines mark different
zones of particle precipitation and Birkeland current
(current direction marked by arrows).

nm is ˜ 0.2 at 061448 UT. The electron flux shows a corresponding local maximum at this time with the peak of the
energy spectrum around 1 keV. At the same location proton
intensity is strongly reduced below ˜ 1 keV. A double
structure is seen in the photometer trace, with a minor
maximum towards the north at 061448 UT. A similar feature
is seen in the electron precipitation. The actual auroral
structure was observed between 0553 and 0618 UT. It moved
poleward during the first part of the interval and then
returned equatorward. This aurora is an east-west aligned
rayed arc-like form. The magnetic deflection in Figure 2B
(lower panel) indicate a pair of oppositely directed Birke-

316

land currents associated with this discrete aurora. The
latitudinal width and the current density of the upward
flowing current, corresponding to the main auroral peak, is
~ 15 km and $2\mu A/m^2$, respectively. This current was traversed during 061449-51 UT. A close inspection of the magnetic
data reveal another slightly smaller structure of upward
current further north, passed during 061454-55 UT, corresponding to the minor peak on the poleward side in the
photometer profile in Figure 1. These upward currents are
embedded in a more large scale downward directed current.

JAN. 23, 1985

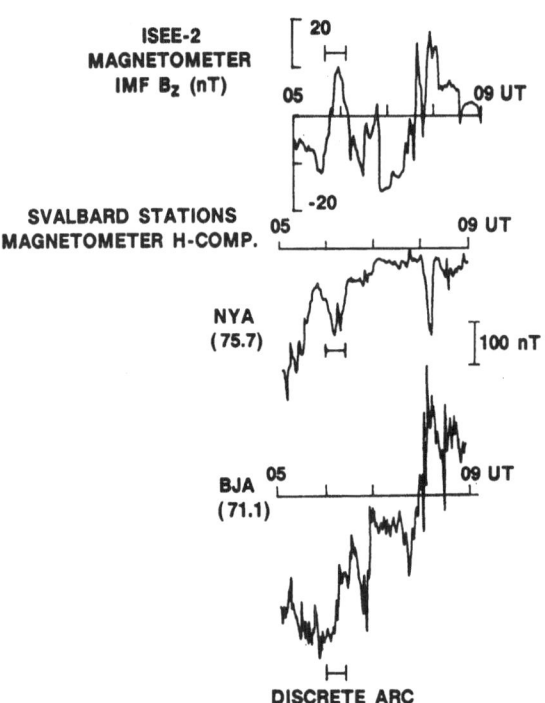

Fig. 3: Upper panel: IMF B_z component from S/C ISEE-2.
Second and third panels: H-component deflections at Ny
Alesund (NYA) and Bear Island (BJA). The time of
occurrence of discrete auroral arc near Ny Alesund
(Figures 1 anda 2) is marked in all three panels.

The discrete aurora on the poleward side of the cleft occurred during a short interval with positive IMF B_z, within
a longer period with negative B_z, as shown in Figure 3.
Poleward and subsequent equatorward motions of this arc did
correspond closely with the northward and southward IMF
transitions, respectively. IMF B_y was ~ -8 nT within 04-
07 UT.

3. SUMMARY

Simultaneous latitudinal profiles along the same meridian
(~ 09 MLT) of electron and proton precipitation, Birkeland
currents and auroral emissions, have been obtained by
coordinated satellite and ground-based observations. The
total data set provides both the time history of the events
as observed by the groundbased instruments as well as
"snapshot" precipitation profiles obtained from the spacecraft. The auroral emissions within the field of view can
be separated in different latitudinal zones with corresponding structures in the particle precipitation. The
temporal evolution of these structures indicate the presence of both stationary cleft-like emissions produced by
soft electron fluxes (energy few hundred eV) and discrete
arcs and arc-fragments (green line intensity 1-10 kR)
corresponding to the precipitation of 1 keV electrons. The
latter are associated with narrow sheets of upward flowing
Birkeland current (~ 2 μA/m^2) and reduced proton precipitation, suggesting the presence of field-aligned potential
drops above the spacecraft. These structures are observed
at different latitudes within the cleft, including its
equatorward and poleward boundaries, during different cases
/11/.

The large-scale gradients in the east-west component of
magnetic deflection (Figure 2B) indicate the existence of
two pairs of current sheets, corresponding to two very
different plasma regions. Our interpretation is that the
gradients from 68.3 to about 70.5° and from about 70.5 to
72.7° MLAT correspond to the Region 2 and Region 1 current
systems, respectively. The precipitating particles in these
zones are characteristic of the plasma sheet. Further north
the particle precipitation is more characteristic of the
low-latitude boundary layer (LLBL). The northernmost field-
aligned currents may correspond to the so-called B$_y$ currents discussed by Troshichev et al. /12/ (cf. also /13/.

ACKNOWLEDGEMENT

This study was sponsored by NATO collaborative research
grant No 85/0521. Dr. R. Lepping kindly provided the ISEE-
2 IMF data used in this study.

REFERENCES

1. P.E. Sandholt, C.S. Deehr, A. Egeland, B. Lybekk, R. Viereck, and G.J. Romick, Signatures in the dayside aurora of plasma transfer from the magnetosheath, J. Geophys. Res., 91, 10063 (1986).

2. P.E. Sandholt and A. Egeland: Polar cusp electrodynamics, ESA SP-270, p. 255, 1987.

3. R. Lundin, L. Eliasson, and I. Sandahl: First results from the VIKING particle experiment, Physica Scripta, vol. 35 (1987).

4. P.E. Sandholt, B. Lybekk, A. Egeland, R. Nakamura, and T. Oguti, Midday auroral breakup, J. Geomagn. Geoelectr. 41, (1989).

5. P.E. Sandholt, IMF control of polar cusp and cleft auroras, Adv. Space Res., in press (1989)

6. P.T. Newell and C.-I. Meng, The cusp and the cleft/-LLBL, Low altitude identification and statistical local time variation, J. Geophys. Res. in press-(1988).

7. R.B. Torbert and C.W. Carlson, Evidence for parallel electric field particle accelleration in the dayside auroral oval, J. Geophys. Res., 85, 2909 (1980).

8. R. Lundin and D.S. Evans, Boundary layer plasmas as a source for high-latitude, early afternoon, auroral arcs, Planet. Space Sci., 33, 1389 (1985).

9. C.-I. Meng and R. Lundin: Auroral morphology of the midday oval, J. Geophys. Res., 91, 1572 (1986).

10. F.J. Rich, D.A. Hardy, and M.S. Gussenhoven, Enhanced ionosphere - magnetosphere data from the DMSP satellites, EOS, 66, 513 (1985).

11. P.E. Sandholt, B. Jacobsen, B. Lybekk, A. Egeland, C.-I. Meng, P.T. Newell, F.J. Rich, and E.J. Weber, Structure and dynamics in the polar cleft: Coordinated satellite and ground-based observations in the prenoon sector, J. Geophys. Res., in press (1989).

12. O.A. Troshichev, B.D. Bolotinskaya, A.L. Kotikov, and V.O. Papitashvili, B_y dependent current in the southern polar region during positive B_z, Planet. Space. Sci., 36, 523, 1988.

13. E. Friis-Christensen, Ground maganetic perturbations in the polar cap and cleft: Structure and dynamics of ionospheric currents, this volume.

6300-Å AURORAL EMISSIONS AT SOUTH POLE: DAYSIDE POLE-WARD MOTION AND SUN ALIGNED ARCS

R. L. Rairden and S. B. Mende

Lockheed Palo Alto Research Laboratory, Dept 91-20, Building 255, 3251 Hanover Street, Palo Alto, CA 94304

ABSTRACT.

Optical (6300 Å) observations of the night sky in the polar cap region were recorded during the April to August austral winters of 1983 through 1986 with an image-intensified all-sky camera located at South Pole Station. Auroral structures seen in these all-sky images were correlated with the signatures registered by a meridian slit (keogram) imager. The keograms provide a convenient survey of auroral activity, but the all-sky images are required in order to determine the precise auroral morphology associated with the commonly observed keogram traces. During the midday interval discrete poleward-moving arcs occasionally appear, likely associated with flux transfer events or similar changes in the plasma flow pattern. Nightside keograms show the characteristic signatures of sun-aligned arcs, which are well-defined in the associated all-sky images. Two thirds of the sun-aligned arcs are observed on the dawn side of local midnight. The arcs drift overhead at velocities typically less than 250 meters/sec with reference to a sun-oriented non-corotating frame. No significant preference is shown for easterly (dawnward) or westerly (duskward) drift directions. These sun-aligned arcs occur with greater frequency than previously thought: we observe a total of 114 such events during 219 diurnal periods of image collection. The available interplanetary magnetic field measurements, though limited, indicate $B_z \geq 0$ for two thirds of the sun-aligned arcs. While no correlation is found between B_y and their local time distribution, there is a suggestion that dawnward motions appear preferentially with $B_y > 0$ and duskward motions with $B_y < 0$.

P. E. Sandholt and A. Egeland (eds.), Electromagnetic Coupling in the Polar Clefts and Caps, 319–342.
© *1989 by Kluwer Academic Publishers.*

1. INTRODUCTION

SOUTH POLE STATION FIELD OF VIEW
RELATIVE TO AURORAL OVAL

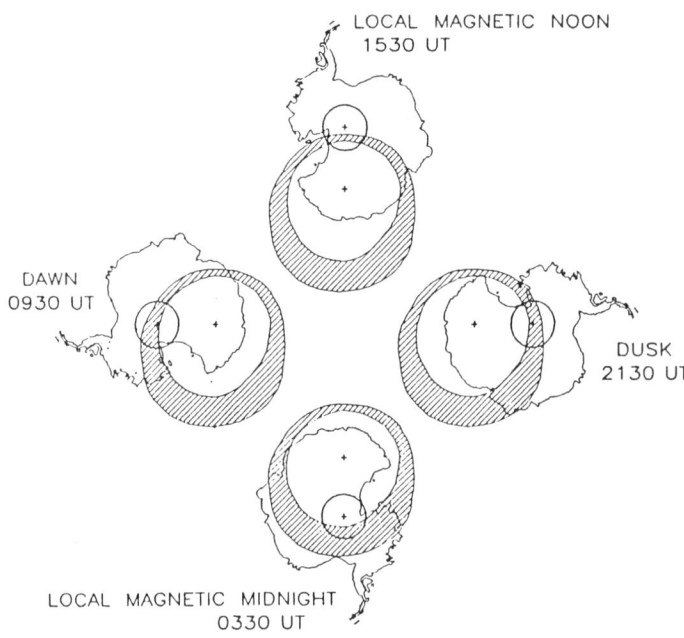

Fig. 1. Size and location of the all-sky camera field of view (circle) relative to the Antarctic continent and the nominal auroral oval. Orientations are shown at four times of day, indicated in magnetic local time and the corresponding UT. The solar direction is towards the top of this figure. South Pole Station, in continual darkness April through August, is ideally situated for the observation of auroral oval and polar cap activity.

Faint optical images of the polar cap region in 6300-Å emissions were recorded during the austral winters of 1983 through 1986 with the Lockheed image-intensified all-sky camera located at South Pole Station. A meridian slit (keogram) camera at the same site provides a well-correlated alternate rendition of the auroral activity in this wavelength. The high magnetic latitude and the continuous optical darkness from April to August allow the observation of faint auroral signatures at all magnetic local times in the high-latitude auroral zone and the polar cap. This region is under intensive study by several investigators due to the diversity of phenomena taking place. When the interplanetary magnetic field (IMF) is southward ($B_z < 0$) enhanced merging occurs, and on the dayside freshly reconnected flux tubes move antisunward causing observable changes in the local ionosphere.

When the IMF is northward ($B_z > 0$) the normally quiescent precipitation in the polar cap, the polar rain, gives way to polar shower precipitation [Hardy, 1984] and the aurora shows discrete patterns in the form of poleward expanded ovals and sun-aligned arcs [Lassen and Danielsen, 1978; Ismail and Meng, 1982; Gussenhoven, 1982]. During such times the polar cap shows structured plasma signatures, the presence of accelerated plasma sheet-like particles [Burch et al, 1979], energetic ions [Peterson and Shelley, 1984], and intricate field-aligned current patterns [Iijima and Shibaji, 1987]. From topside imaging Frank et al. [1982] have shown that the polar cap can be seen divided by a transpolar "theta" arc during such times. Much of this complex behavior can be viewed in terms of the convection patterns proposed by Reiff and Burch [1985] for northward B_z. Ground based intensified images record the faint footprints of all these phenomena. The images provide information regarding the spatial extent of these footprints, their morphological appearance, and their time development. There have been numerous studies of white-light all-sky camera images taken at polar cap stations, however such an all-sky camera produces sporadic data because it does not detect the faint optical signatures which continuously characterize these regions. Eather et al. [1979] used photometers, and Eather and Mende [1980] and Eather [1981] used image intensifiers in their keogram camera at South Pole to show that the fainter auroral signatures produce an uninterrupted record. A keogram by itself provides only a one-dimensional view of the auroral phenomena. In this paper we discuss the two-dimensonal intensified imaging data taken at South Pole and intercompare it with the simultaneous keogram data.

This paper is organized as follows: First we describe the instrument and its daily coverage of geomagnetic phenomena at South Pole, briefly explaining the observing technique and methods of data presentation. Three subsequent sections describe auroral features characteristic of differing magnetic local times. These comprise morning and afternoon auroral oval observations; dayside (cusp) features during the more active times; and nightside polar cap region observations with special emphasis on sun-aligned polar cap arcs. Each section is introduced with a summary of the present understanding of the observed phenomena, then continues with a display of our imaging data, and ends with a discussion of the observations.

2. INSTRUMENTATION

The Lockheed image-intensified all-sky camera, a modified version of the instrument described by Mende et al. [1977], photographs the sky through a 30-Å wide filter at 6300 Å. This filter eliminates much of the unwanted sky background and emphasizes the soft electron precipitation or regions of high ion density. Our field of view is 160 degrees full angle centered at zenith, which is a circle of diameter 1620 km at the assumed emissions height of 200 km altitude. The size and location of this circular field of view relative to the auroral oval are indicated in Figure 1. Four times of day are represented, with the solar direction towards the top of the figure. The upper illustration shows South Pole Station located

322

INTENSIFIED ALL-SKY CAMERA ORIENTATION

FIELD OF VIEW AT 200 KM IS 1620 KM DIAMETER

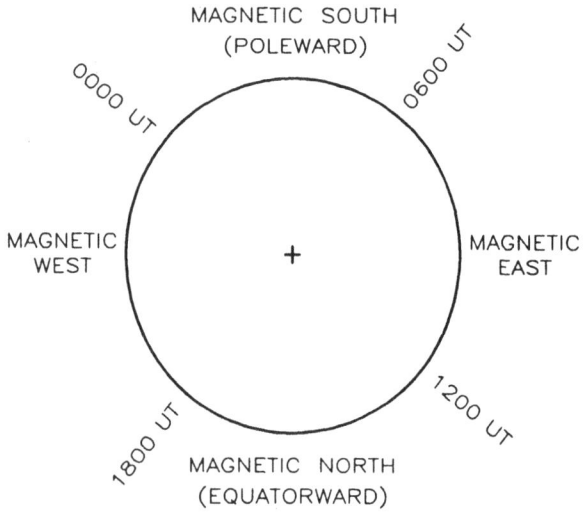

WAVELENGTH 6300 Å   SENSITIVITY 200 R

Fig. 2.   All-sky camera field of view, looking upward
with zenith centered. Subsequent presentations of all-
sky images are oriented thusly, with the south magnetic
poleward direction toward the top of each frame. The
solar direction then rotates as a function of UT, as indicated around the perimeter.

at magnetic midday. At this time the auroral oval is often magnetic poleward
but still within the field of view. This configuration is ideal for studying auroral dayside phenomena. During the austral winter there is continuous darkness
at South Pole permitting the optical investigation of polar cap activity over the
entire diurnal period. The right-hand illustration of Figure 1 represents the situation six hours later at magnetic dusk. The auroral oval is generally overhead,
as it crosses from the poleward side to the equatorward side of the station. In
the bottom picture our location at magnetic midnight is shown. At this time the
quiescent auroral oval is invariably equatorward of the geographic South Pole.
The all-sky images in the remaining figures of this paper are also displayed in
the upward-looking perspective and are rendered as photographic negatives oriented as shown in Figure 2. The cardinal directions are indicated. The magnetospheric dawnward and duskward directions depend on the time of day as the
sun (several degrees below the horizon) travels around this figure clockwise from
the upper left at 0000 UT. Local magnetic midnight occurs at 0330 UT, with the

sun off the top of the figure, or poleward. The sun is off to the right at local dawn, off the bottom or equatorward at local noon (1530 UT), and toward the left at local dusk. The camera records emission rates down to ~ 200 rayleighs. It should be noted that while the angular distance scale from zenith is approximately linear on these images, the spatial extent and velocity of auroral features at the larger zenith angles are subject to the distortion inherent to this geometry.

During the observation periods exposures of one-second duration are made at two- or three-minute intervals on 35-mm Kodak Tri-x panchromatic film (the 100-ft film cannister is reloaded by on-site personnel every ~ 2 days). This provides good time-resolution of the overhead auroral activity throughout the April to August period of antarctic night, with the exception of a few days each month when the moon is near full and above the horizon. Weather is generally not a problem at South Pole.

Our all-sky images add another dimension to the keogram data collected simultaneously by a meridian slit camera also at South Pole. The keogram camera and some of the puzzling auroral phenomena it monitors are described by *Eather* [1984; 1985]. The South Pole Station magnetic latitude is ~ 74 degrees, a point over which the average auroral oval crosses poleward near local dawn and returns equatorward near local dusk. The keogram meridian is aligned such that the top edge of the film is toward the magnetic pole and the bottom edge toward the equator, while a motor drives the film across the slit at a rate of one inch per hour. The keogram field of view is essentially a vertical line up the center of the all-sky camera field of view. The keograms record the motion of the intersection of auroral features with this vertical slit. From day to day the auroral oval crossings exhibit a large degree of variability, but some characteristic features persist.

The keograms graphically display a good time-resolution record of the poleward and equatorward movement of arcs within the auroral oval. Each day's activity is conveniently summarized on 24 inches of film, and several of these strips can be aligned to readily show the variations in activity from day to day. The correlation of all-sky camera images is necessary, however, to determine the longitudinal extent of these features and the component of their motions perpendicular to the meridian.

3. DAYSIDE POLEWARD MOTIONS

3.1 Plasma Convection Models

One of the central questions in magnetospheric physics is the nature of the driving forces responsible for plasma convection. The interaction between the boundary layer flow and the solar wind can be maintained by viscous interaction [*Axford and Hines*, 1961], by reconnection [*Dungey*, 1961] or by impulsive injection of large plasmoids into the magnetosphere [*Lemaire*, 1977]. Although there is a great deal of indirect observational evidence for reconnection there is still some disagreement whether that is the most important process [*Heikkila*, 1983]. It appears that reconnection can take the form of quasi-steady flow [*Paschmann et al.*, 1979] or it can be impulsive with large region of flux suddenly changing from "closed" to "open" field lines. These impulsive unsteady forms of reconnection have became known as flux transfer events (FTEs) [*Russell and Elphic*, 1978; 1979]. The basic difference between the steady-state reconnection processes and impulsive flux transfer events is in the timing and the time scales associated with impulsive flux transfer events is on the order of a few minutes while the time scales of the continuous reconnection should proceed with time scales more in the 20 minutes range [*Goertz et al.*, 1985]. The reconnection process would be strongly enhanced under IMF B_z negative conditions. From satellite observations *Goertz et al.* [1985] estimate that sporadic and spatially isolated flow across the convection boundary occurs on the time scale of a few minutes and has an approximate spatial extent of 50 to 300 km when projected down to ionospheric heights.

Lee and Fu [1985] have proposed a model in which the FTEs represent magnetic islands generated in the magnetopause current layer by the tearing mode. Apparently the growth rate of these is substantial, and islands of large size which involve extensive longitude segments are generated. The *Russell and Elphic* [1979] model produces short segments which can be described as slanted holes in the magnetopause. Hence the spatial extent of such regions is significant in determining which of the two mechanisms is responsible. Optical monitoring of the magnetic field tubes might provide a good description of the spatial extent of the flux tube regions.

It would be of considerable importance to find an appropriate ground signature for these reconnection events. Unfortunately the magnetosheath field lines do not ususaly connect to the ground. However the convection boundary which is the projected image of the boundary layer between sunward and antisunward convection is observable with some ground-based techniques and we would expect that the freshly reconnected plasma would move across the boundary. Thus a radar system such as the STARE radar which measures the ion convection motions and the convection boundary by doppler techniques is used to detect plasma motion across the boundary by *Goertz et al.* [1985].

A newly reconnected flux tube contains a mixture of magnetosheath and magnetospheric plasmas *Thomsen et al.* [1987] show that the ion and electron distributions are consistent with two superimposed populations. The two interacting plasmas would appear to cause pitch-angle scattering and perhaps some enhancement in precipitation. Thus it is possible that there is some optically detectable precipitation signature which is associated with the freshly reconnected flux tubes.

Intense precipitation is usually the result of field-aligned currents and the freshly reconnected flux tubes are expected to carry such field-aligned currents [*Cowley*, 1982; *Paschmann et al.*, 1982]. According to *Sonnerup* [1987] the net field-aligned current is parallel to B (downward in the northern hemisphere) for $B_y > 0$ and antiparallel to B for $B_y < 0$.

Ground-based magnetic signatures for flux transfer events are modeled by *McHenry and Clauer* [1987]. The freshly reconnected flux tube would be pulled tailward and the stresses exerted on the field line would have to be transmitted to the ionosphere. This must be accomplished through the action of field-aligned currents [*Cowley*, 1982; *Paschmann et al.*, 1982] which are closed through the iono-sphere. The ground magnetic signature of the field-aligned current depends on the model of the current configuration. One of the models consists of a central core with a diffuse return current while the other model is asymmetric with two oppositely directed field-aligned currents on opposite edges of the FTE tubes [*Southwood*, 1987]. If the field-aligned currents were to stimulate precipitation then the resultant auroras would be expected to "light up" a symmetrical inner core, or an asymmet-ric pattern on either side of the flux tube, dependent of the applicable model.

Using STARE radar data *Goertz et al.* [1985] examine the motion of flux tubes across the region which is the ground-based signature of the convection boundary and find a unique signature which they associate with flux transfer events. The STARE data has the advantage that it can see flow reversal boundaries. Although these boundaries cannot be delineated without radar support, we follow the exam-ple of *Sandholt et al.* [1986] and discuss plasma transfer from the magnetosheath into the magnetosphere in terms of our optical auroral signatures alone. The fea-ture which we associate with plasma transfer is a poleward moving signature in the optical 6300 Å emission on the dayside. The dayside optical aurora is most probably created by the parallel electric fields at the flow reversal boundary and, similar to nighttime auroras, the oval arcs represent the shear flow produced by viscous interaction between the magnetosheath flow and the sunward convecting magnetospheric plasma [*Reiff and Burch*, 1985]. Thus any auroral form on the dayside which is rapidly moving poleward in a direction essentially perpendicular to the dayside aurora can be regarded as a candidate for the auroral signature of recently merged anti-sunward convecting flux tubes. With this assumption we use our optical techniques to ascertain the properties of convecting flux tubes.

3.2 Observations of Dayside Poleward Motions

At South Pole Station the observation of poleward moving features requires the polar cap to be in an expanded condition such that at local noon (~ 1530 UT) we are still at the latitude of the auroral oval. This is the case on May 24, 1984 during acquisition of the upper keogram in Figure 3, which shows several bright features moving poleward from the auroral oval and fading toward zenith after five or ten minutes.

The repetition of these events is somewhat irregular, appearing from once to perhaps three to four times per hour. On the interval from 1430 to 1545 UT on the lower keogram trace of Figure 3 (May 25, 1984) we see less distinct but far more frequently repeating events. The keogram has a time resolution limit of about 30 sec-onds, so signatures on the minutes time-scale are observable. The nominal exposure repetition rate for the intensified all-sky camera is set at either two or three minutes.

As an example we show the all-sky image sequence corresponding to the event at ~ 1555 UT in Figure 3. These consecutive images at two-minute intervals show the arc breaking away from the oval and moving zenithward into the polar cap until it fades. Its poleward velocity is estimated at 600 m/s initially, slowing to

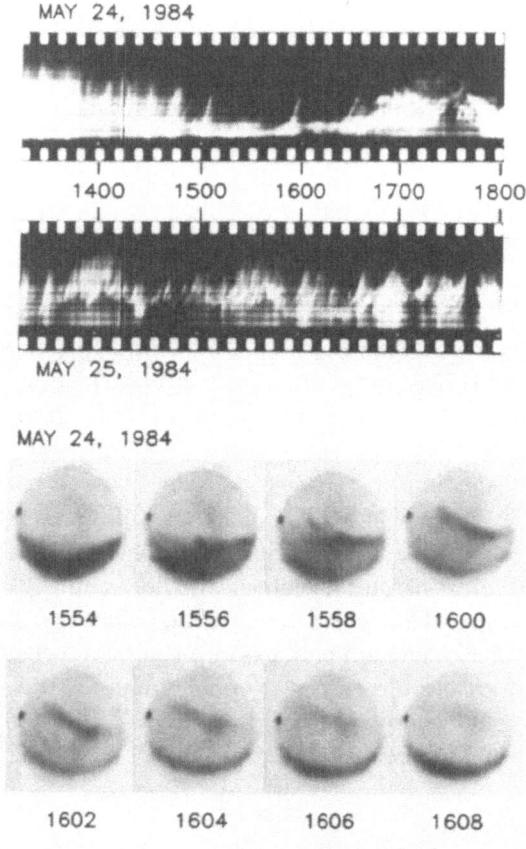

MAY 24, 1984

1400   1500   1600   1700   1800

MAY 25, 1984

MAY 24, 1984

1554   1556   1558   1600

1602   1604   1606   1608

Fig. 3.   Keogram signatures of dayside poleward motions recorded on May 24, 1984 (upper trace) and May 25, 1984 (lower trace).   Prominent poleward-moving features are seen on the May 24 keogram at 1430, 1445, 1500 and 1555 UT. The May 25 keogram shows less distinct but far more frequent poleward-moving events, on the 1430 to 1545 UT interval. The sequence of eight consecutive all-sky images corresponds to the 1555 UT May 24 event. The arc is displayed breaking away from the auroral oval and proceding poleward at ~ 600 m/sec. The longitudinal extent of this poleward-moving form exceeds 500 km, while its latitudinal width is ~ 70 km.

200 m/s as it crosses zenith. The longitudinal extent of this poleward-moving form is ∼ 500 km, while its latitudinal width is ∼ 70 km.

3.3 Discussion of Dayside Poleward Motions

For the enhanced reconnection on the dayside most models require that the interplanetary field B_z should be negative. Although no IMF measurements are available for our May 24, 1984 examples, we know that the position of the dayside cusp equatorward of South Pole Station signifies a disturbed period with an enlarged polar cap and therefore tends to indicate long periods of B_z negative [*Eather et al.*, 1979; *Carbary and Meng*, 1986].

Inspection of the keogram of May 24, 1984 shows from one to perhaps four poleward-moving features per hour. Thus the timing is more consistent with quasi-steady reconnection than with rapid impulsive FTE events. While the keogram trace for May 25, 1984 shows faster repetition rates, the duration of each event remains in the 5 to 10 minute range. The 30-second time resolution of the keogram would have recorded faster-moving events had they occurred.

The longitudinal extent of the features, 300–500 km, agrees with the predicted size for flux transfer events [e.g. *Goertz et al.*, 1985], however the asymmetrically narrow latitude extent and the observable turbulent appearance at 1604 and 1608 UT on May 24, 1984 point perhaps towards the *Lee and Fu* [1985] model. In this model FTEs represent magnetic islands, generated in the magnetopause current layer by the tearing mode, which cover substantial longitude segments. The *Russell and Elphic* [1979] model would have produced more symmetrical and perhaps shorter segments.

The whole question regarding the relationship between the location of the field-aligned current and the observable precipitation is difficult to investigate using the optical data by itself. With most arc-like features one would assume that the downward electron flow, i.e. the precipitation carrying the upward field-aligned current, is on one side of the moving flux tube. However according to some models the bulk of the reconnected flux carries enhanced field-aligned current. If the observed precipitation enhancement is due to the upward field-aligned current, then B_y must be positive for upward current in the southern hemisphere [*Sonnerup*, 1987].

The above discussion relies upon the assumption that the optically visible features are the footprints of reconnected plasma tubes convecting in the antisunward direction. Although we cannot offer conclusive proof that this assumption is valid, it appears that the gross characteristics of these features agree well with some of the model predictions.

4. SUN-ALIGNED ARCS

4.1 Introduction to Sun-Aligned Arcs

Early ground-based optical observations leading to the definition of a statistical auroral oval also found occasional higher-latitude arcs, within the polar cap, which tended to be aligned with the sun-earth direction. Systematic studies revealed certain qualitative aspects of the appearance of these "sun-aligned" arcs and their correlation with the interplanetary magnetic field. The preference of sun-aligned arcs for the dawn side of the polar cap is shown by *Lassen and Danielsen* [1978] in their analysis of data from a chain of all-sky camera stations in Greenland. The diminishing occurrence rate of sun-aligned arcs with the expansion of the auroral oval as B_z turns negative (southward) is also demonstrated.

Space-based imaging has contributed as well. Reporting on observations of northern hemisphere sun-aligned arcs from the DMSP and ISIS-2 spacecraft, *Berkey et al.* [1976] show a strong correlation between these arcs and a northward directed IMF ($B_z > 0$). In another analysis of the ISIS-2 observations, *Ismail et al.* [1977] infer duskward velocities of sun-aligned arcs, measuring from 70 to 250 m/s in a non-corotating reference frame. Their intensity ratios of 5577-Å and 3914-Å emissions indicated average precipitating electron energies in the 0.3 to 1.25 keV range.

Gussenhoven [1982] catagorizes the polar cap auroral forms seen in over 300 northern hemisphere DMSP images. Among these distinguishable catagories are the polar cap sun-aligned arcs, the extremely high latitude morning arcs, and the extremely high latitude evening arcs, designated in that work as P(1), P(2), and P(3), respectively. These types of arcs are all more or less sun-aligned, but differ in their origins and evolution. For these features *Gussenhoven* [1982] finds B_z usually directed northward. Additionally for P(2) and P(3) arcs the B_y component is predominantly negative or positive, respectively. In a similar study of DMSP images *Ismail and Meng* [1982] achieve comparable results, though using a different classification scheme. They view both hemispheres, and, finding B_x usually negative for northern polar cap arcs and positive for southern, suggest that sun-aligned arcs may be non-conjugate.

Potemra et al. [1984] demonstrate that spacecraft measurements of polar cap convective electric fields and Birkeland currents agree with the auroral patterns observed. During periods of northward IMF, the polar cap convective flow pattern for $B_y \approx 0$ exhibits two symmetric cells, producing sunward plasma flow across the cap. As B_y turns positive (negative) the dawnside (duskside) cell expands, compressing the opposite cell and increasing the Birkeland current density in this opposite cell [*Iijima et al.*, 1984]. This may be the mechanism through which the location of sun-aligned arcs is affected by the sign of B_y, if indeed the arcs occur in the regions of increased current density gradients [*Potemra et al.*, 1984]. These convection patterns are consistent with the merging of interplanetary magnetic field lines to geomagnetic field lines in the tail lobe. The resulting closed field lines divide the magnetotail lobe and provide a convection path and field-aligned transport from the plasma sheet into the polar cap. Perhaps the most striking manifestation of such a polar cap signature is the "theta aurora" as seen in DE-1 images [*Frank et al.*, 1982; 1985; 1986]. *Peterson and Shelley* [1984] support this polar cap topology with theta aurora ion composition data from DE-2 which indicate a source in the distant plasma sheet. The precipitating electron spectra from sun-aligned arcs are also shown to be consistent with a source 5 to 8 R_E tailward in the plasma sheet [*Hoffman et al.*, 1985].

The various convection models and magnetic field line merging schemes offered by *Lyons* [1985], *Kan and Burke* [1985], *Cornwall* [1985], and *Chiu et al.* [1985] serve to illustrate the accelerated interest in the complexities of polar cap activity. *Burch et al.* [1985] and *Reiff and Burch* [1985] together present a generalized global convection model, describing the viscous cells, merging cells, and lobe cells at all IMF orientations. Their configuration for northward B_z can show the development of single lobe cells filling the polar caps, rotating oppositely in the opposite hemispheres. Several predictions are then made with respect to antisymmetries between sun-aligned aurora in the two polar caps.

Enough polar cap imaging has been recorded at South Pole Station during the past few years to make herewith a substantial contribution to the statistical data base from which the abovementioned theories may derive their support.

4.2 Keogram and All-Sky Signatures

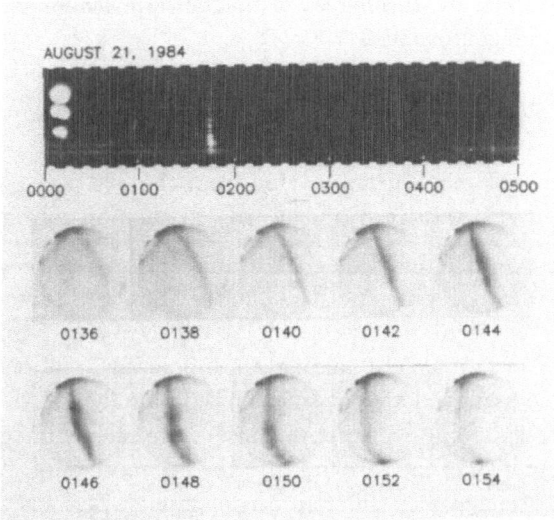

Fig. 4. Keogram signature of a sun-aligned arc at 0144 UT on August 21, 1984 and the corresponding sequence of consecutive all-sky images. The arc is very nearly aligned with the meridian slit and thus transits the keogram field of view on a brief interval. Keograms provide a convenient survey of nightside polar cap activity, but the all-sky sequences are required in order to derive the extent, orientation, and velocity of the arcs.

Sun-aligned arcs near local midnight are oriented vertically in our all-sky camera field of view, with the sunward direction off the top of the frame. We find the motion of these arcs to be usually duskward, or toward the left across the images at this time of day. Thus while a sun-aligned arc may remain in the all-sky image field for an extended period of time, it moves across the similarly-aligned keogram slit very quickly. Note that slightly before or after local midnight the sunward direction is slightly to the left or right of vertical, respectively, and the sun-aligned arcs are tipped a little one way or the other. This causes their passage through the keogram field of view to appear moving rapidly poleward or equatorward, tracing the intersection of this oblique crossing.

The keogram in Figure 4 shows a sharp vertical signature within the polar cap at 0145 UT on August 21, 1984. The auroral morphology associated with this momentary keogram event is revealed in the corresponding sequence of all-sky images also displayed in Figure 4. These ten consecutive images show a very well defined sun-aligned arc, with the sun off the top and slightly to the left at this pre-midnight local time. The arc moves duskward, which at this local time is westward, at a velocity of ~ 320 m/s relative to our field of view. In magnetospheric coordinates, relative to the sun-earth line, our field of view rotates eastward about the south magnetic pole displaced by 16 degrees in magnetic latitude. So the instrument's velocity under the magnetosphere is

$$\frac{\sin(16°) \times 6371 \, \text{km} \times 2\pi}{86400 \, \text{sec}} \approx 130 \, \text{meters/sec}$$

eastward. This amount must be subtracted from the 320 m/s giving the arc a net velocity of 190 m/s, still westward (duskward), in the non-corotating frame.

Another example of a sun-aligned arc appears on the keogram at 0550 UT May 25, 1984 shown in Figure 5. It is fainter than the previous example, but still distinct, and this signature sweeps poleward over a $\sim$ 5-minute period. Being two hours after local magnetic midnight the poleward motion is an artifact of a duskward-moving sun-aligned arc crossing the keogram field of view at a slanted angle. Again the corresponding sequence of consecutive all-sky images is shown below the keogram. These images likewise indicate the arc more faintly than in the previous example, and oriented at a different angle, though perhaps not as sun-aligned until the final five frames here. (The initial five frames show it more poleward-aligned than sunward-aligned.) By the middle frames we see that it does cross the keogram field of view at an angle that generates a poleward moving intersection. The duskward velocity here is similar to that of the previous example.

In Figure 6 and Figure 7 we show two more examples of pre-dawn sun-aligned arcs as they appear on the keograms and all-sky images. The June 26, 1984 case (Figure 6) exhibits a bright poleward-moving intersection at 0650 UT. A fainter poleward-moving feature at 0710 UT is recorded on the June 30, 1984 keogram (Figure 7). The corresponding sequence of consecutive all-sky images in Figure 6 shows a bright but knotty-looking polar cap arc moving not exactly leftward but more toward the lower left in the frames, duskward and anti-sunward. The all-sky sequence of Figure 7 shows an arc that is more sun-aligned, with the sun off to

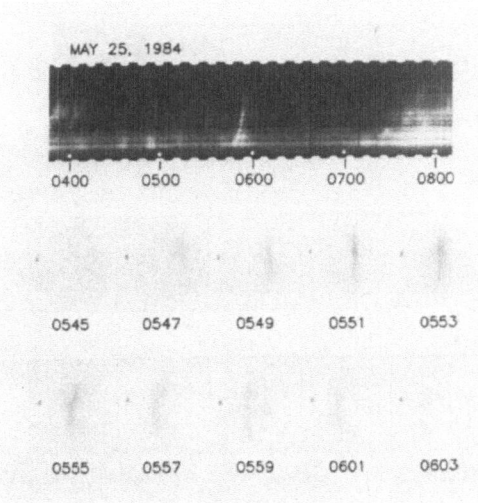

MAY 25, 1984

0400   0500   0600   0700   0800

0545   0547   0549   0551   0553

0555   0557   0559   0601   0603

Fig. 5. Keogram signature of a sun-aligned arc at 0550 UT on May 25, 1984. Leftward motion in these all-sky images is duskward at this time of day.

the upper right. The junction of this arc with the auroral oval can be seen, particularly in the 0707 and 0709 UT images where it appears to hook toward the right into the oval. This arc seems to spread and diffuse as it moves duskward. The relative velocity calculations at these local times are not as simple as near local magnetic midnight, because duskward is no longer directly toward the left of the frame but more toward the upper left, while the frame's motion in magnetospheric coordinates is still directly rightward. Thus we must multiply the 130 m/s instrument motion by the cosine of the sun-angle from the poleward direction, or by $\cos(53°)$ near 0700 UT, before subtracting it from the duskward velocity relative to the field of view. In the June 30, 1984 sequence the relative duskward velocity is 250 m/s, from which we subtract 80 m/s to get a net duskward velocity of 170 m/s in the non-corotating frame. The result is very close to the 190 m/s duskward velocity found in our pre-midnight first example.

The accuracy of these velocity measurements, especially at large zenith angles, is dependent on the assumed altitude of the emissions, taken here as 200 km. The velocities of well-defined arcs which are tracked directly overhead are determined to within ±20 m/s, while those of faint, brief, or more diffuse arcs must be assigned uncertainties of up to ±100 m/s.

Through the four austral winters of image collection we record a total of 114 sun-aligned arcs. The cumulative number of dusk to dawn periods of useful observation is 219. The velocity and magnetic local time distribution of these sun-

332

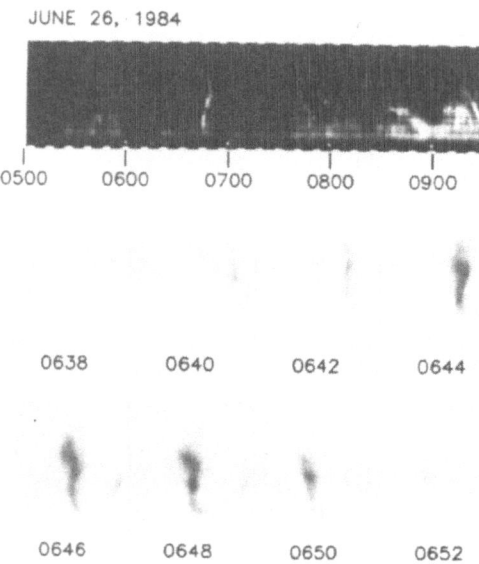

JUNE 26, 1984

0500   0600   0700   0800   0900

0638   0640   0642   0644

0646   0648   0650   0652

Fig. 6. Keogram signature of a sun-aligned arc at 0645 UT on June 26, 1984. The sunward direction is toward the upper right in this all-sky image sequence, so this polar cap arc is not precisely "sun-aligned," though it does exhibit the usual net duskward component of motion.

aligned arcs is illustrated in the scatter-plot and associated histograms given in Figure 8. Arc drift velocities in the non-corotating frame are seen to range from nearly 500 m/s dawnward to 500 m/s duskward. The curved line on the scatter-plot shows the relative dawnward velocity of South Pole Station as it varies with the cosine of magnetic local time, reaching a maximum of 130 m/s at magnetic midnight. Symbols near this curved line represent arcs which appear relatively motionless in the field of view, while those points above or below denote arcs with a relative drift across the images dawnward or duskward, respectively. The velocity histogram at the left in Figure 8 indicates the number of sun-aligned arcs with velocity values falling into each 100 m/s interval in the non-corotating frame. For these 114 arcs, duskward velocities appear to predominate over dawnward by a 68 to 46 margin. Care must be taken in this interpretation, however. Although these velocities have been corrected to a non-corotating reference frame, the dawnward motion of the station continues to induce a preference for the observation of duskward-drifting arcs. For example we clearly have a better chance

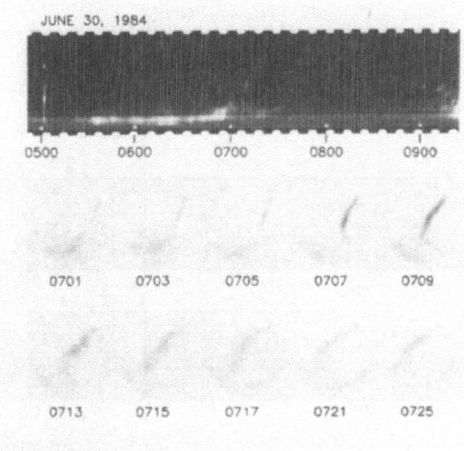

Fig. 7. Keogram signature of a sun-aligned arc at 0710 UT on June 30, 1984. The junction of the sun-aligned arc with the auroral oval is seen in this pre-dawn (magnetic local time) image sequence.

of intercepting an arc which sweeps duskward across the polar cap at 130 m/s than one which sweeps across dawnward at that velocity, though when we do see the dawnward-moving arc it may remain within the field of view for a much longer time. The overall excess of duskward-moving arcs indicated in Figure 8 becomes statistically insignificant when compensated for by the effects of this observational bias. And while the population of the upper left quadrant (pre-midnight dawnward-moving arcs) appears relatively sparse, we claim no convincing evidence of any correlation between an arc's velocity and the local time of its appearance.

The local time distribution summed into one-hour bins across the bottom of Figure 8 shows a nearly 2 to 1 asymmetry towards the morning hours. At the dusk or dawn extremities it becomes difficult to distinguish between a sun-aligned arc and a fragment of the auroral oval, due to the similar alignment. This factor is responsible for the lack of arcs identified later than 4 hours MLT or earlier than 19 hours MLT.

The frequency of occurrence of sun-aligned arcs found in our survey gives a ~ 40% probability on the average of seeing one or more such arcs at South Pole Station during any given diurnal period. This is a somewhat higher percentage than inferred by *Gussenhoven* [1982] in a statistical study of high-latitude aurorae photographed by a DMSP satellite. It may be expected that an increasing number of fainter arcs will be observed as instrument sensitivities are improved. Initial observations with an image-intensified video camera of greater sensitivity from Sondre-Stromfjord,

334

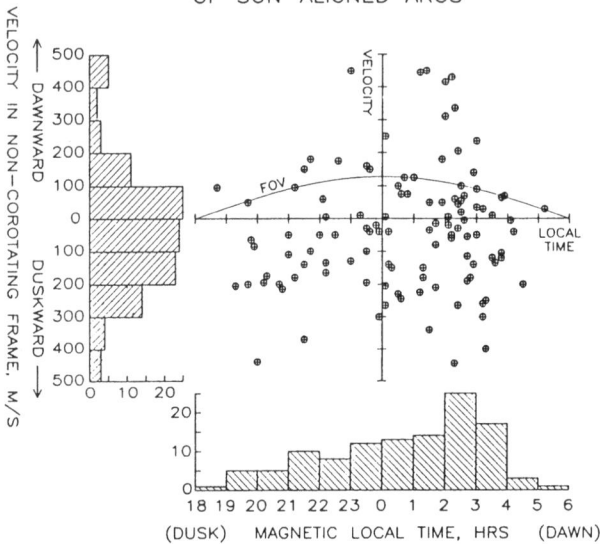

LOCAL TIME AND VELOCITY DISTRIBUTION
OF SUN-ALIGNED ARCS

Fig. 8. Local time and velocity distribution of 114 sun-aligned arcs discerned in the 1983–1986 nightside all-sky image data. Observations are binned into 100 m/s intervals for the velocity histogram at the left, and one hour intervals for the magnetic local time histogram across the bottom. The sun-aligned arcs occur preferentially on the dawn side of midnight by about a 3 to 2 ratio. The apparent excess of duskward-moving arcs is an artifact of the relative dawnward motion of our field of view. Velocity values refer to a non-corotating frame; velocity uncertainties range up to ±100 m/s. The curve on the scatter-plot shows the dawnward component of the imager field of view as a function of local time. (Units and scale of the scatter-plot axes correspond to the histogram labels.)

Greenland show this to be the case [*Mende et al.*, 1988].

4.3 IMF Direction

A preliminary survey of the IMP-8 IMF hourly averages presently available from the National Space Science Data Center (NSSDC) gives 81 instances of overlap with our nightside all-sky image acquisition. Values of $\bar{B}_z$ on these intervals are about evenly divided between $\bar{B}_z \geq 0$ and $\bar{B}_z < 0$. During these 81 nightside sequences we record 26 sun-aligned arcs, of which 17 appear with $\bar{B}_z \geq 0$ and 9 appear with $\bar{B}_z < 0$. Similarly, the distribution with $\bar{B}_y$ is found to be 10 sun-aligned arcs with $\bar{B}_y \geq 0$ and 16 with $\bar{B}_y < 0$. These somewhat subjective results are summarized in Table 1, giving the IMF statistics for the 55 nightside periods without polar-cap arcs as well as the 26 periods with arcs. It is curious that $\bar{B}_y$ is positive in sign for the preponderance of polar cap activity when $\bar{B}_z$ is southward. Such an asymmetry is not apparent in the plots of *Lassen and Danielsen* [1978]. Closer scrutiny is necessary to assess its statistical significance.

Table 1.   Summary of all-sky imaging nightside coverage
and sun-aligned arc occurrence rates.

Year	Number of dusk to dawn periods observed	Number of sun-aligned arcs observed
1983	26	17
1984	74	35
1985	83	42
1986	36	20
4-yr total	219	114

The polar cap activity preference seen here for $\bar{B}_z \geq 0$ is not as strong as might be expected from the models or from the earlier results of *Lassen and Danielsen* [1978], but hourly averages may insufficiently represent the IMF behavior. Limited sets of ISEE-1 one-minute time resolution IMF measurements from the UCLA magnetometer [R. C. Elphic, private communication], however, do not immediately indicate any short-timescale excursions that would alter the above result.

Five nightside periods of the ISEE-1 B_z data are plotted in Figure 9. Intervals are chosen for the well-defined sun-aligned arcs present in the concurrent all-sky images at the times marked on the plots. For August 18, 1984 and August 16, 1985 (first and third traces in Figure 9), B_z appears relatively stable and at ~ 0 nT. On August 12, 1985 (second trace), B_z has been clearly northward for at least three hours before the sun-aligned arc appears at 0040 UT. On August 18, 1985 (fourth trace) a sun-aligned arc appears where an hourly average B_z would be southward (0230 UT), but B_z is seen to have just turned southward from a large positive value. No allowances have been made on these plots for time lags due to the ISEE-1 to Earth distance. During the August 18–19, 1985 nightside interval (bottom trace), B_z fluctuates between -7 and $+7$ nT before the first sun-aligned arc at 2340 UT. Northward values then occur before the arcs following at 0120 and 0155 UT.

Figure 9 shows that hourly averages are fairly representative of the B_z direction. The statistical results given in Table 1, however, comprise a variety of IMF circumstances which would be more properly evaluated on a case by case basis with more complete magnetic field data. IMF correlations are dependent on the behavior of the field direction during the few hours prior to each sun-aligned arc, because of the time lags required for the convection patterns to form and propagate.

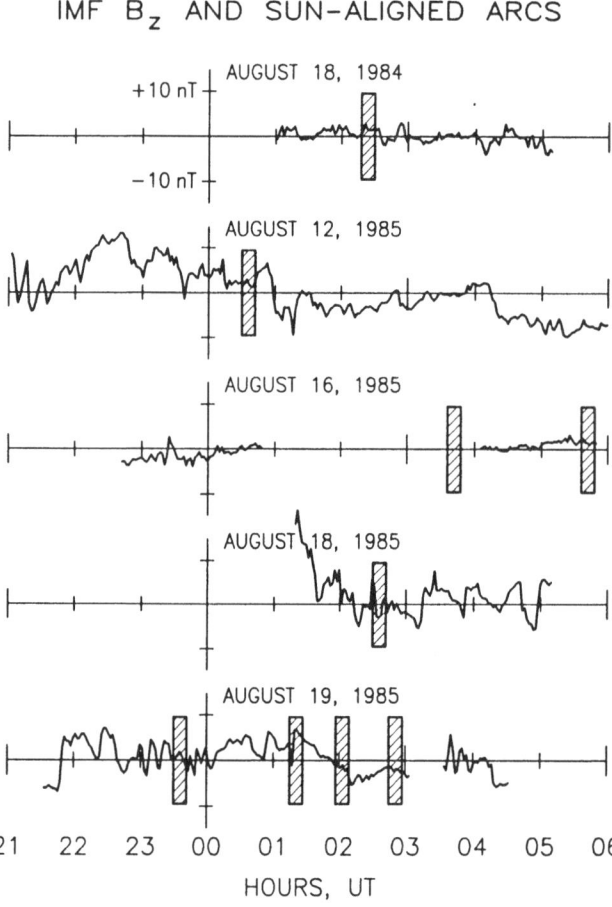

Fig. 9. Selected one-minute resolution IMF B_z measurements from the ISEE-1 UCLA magnetometer. Occurrences (UT) of sun-aligned arcs over South Pole Station are indicated. No adjustments have been made for spacecraft distance from Earth.

4.4 Discussion of Polar Cap Arc Observations

The characteristics of the sun-aligned arcs found in our study are similar to those of earlier work. Most stretch from horizon to horizon in our all-sky images, consistent with the arc lengths seen from DMSP [*Gussenhoven, 1982; Ismail and Meng, 1982*]. The local time distribution is in general agreement with the post-midnight preference shown in the statistical studies of *Ismail et al.* [1977] and *Lassen and Danielsen* [1978]. Most of these arcs exhibit drift velocities of absolute value less than ~ 250 m/s, consistent with past measurements [*Ismail et al., 1977; Hoffman et al., 1985*]. Such a variety of behavior is seen that we have not attempted to catagorize our sun-aligned arc observations into morphological subgroups as have previous investigators, though each of the previously found characteristics is evident. Our Figure 7, for example, fits *Ismail and Meng's* [1982] definition of a "hook-shaped" arc.

The magnetospheric phenomena responsible for the appearance and motion of sun-aligned arcs are related to the convection cells and magnetic merging in the polar cap region as driven by variations in the IMF. Models of the magnetospheric currents and plasma flow are presented by *Burch et al.* [1985] and *Reiff and Burch* [1985]. Though only a limited amount of IMF data has been correlated to our periods of observation, we find a statistically significant 2 to 1 preference for northward B_z among sun-aligned arcs. Our Table 1 shows also the B_z direction during periods without sun-aligned arcs, in order to demonstrate that our results are not due to a sampling bias.

Changes in the IMF B_y component are thought to cause the growth or disappearance of convective regions, driving the duskward or dawnward motions of polar cap arcs. The few arcs for which we have IMF data indicate a $\sim 75\%$ preference for dawnward motion when $\bar{B}_y \geq 0$ and a similar preference for duskward motion when $\bar{B}_y < 0$. There is no local time correlation with $\bar{B}_y$. And whether B_y is increasing or decreasing at these times is not determined. The *Reiff and Burch* [1985] model predicts dawnward motion in the southern polar cap for B_y greater than zero and increasing (see their Figure 5, which depicts the growth of polar cap convection patterns during various IMF conditions). In addition their model predicts non-conjugate behavior in the northern hemisphere. *Potemra et al.* [1984] model such B_y-dependent convection patterns as well, also predicting non-conjugate behavior between hemispheres. We note here that our best example of a dawnward drifting (~ 450 m/s) sun-aligned arc occurred at 0230 UT on June 3, 1983, and is correlated to a $\bar{B}_y \gtrsim +3$ nT.

For local dayside observations of sun-aligned arcs South Pole is not a suitable location, because a much expanded oval is required in order that the site be within the polar cap on the dayside. Such a condition necessitates a large southward B_z, which is generally mutually exclusive to sun-aligned arcs.

5. SUMMARY AND CONCLUSIONS

The consistent long-term monitoring of the auroral oval by the image-intensified all-sky camera at South Pole Station provides a valuable set of observations for determining the auroral position and morphology over the complete cycle of local times. The months of continuous image collection allow sufficient statistics to characterize the behavior of intermittent features, and to correlate these features with parameters that are available on a less than continuous basis such as the IMF.

We have described our instrumentation and data collection methods for the 1983–1986 austral winters at South Pole Station. Sample images from different local times have been presented and compared with simultaneously collected keogram data. The distinctive characteristics of the morning and afternoon auroral oval have been shown and contrasted. Discrete poleward moving arcs from the dayside appear in keograms and image data, and can be described as the footprints of reconnected plasma tubes convecting toward the antisunward direction.

In our nightside polar cap observations we have concentrated on the shapes and motions of sun-aligned arcs. Their local time distribution and velocities are comparable with the results of previous studies. Our statistics show that roughly two thirds of the sun-aligned arcs appear in the dawnside half of the polar cap. After compensation for the relative motion of our observation site and the concomitant bias toward viewing duskward moving arcs, we assert that the arcs exhibit no significant preference in their direction of drift velocity in a non-corotating frame. Correlations with a limited amount of IMF measurements indicate that more sun-aligned arcs occur when B_z is positive valued, as found by earlier investigators. Insufficient IMF data is available for our study to confirm any effects of the B_y component on the direction of arc drift. Further study will include the correlation of auroral electrojet (AE) indices as done by Ismail and Meng [1982]. Questions concerning polar cap conjugacy may be addressed with imaging data of the northern hemisphere obtained by the DE-1 spacecraft.

Intensified all-sky image collection at South Pole Station continues, and beginning with the 1987 austral winter our recording medium has been changed from 35-mm film to VHS video cassettes. This greatly facilitates data analysis and image reproduction. Time resolution has also been increased to 1-minute intervals between images. An initial survey of the 1987 data indicates a successful transition of format.

Acknowledgments. Much credit for the success of the Lockheed South Pole operations is due to J. H. Doolittle for his management and logistical efforts. The contributions of S. P. Geller and E. K. Aamodt to the design and development of the Lockheed all-sky camera are greatly appreciated. We thank R. H. Eather for supplying the keogram data and for helpful discussions. ISEE-1 IMF measurements were kindly provided by C. T. Russell and R. C. Elphic of the University of California at Los Angeles. We also thank the South Pole winter-over technicians who actively monitored the all-sky camera performance and kept the film rolling. This work has been supported by the National Science Foundation under grants DPP-86-00018 and ATM-87-13214, and by a Lockheed Independent Research Program.

REFERENCES

Axford, W. I., and C. O. Hines, A unifying theory of high-latitude geophysical phenomena and geomagnetic storms, *Can. J. Phys.*, **39**, 1433–1464, 1961.

Berkey, T., L. L. Cogger, S. Ismail, and Y. Kamide, Evidence for a correlation between sun-aligned arcs and the interplanetary field direction, *Geophys. Res. Lett.*, **3**, 145–147, 1976.

Burch, J. L., S. A. Fields, and R. A. Heelis, Polar cap electron acceleration regions, *J. Geophys. Res.*, **84**, 5863–5874, 1979.

Burch, J. L., P. H. Reiff, J. D. Menietti, R. A. Heelis, W. B. Hanson, S. D. Shawhan, E. G. Shelley, M. Sugiura, D. R. Weimer, and J. D. Winningham, IMF B_y-dependent plasma flow and Birkeland currents in the dayside magnetosphere 1. Dynamics Explorer observations, *J. Geophys. Res.*, **90**, 1577–1593, 1985.

Carbary, J. F., and C.-I. Meng, Correlation of cusp latitude with B_z and AE (12) using nearly one year's data, *J. Geophys. Res.*, **91**, 10,047–10,054, 1986.

Chiu, Y. T., N. U. Crooker, and D. J. Gorney, Model of oval and polar cap arc configurations, *J. Geophys. Res.*, **90**, 5153–5157, 1985.

Cornwall, J. M., Idealized model of polar cap currents, fields, and auroras, *J. Geophys. Res.*, **90**, 3541–3544, 1985.

Cowley, S. W. H., The causes of convection in the Earth's magnetosphere — A review of developments during the IMS, *Rev. Geophys. Space Phys.*, **20**, 531–565, 1982.

Dungey, J. W., Interplanetary magnetic field and the auroral zones, *Phys. Rev. Lett.*, **6**, 47–48, 1961.

Eather, R. H., Dayside aurora studies with a color keogram camera, *Antarct. J. U.S.*, **16**, 218, 1981.

Eather, R. H., Dayside auroral dynamics, *J. Geophys. Res.*, **89**, 1695–1700, 1984.

Eather, R. H., Polar cusp dynamics, *J. Geophys. Res.*, **90**, 1569–1576, 1985.

Eather, R. H., and S. B. Mende, Dayside aurora studies with a keogram camera, *Antarct. J. U.S.*, **15**, 203, 1980.

Eather, R. H., S. B. Mende, and E. J. Weber, Dayside aurora and relevance to substorm current systems and dayside merging, *J. Geophys. Res.*, **84**, 3339–3359, 1979.

Frank, L. A., J. D. Craven, J. L. Burch, and J. D. Winningham, Polar views of the Earth's aurora with Dynamics Explorer, *Geophys. Res. Lett.*, **9**, 1001–1004, 1982.

Frank, L. A., J. D. Craven, and R. L. Rairden, Images of the Earth's aurora and geocorona from the Dynamics Explorer mission, *Adv. Space Res.*, **5**, 53–68, 1985.

Frank, L. A., J. D. Craven, D. A. Gurnett, S. D. Shawhan, D. R. Weimer, J. L. Burch, J. D. Winningham, C. R. Chappell, J. H. Waite, R. A. Heelis, N. C. Maynard, M. Sugiura, W. K. Peterson, and E. G. Shelley, The theta aurora, *J. Geophys. Res.*, **91**, 3177–3224, 1986.

Goertz, C. K., E. Nielsen, A. Korth, K. H. Glassmeier, C. Haldoupis, P. Hoeg, and D. Hayward, Observations of a possible ground signature of flux transfer events, *J. Geophys. Res.*, **90**, 4069–4078, 1985.

Gussenhoven, M. S., Extremely high latitude auroras, *J. Geophys. Res.*, **87**, 2401–2412, 1982.

Hardy, D. A., Intense fluxes of low-energy electrons at geomagnetic latitudes above 85°, *J. Geophys. Res.*, **89**, 3883–3892, 1984.

Heikkila, W. J., Comment on "The causes of convection in the earth's magnetosphere: A review of developments during the IMS" by S. W. H. Cowley, *Rev. Geophys. Space Phys.*, **21**, 1787–1788, 1983.

Hoffman, R. A., R. A. Heelis, and J. S. Prasad, A sun-aligned arc observed by DMSP and AE-C, *J. Geophys. Res.*, **90**, 9697–9710, 1985.

Iijima, T., and T. Shibaji, Global characteristics of northward IMF-associated (NBZ) field-aligned currents, *J. Geophys. Res.*, **92**, 2408–2424, 1987.

Iijima, T., T. A. Potemra, L. J. Zanetti, and P. F. Bythrow, Large-scale Birkeland currents in the dayside polar region during strongly northward IMF: A new Birkeland current system, *J. Geophys. Res.*, **89**, 7441–7452, 1984.

Ismail, S., and C.-I. Meng, A classification of polar cap auroral arcs, *Planet. Space Sci.*, **30**, 319-330, 1982.

Ismail, S., D. D. Wallis, and L. L. Cogger, Characteristics of polar cap sun-aligned arcs, *J. Geophys. Res.*, **82**, 4741–4749, 1977.

Kan, J. R., and W. J. Burke, A theoretical model of polar cap auroral arcs, *J. Geophys. Res.*, **90**, 4171–4177, 1985.

Lassen, K., and C. Danielsen, Quiet time pattern of auroral arcs for different directions of the interplanetary magnetic field in the Y-Z plane, *J. Geophys. Res.*, **83**, 5277–5284, 1978.

Lee, L. C., and Z. F. Fu, A theory of magnetic flux transfer at the Earth's magnetopause, *Geophys. Res. Lett.*, **12**, 105–108, 1985.

Lemaire, J., Impulsive penetration of filamentary plasma elements into the magnetosphere of the Earth and Jupiter, *Planet. Space Sci.*, **25**, 887–890, 1977.

Lundin, R., and D. S. Evans, Boundary layer plasmas as a source for high-latitude, early afternoon, auroral arcs, *Planet. Space Sci.*, **33**, 1389–1406, 1985.

Lyons, L. R., A simple model for polar cap convection patterns and generation of Θ auroras, *J. Geophys. Res.*, **90**, 1561–1567, 1985.

McHenry, M. A., and C. R. Clauer, Modeled ground magnetic signatures of flux transfer events, *J. Geophys. Res.*, **92**, 11,231–11,240, 1987.

Mende, S. B., R. H. Eather, and E. K. Aamodt, Instrument for the monochromatic observation of all sky auroral images, *App. Opt.*, **16**, 1691–1700, 1977.

Mende, S. B., J. H. Doolittle, R. M. Robinson, R. R. Vondrak, and F. J. Rich, Plasma drifts associated with a system of sun-aligned arcs in the polar cap, *J. Geophys. Res.*, **93**, 256–264, 1988.

Paschmann, G., B. U. Ö. Sonnerup, I. Papamastorakis, N. Sckopke, G. Haerendel, S. J. Bame, J. R. Asbridge, J. T. Gosling, C. T. Russell, and R. C. Elphic, Plasma acceleration at the Earth's magnetopause: Evidence for reconnection, *Nature*, **282**, 243–246, 1979.

Paschmann, G., G. Haerendel, I. Papamastorakis, N. Sckopke, S. J. Bame, J. T. Gosling, and C. T. Russell, Plasma and magnetic characteristics of magnetic flux transfer events, *J. Geophys. Res.*, **87**, 2159–2168, 1982.

Peterson, W. K., and E. G. Shelley, Origin of the plasma in a cross polar cap auroral feature (theta aurora), *J. Geophys. Res.*, **89**, 6729–6736, 1984.

Potemra, T. A., L. J. Zanetti, P. F. Bythrow, and A. T. Y. Lui, B_y-dependent convection patterns during northward interplanetary magnetic field, *J. Geophys. Res.*, **89**, 9753–9760, 1984.

Reiff, P. H., and J. L. Burch, IMF B_y-dependent plasma flow and Birkeland currents in the dayside magnetosphere 2. A global model for northward and southward IMF, *J. Geophys. Res.*, **90**, 1595–1609, 1985.

Russell, C. T., and R. C. Elphic, Initial ISEE magnetometer results: Magnetopause observations, *Space Sci. Rev.*, **22**, 681–715, 1978.

Russell, C. T., and R. C. Elphic, ISEE observations of flux transfer events at the dayside magnetopause, *Geophys. Res. Lett.*, **6**, 33–36, 1979.

Sandholt, P. E., C. S. Deehr, A. Egeland, B. Lybekk, R. Viereck, and G. J. Romick, Signatures in the dayside aurora of plasma transfer from the magnetosheath, *J. Geophys. Res.*, **91**, 10,063–10,079, 1986.

Sonnerup, B. U. Ö., On the stress balance in the flux transfer events, *J. Geophys. Res.*, **92**, 8613–8620, 1987.

Southwood, D. J., The ionospheric signatures of flux transfer events, *J. Geophys. Res.*, **92**, 3207–3213, 1987.

Thomsen, M. F., J. A. Stansberry, S. J. Bame, S. A. Fuselier, and J. T. Gosling, Ion and electron velocity distributions within flux transfer events, *J. Geophys. Res.*, **92**, 12,127–12,136, 1987.

PLASMA TRANSPORT THROUGH THE DAYSIDE CLEFT: A SOURCE OF IONIZATION PATCHES IN THE POLAR CAP

John C. Foster
M.I.T. Haystack Observatory
Westford, MA 01886
U.S.A.

ABSTRACT

Rapid sunward convection from the post-noon ionosphere carries high-density solar-produced F region plasma through the dayside cleft and into the polar cap. This plasma is swept through the noontime cleft and enters the polar cap as a tongue of ionization which delineates the convection trajectory and its dynamics and provides the source for enhanced F region plasmas and their effects which are observed at high polar latitudes away from noon.

1. INTRODUCTION

The general pattern of ionospheric convection, driven by magnetospheric electric fields, features sunward flow towards noon at auroral latitudes and anti-sunward velocities at high latitudes in the polar cap. During disturbed conditions the region of strong sunward convection expands towards lower latitudes resulting in appreciable ionospheric plasma transport towards the noon meridian from the afternoon local time sector where solar-produced densities are high. The convection pattern turns poleward near noon and enters the polar cap in the vicinity of the ionospheric cusp or cleft. The solar-enhanced plasma convected towards noon from the post-noon F region is swept through the noontime cleft and enters the polar cap as a tongue of ionization which delineates the convection trajectory and its dynamics.

The large-scale characteristics of the noontime convection pattern respond strongly to the interplanetary magnetic field (IMF) orientation and strength. Average patterns of dayside convection electric field for disturbed geomagnetic conditions (Kp > 4) have been prepared from the large synoptic data base obtained with the Millstone Hill incoherent scatter radar (Holt et al., 1987). Patterns for toward and away inferred IMF sectors, presented in Figure 1, display convection streamlines associated with plasma transport toward noon from lower latitudes in the afternoon sector and entry into the polar cap through a rather broad region (> 3 hours of LT) whose large-scale features differ significantly with the orientation of the IMF.

A number of previous studies, using the incoherent scatter technique to provide simultaneous observations of plasma transport velocity and plasma density, have concluded that the large-scale convection of high-density solar-produced F region plasma through the noontime cleft constitutes a major source of enhanced ionospheric density features at high polar latitudes. This paper summarizes these previous observations and highlights the characteristics and dynamics of the F region plasma observed during one such event.

343

P. E. Sandholt and A. Egeland (eds.), Electromagnetic Coupling in the Polar Clefts and Caps, 343–354.
© 1989 by Kluwer Academic Publishers.

344

MILLSTONE MODEL   TOWARD IMF   KP > 4
In Magnetic LAT vs. LT

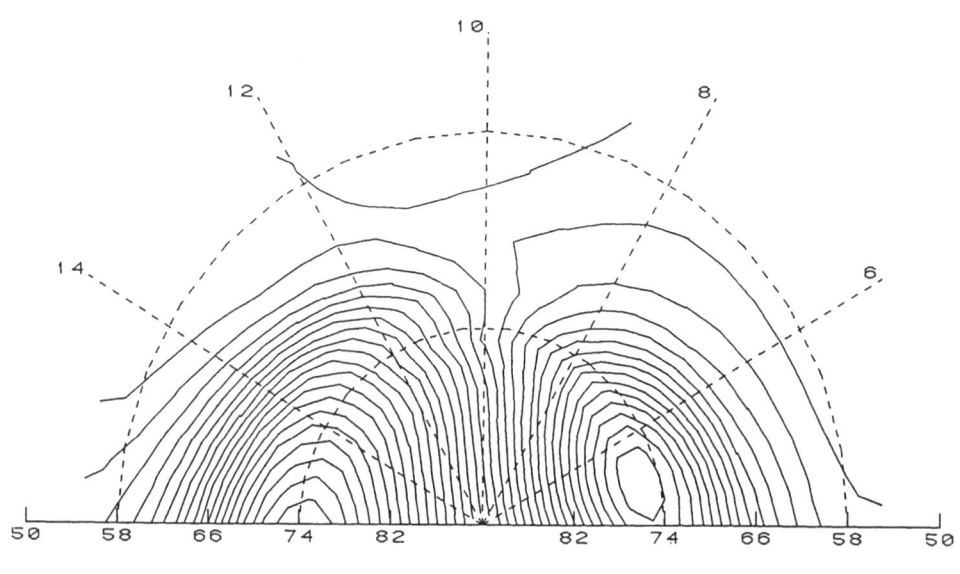

MILLSTONE MODEL   AWAY IMF   KP > 4
In Magnetic LAT vs. LT

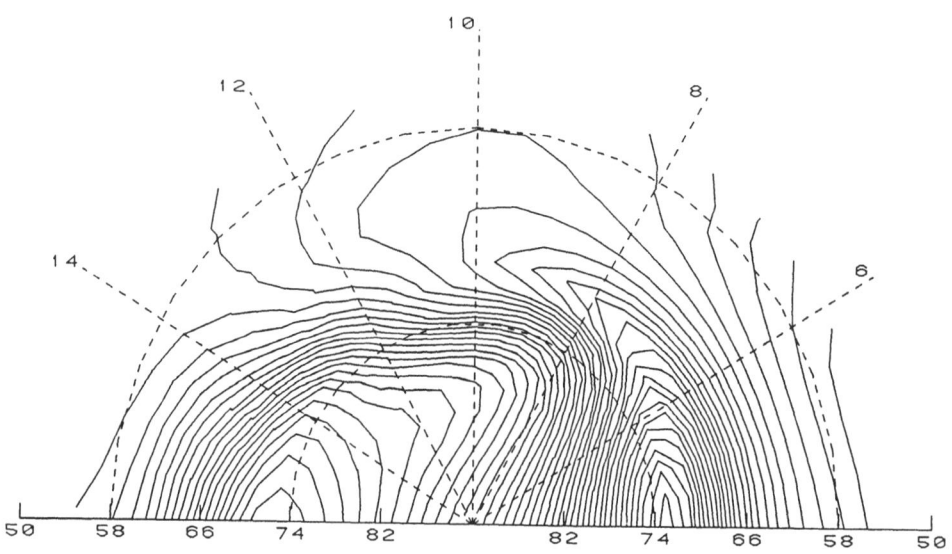

Figure 1. Inferred IMF-dependent average dayside convection patterns (2 kV equipotential contour spacing) derived from Millstone Hill radar observations (Holt et al., 1987).

345

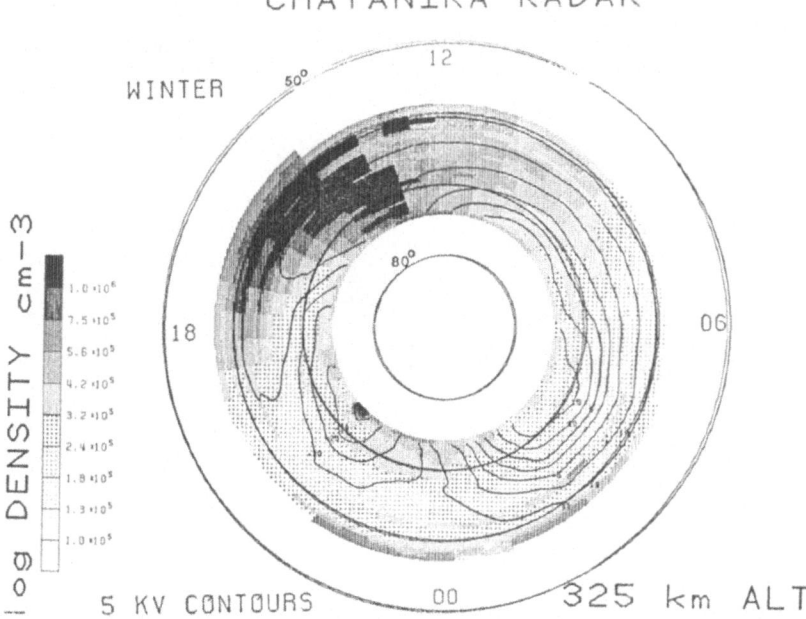

Figure 2. Convection equipotential contours and average F region density observed during winter experiments with the Chatankia radar (Foster, 1984).

2. RADAR STUDIES OF DAYSIDE PLASMA TRANSPORT

Synoptic observations of the auroral-latitude ionospheric convection pattern were made with the Chatanika, Alaska incoherent scatter radar between 1978 and 1981 (Foster et al., 1981). These observations were combined to produce averaged convection patterns for summer and winter conditions (Foster, 1983) which were then compared with features of the ionospheric F region plasma density by Foster (1984) in an investigation of ionospheric signatures of large-scale magnetospheric boundaries and convection. Figure 2, taken from that latter work, intercompares the winter convection streamlines, presented in non-corotating coordinates, with the average F region density distribution observed during experiments which contributed to the convection model. In these inertial coordinates, convection stagnates in the post-noon region of strong solar production before turning sunward toward the high-latitude noontime cleft. A tongue of enhanced density is seen to follow the convection contours towards the cleft and the polar cap. Foster (1984) conjectured that the high-density plasma which enters polar latitudes at noon is convected rapidly anti-sunward across the polar cap where it contributes to the enhancement of F region density seen above 70°Λ near midnight. In a study of the nighttime F region density enhancements which were consistently seen at the different radar sites during solar maximum, de la Beaujardiere et al. (1985) concluded that this plasma was of solar-produced origin and had been convected across the polar cap from a sunlit source region on the dayside.

A more detailed study of an individual noontime event observed from Chatanika was presented by Foster and Doupnik (1984) who observed high-density F region plasma convecting poleward through the cleft from a source at lower latitudes in the afternoon sector. Figure 3 presents a synopsis of their observations which identified a tongue of ionization which followed the

346

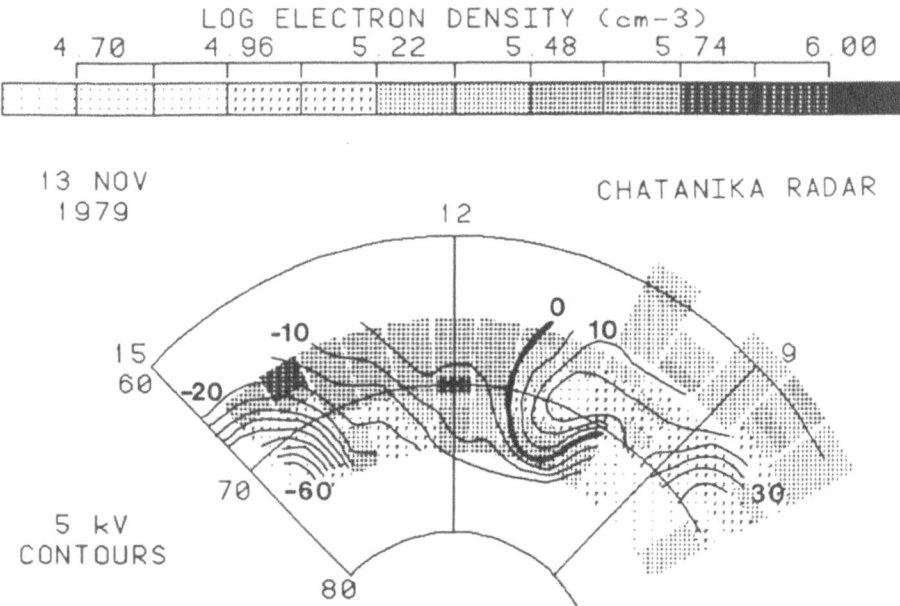

Figure 3. Chatanika radar observations of convection streamlines and F region plasma density in magnetic latitude - local time coordinates (Foster and Doupnik, 1984).

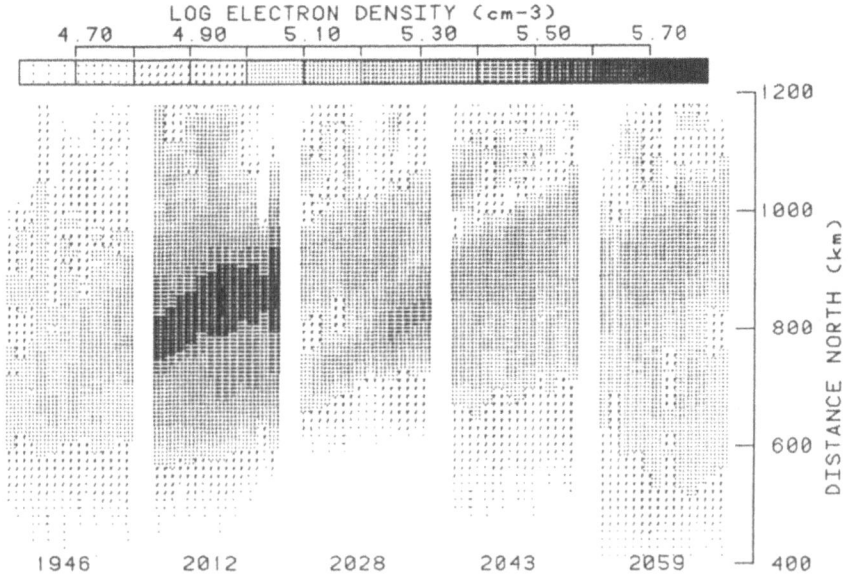

Figure 4. High time resolution (15 s) density observations made looking poleward into the cleft ionosphere show plasma density patches streaming poleward in the convection pattern shown in Figure 3 (Foster and Doupnik, 1984).

convection equipotentials through the cleft to polar latitudes. Although the pre-noon convection cell, bringing plasma sunward from the pre-dawn darkness, was associated with low F region densities, the post-noon convection cell was clearly marked by the higher densities of the lower latitude afternoon sector F region. While the radar probed poleward in the vicinity of the convection convergence and the ionization tongue, high temporal resolution observations, presented in Figure 4, revealed discrete patches of topside plasma moving with the observed convection velocity into the polar cap. The study of Foster and Doupnik (1984) suggested that convection through the cleft can result in spatially discrete patches of enhanced F region density and that the motion of these patches can be used to delineate dayside convection trajectories. Weber et al. (1986) tracked such ionization patches flowing in the anti-sunward direction from the center of the polar cap to the edge of the nightside auroral oval.

In 1983, the former Chatanika radar began operations in the near vicinity of the noontime cleft at Sondrestrom, Greenland. From this much closer vantage point the features of the cusp and cleft convection patterns and their ionospheric signatures could be examined in detail. Foster et al. (1985) used radar azimuth scans to determine the noontime convection pattern over a span of two hours of local time and 10° of latitude with 20 minute time resolution. The resultant detailed patterns were fully consistent with the large-scale two-cell convection patterns which had been synthesized from many radar and satellite observations and which are characterized by our Figure 1. Figure 5 presents the averaged convection and F region density patterns observed from Sondrestrom on a day when the cleft convection reversal was overhead at Sondrestrom at 12 MLT. As had been seen in the Chatanika patterns presented above, a region of solar-enhanced ionization was carried through the cleft and into the polar cap by the post-noon convection cell. Detailed radar observations of E region density enhancements and F region electron temperatures and direct satellite observations of precipitating particles revealed that precipitation effects were confined to the near vicinity of the convection reversal and verified that the high-density plasma being carried into the polar cap was not locally produced by particle precipitation.

3. MILLSTONE HILL OBSERVATIONS: 31 JANUARY 1982

A large geomagnetic disturbance (Kp ranged from 4 to 6) took place on 31 January 1982 while the Millstone Hill radar surveyed the daytime ionosphere as a part of the MITHRAS (de la Beaujardiere et al., 1984) observing program. The radar was performing low elevation angle azimuth scans which permit the determination of the plasma convection velocity pattern over a wide span of latitude and local time with approximate 30 minute temporal resolution (Holt et al., 1984). The regions of cusp and cleft precipitation and the dayside convection convergence moved to relatively low latitudes (69°Λ) during this event (Foster et al., 1989), well within the radar field of view. The high topside plasma density which characterizes solar cycle maximum further enhanced the radar's effective range and sensitivity and extended its high-latitude coverage to 75°Λ for spectral measurements and to 80°Λ for densities. We have constructed the "averaged" convection pattern observed during this event by combining all the line of sight velocity observations obtained at a given local time and latitude following the procedure of Holt et al. (1985) for determining the "best fit" convection pattern. Although temporal averaging somewhat smears detailed features, the general characteristics of the large-scale pattern are well-represented by the averaging technique. Large-scale plasma transport and the resultant redistribution of ionospheric plasma at high latitudes are driven by such a time-averaged convection electric field. The dayside convection pattern for the 31 January 1982 event is presented in Figure 6 along with the bin-averaged electron density normalized to the F region peak value. Again these data clearly depict a plasma density tongue extending along the convection trajectories from the mid-latitude post-noon sector and into the pre-noon polar cap. In the following sections the characteristics of this plasma tongue are examined in detail using high-resolution radar azimuth scan and elevation scan data.

348

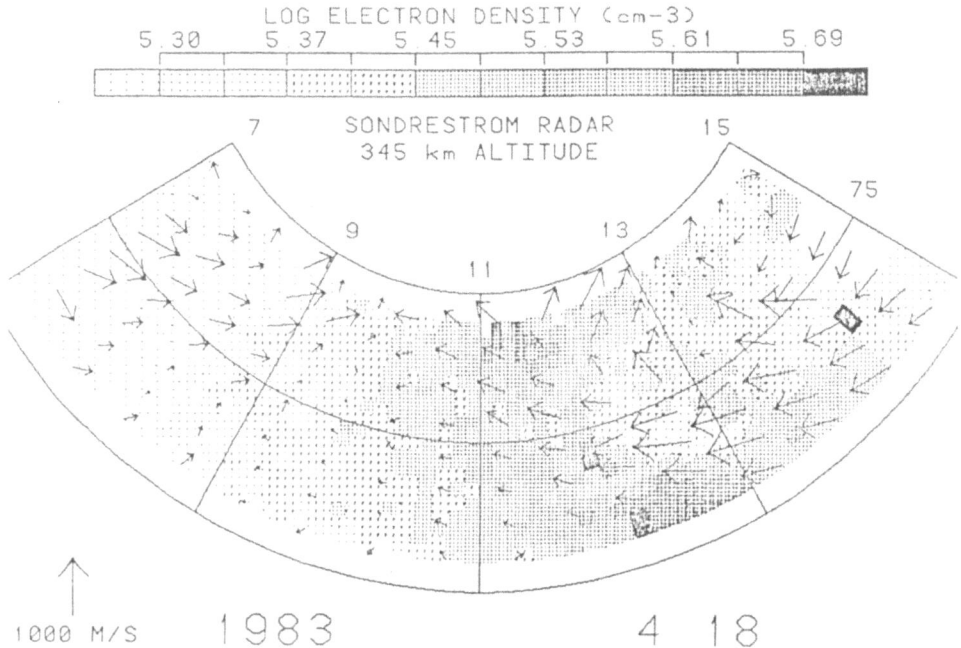

Figure 5. Plasma density and convection streamlines observed from Sondrestrom in the vicinity of the cleft (Foster et al., 1985). Magnetic latitude - LT coordinates are used with latitude increasing toward the top of the figure.

Figure 6. Average dayside convection equipotential contours (5 kV spacing) and F region peak density observed from Millstone Hill on 31 January 1982.

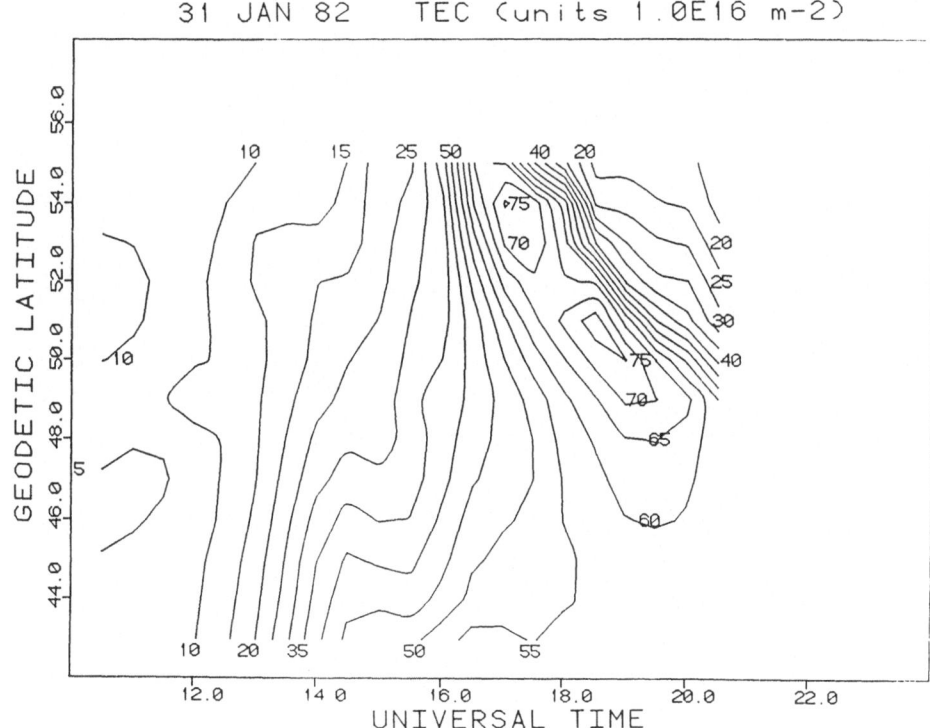

Figure 7. Total electron content (TEC) derived from density observations between 200 km and 650 km altitude along the Millstone Hill meridian reveal a factor of 3 increase associated with the sunward convecting plasma density feature. Local noon occurs at 17:00 UT.

The sunward-convecting ionization feature which extends along the equatorward edge of the post-noon trough and into the polar cap during disturbed conditions represents a large enhancement in the total ionization at a given latitude. Millstone Hill radar elevation scans intersect this feature to the north and provide both density/altitude profiles through the plasma tongue and map its local time - latitude behavior. Total electron content (TEC) as a function of latitude was determined from field-aligned altitude profiles of electron density between 200 km and 650 km altitude for the 31 January 1982 event. Contours of TEC (in units of 10^{16}m^{-2}) are presented in Figure 7 as a function of universal time between 43° and 55° geodetic latitude along the Millstone Hill meridian (75° W longitude). (Local noon occurs at 17:00 UT and 52° geodetic latitude corresponds to 64°Λ along this meridian.) Limited radar coverage of the lower F region (due to the curvature of the earth) determines the latitude range over which TEC can be calculated. For the latitudes shown in this figure only the sunward convecting portion of the ionization feature, equatorward of the post-noon cleft, can be seen. Direct solar production accounts for up to 25 TEC units before noon at these latitudes while the rapidly convecting plasma tongue is seen at progressively lower latitudes after the noon meridian is crossed and accounts for an enhancement of TEC by a factor of 3 to about 75 TEC units. A deep ionization trough is seen on the poleward side of the TEC enhancement.

350

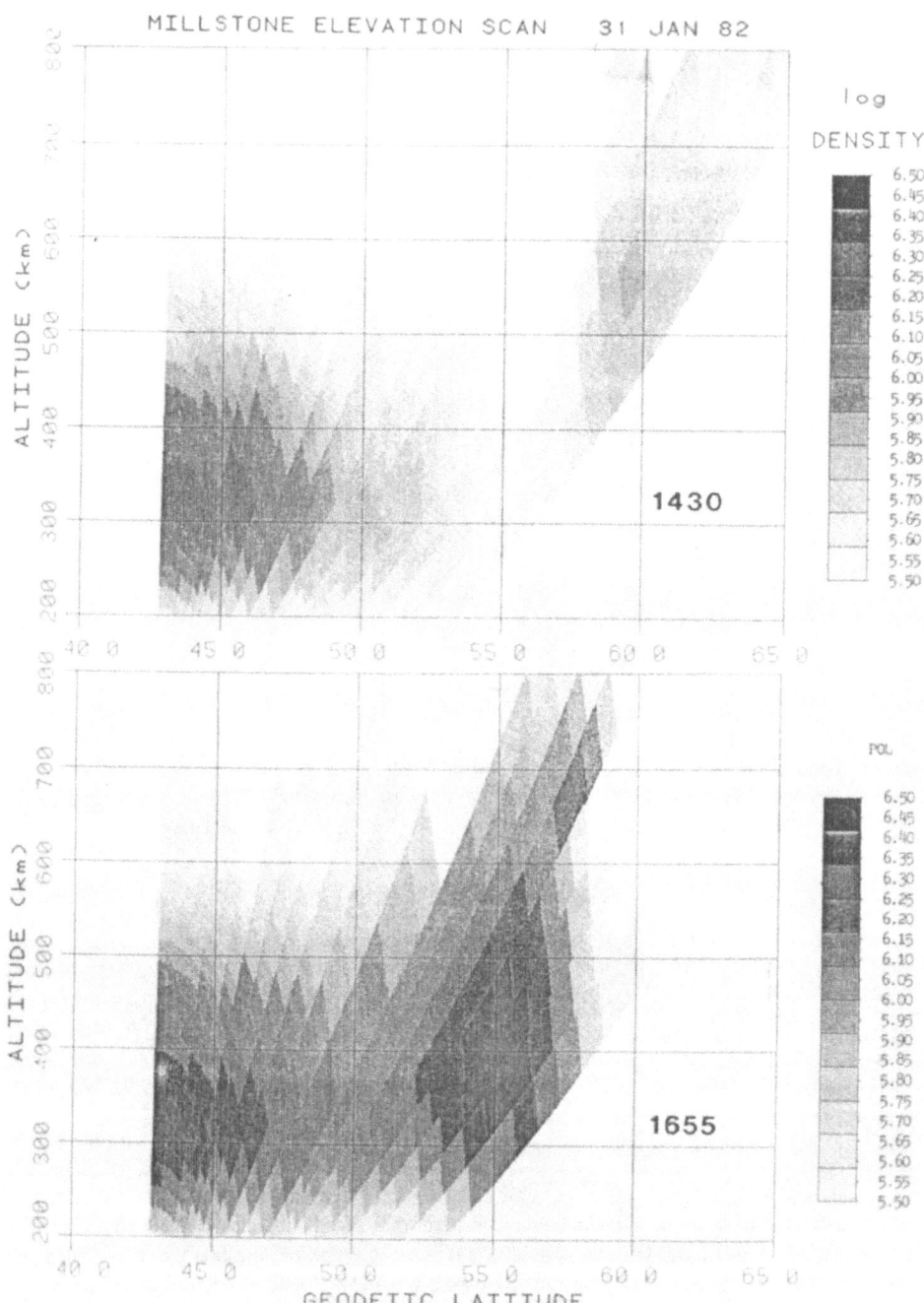

Figure 8. Millstone Hill elevation scans along the local meridian (75° W) crossed the poleward convecting ionization tongue.

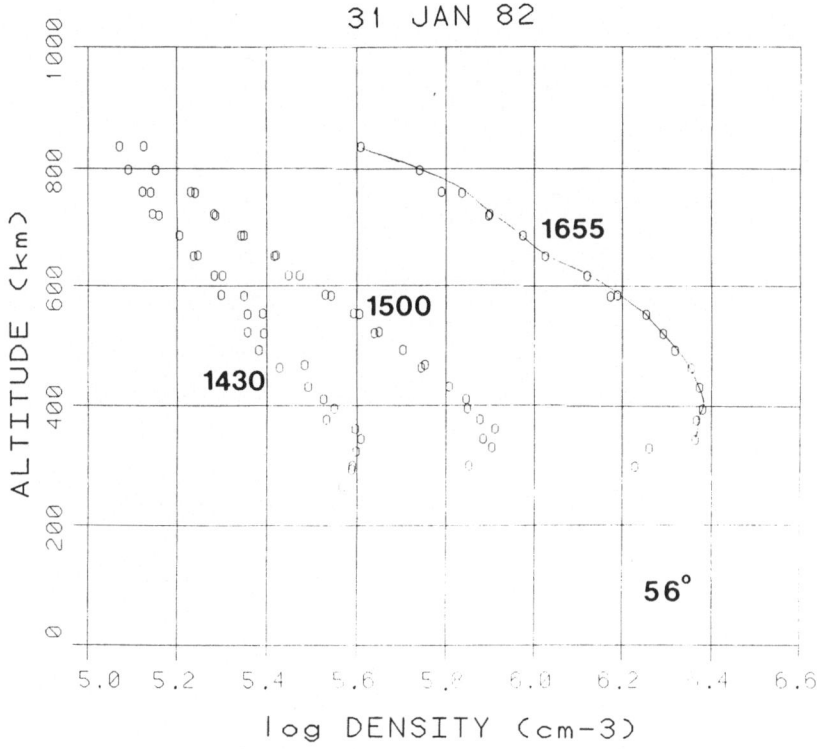

Figure 9. Density-altitude profiles at 56° geodetic latitude (68°Λ) derived from the elevation scan data of Figure 8 indicate a factor of 5 to 8 increase in topside density within the plasma tongue convecting through the cleft into the polar cap.

The altitude/latitude distribution of F region and topside plasma density observed by radar elevation scans from Millstone Hill which intersect the sunward-convecting ionization enhancement are presented in Figure 8. At 14:30 UT (09:30 MLT) the plasma tongue appears as a wall of ionization poleward of 58° geodetic latitude (70°Λ). Radar coverage is limited to altitudes above 450 km at these latitudes but densities of $6*10^5 cm^{-3}$ at 600 km - 700 km altitude characterize the plasma entering the polar cap at this time. By 16:55 UT the radar had rotated to the noon MLT meridian where the elevation scan crossed the ionization tongue at 55° geodetic latitude (67°Λ). Peak densities of $2.5*10^6 cm^{-3}$ at 400 km altitude and a sharp poleward edge of the enhanced density feature were observed. Plasma poleward of 54° was convecting into the polar cap at speeds in excess of 1.5 km/sec (see Figure 10 below). The apparent density enhancement near 700 km altitude at 58° could be either a satellite echo or the radar signature of a topside plasma instability (Foster et al., 1988).

Density/altitude profiles have been derived from the elevation scan data in order to characterize the plasma enhancement associated with the poleward-convecting plasma tongue. Figure 9 presents profiles for altitudes between 250 km and 850 km for 56° geodetic latitude (68°Λ) corresponding to the scan data shown in Figure 8. At 14:30 UT this latitude lay at the poleward edge of the solar-produced ionization and equatorward of the plasma tongue. The F region peak altitude and density were 325 km and $4*10^5 cm^{-3}$. The topside e-folding density scale height was approximately 300 km. At 16:55 UT the profile latitude sampled the ionization tongue and the F

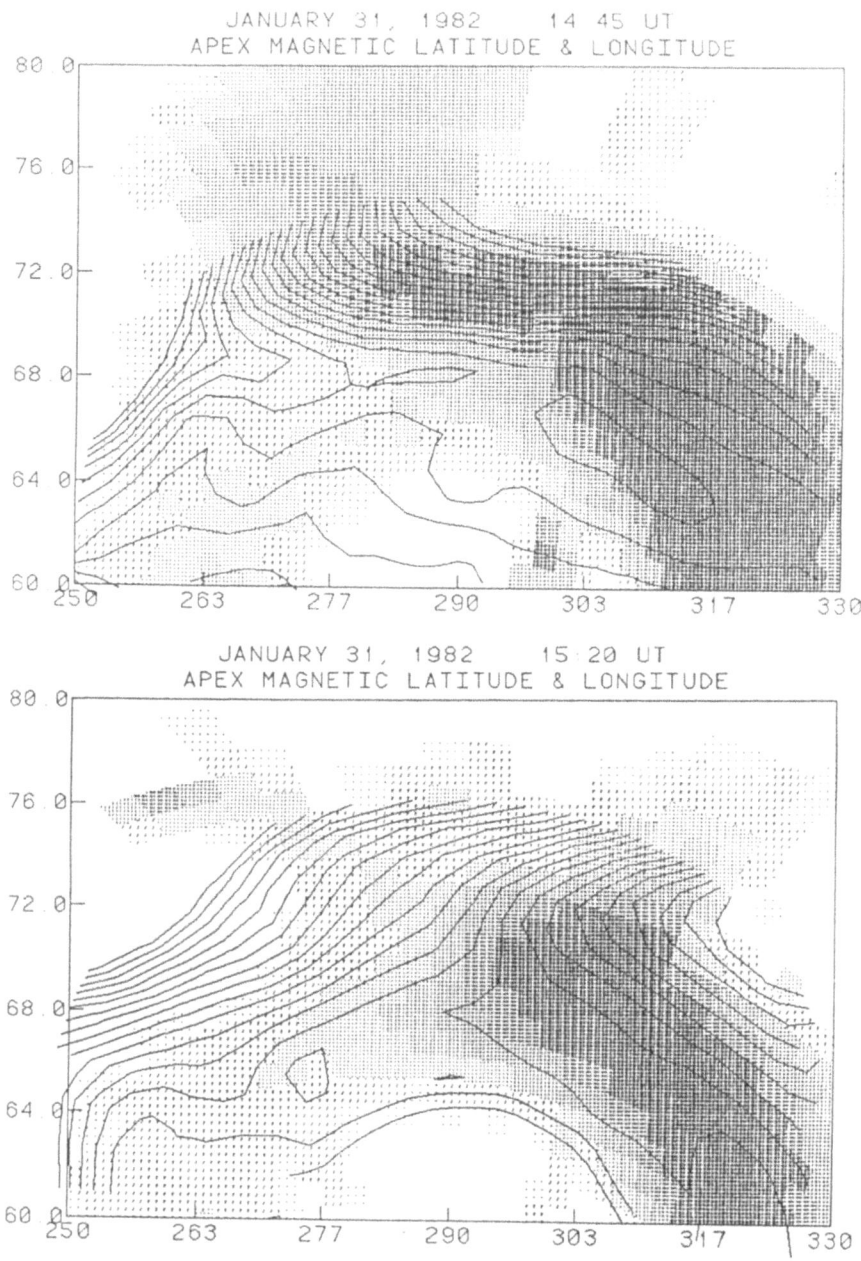

Figure 10. Azimuth scan maps of plasma density and convection provide "snapshots" of the dayside region of plasma entry into the polar cap with 30 minute temporal resolution. An ionization tongue follows the convection pattern through the cleft into the polar cap.

region peak altitude and density and topside scale height were observed to be 400 km, $2.5*10^6 cm^{-3}$, and 200 km, respectively. Plasma densities convected through the noontime cleft and into the polar cap on this day were enhanced by a factor of 5 to 8 at all altitudes in the topside.

Radar azimuth scans were made each 30 minutes throughout the event and these produced a series of "snapshots" of the large-scale convection and density patterns over a span of 6 hours of local time and 20° of latitude. Whereas the several-hour averaged convection pattern displayed in Figure 6 depicts rather orderly two-cell convection with polar cap entry over a broad span of local times centered around 10 MLT, the individual scan maps reveal the large-scale dynamics of the convection pattern near the noontime cusp and cleft. Figure 10 presents two "snapshots" of the region of dayside plasma entry into the polar cap, separated in time by about 30 minutes. Each picture was derived from data taken over a 20 minute interval centered on the indicated time and magnetic latitude and longitude coordinates are used. Equipotential contours of the ionospheric electric field (2 kV spacing) indicate plasma convection streamlines and in each case a potential difference of about 45 kV is seen across a plasma entry region which spans 2 to 3 hours of MLT. In this format plasma density, uncorrected for the increase of altitude with range from the radar, is indicated by the intensity of the shading. (The altitude variation with geodetic latitude when looking to the north is indicated by the low altitude cutoff of data coverage seen in the elevation scan at the top of Figure 8.) The region of solar-enhanced F region density at mid latitudes is seen at the lower right of each picture and these densities are observed to follow the convection streamlines to higher latitudes into the polar cap. Particularly striking is the distinct plasma tongue at 14:45 UT extending from the post-noon region of solar-produced densities to polar latitudes. This density tongue closely follows the observed convection streamlines up to 75°Λ and delineates the instantaneous pattern beyond the radar convection field of view at higher latitudes. In the 30 minute interval between the radar scans the convection convergence region shifted eastward by 30° sweeping the convecting plasma tongue with it. Throughout the event the high density plasma tongue followed the dynamics of the dayside convection electric field and served as an accurate tracer of its motion.

SUMMARY

During disturbed conditions rapid sunward convection from the post-noon mid-latitude ionosphere carries high-density solar-produced F region plasma through the dayside cleft and into the polar cap. Radar observations from a number of sites have indicated the repeatability of this feature of the dayside ionosphere both in average studies and individual cases. Within the poleward convecting feature, plasma densities are enhanced throughout the topside by a factor of 5 or more and the total electron content is increased by a factor of 2 to 4. This plasma is observed to spread out along convection trajectories within the polar cap where it constitutes a source for the observed polar cap F region density enhancements and their effects. The plasma tongue carried through the cleft from lower latitudes serves as a tracer of polar cap convection away from the cusp and cleft.

ACKNOWLEDGEMENTS

Radar observations at Chatanika, Sondrestrom, and Millstone Hill were supported by the U. S. National Science Foundation. The contributions of J. M. Holt and H.-C. Yeh are gratefully acknowledged. The analyses leading to this report were supported by the U. S. Air Force Office of Scientific Research through grant AFOSR-86-0023 and by National Science Foundation Cooperative Agreement ATM-88-08137 with the Massachusetts Institute of Technology.

REFERENCES

de la Beaujardiere, et al., Universal Time Dependence of Nighttime F Region Densities at High Latitudes, J. Geophys. Re., 90, 4319, 1985.

de la Beaujardiere, et al., MITHRAS: a brief description, Radio Science, 19, 665, 1984.

Foster, J. C., J. R. Doupnik and G. S. Stiles, Large-scale patterns of auroral ionospheric convection observed with the Chatanika radar, J. Geophys. Res., 86, 11357, 1981.

Foster, J. C., An empirical field model derived from Chatanika radar data, J. Geophys. Res., 88, 981, 1983.

Foster, J. C., Ionospheric signatures of magnetospheric convection, J. Geophys. Res., 89, 855, 1984.

Foster, J. C. and J. R. Doupnik, Plasma convection in the vicinity of the dayside cleft, J. Geophys. Res., 89, 9107, 1984.

Foster, J. C., J. M. Holt, J. D. Kelly, and V. B. Wickwar, High resolution observations of electric fields and F-region plasma parameters in the cleft ionosphere. The Polar Cusp, J. A. Holt et and A. Egeland (eds.), 349, 1985.

Foster, J. C., C. del Pozo, K. Groves, and J.-P. St. Maurice, Radar observations of the onset of current driven instabilities in the topside ionosphere, Geophys. Res. Lett., 15, 160,1988.

Foster, J. C., H.-C. Yeh, J. M. Holt, and D. S. Evans, Two-dimensional mapping of dayside convection, this volume, 1989.

Holt, J. M., R. H. Wand, and J. V. Evans, Millstone Hill measurements on 26 February 1979 during the solar eclipse and formation of a midday F region trough, J. Atmos. Terr. Phys., 46, 251, 1984.

Holt, J. M., J. V. Evans, W. L. Oliver, and R. H. Wand, Millstone Hill observations of ionospheric convection, in Physics of Space Plasmas, p. 53, Scientific Publishers, Cambridge, Mass., 1985.

Holt, J. M., R. H. Wand, J. V. Evans, and W. L. Oliver, Empirical models for the plasma convection at high latitudes from Millstone Hill observations, J. Geophys. Res., 92, 203, 1987.

Weber, E. J., et al., Polar cap F layer patches: structure and dynamics, J. Geophys. Res., 91, 12121, 1986.

GEOMAGNETIC RESPONSE OF THE POLAR THERMOSPHERE AND IONOSPHERE.

D. REES and T.J. FULLER-ROWELL,

DEPARTMENT OF PHYSICS AND ASTRONOMY, UNIVERSITY COLLEGE LONDON, GOWER ST., LONDON WC1E 6BT, UK.

ABSTRACT

A self-consistent coupled thermospheric / ionospheric model has been developed by merging the University College London Global Thermospheric Model and the Sheffield University Ionospheric Model. The neutral thermospheric wind velocity, composition, density, and energy budget are computed, including their full interactions with the high-latitude ion drift, precipitation, Joule heating and plasma density. This model has been used to examine thermospheric and ionospheric coupling within the polar cap, polar cusp and auroral oval. Simulations have been performed corresponding to high solar activity, moderate geomagnetic activity (Kp = 3), for the June and December solstices, and for convection electric field patterns corresponding to positive and negative values of the IMF-BY component to examine variations with season, and responses to the interplanetary magnetic field. In the winter polar region, ion transport and the diurnal migration of the polar convection pattern into and out of sunlight, play a major role in the plasma density structure at F-region altitudes. Regions of intense Joule heating, and field-aligned currents and locations of high ion temperatures are very dependent on convection field distributions, so that regions of strong neutral-ionospheric interactions are dependent on the IMF-BY component. In the summer polar region, the proportion of molecular to atomic species increases sharply, driven by the summer to winter seasonal thermospheric circulation, augmented by additional geomagnetic forcing. In the winter polar region at 300 km the dominant ion is O^+. As a consequence of the seasonal neutral composition change, at levels of moderate geomagnetic activity, molecular (NO^+ and O_2^+) and O^+ ions are of similar number densities in the summer polar cap. The increased destruction of F-region ions in the summer polar region reduces the mean level of ionization to similar mean winter levels, despite increased solar insolation and ion production. The summer ion temperature at 300 km exceeds the winter values by $500°K$, due to the underlying change in neutral temperature. In the lower thermosphere auroral oval the ion density is dominated by auroral precipitation in summer and in winter. Overall, there is a seasonal dependence in the height-integrated Joule heating rate and field-aligned currents (FAC) of about a factor of 2 - 3. Within the polar cusp, extra solar ionization in summer increases the conductivity to produce a threefold increase in peak Joule heating rates. There is a corresponding increase in the summertime cusp ionospheric currents and related FAC distributions. The intense neutral winds significantly modify the distribution of ionospheric currents, Joule heating and FAC, particularly in the dusk sector. Most of the neutral and electrodynamic parameters considered have strong IMF-BY dependence.

P. E. Sandholt and A. Egeland (eds.), Electromagnetic Coupling in the Polar Clefts and Caps, 355–391.
© *1989 by Kluwer Academic Publishers.*

1. INTRODUCTION.

In keeping with the general theme of this symposium, this review will concentrate on the behaviour of the thermosphere and ionosphere within the polar cap and polar cusp, and on phenomena within adjacent regions which affect the polar cap and cusp. At this point in time, the development of coupled numerical models of the terrestrial thermosphere and ionosphere system are sufficiently advanced that it is possible to review many of the most important interactions using simulations performed with these numerical models. Obviously, these simulations represent situations which are much less complex than the real world, and deal with climatology of the thermosphere and ionosphere, rather than the complex and rapidly evolving day by day and hour by hour meteorological variations which are actually observed, particularly at times of high geomagnetic activity.

The magnetosphere always imprints unmistakeable signatures upon the high-latitude thermosphere and ionosphere. These thermospheric and ionospheric signatures are always associated with energy and momentum deposition by energetic particle precipitation and convective electric fields, or else represent "fossils" of strong recent geomagnetic forcing. Under quiet magnetospheric conditions, the affected regions contract poleward, away from areas of historical observations. However, recent global observations from satellites, and new ground-based polar cap observatories have shown clearly that ionospheric structures, and thermospheric winds, temperature and composition are persistently disturbed in the vicinity of the auroral oval and within the geomagnetic polar cap, even under the most quiet conditions.

As geomagnetic activity levels increase, the regions of magnetospheric inputs expand away from the geomagnetic poles. The energy and momentum deposition rates increase greatly, and are strongly space and time-dependent. This intensification of auroral precipitation is well shown in statistical surveys and analyses of the energetic electron precipitation [1,2]. These statistical surveys complement the impression obtained from individual observations. As shown by analyses of polar plasma convection [3-5], the regions of strong magnetospheric convection electric fields imprinted on the polar ionosphere undergo a similar and closely related equatorward expansion and intensification as geomagnetic activity increases. From the point of view of numerical modelling, it is critically important that convection and precipitation boundaries match realistically, particularly if time-dependent simulations for variable geomagnetic activity are to be performed.

Convection electric fields drive ionospheric plasma of the auroral oval and polar cap to velocities of the order of 1 km/sec. The ions impart momentum to the neutral gas via 'ion drag', at the same time losing a little of their net ($\underline{E} \wedge \underline{B}$) velocity, creating the dissipative

Pedersen ionospheric current component, which causes Joule heating. For effective energy or momentum transfer from the solar wind, via the magnetopause and magnetosphere to the ionosphere and thermosphere, A/C and D/C components of the field-aligned current (FAC, or Birkeland current) are required. Such magnetospheric currents are associated with the dissipative Pedersen current within the auroral ionosphere (component parallel to the ionospheric electric field). Both the FAC and the Pedersen currents within the auroral ionosphere intensify sharply as geomagnetic activity increases. The efficiency of momentum transfer and the Joule heating both increase linearly with ionospheric plasma density. Since Joule heating almost always considerably exceeds direct particle precipitation, knowledge of the ionospheric plasma response to precipitation is particularly important /6/.

The polar regions which display the imprints of these important magnetospheric phenomena also show a wide range of other disturbances. These disturbances produce a range of characteristic signatures in the charged and energetic particle populations, the AC and DC electric and magnetic fields, and the optical aurora. However, for the moment, we will concentrate on those phenomena which have the most direct connection to excitation of the thermosphere and ionosphere.

The classical ground-based signatures used to identify intense auroral substorms are typically short, strong, negative magnetic excursions, brilliant aurorae, and riometer absorption events. These are only a small part of the sequence of phenomena which cause the ionospheric and thermospheric response to geomagnetic disturbances. Indeed the signatures of the most important momentum and energy sources for the thermosphere are difficult to sense except by satellite or ground-based remote sensing. For these reasons, major thermospheric and ionospheric disturbances are often poorly related to classical magnetospheric activity indices such as Kp and AE, which are dominated by the magnetic effects of Hall currents during auroral substorms. Pedersen currents more accurately reflect the intensity of momentum and energy transport from the magnetosphere. However, the Pedersen current and the FAC system are poorly reflected in ground-based magnetic perturbations within the auroral oval, and are only extracted with some difficulty from global chains of magnetometers, by assuming a relationship between Hall and Pedersen conductivity.

A signature of increasing geomagnetic activity is the intensification of convective electric fields and auroral precipitation, with an equatorward expansion of the auroral oval. The FAC, Joule heating and ion-drag acceleration of thermospheric winds all increase. The temporal and spatial variability of all of the previously-mentioned terms - 'geomagnetic forcing' - increases sharply, particularly at very high activity levels ($6<Kp<9$). Disturbances of the lower ionospheric regions, in the E-region up to 150 km, respond directly to 'auroral' inputs and ion production.

At higher thermospheric and ionospheric altitudes, the situation becomes much more complex. Large scale advection and convection forced upon the thermosphere by geomagnetic heating causes the polar F-region neutral gas composition to change dramatically /7,8/. Very strong enhancement of molecular nitrogen density, and a corresponding depletion of atomic oxygen density occurs, particularly in the disturbed summertime polar cap /8/. Enhanced concentrations of molecular nitrogen cause significant depletions of F-region plasma density by greatly increasing the effective recombination coefficient, while ionisation rates, due to the combination of solar photoionisation and auroral precipitation, are only slightly changed. The combination of induced ionospheric chemistry changes and dynamical effects on the F-region plasma, resulting from strong induced horizontal winds, cause the ionospheric response to magnetospheric forcing to be very complex and non-linear.

Global-scale disturbances within the thermosphere follow the initial high-latitude geomagnetic forcing. Propagating waves, and the consequences of gross wind-driven compositional changes have truly global consequences for the thermosphere, and force some very large, and long-lasting, disturbances of the ionosphere - the ionospheric F-region storm /9/. During major geomagnetic disturbances, the decay of energetic particles from the ring current, probably mainly energetic ions /10/, may directly cause the negative phase of the low-latitude ionospheric storm. This process is, however, still rather difficult to include in a coherent and self-consistent model of the entire coupled Solar - Terrestrial system.

Global numerical models provide a framework incorporating the basic, well-understood, mechanisms and phenomena. Numerical models provide a means of predicting the mean structure, and of the qualitative and quantitative variations caused by seasonal changes and by solar activity variations. The general form of large-scale thermospheric disturbances resulting from magnetospheric activity can also be simulated quite well. Localised and short-lived features will not be predicted in detail.

Numerical simulation of the thermosphere from first principles requires that the most important physical processes are properly treated /11-14/. It is assumed that most of the energy and momentum sources driving the thermosphere are predetermined, and invariant to the response of the thermosphere. The thermosphere does not determine the nature of the solar UV and EUV inputs which provide important heat and ionisation sources. However, the thermosphere does react strongly to forcing. The major responses in wind, temperature and composition of the polar thermosphere to ion convection and heating within the auroral oval and polar cap are now well documented by ground-based and spaceborne observation /15-17/. Some of these thermospheric responses may change the nature or magnitude of the forcing itself.

While this review will concentrate on external thermospheric forcing from the magnetosphere, it should be noted that significant effects, particularly in the lower thermosphere, occur from internal forcing from the lower atmosphere, as a result of the combination of tidal, gravity and planetary waves. The first can be handled numerically within a thermospheric model /18,19/ by introducing a 'flexible' lower boundary. Self-consistent wind and temperature amplitude and phase changes corresponding to specific propagating tidal modes can be adjusted, by numerical experiment, until the tides within the lower thermosphere correspond to observed tidal variations as functions of altitude, season and latitude. The relatively large amplitudes of observed tidal winds in the lower thermosphere can be successfully simulated by introducing such propagating tides. This is not possible, if only the in-situ generated tides are considered.

Ion-neutral frictional drag /20/ in regions of rapid convective ion flow causes direct heating of both ions and neutrals, commonly known as Joule heating /21,22/. Induced winds may increase or decrease (but generally decrease) the ion drag, and the resulting frictional heating. The induced winds (or more correctly, changed winds, since there is always a complex wind system in existence prior to a given geomagnetic disturbance) may induce a 'back-EMF', opposing the initial magnetospheric convective electric field, and decreasing the electromotive force e.$(\underline{E} + \underline{V}_n \wedge \underline{B})$. This also limits the maximum induced winds, the local electrojet current (at all heights) and the Joule heating. These processes are independent of any plasma density and conductivity modifications, however, the thermosphere will respond to the reduction in both electromotive force and electrojet current and consequent Joule heating.

The induced wind system (subtly modified by gas pressure changes due to neutral heating) will also induce ion drifts (or changes in ion drifts) parallel to the local magnetic field. Such 'parallel' ion drifts will also induce a field-aligned electron flow, to maintain quasi-charge neutrality. Thus the entire vertical plasma distribution will respond to wind changes, an effect which becomes increasingly important at greater altitudes. This change of ion density distribution will modify the consequent ion drag on the neutrals, and thus the wind acceleration terms, and finally the winds themselves. Since, in the vicinity of the auroral oval and polar cap, there are always various contra-flowing streams of field-aligned thermal and suprathermal particles, it is difficult to identify the net FAC by direct observation. Yet it is the net flow which powers the magnetosphere - thermosphere forcing process, and variations of this field-aligned flow caused by feedback processes are important, but necessarily second order changes, and thus difficult to observe directly.

Modifications of the horizontal current (usually decreases) due to the induced winds, affect the capacity of the ionosphere to carry FAC connecting to the magnetosphere. Intuitively, the feedback effects of the induced winds on the total electromotive force, and the modified capacity of the auroral ionosphere to transmit or connect the FAC might be expected to cause some significant effects on the magnetosphere at times of large disturbances, when the E-region winds are known to reach 50 % of the $\underline{E} \wedge \underline{B}$ ion drift velocity, driven by magnetospheric electric fields /23-25/. Although some numerical experiments in these areas are in progress /26,27/, theoretical and experimental exploration of these problems is still at a very preliminary phase.

2. THE COUPLED GLOBAL THERMOSPHERE / POLAR IONOSPHERE MODEL.

The development of the UCL Three Dimensional Thermospheric Model is well documented in previous publications /11,12/, as is the Sheffield ionospheric model /28/. The development of this coupled model is described in /28-33/.

The UCL Three-Dimensional Time-Dependent Thermospheric Model (or GCM) simulates the time-dependent structure of the vector wind, temperature, density and composition of the neutral atmosphere by numerically solving the non-linear equations of momentum, energy and continuity /11/, and a time-dependent mean mass equation /12/. The global atmosphere is divided into a series of elements in geographic latitude, longitude and pressure. Each grid point rotates with the earth to define a non-inertial frame of reference in a spherical polar coordinate system. The latitude resolution is 2°, the longitude resolution is 18°, and each longitude slice sweeps through all local times, with a 1 min time step. In the vertical direction the atmosphere is divided into 15 levels in log pressure, each layer equivalent to one scale height thickness, from a lower boundary of 1 Pascal at 80km height.

The time-dependent variables of southward and eastward neutral wind, total energy density, and mean molecular mass are evaluated at each grid point by an explicit time stepping numerical technique. After each iteration the vertical wind is derived, together with temperature, heights of pressure surfaces, density, and atomic oxygen and molecular nitrogen concentrations. The data can be interpolated to fixed heights for comparison with experimental data, or with empirical models. The momentum equation is non-linear and the solutions fully describe the horizontal and vertical advection, i.e. the transport of momentum.

The initial versions of the global 3-dimensional time-dependent numerical thermospheric models /11,12/ used theoretical models or the simple empirical Chiu /34/ global model of the ionosphere to calculate ion drag and Joule heating. However, the lack of any response at high latitudes to geomagnetic processes (precipitation, convection) within the Chiu ionospheric model caused a gross underestimate of the magnitude of ion drag and of Joule / frictional heating at E-region altitudes in the auroral oval. In the F-region, the Chiu model did not so seriously underestimate plasma densities. When the Chiu model was used in the 3-D T-D (or GCM) models, it was possible to simulate F-region winds and temperatures within the upper thermosphere which were realistic for quiet and slightly disturbed geomagnetic conditions. However, under disturbed conditions, and in the E-region at all times, simulations using the Chiu /34/ ionospheric model generated winds, currents and heating which were all unreasonably low /16-17/.

In the first interactive model for the polar ionosphere and thermosphere /28,29/, data sets from the UCL global thermosphere and the 'Sheffield' polar ionosphere model (UT-independent) were iteratively exchanged until stability was achieved. The effects of the model iterations showed that significant changes in plasma density were caused by the effects of induced winds. The auroral oval plasma densities were greatly enhanced compared with those of the Chiu model. As a result, induced thermospheric winds and heating were generally greatly increased compared with previous simulations using the global Chiu model.

This 'simple' coupled model could not, however, be universally applied to study UT variations, let alone the effects of variable solar, geomagnetic and seasonal conditions. The next stage, was to develop a fully interactive thermosphere and polar ionosphere model /8,33/. The fully coupled model exchanges ionospheric and thermospheric parameters throughout the region poleward of 40° geomagnetic latitude. At lower latitudes, the numerical model is presently still dependent on empirical ionospheric descriptions. When the physics of the major ionospheric-thermospheric interactions are included within the coupled model, many of the additional 'geomagnetic' energy sources required previously to explain observations, are unnecessary /8,33,35/.

The approach taken in the development of the coupled model described above differs from that employed by the Utah State group /36/. In the latter modelling, the structure, dynamics and composition of the neutral atmosphere is assumed to be invariant to the response of the ionosphere to solar and geomagnetic forcing. In this work, we show that induced neutral atmosphere winds, temperature and composition changes do, in fact, cause major feedback changes of the ionosphere.

The neutral atmosphere numerical model uses an Eulerian approach. The ionospheric code (30-33) is evaluated in a Lagrangian system that is closely tied to the Eulerian grid-point frame. From each grid point the parcels of plasma are traced along their convection paths under the influence of the magnetospheric convection electric field. The effect of co-rotation is implicit within this scheme, and the paths are followed, if necessary, in a series of sub-steps, assuming pure $\underline{E} \wedge \underline{B}$ drift at F-region altitudes. Neutral temperature, composition and wind velocity are interpolated from the neighbouring grid-points for evaluation of plasma density (H^+, O^+, NO^+, N_2^+, O_2^+) and ion temperature. After each complete time-step (1 min) final values are interpolated back to the fixed grid scheme for use in the neutral thermosphere code, and as start-up for the subsequent time steps. In such a way the evolution of the coupled ionosphere/thermosphere system proceeds through the UT day.

In the ionospheric code, atomic (H^+ and O^+) and molecular ion concentrations are evaluated over the height range from 100 to 1500 km, and used in the thermospheric code poleward of 40° magnetic latitude. The use of the self-consistent ionosphere at high- and mid-latitudes and an empirical description at low-latitudes can result in a discontinuity at the boundary. The ionospheric code is being extended to include the self-consistent calculation at low-latitudes, including computation of the equatorial anomaly, and allowance for inter-hemispheric flow, but these new results will not be discussed here.

3. COUPLED MODEL SIMULATIONS.

Two simulations of the coupled thermosphere and ionosphere for December solstice and two for the June solstice, using the UCL / Sheffield coupled model, will be presented. These simulations have been generated for conditions when the IMF BY component was either strongly positive, or strongly negative, for a geomagnetic activity level corresponding to approximately Kp = 3 to 4, and for moderately high solar activity ($F_{10.7}$ cm = 185). The simulations are time-dependent, that is they are UT-dependent, and the results are diurnally reproducible. However, the external solar and geomagnetic inputs are time-independent. These two simulations use an offset dipole representation of the geomagnetic field.

The characteristic UT variations of the summer and winter polar regions are dependent on the offset of the geomagnetic poles from the geographic poles. During the UT day, at all seasons, the geomagnetic polar caps are carried into and out of sunlight. There is, therefore, a large diurnal modulation of the solar photoionisation and UV/EUV heating of the geomagnetic polar regions which also causes large UT variations in plasma density, conductivity, ion drag and Joule and solar heating of the polar thermosphere. There are consequent large UT modulations of the thermospheric and ionospheric response. These characteristic UT variations of thermospheric and ionospheric structures, and the associated thermospheric-ionospheric interactions are discussed in /8/.

Figure 1 illustrates the global wind and temperature structure for the December solstice, at pressure level 12, close to 300 km altitude, at 18 Universal Time (UT) and moderate geomagnetic activity ($K_p=3^+$). A large temperature gradient has developed from the summer to winter hemisphere of over 500K, due to the differential heating by the solar UV and EUV radiation. Superimposed on this latitude structure is a smaller amplitude diurnal variation of about 250°K range at equatorial latitudes. At high latitudes the magnetospheric momentum and energy sources resulting from the convection electric field, auroral particle precipitation and Joule heating have increased the neutral wind velocities, with the additional heat source coming primarily from Joule heating. In the winter hemisphere this additional heating from Joule and auroral particle sources reverses the global summer to winter temperature gradient, so that minimum temperatures occur at winter mid-latitudes.

Near solstice, the global temperature and pressure fields create a prevailing summer to winter flow at mid and high altitudes, which is particularly pronounced under quiet geomagnetic conditions. For average or high geomagnetic activity, the geomagnetic heating in the summer hemisphere enhances this seasonal flow, whereas in winter it opposes the solar forcing. At upper thermospheric altitudes, a mean wind then flows from the summer pole and from the winter pole to winter mid latitudes. The global pattern of mean molecular mass in the upper thermosphere, shown in Figure 2 for the same simulation as Figure 1, is largely a result of the convergence and divergence of these large-scale horizontal, and the associated vertical, wind fields. High values of mean molecular mass (above 20 amu at F-region altitudes) are found over the summer geomagnetic polar region, even under quiet conditions. As the geomagnetic activity rises, the summer polar values increase further, as high as 24-25 amu, while values of above 20 amu can be found over the winter geomagnetic pole under active geomagnetic conditions. At winter mid-latitudes, the mean molecular mass is close to 16 amu, nearly pure atomic oxygen.

Figure 3 illustrates the equivalent global structure of wind and temperature in the lower thermosphere, at pressure level 7 (around 125 km), and again at 18 UT. At mid and low latitudes, a large part of the response is due to the influence of semi-diurnal tidal modes propagating from the middle and lower atmosphere. At high latitudes, for this case of moderately disturbed geomagnetic activity, the wind pattern is controlled more by the electrodynamic forcing from the magnetosphere, than either in-situ solar forcing or propagating tides. At high latitudes, the lower thermosphere wind magnitudes are about a factor of two smaller than those shown in Figure 1 for the upper thermosphere. Note that the wind scale differs by a factor of two compared with the previous Figures. However, it is interesting that the E- and F-region polar wind patterns are remarkably similar. The ion drag momentum forcing is different, not only due to reduced velocities, but also due to rotation of the ion drift velocity vector from the $E \wedge B$ direction toward the electric field vector direction (Hall component). However, steady-state pressure gradients constrain the divergent or convergent winds induced by the Hall component of the ion drift at these E-region altitudes. At mid and low latitudes the influence of the propagating tides has brought the magnitudes of the winds to levels similar to those seen in the upper thermosphere, of the order of 100 m/s. At the lower levels the winds are predominantly semi-diurnal, however, compared with a mainly diurnal variation in the upper levels, so that wind patterns of the upper and lower thermosphere are quite different.

4. POLAR F-REGION WINDS, COMPOSITION AND ION DENSITY.

Figure 4 illustrates, in four panels, the response of the upper thermosphere wind and the mean molecular mass, in the northern polar regions (50-90°) to the orientation of the IMF-BY in summer and in winter. Figure 4a and b are for winter, IMF-BY negative and positive respectively, 4c and d are the equivalent for the summer.

At F-region altitudes, the major changes of wind patterns are IMF BY-related, and seasonal effects are slight. In the winter, for BY positive, the winds in the dawn auroral oval show a slight tendency to turn sunward with the relatively weak sunward ion convection. In the summer there is no significant indication of sunward winds in the dawn cell for BY positive. Within the strong clockwise circulation wind cell which follows the strong sunward ion convection in the dusk auroral oval and antisunward ion flow over the dawn side of the polar cap, only minor seasonal differences occur. Equatorward of the nightside auroral oval, the wind flow is more strongly equatorward in the June simulation (200 m/sec) than in the December simulation (100 m/sec).

There are two major responses of high-latitude F-region winds to changes of IMF BY sense. Firstly, a region of high-velocity anti-sunward flow shifts from the dusk (BY negative) to dawn (BY positive) side of the polar cap. This has been previously reported /17,38/ in observations from the DE-2 spacecraft and from the ground. Secondly, the winds of the dawn auroral oval also show a characteristic change. For IMF BY positive, there is very little indication of sunward wind flow in response to the weak dawn auroral oval / polar cap convection cell, while for IMF BY negative, sunward winds of 100 - 200 m/sec occur. The sunward winds of the dawn auroral oval are always weaker than the strong sunward winds of the dusk auroral oval. The primary reasons for this dusk / dawn asymmetry have been discussed previously /16,17,29,37/.

The asymmetric response is caused by a natural atmospheric 'resonance' in which the clockwise wind vortex, forced by sunward ion convection in the dusk auroral oval and antisunward ion convection over the polar cap, is preferentially excited, given the sense and rate of rotation of the Earth. Coriolis and curvature accelerations balance within the clockwise vortex, of which the dusk auroral oval is part. In the dawn auroral oval, the conditions for this resonance do not exist, since the Coriolis and curvature accelerations act in the same sense, causing a 'spin-out' of the anticlockwise vortex. As a result, for the same plasma densities and convection velocities, there is a much smaller effective sunward wind acceleration. This effect was noted in earlier rocket wind measurements /23,24/, and has been well observed by DE-2 /15,16,39/. The effect was initially predicted by rather simple simulations /29,37/, however, the asymmetric dusk / dawn wind response is a major feature of the present relatively sophisticated simulations. In principle, a complex feedback process between the thermosphere and ionosphere might have changed the nature or magnitude of the wind acceleration and asymmetric response. However, the coupled model accounts for all the major feedback mechanisms, with the possible exception of feedback processes involving magnetosphere / ionosphere coupling.

The increased geomagnetic heating in the northern polar cap creates, in addition to the high neutral gas temperature, a region of raised mean molecular mass, caused by systematic upwelling and outflow to the high mid-latitudes. The value of mean molecular mass increases steadily with sustained geomagnetic forcing and heating (enhancement of molecular nitrogen, depletion of atomic oxygen).

The seasonal variation of neutral thermospheric composition at a fixed height level is well established by experimental observations /40/. At middle latitudes, enhanced concentrations of $[N_2]$ relative to $[O]$ in the summer hemisphere between 250 and 350 km feeds back into the ionospheric chemistry by increasing the recombination of the dominant ion O^+, decreasing the N_mF_2 (peak electron density at the peak of the F_2 layer). Conversely, in the winter hemisphere, the enhanced $[O/N_2]$ ratio decreases the recombination rate of the F region O^+ ions, and causes a general increase in N_mF_2, despite the overall decrease of solar insolation.

These general features are well illustrated by comparisons between Figures 4a and 4c, for the F-region (IMF BY negative) or Figures 4b and 4d (IMF BY positive). The highest values of mean molecular mass which occur poleward of 50° latitude in the winter polar F-region equal the lowest value in the equivalent summer polar region. At high winter mid-latitudes, the mean molecular mass is close to 16, indicating a composition which is nearly pure atomic oxygen. There is then a winter-time plateau around the geomagnetic pole, due to heating, upwelling and outflow, where the mean molecular mass reaches 20.

This value of mean molecular mass within the winter plateau is very dependent on geomagnetic activity. At very low activity levels the mean molecular mass may be as low as 16 - 17. During extended geomagnetic storm conditions, the mean molecular mass may reach values as high as 22.

In the summer polar cap, the lowest values of mean molecular mass above 50° are 20, and the highest values within the geomagnetic polar cap reach 24 to 25. At this constant pressure level (12), this implies a 4 fold reduction in the density of atomic oxygen between high summer mid-latitudes and the pole. Atomic oxygen concentrations at the summer pole are a factor of 10 lower than those found at high winter mid-latitudes. The variation of molecular nitrogen density is in direct anti-phase, and compensates for the atomic oxygen changes.

There are significant changes in the compositional variations caused by changes in the sense of IMF BY. These detailed changes in mean molecular mass are modest (1 - 2 units) compared with the larger changes (3 - 4 units) induced by the combination of seasonal solar insolation changes and the heating and consequent overturning of the thermosphere resulting from geomagnetic forcing. However, the F-region plasma density is highly responsive to the combination of BY-induced modulation of neutral composition and ion transport.

Figure 5 illustrates the response of the F-region plasma density for the same conditions as described in Figure 4. Comparing Figures 5a and 5c, the high-latitude seasonal F-region ionospheric anomaly is apparent: in the sunlit regions on the dayside, winter plasma densities at high latitudes exceed those of the equivalent summer sunlit regions. For example, peak densities in the winter sunlit cusp region (which is just in sunlight at 18 UT) are about a factor of 2 - 3 higher than the equivalent summer region. The mean levels over the high latitude region are similar in summer and winter at this UT.

The winter polar cap, at 18 UT, is filled with high density plasma which is rapidly convected from the dayside through the cusp region. There is a well-developed sub-auroral trough, from 16 LST to 08 LST at this UT. At UT times when the cusp is not in sunlight, the plasma density within the polar cap tends to fall to significantly lower values /9/. The discontinuity at the equatorward edge of the sub-auroral trough in ion density is an artifact of using the Chiu model for the lower latitude region. A more realistic picture of the mid-latitude ion density can be seen later in Figure 6, since O^+ is the dominant ion in this case.

In contrast, the summer sunlit polar cap is a region where the electron recombination rate is high, so that plasma densities are low, despite the combination of solar photoionisation and auroral electron precipitation. The regions of lowest plasma density have a one-to-one relationship with regions of the highest mean molecular mass and highest Joule heating.

The sub-auroral trough, caused by the sunward convection of low-density plasma from the nightside encircles the entire auroral oval at the winter solstice for much of the UT day except within about +/- 3 hours of 18 UT. In another study /41/ we have shown that the sub-auroral trough migrates equatorward with increasing activity, remaining on and equatorward of the equatorward edge of the auroral oval. The sub-auroral trough is present in the winter hemisphere and survives equinox. It usually disappears in the summer hemisphere near the full solstice, due to photoionisation throughout the summer polar region.

The sub-auroral trough in the winter hemisphere has relatively little response to the sense of the IMF BY component. When IMF BY is positive, the dusk trough is somewhat broader in latitudinal extent than when BY is negative. This is due to the stronger sunward convection in the dusk oval, but the effect is relatively minor, and must be dependent on the coincidence (or non-coincidence) of convection and precipitation boundaries.

Sub-auroral troughs which develop in the summer hemisphere are likely to be the fossils of a much-expanded auroral oval during a recent period of intense geomagnetic activity. Intense heating, forcing enrichment of molecular species by convective overturning and advection, will lead to the rapid destruction of plasma once the auroral source diminishes at the end of the disturbance.

Comparing Figures 5a (BY negative) and 5b (BY positive), the major response of the winter polar F-region plasma density in response to IMF BY changes can be seen. For BY negative (5a), a long extended tongue or plume of plasma is rapidly transported antisunward from the dayside and cusp. This rapid transport is produced by the combination of strong antisunward plasma convection on the dusk side of the polar cap reinforced by the effect of co-rotation in that region. On the dawn side of the polar cap, the weak antisunward plasma convection is opposed by co-rotation so that a deep plasma hole develops (i.e. no transport, no sunlight, little precipitation).

When IMF BY is positive, the rapid antisunward plasma transport on the dawn side of the polar cap is opposed by co-rotation. The corresponding tongue or plume of anti-sunward moving plasma is thus weaker and shorter, and tends to lie in the central polar cap, rather than the dawn side. The co-rotation velocity is very small in the central polar cap region and thus does not oppose convection. Co-rotation also aids the otherwise weak antisunward transport on the dusk side of the polar cap, another effect which broadens the width of the antisunward plasma tongue or plume when IMF BY is positive.

When the IMF BY component is negative, the combined effects of plasma convection and co-rotation are sufficiently strong that, for the same precipitation, plasma densities within the dusk auroral oval are roughly double those when the IMF BY is positive. This is entirely due to the shorter time required to transport plasma from the dayside cusp through the polar cap and into the dusk auroral oval for IMF BY negative.

In the summer polar cap F-region, there are some quite dramatic variations with IMF BY component. Troughs form in regions where there is strong photoionisation and some particle precipitation, caused by the neutral compositional response to strong Joule heating: the stronger the ion convection, the stronger the Joule heating and, generally, the stronger the convective / advective response of the neutral atmosphere. The rapid convective upwelling, combined with horizontal advection causes the very large increase in F-region mean molecular mass, which then causes the plasma troughs.

The most important conclusion which comes from this study, related to F-region behaviour and ionosphere-thermosphere coupling, is the distinctive seasonal behaviour. In the winter polar region, it is primarily plasma convection which determines the distribution of plasma density and so modulates the local frictional heating and the acceleration of winds by ion drag. The transport of F-region plasma by the combination of convection and co-rotation control the existence and location of tongues of high plasma density within the polar cap and the formation of the sub-auroral trough. At higher levels of geomagnetic

368

activity, strong convection and intense Joule heating may create localised troughs of low plasma density within the polar cap. Neutral wind effects are significant through the vertical transport of plasma. The convective overturning of the neutral atmosphere and associated advection causes significant compositional changes which cause noticeable, but not dominant, effects on the winter-time F-region plasma distribution, except under very disturbed conditions.

In the summer hemisphere, the effects of combined solar insolation and geomagnetic heating force a much greater convective / advective overturning of the thermosphere. In the summer polar cap, at levels of moderate geomagnetic activity, the greatly increased proportion of molecular species causes a fundamental change in the F-region ionospheric chemistry. As a result, the neutral composition distribution dominates the detailed ionospheric response. Regions where, in the winter pole, the plasma density is enhanced by rapid convection from the dayside, tend to become regions of low plasma density in summer. The combination of plasma convection and co-rotation are still important.

While in the course of this paper, we have only considered two, asymmetric, convection patterns, the general conclusions described above are true for other representations of the convection field, such as the Foster et al (1986) models. Detailed inter-comparisons with those described in the present study confirm the conclusions of the previous two paragraphs /41/.

5. ION TEMPERATURE AND ION COMPOSITION.

Figures 6 and 7 illustrate the ion temperature, O^+, molecular ion, and H^+ number density distributions for the winter and summer northern polar regions, respectively. Both figures are from the simulations with IMF-BY negative, at 18 UT, and the altitude is 301 km. For the solar activity depicted here ($F_{10.7} = 185$), this altitude is close to the F2 peak in winter but significantly below it for the summer case.

The general patterns of ion temperature in summer and winter are very similar, given that the same convection pattern has been used in both these two simulations. The ion temperature structure is highly dependent on Joule heating, and thus on the magnitude of the relative ion to neutral drift vectors. These, in turn, are highly convection-pattern dependent - in our simplistic modelling, dependent on the orientation of the IMF-BY. The summer ion temperatures exceed those of the winter by about 500 to 600°K, approximately equal to the difference in the neutral temperature (Figure 1). The influence of the neutral wind will not be addressed specifically in this paper.

A further consequence of the global seasonal composition difference is the effect on the relative proportion of atomic and molecular ions in the polar regions. The patterns of ion composition are substantially different in the two seasons. The general magnitudes of the O^+ number density in the polar regions are very similar in summer and winter at 300 km altitude, although the reasons for their individual structure are very different. The winter polar F-region, although in darkness, is maintained by the transport of O^+ from the dayside sunlit sub-auroral ionosphere, where the winter plasma densities are consistently larger than corresponding summer-time values. The very slow recombination rates in the neutral oxygen-rich atmosphere is sufficient to keep the winter polar O^+ density above the summer values. In fact, the dominant ion is O^+ in the winter at these altitudes, the mechanisms creating the pattern are therefore identical to the previous discussion for the total ion density. The molecular ions in winter at this altitude generally contribute less than 5% to the total ion density. The pattern of molecular ions, predominantly NO^+, has a similar trend to that of the pattern of ion temperature (and Joule heating). The regions of upwelling are related to regions of enhanced molecular ions. The regions of upwelling expand and intensify with increased activity, and the general proportion of molecular ions increases at higher activity levels, even within the winter polar region.

In summer the ratio of molecular to atomic ions is greatly increased, approximately 10 fold within the geomagnetic polar cap (Figure 7). For example, within the summer polar cap, as the geomagnetic activity level increases, there is a steady increase of molecular ion density. From a series of simulations for early May the ratio of molecular to atomic ions increased from about 5 % to 30 %, as Kp increases from 1 to 4 /41/. At times of high geomagnetic activity, even at this altitude of 301 km, in some regions of the summer polar thermosphere, the concentration of molecular ions exceeds that of atomic ions. Generally the levels are very similar over much of the summer polar regions. Outside the summer polar region of enhanced molecular ions, associated with the region of high mean molecular mass found in the neutral thermosphere, the proportion of molecular to atomic ions decreases rapidly toward mid-latitudes.

Atomic oxygen ion number densities are little changed in the summer polar cap and auroral oval at 300 km as geomagnetic activity increases, except for depletions in the dawn and dusk auroral oval. These depletions are created by the sunward ion drift, transporting low density plasma from the nightside toward the dayside. The location of this 'sub-auroral' trough follows the equatorward expansion of the convection features.

The H^+ number density is typically a few orders of magnitude less in concentration than the total ion density, although it does become significant at much higher altitudes, in excess of 500 km.

6. LOWER THERMOSPHERE WINDS AND ION DENSITY.

Figure 8 illustrates the neutral wind vector and plasma density in the lower thermosphere, at pressure level 7, about 125 km altitude, close to the peak in the Pedersen conductivity. The figure shows the same four seasonal and IMF BY conditions described in Figures 4 and 5. There are major changes of the E-region winds resulting from seasonal factors and from the response to the sense of the IMF BY component.

Firstly, there is a distinctly different pole to pole thermal structure at the June and December solstices. There is a pronounced cyclonic vortex around the winter polar region, which is observable at high mid-latitudes, superimposed on a generally weak flow away from the sub-solar point. This circulation is replaced by a pronounced anti-cyclonic vortex around the summer polar region (again best seen at high mid-latitudes).

Within the auroral oval and polar cap, the wind patterns show little change with season for the same convection and precipitation patterns. The major wind features result from ion drag, and are thus responsive to both convection pattern changes and plasma density enhancements (resulting from combined particle precipitation and solar photoionisation). The basic auroral oval wind pattern follows that of ion convection, as in the F-region. The E-region wind velocities are typically about a factor of 2 - 3 lower than at F-region altitudes.

The self-consistent simulations show E-region winds, at times of moderate geomagnetic disturbances, which are of similar magnitudes and structures to those reported in experimental studies of E-region winds /23-25,42,43/. Until the present self-consistent thermosphere - ionosphere code was developed, such correspondence of simulated E-region winds with the real world could only be obtained by extensive and artificial tuning of E-region plasma density distributions (even in earlier coupled codes /28,29/).

The response of the auroral and polar cap E-region winds to the IMF / plasma convection pattern changes is very marked. Obviously, this response is also modulated by the precipitation, which causes the large increases of ion drag coupling via enhanced E-region plasma density.

For IMF BY positive, there is a large clockwise swirl, sunward around the dusk auroral oval and antisunward over the dawn side of the polar cap. The peak E-region wind speeds in the dusk auroral oval reach 200 - 250 m/sec, which is quite consistent with observations /28/. There is a relatively weak sunward wind flow in the dawn auroral oval (less than 50 m/sec), combined with a significant equatorward flow component (50 - 100 m/sec).

For IMF BY negative, the sunward wind flow in the dusk auroral oval is still strong, except that it only extends to about 14 LST. The antisunward flow over the polar cap is more generally spread over the entire width of the polar cap, rather than being confined to the dusk side, as tends to happen in the F-region (see earlier discussion). The sunward winds of the dawn auroral oval, for IMF BY negative, are significantly increased, to nearly 100 m/sec, and there is a generally more coherent flow, following the sunward ion convection of the dawn auroral oval.

The asymmetry of the wind response in the dusk and dawn parts of the auroral oval is due to the conditions required for resonance being met in the dusk, but not the dawn sectors of the auroral oval, as in the F-region. The E-region winds follow the $\underline{E} \wedge \underline{B}$ direction more than would be expected, considering the rotation of the ion drift vector toward the $\underline{E}$ (electric field) direction at these altitudes (Hall component). This is the result of steady-state pressure gradients being induced to oppose the divergent or convergent winds tending to follow the Hall component of ion drift at these altitudes. A short-period impulse would have rather different consequences. The initial winds will follow more closely the mean ion vectors, before the pressure gradients are generated.

These four simulations show that E-region ionospheric chemistry is not fundamentally changed, despite a significant seasonal modulation of mean molecular mass, even at E-region altitudes. Plasma densities in regions outside the auroral oval show an expected increase in the summer hemisphere, as the result of increased photoionisation. E-region plasma density values on the nightside in the winter hemisphere are predicted to be low (2 - 5x10^9 m^{-3}) in places. Within the geomagnetic polar cap, E-region plasma densities within the winter hemisphere are lower than in the summer hemisphere. This is the direct result of additional solar photoionisation within the summer polar cap.

What we see from this study is that E-region plasma densities around the auroral oval are primarily enhanced as the result of precipitation. The seasonal modulation and the effects of plasma wind transport are generally of little importance. Vertical plasma transport due to the horizontal winds however, may be particularly important in connection with the formation and subsequent behaviour of sporadic E layers, resulting from metallic ions. Such long-lived ions are not subject to the rapid destruction of the E-region molecular ions which alone are considered in this study. For metallic E-region ions, horizontal as well as vertical transport effects will be much more significant /44/.

The actual values of ion drag wind acceleration terms, ionospheric conductivities, electric currents, and the implicit distribution of magnetospheric FAC, which power the entire high-latitude system, are therefore determined by the effects of precipitation, as well as by the convection field. Typically, in summer time, peak E-region ion density values within the moderately disturbed auroral oval are a factor of 3 greater than those of the surrounding sunlit, non-auroral, ionosphere. In winter time, the corresponding enhancement factor due to electron precipitation is between a factor of 10 and 100 times above the non-sunlit E-region ionosphere.

7. ELECTRODYNAMICS.

In the following section, the key electrodynamic parameters - the height-integrated Joule heating rate, the horizontal ionospheric currents and the FAC - obtained from the numerical simulations will be discussed and compared. Unlike the previous figures which have been referenced to a fixed pressure or height level, the electrodynamic parameters are height-integrated through all the levels. Generally, the largest contribution to both these quantities comes from the lower thermosphere. However, there are times and locations when this is not the case, and a significant contribution can come from the upper levels.

The height-integrated Joule heating rate is very sensitive to model inputs. This is due to its dependence on the second power of the electric field, and also due to its critical dependence on the co-location of boundaries of conductivity enhancements through electron precipitation, and maximum ion drifts. Obviously, the use of the product of two independently produced models, anticipating a strong causal relationship between the two sources via their magnetospheric origin, must be treated carefully. Ideally, the models of precipitation and convection should be assembled 'self-consistently'. Any correlation or anti-correlations (as have been reported) between the two data sets can then be maintained. However, no convenient empirical statistical model of the combined magnetospheric sources is available, so we must rely on using, with some caution, those currently at our disposal.

Figure 9 illustrates the height-integrated Joule heating rate for the four cases. The patterns are very strongly dependent on the orientation of the IMF-BY (and thus on any specific convection field), while the magnitudes are seasonally-dependent. Due to the second power dependence on the neutral wind - ion velocity difference, the peak Joule heating rates tend to be associated with the peaks in the ion drift vectors. A particular characteristic of using electric field models with two extreme orientations of the IMF-BY, is that either the dusk or dawn boundaries between polar cap and auroral oval have a double-maximum feature in Joule heating. These double features are created by the combination of the expected peak that occurs in the dawn and dusk sector auroral oval (maximum sunward ion drift), and a second feature associated with the enhanced antisunward ion flow over either the dusk or dawn side of the polar cap. For IMF-BY negative the twin maximum feature is on the dusk side, for IMF-BY positive it is on the dawn side.

We should note two important features of the Joule heating which are related to seasonal conductivity variations which are not obvious from Figure 9. Firstly, there is a large difference in the lower thermosphere conductivity in summer and winter (Figure 8 and /41/). Secondly, in winter the conductivity of the lower thermosphere, except where enhanced by auroral ionisation, is very small. A significant contribution to the height-integrated conductivity, and consequently also the Joule heating rate, comes from the upper thermosphere. As was shown before, the high latitude winter F-region maintains as much if not more plasma density, as the equivalent summer regions. The peak Joule heating values outside the auroral oval therefore reflect these seasonal differences in the distribution of conductivity within the thermosphere. The hemispheric integrated Joule heating rates, however, maximise in the summer hemisphere, where the total input is generally between a factor of 2 and 3 higher than in winter.

The seasonal difference is most clearly seen by comparing Figure 9a and c, for winter and summer conditions respectively (IMF BY negative). Within the auroral oval the dusk or dawn auroral oval peak Joule heating values have increased by about 50% due to the additional solar ionization. On the dusk side of the polar cap, however, the Joule heating values are considerably increased in summer, in some areas by more than a factor of three. The extreme difference in magnitude of this polar cap feature is because, in winter, at this time, the location of the peak (which is associated with the fast ion velocities near 0200 LT) is just in darkness, and experiences only modest fluxes of auroral precipitation. In summer this same location has a zenith angle close to 50°, at this time, and the additional photo-ionisation greatly increases the conductivity and Joule heating.

Peak Joule heating values for both simulations with IMF-BY negative occur in the early dusk magnetic sector. In winter and summer, dusk auroral oval values exceed the dawn values by a factor of two. In summer the extreme polar cap maximum exceeds the dawn sector values by a factor of six.

For IMF-BY positive (Figure 9b and d) the seasonal differences are less dramatic, and both patterns are very different from Figures 9a and c. The double feature is now in the dusk sector, and the peak values are near midnight, within the auroral oval, and towards the dayside for the polar cap feature. The polar cap feature again relies on solar produced plasma, and again increases in magnitude by a factor of three. The midnight sector peak increases by about 50% from winter to summer.

It is also possible to use these simulations to examine the effect of ignoring the influence of the neutral wind on the height-integrated Joule heating rate. Values in the dawn and midnight sector auroral oval change little, however, Joule heating in the dusk auroral oval is increased by 30 to 40% when effect of neutral winds are excluded. This result is due to the neutral wind resonance discussed earlier. The sunward ion motion in the dusk auroral oval naturally drives the larger winds, and thus when the $(\underline{V}_n \wedge \underline{B})$ component is removed from the full expression for the Joule heating rate, the main effect is an increase in the dusk sector. In regions of high Joule heating within the polar cap, on the dusk (BY negative) and dawn (BY positive) sides, the peak values are also reduced when neutral winds are excluded.

The ion temperature (F-region) distributions shown in Figures 6 and 7 correspond roughly to the Joule heating patterns, when the baseline F-region neutral gas temperature is considered (Figure 1). As the general level of geomagnetic activity rises, this baseline neutral gas temperature also increases. In these models, using simple and smooth functions for convection patterns and precipitation, the neutral gas temperature distribution is relatively smooth. However, there are large localised increases of neutral temperature in the polar region which are not represented by semi-empirical thermospheric models such as MSIS 1986 /40/. Thus some care is required to properly interpret ion temperature data alone when attempting to infer either Joule heating rates, or the neutral gas temperature. The Joule heating rates in the real world are considerably more structured, and with locally much higher values, than these model simulations for mean conditions would indicate.

The horizontal ionospheric current and FAC distributions are shown in Figure 11. Even using the fully-coupled thermosphere-ionosphere model (with statistical models of convection and precipitation) such results have to be used cautiously. The height-integrated horizontal ionospheric current system is obtained from a product of the conductivity and total electric field, and the FAC system is calculated from the divergence of the horizontal current. This condition then satisfies current continuity. Combining this calculation with the previously-derived quantities, we then have an indication of the total demand made by the ionosphere on the magnetospheric sources of plasma, charge, momentum and energy. For the horizontal current to flow in the ionosphere, current continuity demands that either the magnetosphere is able to supply the necessary FAC, or polarization charges build up within the ionosphere, (or via instabilities triggered by large FAC or 'double-layers' etc), which tend to reduce the magnitude of the electric field imposed upon the upper atmosphere. The sensitivity of the magnetosphere to the feedback from the ionospheric current system is as yet difficult to quantify experimentally or theoretically.

The region 1 and region 2 Birkeland (FAC) current system are clearly defined in Figure 11, even at the lowest activity level, and generally expand and intensify as the activity increases.

In winter the peak upward FAC are larger in magnitude than the downward FAC, with peak values a factor 1.5 to 2 larger. The regions of convergence of the height-integrated horizontal current, which drive the upward FAC, are within the auroral oval. The regions of divergent horizontal current tend to lie outside those of auroral precipitation and therefore rely on solar-produced conductivities. The upward FAC is confined to a smaller area. Globally-integrated, the differences between total upward and downward FAC appear to be small, within physical and numerical uncertainties, as might be expected.

In summer the upward current increases only by about 50%. In contrast the downward current increases by a factor of 2-3, reflecting the importance of the increase in solar produced conductivities. The result is that in summer the peak in the upward and downward FAC are roughly equal.

A number of differences occur in the FAC due to the orientation of the IMF-BY. The magnitudes for IMF-BY positive exceed those for BY negative by 50%. For BY negative the upward FAC closes on the dayside, to just form a continuous circle, surrounding the region of downward FAC. For IMF-BY positive the continuity of the upward current on the dayside is split by a region of downward current. The region of upward FAC is now diverted over the polar cap, but in the midnight sector the continuity of the pattern is much smoother.

8. SUMMARY

Data is presented from a series of four numerical simulations of the coupled thermosphere and ionosphere. The data from high latitudes is shown, at high (300km) and low (125km) thermospheric altitudes, to illustrate the response of the system to changes in the orientation of the IMF By component at the summer and winter solstices. Each simulation is at high solar activity and moderate geomagnetic activity, and data are presented for 18 UT.

In the winter F-region, the plasma distribution is strongly controlled by transport. As the convection electric field moves into sunlight, solar produced ionization is 'picked-up' and carried over the polar cap. For By negative, a tongue of plasma extends over the dusk side of the polar cap, the convection electric field being assisted by the natural tendency of co-rotation. A polar ionisation hole is created by the lack of transport from regions of plasma generation, no photoionisation and negligible particle ionization. Plasma carried anti-sunward over the dusk side of the polar cap continues into the dusk auroral oval, where number densities exceed those in the dawn sector auroral oval by at least a factor of two. For By positive the tendency of the convection electric field is to transport the plasma to the dawn side of the polar cap. The opposing influence of co-rotation in this case restricts the flow, resulting in a much smaller tongue over the central polar cap. Auroral oval ion densities in this case remain symmetric. In both cases a sub-auroral trough encircles the nightside.

There is also a characteristic neutral wind response in the upper thermosphere to the changing sense of IMF By. For By negative, enhanced anti-sunward winds occur on the dusk side of the polar cap, and modest sunward winds are generated in the dawn auroral oval. For By positive the peak anti-sunward winds move to the dawn side of the polar cap, and the sunward winds in the dawn auroral oval are virtually eliminated. In both cases substantial sunward winds occur in the dusk auroral oval, but are stronger for By negative due to the higher ion densities, resulting from fast ion transport from the dayside, across the dusk side of the polar cap and into the dusk auroral oval.

In summer the plasma distributions are flatter, are generally lower in concentration, and lack the sub-auroral trough, at least at 18 UT, as presented here. Transport is much less important in defining the plasma distribution which, in summer, at this level of geomagnetic activity, is controlled by the peaks and troughs in neutral composition. Plasma density minima correlate with peaks in the mean molecular mass, and contrast with the peaks in ion density that occurred at similar locations in winter. The generally higher mean molecular mass (i.e. enhanced molecular species) in summer is created by the global seasonal neutral circulation pattern. At high latitudes the pattern is modulated by regions of enhanced upwelling due to Joule heating, and the characteristic By-dependent composition distribution is finally imposed by the transport effect of the horizontal neutral wind. The neutral wind maintains its By dependence in summer, and changes very little with season.

The ion temperature in the F-region is higher in summer due to the seasonal difference in the temperature of the neutral atmosphere. The overall pattern of ion temperature is primarily dependent on the ion-neutral velocity difference, which will be a function of the IMF-BY direction, at least, and will depend little on season. Ion temperature distribution, as with many other features, will be highly dependent on the actual ion convection pattern.

The ratio of atomic to molecular ions at 300 km altitude has a strong seasonal dependence, virtually reflecting the strong seasonal difference in the neutral composition. In winter O^+ dominates, in summer the concentration of molecular ions increases dramatically, such that in some areas of the summer polar region they exceed the atomic ion concentration.

In the E-region, at this level of geomagnetic activity, the ion density in the auroral oval is dominated by auroral precipitation, which for these simulations has no By or seasonal dependence. Small changes do occur due to the seasonal changes in solar ionization. Outside the auroral oval, increased solar illumination in summer generates ion densities exceeding those in winter by about a factor of three.

The neutral wind in the lower thermosphere maintains the signature of the IMF By that was the hall-mark of the F-region circulation. It has a noticeable but not dominant seasonal variation due to the increase of ion density by the background solar illumination.

The height-integrated Joule heating rate shows a strong seasonal and BY dependence. A double-peak feature develops in the dusk sector for BY negative, and in the dawn sector for BY positive. Within the auroral oval, the seasonal increase is small, 50%, due to primary dependence of conductivity on auroral precipitation. In the polar cap the seasonal response is large, due to dependence on solar produced conductivity.

In the dusk auroral oval, Joule heating exceeds that in the dawn sector by a factor of two. Ignoring neutral winds enhances Joule heating, particularly in the dusk sector, thus increasing the ratio of Joule heating in the dusk/dawn sectors. The conductivity changes are also reflected in the patterns of FAC.

ACKNOWLEDGMENTS

We are indebted to Dr. Fred Rich for provision of the Heppner and Maynard polar electric fields in the form of harmonic coefficients, and to Drs. Shaun Quegan, Roy Moffett and Graham Bailey for their help in the development of the coupled ionosphere/thermosphere model. We would like to express our particular thanks to John Harmer and Hilary Hughes for their assistance in preparing, running and processing the computer simulations using the UCL / Sheffield coupled ionospheric / thermospheric model. Computer time was made available by the University of London Computer Center (CRAY 1-S), and the CRAY-XMP-48 at the Rutherford Appleton Laboratory. The research was supported by grants from the UK Science and Engineering Research Council, and from the European Office of Aerospace Research and Development (AFOSR-86-341).

REFERENCES:

1.   FULLER-ROWELL T.J. and D.S. EVANS, (1987), Height-Integrated Pedersen and Hall Conductivity Patterns Inferred from the NOAA/TIROS Satellite Data. J. Geophys. Res. 92, 7606-7618.
2.   HARDY D., M.S. GUSSENHOVEN, E. HOLEMAN, (1985), A statistical model of auroral electron precipitation. J Geophys. Res., 90, 4229-4248.
3.   HEPPNER J.P., (1977), Empirical Models of High Latitude Electric Field, J. Geophys. Res. 82, 1115-1125.
4.   HEPPNER J.P. and N.C. MAYNARD, (1987), Empirical High-Latitude Electric Field Models, J. Geophys. Res. 92, 4467-4490.
5.   FOSTER J.C., J.M. HOLT, R.G. MUSGROVE and D.S. EVANS, (1986), Ionospheric Convection associated with Discrete Levels of Particle Precipitation, Geophys. Res. Lett., 13, 656-659.
6.   REES M.H., B.A. EMERY, R.G. ROBLE, and K. STAMNES, (1983), Neutral and ion gas heating by auroral electron precipitation, J. Geophys. Res., 88, 6289-6300.
7.   REES D., R. GORDON, T.J. FULLER-ROWELL, M.F. SMITH, G.R. CARIGNAN, T.L. KILLEEN, P.B. HAYS, and N.W. SPENCER, (1985), The composition, structure, temperature and dynamics of the upper thermosphere in the polar regions during October to December 1981, Planet. Space. Sci. 33, 617-666.
8.   FULLER-ROWELL T.J., D. REES, S. QUEGAN, R.J. MOFFETT, G.J. BAILEY, (1988), Simulations of the seasonal and universal time variations of the thermosphere and ionosphere using a coupled, three-dimensional, global model. PAGEOPHYS, 127, 189-217.

9.   MARTYN D.F., Geo-morphology of F2-region ionospheric storms, (1953), Nature, 171, 14-16.
10.   TINSLEY B.A., Y. SAHAI, M.A. BIONDI and J.W. MERIWETHER, (1988), Equatorial particle precipitation during geomagnetic storms and relationship to equatorial thermospheric heating, J. Geophys. Res., 93, 270-276.
11.   FULLER-ROWELL T.J. and D. REES, (1980), A three-dimensional, time-dependent, global model of the thermosphere, J. Atmos. Sci. 37 2545-2567.
12.   FULLER-ROWELL T.J. and D. REES, (1983), Derivation of a conservative equation for mean molecular weight for a two constituent gas within a three-dimensional, time-dependent model of the thermosphere, Planet. Space Sci. 31, 1209-1222.
13.   ROBLE, R.G., R.E. DICKINSON and E.C. RIDLEY, (1982), The global circulation and temperature structure of the thermosphere with high-latitude plasma convection, J. Geophys. Res. 87, 1599-1614.
14.   DICKINSON R.E., E.C. RIDLEY and R.G. ROBLE, (1984), Thermospheric general circulation with coupled dynamics and composition, J. Atmos. Terr. Phys. 41, 205-219.
15.   HAYS P.B., T.L KILLEEN. N.W. SPENCER, L.E. WHARTON, R.G. ROBLE, B.A. EMERY, T.J. FULLER-ROWELL, D. REES, L.A. FRANK, and J.D. CRAVEN, (1984), Observations of the dynamics of the polar thermosphere, J. Geophys. Res. 89, 5547-5612.
16.   REES D., T.J. FULLER-ROWELL, R. GORDON, T.L. KILLEEN, P.B. HAYS, L.E. WHARTON and N.W. SPENCER, (1983), A comparison of the wind observations from the Dynamics Explorer satellite with the predictions of a global time-dependent model, Planet. Space Sci. 31, 1299-1314.
17.   REES D., T.J. FULLER-ROWELL, R. GORDON, M.F. SMITH, J.P. HEPPNER, N.C. MAYNARD, N.W. SPENCER, L.E. WHARTON , P.B. HAYS, and T.L. KILLEEN, (1986), A theoretical and empirical study of the response of the high-latitude thermosphere to the sense of the "Y" component of the interplanetary magnetic field, Planet. Space Sci. 34, 1-40.
18.   FESEN C., R.G. ROBLE, E.C. RIDLEY, (1986), Simulations of thermospheric tides at equinox with the NCAR Thermospheric General Circulation Model. J. Geophys. Res., 91, 4471-4489.
19.   PARISH H. T.J. FULLER-ROWELL, D. REES, T.S. VIRDI and P.S.J WILLIAMS, Numerical simulations of the seasonal response of the thermospheric to propagating tides, (1989), Adv. Space Res, (in press).
20.   RISHBETH H. and O.K. GARRIOT, (1969), Introduction to Ionospheric Physics, Academic Press, New York and London.
21.   COLE K.D., (1962), Joule heating of the upper atmosphere, Aust. J. Phys., 15, 223-235.
22.   COLE K.D., (1971), Electrodynamic heating and movement of the thermosphere, Planet. Space Sci. 19, 59-75.
23.   REES D., (1971), Ionospheric winds in the auroral zone, J. Brit. Interplan. Soc. 24, 233-346.
24.   REES D., (1973), Neutral wind structure in the thermosphere during quiet and disturbed geomagnetic periods, in Physics and Chemistry of Upper Atmospheres (Edited by B.M. McCormac), 11-23, Reidel, Dortrecht.

25. PEREIRA E., M.C. KELLEY, D. REES, I.S. MIKKELSON, T.S. JORGENSEN and T.J. FULLER-ROWELL, (1980), Observations of neutral wind profiles between 115 and 176 km altitude in the dayside auroral oval. J. Geophys. Res. 85, 2935-2940.
26. HAREL M., R.A. WOLF, P.H. REIFF. R.W. SPIRO, W.J. BURKE, F.J. RICH and M. SMIDDY, (1981), Quantitative Simulation of a Magnetospheric Substorm, 1, Model Logic and Overview, J. Geophys. Res. 86, 2217-2241.
27. FULLER-ROWELL T.J., D. REES, S. QUEGAN, R.J. MOFFETT, and G.J. BAILEY, (1987), The thermospheric response and feedback to magnetospheric forcing. Extended Abstract, Symposium on Quantitative Modeling of Magnetosphere-Ionosphere coupling processes. Convenors: Y. Kamide and R.A. Wolf, March 9-13, 1987 Kyoto Sangyo University. p20.
28. QUEGAN S., G.J. BAILEY, R.J. MOFFETT, R.A. HEELIS, T.J. FULLER-ROWELL, D. REES and R.W. SPIRO, (1982), Theoretical study of the distribution of ionization in the high-latitude ionosphere and the plasmasphere : First results on the mid-latitude trough and the light-ion trough, J. Atmos. Terr. Phys. 44, 619-640.
29. FULLER-ROWELL, T.J., D. REES, S. QUEGAN, G.J. BAILEY AND R.J. MOFFETT, (1984), The effect of realistic conductivities on the high-latitude thermospheric circulation, Planet. Space Sci. 32, 469-480.
30. QUEGAN S., G.J. BAILEY, R.J. MOFFETT and L.C. WILKINSON, (1986), Universal time effects on the plasma convection in the geomagnetic frame, J. Atmos. Terr. Phys. 48, 25-40.
31. WATKINS B.J., (1978), A numerical computer investigation of the polar F-region, Planet. Space Sci., 26, 559-569.
32. ALLEN B.T., G.J. BAILEY and R.J. MOFFETT, (1986), Ion distributions in the high-latitude topside ionosphere, Ann. Geophysicae 4 A, 97-106.
33. FULLER-ROWELL T.J., S. QUEGAN, D. REES, R.J. MOFFETT, G.J. BAILEY, (1987b), Interactions between neutral thermospheric composition and the polar ionosphere using a coupled ionosphere-thermosphere model, J. Geophys. Res. 92, 7744-7748.
34. CHIU Y.T., (1975), An improved phenomenological model of ionospheric density, J. Atmos. Terr. Phys. 37, 1563-1570.
35. REES D., and T.J. FULLER-ROWELL, Invited paper presented at the AGARD / NATO meeting in Munich, May 1988, to be published in the proceedings.
36. SOJKA J.J. and R.W. SCHUNK, (1983), A theoretical study of the high-latitude F-region response to magnetospheric storm inputs, J. Geophys. Res. 88, 2112-2122.
37. FULLER-ROWELL T.J. and D. REES, (1984), Interpretation of an anticipated long-lived vortex in the lower thermosphere following simulation of an isolated substorm. Planet. Space Sci. 32, 69-85, 1984.
38. MCCORMAC F.G. and R.W. SMITH, (1984), The influence of the Interplanetary Magnetic Field Y component on ion and neutral motions in the polar thermosphere, Geophys. Res. Lett., 11, 935-938.
39. REES D. T.J. FULLER-ROWELL, M.F. SMITH, R. GORDON, T.L. KILLEEN, P.B. HAYS, N.W. SPENCER, L.E. WHARTON and N.C. MAYNARD, (1985), The Westward Thermospheric Jet Stream of the evening auroral oval. Planet Space Sci., 33, 425-456.

40. HEDIN A.E., (1987), MSIS-86 thermospheric model, J. Geophys. Res. 92, 4649.

41. REES D., T.J. FULLER-ROWELL, Invited paper presented at the Chapman Conference on Auroral Physics, Cambridge, June 1988, to be published in the proceedings.

42. KILLEEN T.L., B. NARDI, F.G. MCCORMAC, J.W. MERIWETHER Jnr., J.P THAYER., R.G. ROBLE, T.J. FULLER-ROWELL and D. REES, (1989), Lower Thermospheric Structure and Dynamics Inferred from Satellite and Ground-based Fabry-Perot Observations of the O(1S) Green line Emission. Adv. Space Res. (in press).

43. JOHNSON R.M., V.B. WICKWAR, R.G. ROBLE and J.G. LUHMANN, Lower thermospheric winds at high latitudes: Chatanika radar Observations. (1987), Annal. Geophsyics., 5, 383-404.

44. NYGREN, T., L. JALONEN, J. OSKMAN and T. TURINEN, The role of electric field and neutral wind direction in the formation of sporadic-E layers, (1984), J. Atmos. Terr. Phys., 46, 373.

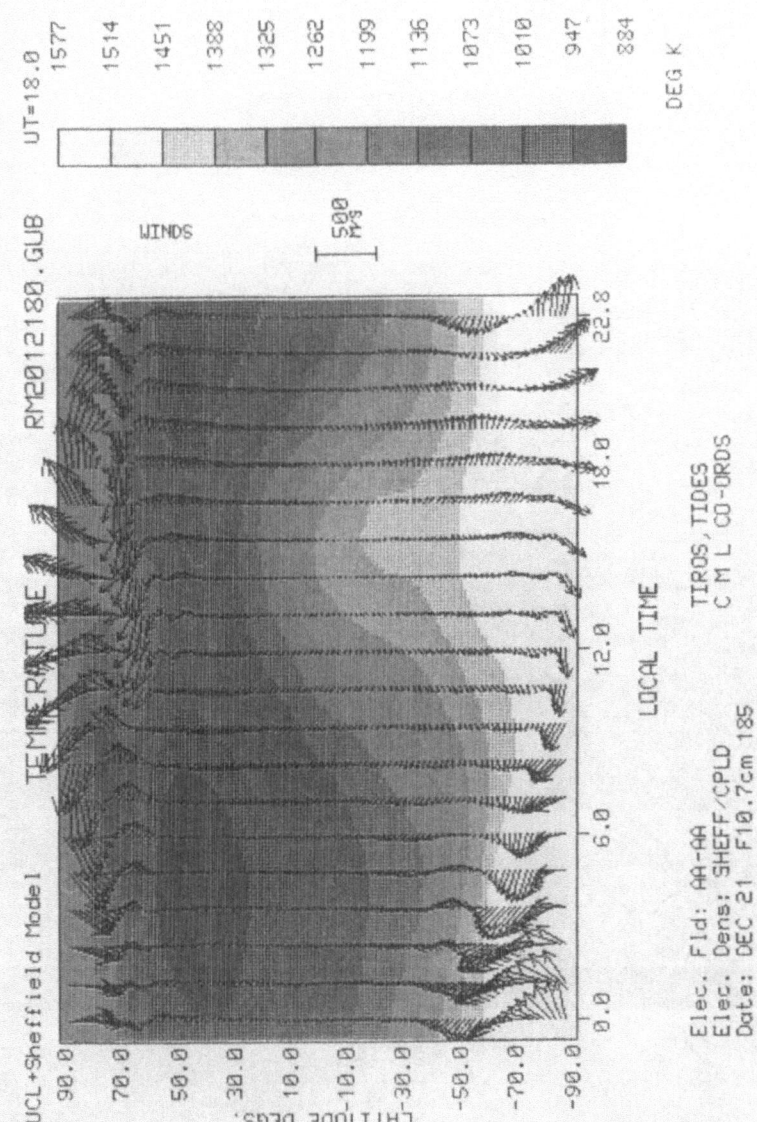

Figure 1. Global distributions of upper thermospheric temperature and wind velocity, pressure level 12, computed for the June solstice, by the Coupled Ionosphere-thermosphere Model, using the Heppner and Maynard (1987) Model 'A' convection field, High solar activity, $F_{10.7} = 185$; Moderate geomagnetic activity, Kp = 3; lower atmosphere propagating semi-diurnal tides.

382

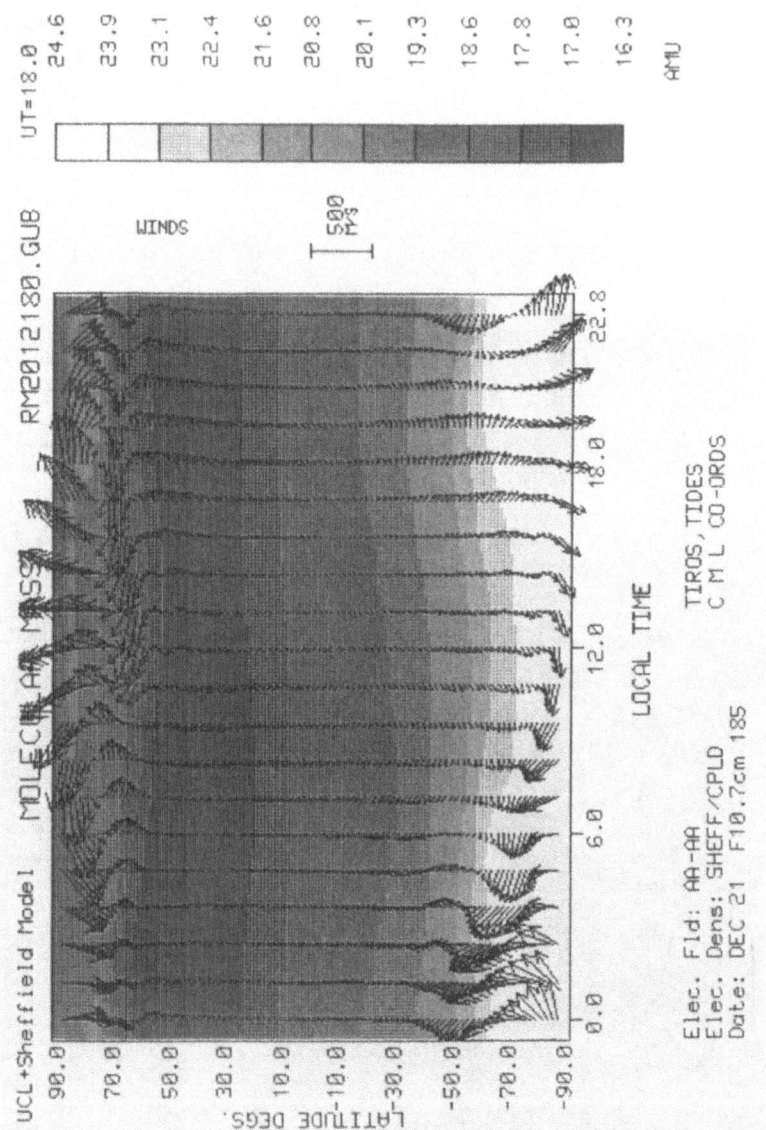

Figure 2. As for Figure 1, except showing the mean molecular mass distribution.

383

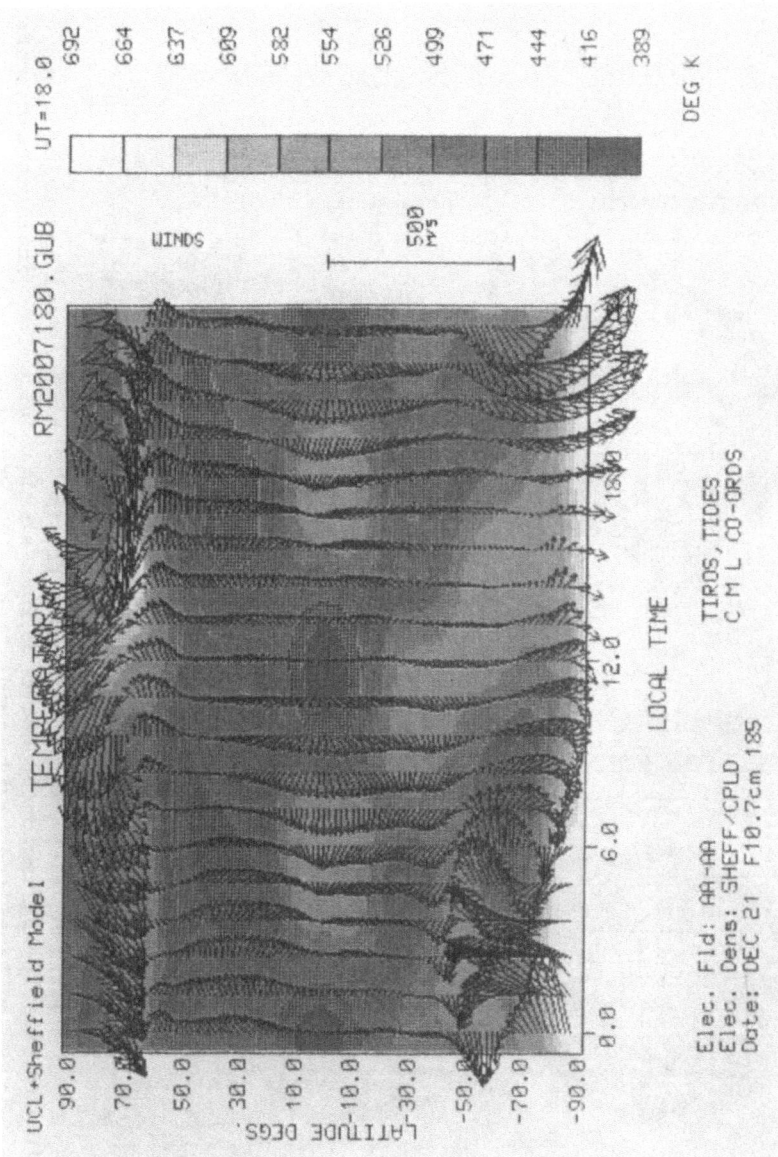

Figure 3. Same as Figure 1, showing the global E-Region (125km) response.

384

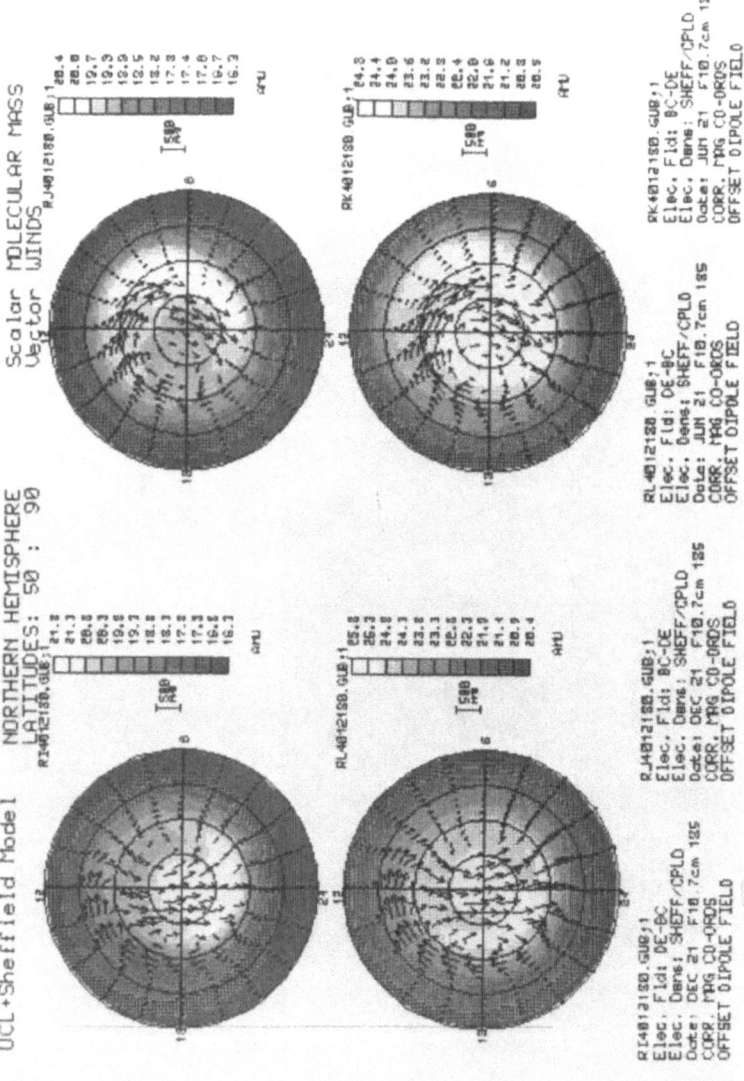

Figure 4. A comparison, at F-Region altitudes (Pr Lv 12) of neutral composition and winds in the North polar region produced in four simulations. The simulations use the same solar and geomagnetic activity conditions as in Figure 1, but the DE/BC convection fields are used to simulate extreme asymmetric values of the IMF BY component, and also contrast northern summer and winter. The simulations include lower atmosphere semi-diurnal tides.

385

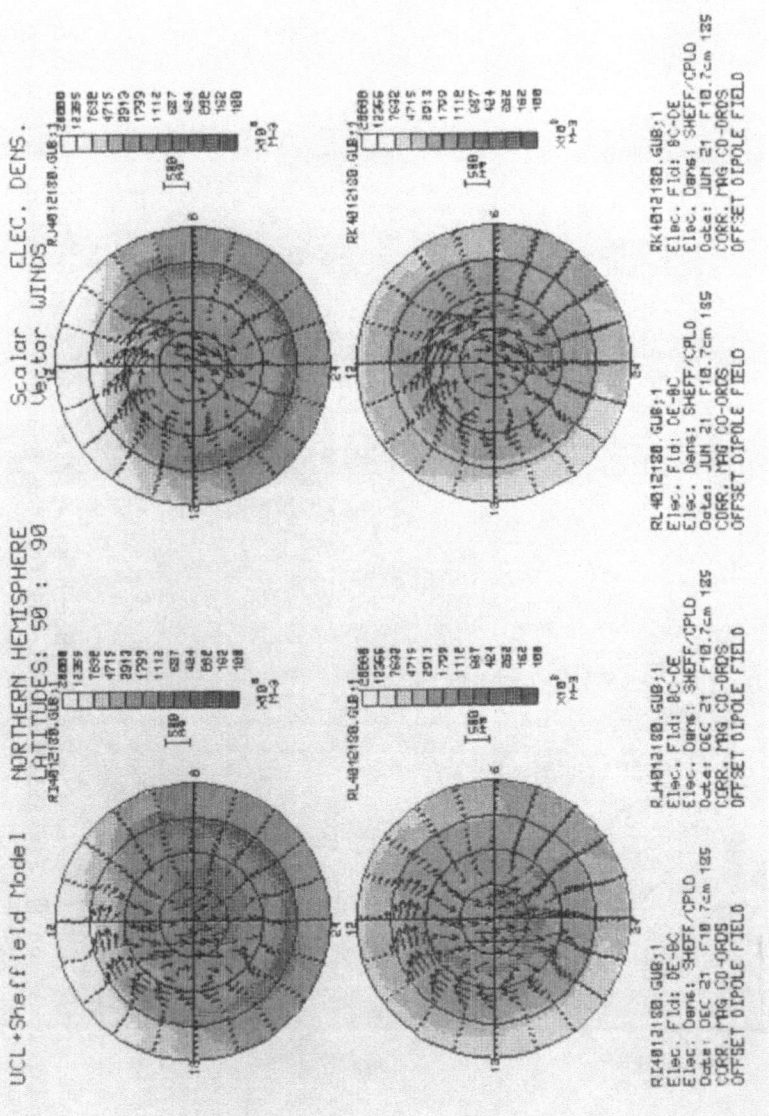

Figure 5. A comparison, for the same four sets of seasonal and IMF BY component conditions as shown in Figure 4, of the corresponding plasma density distributions at F-region altitudes (301 km). There is a contrast between plumes, carried by convection, in the winter simulations, and polar 'holes', corresponding to very high values of neutral mean molecular mass, for the summer simulations.

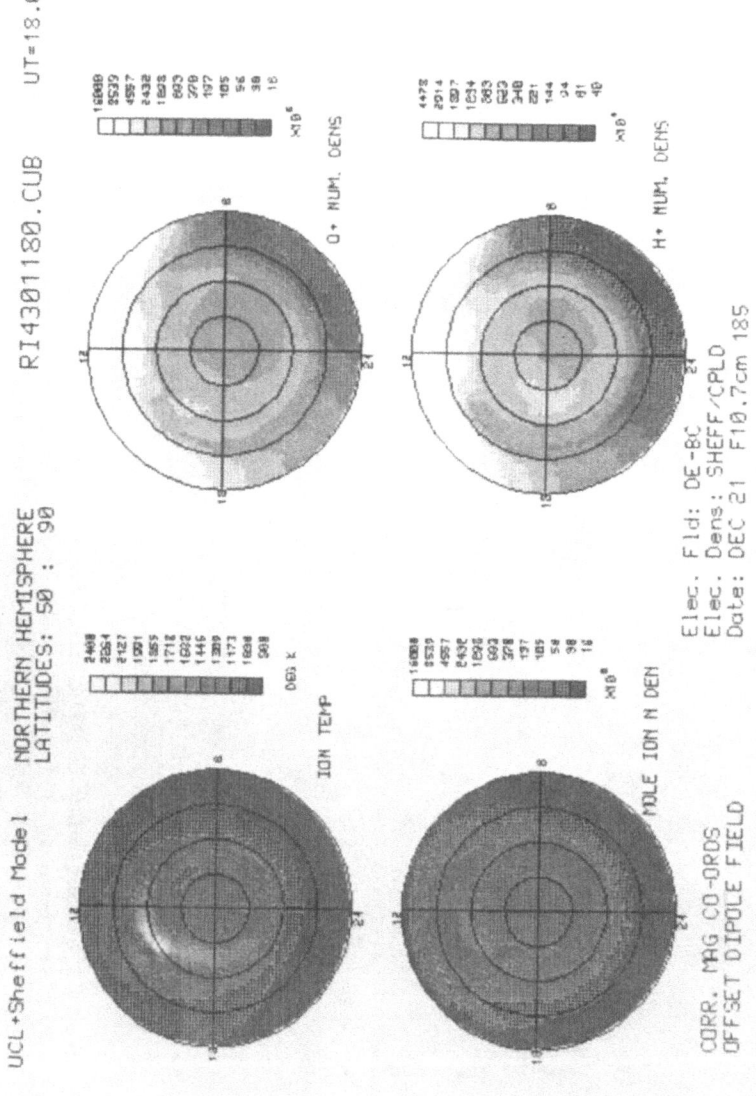

Figure 6. The four sections show the North Polar distributions of Ion Temperature, O^+, Molecular Ions and H^+ densities at 301 km, for the December simulation, IMF BY -ve, (otherwise as Figure 1)

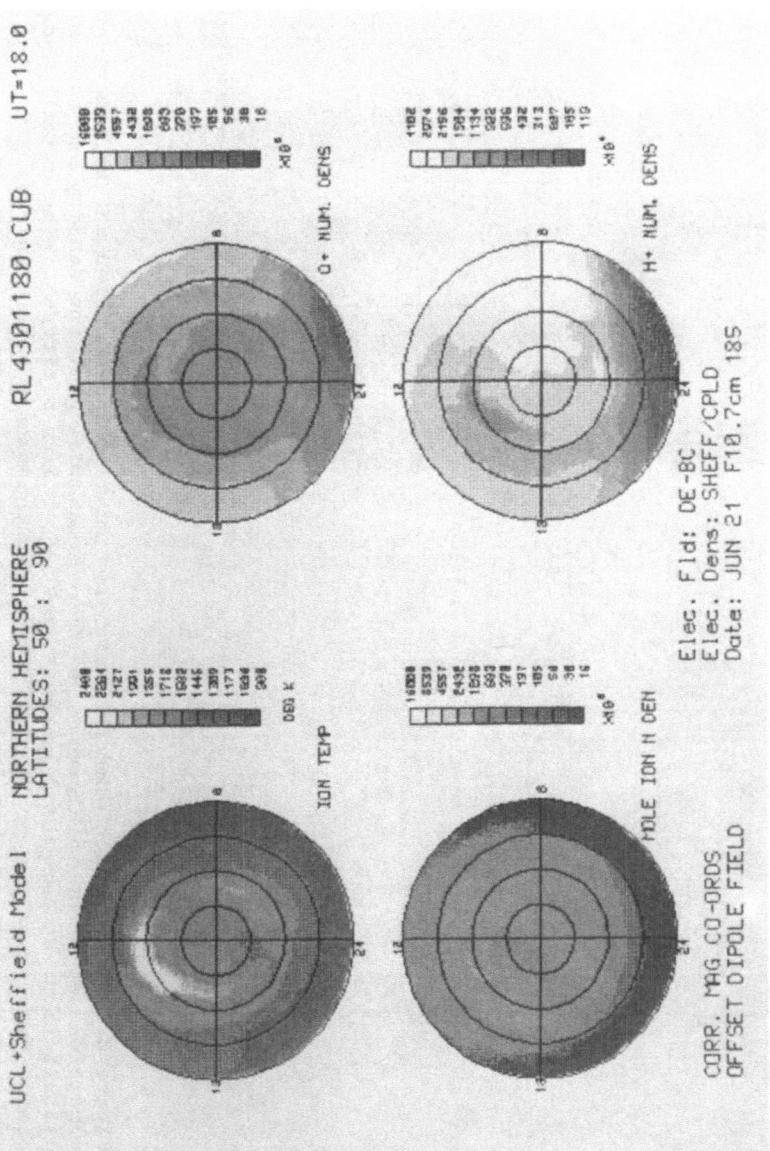

Figure 7. The four sections show the North Polar distributions of Ion Temperature, O⁺, Molecular Ions and H⁺ densities at 301 km, for the June simulation, IMF BY -ve, (otherwise as Figure 1)

388

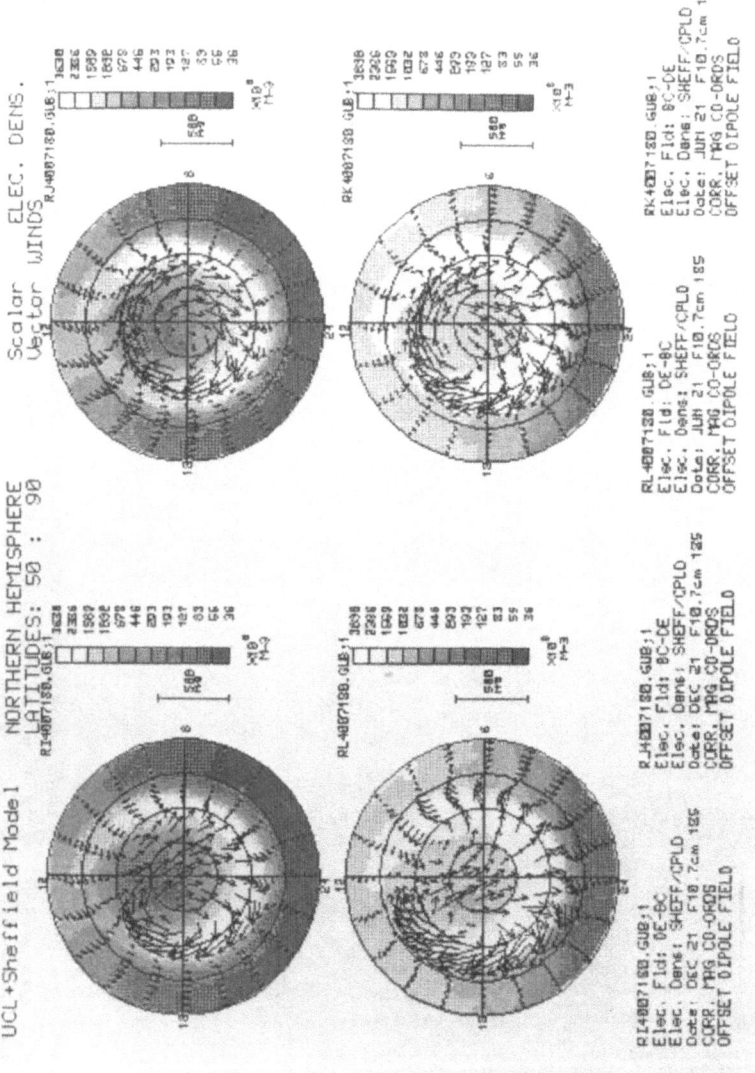

Figure 8. A comparison, for the same four sets of seasonal and IMF BY component conditions as shown in Figure 4, of the corresponding plasma density and neutral wind distributions at E-region (pressure level 7, 125 km altitude). The overall and peak plasma density values are higher throughout the summer polar regions, given the combination of solar photoionisation and energetic electron ionisation.

389

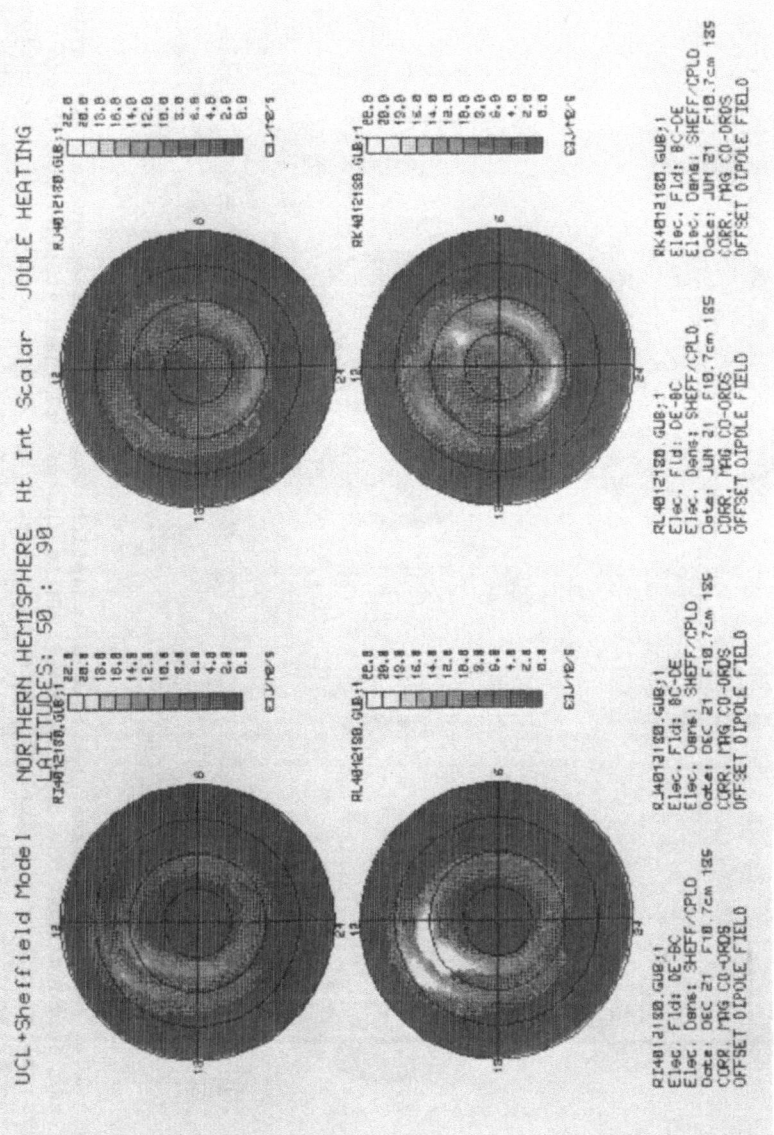

Figure 9. The Height-Integrated Joule Heating rates are shown for the four cases described in Figures 4 - 8. These calculations include the effect of the induced neutral wind dynamo. Sum... ...mf values are generally considerably higher, and there is a marked changes in t... ...ns as the IMF BY component changes sense.

390

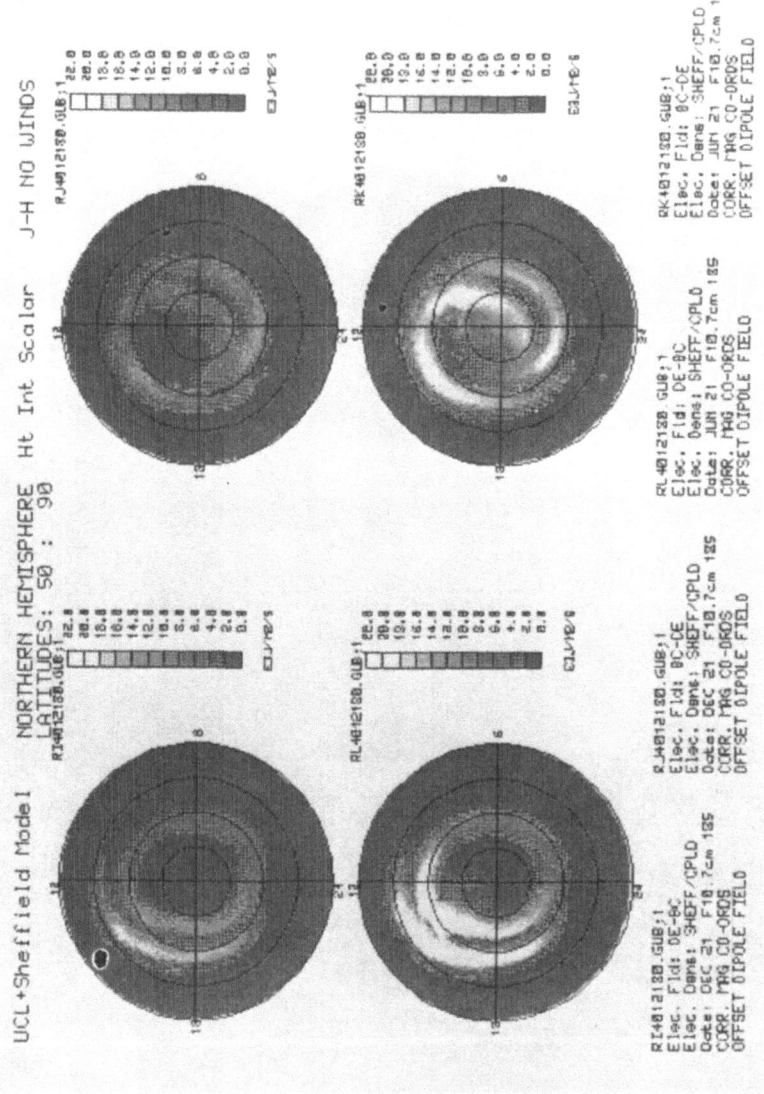

Figure 10. Same as Figure 9, with the exception that the induced neutral wind dynamo has been dropped from the calculation. While the overall patterns do not change dramatically, the peak values increase by a factor of 50 %, and in several locations, the emphasis of the highest values shifts from polar cap to auroral oval, or the converse.

391

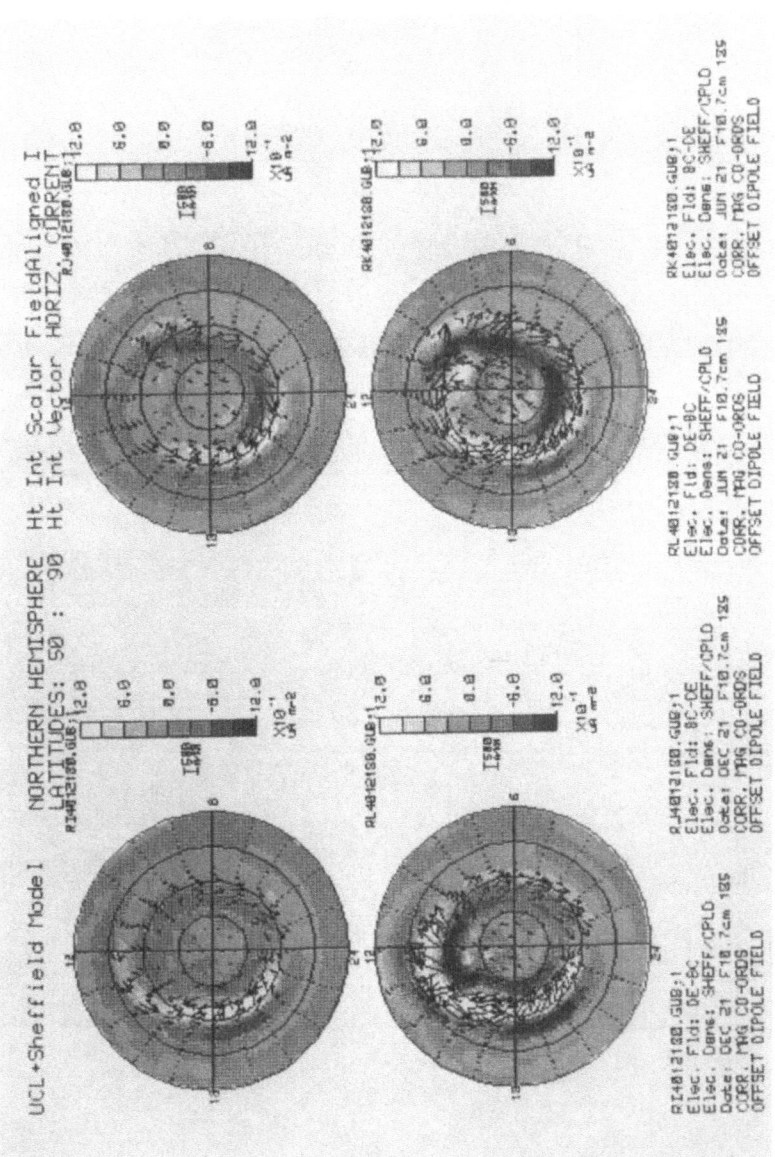

Figure 11. For the same simulations as shown in Figures 4 - 8, the field-aligned currents are displayed, calculated from horizontal current convergence or divergence. The major changes occur in the vicinity of the polar cusp, as a result of dependence on the sense of IMF BY, however, the dawn auroral oval and the region of the Harang discontinuity also show considerable BY-dependent changes.

F-REGION STORMS AND THERMOSPHERIC CIRCULATION

H. Rishbeth
Rutherford Appleton Laboratory,
Chilton, Didcot OX11 0QX, U.K
& Dept of Physics, University,
Southampton SO9 5NH, U.K.

ABSTRACT: This paper outlines current ideas on mechanisms of storm effects in the ionospheric F-layer at mid-latitudes, and of how they depend on the thermospheric circulation. The marked depletions of F2-layer electron density are probably due to changes of neutral composition, which are known to exist but are difficult to explain in terms of energy inputs at auroral latitudes. A promising idea is that the composition changes are caused locally by mid-latitude energy inputs during storms, possibly by precipitation from the ring current.

1. Introduction

Ionospheric storms, though discovered 50 years ago [1], still present many scientific and practical problems. Even in these days of satellite and cable communications, the disruption by storms of radio communications may be serious. The "storm phenomenon" involves many parts of the solar-terrestrial environment: this paper focuses on the response of the thermosphere and F2-layer to the input of energy at high latitudes by electric fields and energetic particles.

The ionospheric F-region, which extends upwards from about 150 km, contains two main layers of ionization: the F1-layer, situated at about 170 km where solar EUV radiation is most strongly absorbed; and the denser, more important and more capricious F2-layer with its peak at 240-400 km. The radio propagation characteristics of the F2-layer depend largely on the peak electron density NmF2 (or the critical frequency foF2) and the peak height hmF2; the total column electron content (TEC) is another important parameter.

Although a first-order theory of the quiet F2-layer was achieved about 30 years ago [Annex A.1], there was no real understanding of F-layer storms. The idea that storm effects are connected with composition changes in the neutral atmosphere originated in the fifties [2], and was followed in the early sixties by a similar hypothesis about the F2-layer seasonal variations [3]. It was only in the late sixties that direct experimental evidence of neutral composition changes began to appear [4]. Several more years elapsed before theory could begin to account for them. Our understanding has been transformed by the huge increase in experimental and theoretical knowledge of the neutral thermosphere. F-layer storm effects, examples of which are shown in Fig. 1, are now at least partially explained, though many problems remain.

P. E. Sandholt and A. Egeland (eds.), Electromagnetic Coupling in the Polar Clefts and Caps, 393–406.

394

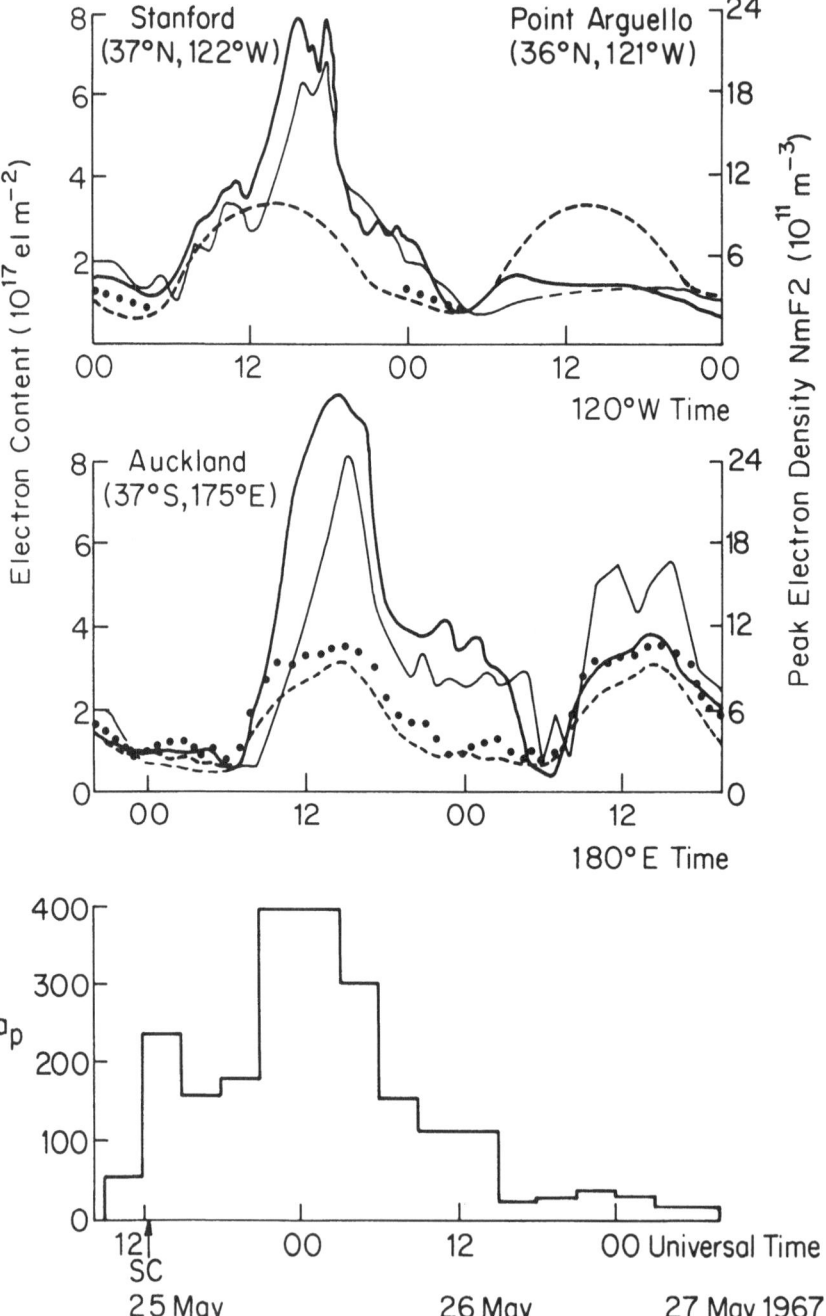

Fig. 1. Total electron content (thick line, monthly mean as ---) and F2 peak electron density (thin line, monthly median as dots, omitted if coincident with ---), storm of 25-27 May 1967 (northern summer, southern winter). Courtesy Rutherford Appleton Laboratory, UK; TEC data from K L Jones, NmF2 from WDC-A (Boulder USA) and WDC-C1 (Chilton UK)

This paper describes results and ideas that have been studied and developed in a collaboration involving the University of Southampton, University College London, the British Antarctic Survey, and the Royal Aerospace Establishment. Some aspects of the underlying theory are reviewed in the Annex [A.1-A.4].

2. Theory of The Thermosphere

The theoretical description of the upper atmosphere is based on the laws of conservation of mass (the continuity equation), momentum (the equation of motion) and energy (the heat-balance equation), applied to each charged or neutral constituent; the equations of state; and the laws of electromagnetism, for charged particles. All these laws are incorporated in the numerical models of the thermosphere, such as the NCAR TGCM [5] and the UCL 3DTD [6], that were developed in the seventies and are still being refined and extended in scope. Aspects of thermospheric structure and dynamics that bear on ionospheric storms include the idea of "constant pressure-levels"; the role of vertical and horizontal winds; how these winds alter the chemical composition of the neutral gas, and consequently affect the F2-layer electron density.

Fig. 2 illustrates the longitude-averaged circulation, as given by numerical modelling [7]. In magnetically quiet conditions, the circulation is mainly driven by solar EUV energy, being symmetrical about the Equator at equinox but with a summer-to-winter bias at the solstices. The high-latitude circulation cells, driven by the magnetospheric energy input, are weak in quiet conditions but are strongly enhanced during magnetic disturbances. Typical horizontal and vertical wind speeds are 100 m s^{-1} and 3 m s^{-1}, though greater vertical winds exist where there is strong heating.

Recent work [8] has given a better understanding of how composition changes are produced by heat inputs, which cause "up-welling" and drive horizontal winds that carry away the energy absorbed by the up-welling. This energy is released by the compressional heating due to the "down-welling" that occurs at a distance from the input. The up-welling, horizontal winds and down-welling form a three-dimensional circulation whose effects on composition - specifically the $[O/N_2]$ ratio that controls NmF2 [A.2] - can be summarized thus:

Close to heat input -> up-welling   -> decreased $[O/N_2]$, NmF2
Far from heat input -> down-welling -> increased $[O/N_2]$, NmF2

To understand the physics of composition changes, one should consider the changes at fixed pressure-levels, as well as at fixed heights [A.3]. It is necessary to separate the effects of simple thermal expansion and contraction, which is one-dimensional (i.e. involving only vertical motion) and does not change the composition at any fixed pressure-level, from those of the up-welling and down-welling that forms part of a three-dimensional circulation [A.4].

3. F-Region Storms

3.1 THE MAIN FEATURES

The "classical" description of F2-layer storms at mid-latitudes is based mainly on statistical studies [9,10], though some individual storms do conform to this pattern.

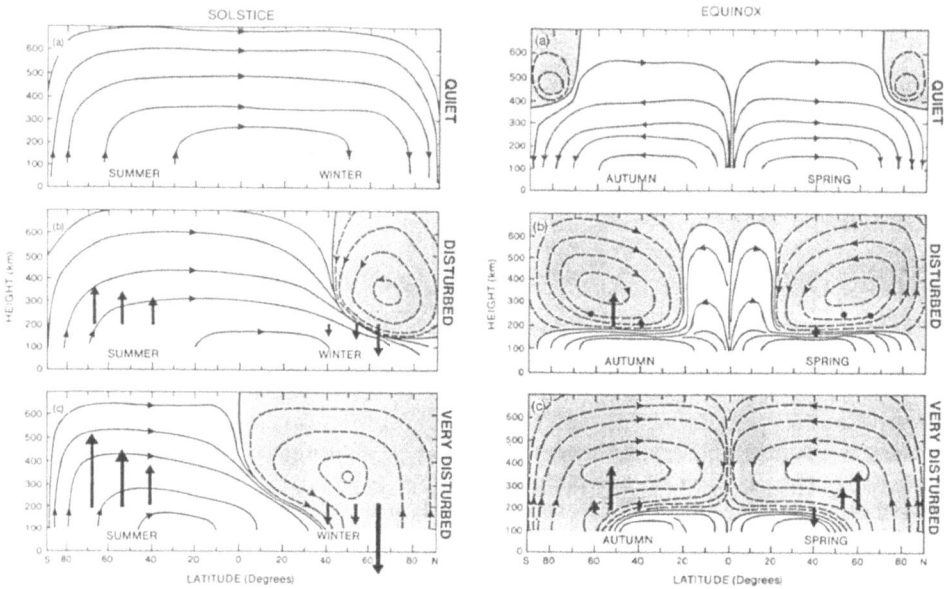

Fig. 2. Schematic longitude-averaged circulation patterns (after R G Roble and R E Dickinson). The superimposed arrows show regions of upwelling and downwelling, deduced from the amplitude of the DC storm effect (section 3.2) at Halley Bay, Argentine Is and Port Stanley (geomagnetic latitudes -66°, -54°, -40°). Courtesy British Antarctic Survey.

Following the geomagnetic SSC (storm sudden commencement), a "positive" phase sets in, in which both NmF2 and TEC are increased (Fig. 1), especially in the evening hours.

After some hours the "main phase" sets in. It frequently brings a severe depletion of NmF2 and TEC, i.e. a "negative storm", which may be regarded as the main feature of F2-layer storms. Sometimes NmF2 is enhanced in the main phase: these "positive storms" principally occur in low latitudes, or in winter at mid-latitudes. The "recovery phase" generally lasts about two days. F2-layer storms show many other features that require explanation: longitude variations, hemispheric differences, influences of the UT or LT at which the storm starts, and so on.

3.2 NMF2 PATTERNS IN THE ANTARCTIC

New ideas about F2-layer storm phenomena have come from studying NmF2 and hmF2 data from the South Atlantic region [11]. One of the many problems of storm studies is to devise a proper index for measuring the effect of the storm, given the variability of the "normal" F2-layer. For this purpose, the parameter used is ln (N/No), N being the observed value of NmF2 and No the 30-day running mean at the same local time (LT), centred on that particular day. The magnetic disturbance is characterized by a parameter ap(τ), which is the geomagnetic index ap smoothed over about 12 hours prior to the time of observation [12]. The 30-day averaging seems to match well the characteristic time-scale of F2-layer seasonal variations, while the 12-hour smoothing matches the time-scale of the F2-layer storm phenomena.

The analysis shows that the LT variation of ln (N/No) follows a seasonal pattern of the kind shown in Fig. 3, in which the "negative" and "positive" effects form part of an overall pattern. This pattern comprises a local-time "AC" variation superimposed on a seasonally-varying "DC" variation [13]. The amplitude of the DC and AC effects, increases with ap(τ). Possible interpretations are discussed later.

3.3 STORM BEHAVIOUR OF HMF2

The storm behaviour of hmF2 is less well documented than that of NmF2, because of the difficulty of obtaining reliable heights from ionograms. Ionograms often show huge increases of F2-layer virtual height during storms, but they are mainly due to changes in shape of the electron density N(h) profile; the real changes of height are actually much smaller. As the F2 peak tends to follow a fixed pressure-level [14], hmF2 should increase during storms, simply because of the heating and consequent thermal expansion of the neutral gas. For the Antarctic stations, a preliminary study suggests that hmF2 does generally increase slightly during storms, but this is largely due to thermal expansion; the corresponding "reduced height" zmF2 [A.3] does not change much.

4. Possible Explanations

4.1 EXPLANATIONS OF THE DC EFFECT

The best current explanation of the DC storm effect seems to be the long-standing "composition change" theory [2]. Regions of reduced [O/N$_2$] ratio, as detected by satellite-borne instruments, have been shown to correspond to depletions of NmF2 [4]. The reduced thermospheric [O/N$_2$] ratio during storms is sufficiently well described to be embodied in empirical models such as MSIS [15], and accounts (at least qualitatively)

398

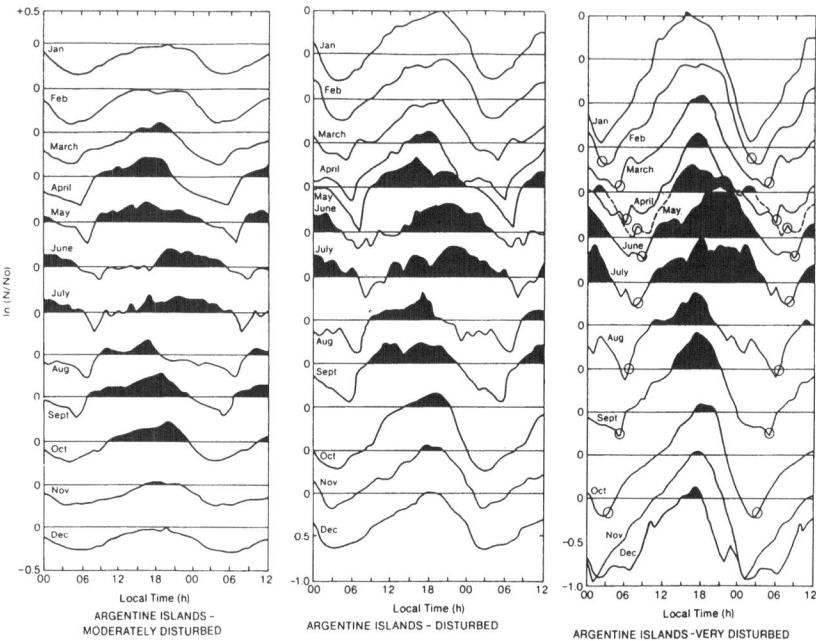

Fig. 3. Local-time variation of ln (N/No) at Argentine Is, 1971-1981, arranged by month, for "moderately disturbed", "disturbed" and "very disturbed" geomagnetic conditions. The period 00-12 LT is repeated to display more clearly the form of the "AC" variation (section 3.2). The black areas, where ln (N/No) > 0, display the seasonal "DC" effect. The circles show the time of sunrise on the 15th of each month. Courtesy British Antarctic Survey.

for the "negative" storm effects, through equation (A1). The difficulties that remain are largely theoretical, but are none the less puzzling.

Important questions about the composition changes are: (i) Are they actually a depletion of [O]; or an increase of $[N_2]$; or a combination of both? (ii) Since they are generally described in terms of changes at given heights, are they essentially the same in pressure coordinates (reduced height)? If so, they could just be due to thermal expansion, which does not affect NmF2 very much [A.1, A.3]. Further investigation shows that the $[O/N_2]$ ratio does actually decrease at fixed pressure-levels [16], so simple thermal expansion is not the only cause, but the answer to (i) is not yet clear.

4.2 MECHANISMS FOR COMPOSITION CHANGES AT MID-LATITUDES

Since the most spectacular energy inputs during storms (particle bombardment and Joule heating) occur at high latitudes [17], they have often been regarded as the probable cause of the mid-latitude storm effects. Several suggestions have been made as to how the $[O/N_2]$ ratio at mid-latitudes can be affected by energy inputs 1000-3000 km away. They include "local" mechanisms, by which the atmosphere is locally disturbed by waves of some kind emanating from the high-latitude energy source; and "convection" mechanisms, by which gas of different chemical composition is transported from higher latitudes (Fig. 4).

None of these mechanisms seems effective. The crux of the matter is that the decreases of $[O/N_2]$ ratio imply local up-welling and a strong local energy input (Sec. 2). For the "local" mechanism illustrated in Fig. 4, the energy must presumably be transported by waves of some kind [18]. This leads to a contradiction; the waves must have low attenuation to travel long distances from the auroral oval, and yet suffer heavy absorption in order to give up the required energy. Furthermore, gravity waves (although certainly generated in auroral disturbances and known to travel long distances) do not seem to perturb the neutral gas composition to any great extent.

The "convection" mechanism might seem more promising, because an auroral-driven "storm circulation" system does exist, and up-welling and composition changes do take place in the auroral oval [19-22]. But modelling shows that these composition changes are localized, and are not transmitted far from the auroral oval [23]. The reason is that, once the gas flows away from the source of heating and up-welling, its composition is quickly restored to normal by down-welling or molecular diffusion. It is therefore necessary to look for direct energy input at mid-latitudes. To illustrate the problem, the locations of up-welling and down-welling that might account for the storm patterns at the South Atlantic stations (Section 3.2) have been marked on Fig. 2.

An important point in favour of the circulation model is that it does seem to account for the "positive" main-phase effects on NmF2, often seen in low latitudes and in the winter hemisphere, at the same time as the "negative" effects in the summer hemisphere. The "down-welling" of the circulation increases the $[O/N_2]$ ratio and thus provides a possible explanation of the increases of NmF2 [23].

4.3 A MID-LATITUDE ENERGY INPUT?

The required source must be effective over the wide latitude range in which negative F2-layer effects are seen. It has to deposit enough energy (perhaps 1 mW m^{-2}) to cause composition changes (say equivalent to $\Delta M \sim 1$) at the F2 peak. To be most effective, the input must probably peak at 120-150 km.

One possibility is that auroral inputs extend to lower latitudes than the conventionally-defined auroral oval. Another possibility might be the precipitation

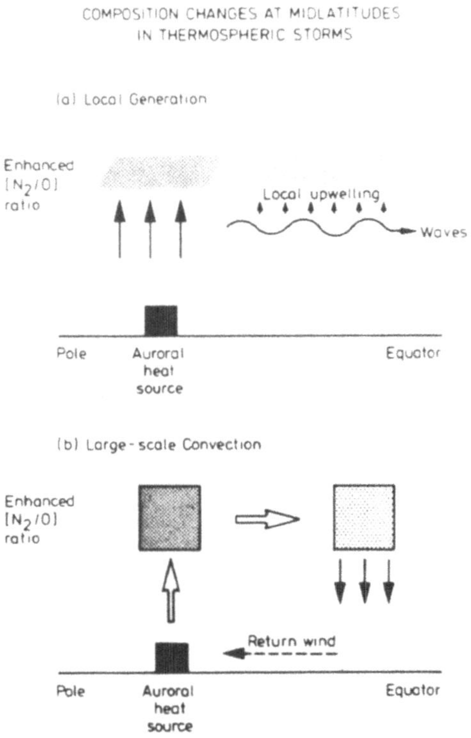

Fig. 4. Sketch to illustrate possible "local" and "remote" mechanisms by which mid-latitude enhancements of the $[N_2/O]$ ratio might be caused by heat input in the auroral oval. (After G.W.Proelss; Rishbeth et al., Physica Scripta 1987)

responsible for stable red arcs. Red arcs are indeed accompanied by depletions and irregularities of F2-layer electron density [24,25], but seem to be more restricted in extent and occurrence than negative storm effects. Another possibility is joule heating caused by the storm electric fields.

At present, the most promising idea is that the mid-latitude energy source is due to precipitation from the Dst ring current [26]; protons of about 20-80 keV seem to be the most likely candidates. This source seems to have the appropriate time-scale and latitude distribution. Its initial modelling is promising [27].

4.4 OTHER FACTORS CONTRIBUTING TO F2-LAYER STORM EFFECTS

(a)   Constituents that destroy O^+ ions (other than N_2) might be produced in situ, or introduced by transport from below or from high latitudes, e.g. vibrationally excited N_2, metastable O_2 or NO. Only the first seems promising.

(b)   Increases of the rate coefficients of O^+ loss reactions due to increases of the neutral gas or electron temperature may make some contribution to negative storm effects.

(c)   Winds and electric fields (although the storm-driven equatorward winds are in the wrong direction to reduce NmF2 at mid-latitudes). Electric fields probably cause only transient changes of hmF2 and NmF2 at mid-latitudes, because of the "ion-drag" effect [29]. However, neutral air winds and electric fields are likely causes of the initial "positive" storm effects mentioned in Sec. 3.1 [30,31].

4.5 EXPLANATIONS OF THE AC EFFECT [13]

(d)   Horizontal thermospheric winds are the most obvious possibility, as they can produce complicated LT and UT patterns of vertical drift, which are considerably affected by the storm changes in thermospheric pressure.

(e)   Composition effects may contribute; the thermospheric circulation (that is thought to account for the DC effect) is three-dimensional, and up-welling and down-welling varies with LT as well as with latitude.

(f)   Chemical rate coefficients may be affected by the changes in neutral and electron temperatures.

(g)   Electromagnetic drifts due to storm electric fields may have some effect, probably not a major one at mid-latitudes.

(h)   The LT variation of the plasma flow between the F2 layer and the protonosphere may change during storms.

5. Conclusion : Key Questions

The last few years have seen advances in the understanding of F2-layer storms. The analysis of storm phenomena in the South Atlantic has led to new ideas about the nature of the storm patterns, and has pointed to promising ideas about their physical causes.

This paper has described the difficulty in explaining the origin of the composition changes (increases of $[O/N_2]$ ratio) that are thought to cause the mid-latitude depletions of NmF2. The alternatives to the composition theory (Section 4.4) do not seem promising, and in any case they do not overcome the basic dilemma: the required composition changes exist, but theory cannot explain how they come about. So it is imperative to confirm that the modelling is correct, and does not contain some basic flaw. Refinements of the UCL model, such as the inclusion of NO cooling [32], have greatly improved the matching to observations of the time-scale of the temperature effects at mid-latitudes; but do not change the conclusion about the limited spatial extent of the composition changes.

As discussed in Sections 4.2 and 4.3, the dilemma could be resolved by establishing that a substantial heat input exists at mid-latitudes during storms. This input could be due to precipitation from the ring-current, probably of energetic protons. Another, but less likely, possibility is joule heating by electric fields.

A practical question is whether ionospheric storms could be forecasted with good real-time monitoring of the solar-terrestrial environment. Even if advance forecasts are unreliable, it would still be useful to detect promptly the onset of a storm and predict its subsequent course. The ideas discussed in this paper might help towards that end.

ACKNOWLEDGEMENTS: This paper describes results obtained from a cooperative project involving T J Fuller-Rowell, D Rees, A S Rodger and G L Wrenn. The work at Southampton was supported by the U K Natural Environmental Research Council through a Special Topic Grant for Antarctic Research, and the computational work at University College, London, by the U.K. Science & Engineering Research Council.

References

[1] Appleton E V & Ingram L J, Nature 136, 548 (1935).
[2] Seaton M J, J. Atmos. Terr. Phys. 8, 122 (1956).
[3] Rishbeth H & Setty C S G K, J. Atmos. Terr. Phys.20, 263 (1956).
[4] Proelss G W, Rev. Geophys. Space Phys. 18, 183 (1980).
[5] Fuller-Rowell T J & Rees D, J. Atmos. Sci. 37, 2545 (1980).
[6] Dickinson R E, Ridley E C & Roble R G, J. Geophys. Res. 86, 1499 (1981).
[7] Fuller-Rowell T J & Rees D, Planet. Space Sci. 31, 1209 (1983).
[8] Rishbeth H, Fuller-Rowell T J & Rees D, Planet. Space Sci. 35, 1157 (1987).
[9] Martyn D F, Nature 171, 14 (1953).
[10] Matsushita S, J. Geophys. Res. 64, 305 (1959).
[11] Wrenn G L, Rodger A S & Rishbeth H, J. Atmos. Terr. Phys. 49, 901 (1987).
[12] Wrenn G L, J. Geophys. Res. 92, 10125 (1987).
[13] Rodger A S, Wrenn G L & Rishbeth H, J. Atmos. Terr. Phys. 51 (1989).
[14] Garriott O K & Rishbeth H, Planet. Space Sci. 11, 587 (1963).
[15] Hedin A E, J. Geophys. Res. 88, 10170 (1983).
[16] Proelss G W, Planet. Space Sci. 35, 807 (1987).
[17] Cole K D, Planet. Space Sci. 19, 59 (1961).
[18] King G A M, J. Atmos. Terr. Phys. 28, 957 (1966).
[19] Duncan R A, J. Atmos. Terr. Phys. 31, 59 (1956).
[20] Mayr H G & Volland H, Planet. Space Sci.20, 379 (1972)
[21] Obayashi T & Matuura N, Solar-Terrestrial Physics, Part IV p.199, Reidel, Dordrecht (1972).

[22] Rishbeth H, J. Atmos. Terr. Phys. 37, 1055 (1975).

[23] Rishbeth H, Gordon R, Rees D & Fuller-Rowell T J, Planet. Space Sci. 33, 1283 (1985).

[24] Rodger A S, J. Atmos. Terr. Phys. 46, 335 (1984).

[25] Rodger A S & Aarons J, Ann. Geophys. 6 (1988).

[26] Tinsley B A, Rohrbaugh R, Rassoul H, Sahai Y, Teixeira N R & Slater D, J. Geophys. Res. 91, 11257 (1986).

[27] Fuller-Rowell T J, Rees D, Tinsley B A, Rishbeth H, Rodger A S & Quegan S, Adv. Space Res. 9 (1989).

[28] Rishbeth H, Fuller-Rowell T J & Rodger A S, Physica Scripta 36, 327 (1987).

[29] Dougherty J P, J. Atmos. Terr. Phys. 20, 167 (1961).

[30] Jones K L & Rishbeth H, J. Atmos. Terr. Phys. 33, 391 (1971).

[31] Tanaka T & Hirao K, J. Atmos. Terr. Phys. 35, 1443 (1973).

[32] Kockarts G, Geophys. Res. Lett. 7, 137 (1980).

[33] Rishbeth H, J. Atmos. Terr. Phys. 48, 511 (1986).

[34] Richards P J & Torr D G, J. Geophys. Res. (19).

[35] Dickinson R E & Geisler J E, Monthly Weather Rev. 96, 606 (1968).

[36] Rishbeth H, Moffett R J & Bailey G J, J. Atmos. Terr. Phys. 29, 1035 (1969).

A. Annex

A.1 CONTROLLING PRINCIPLES OF THE F2-LAYER [33]

(a)   F2-layer ions are mainly O^+, formed by photoionization of atomic oxygen. The F2-layer is well above the level where the production rate q is greatest, which lies in the F1-layer (except at large solar zenith angles), so the F2-layer is optically thin to the ionizing radiation.

(b)   The O^+ ions are destroyed by a two-stage process that involves reactions of O^+ ions with neutral N_2 or O_2; the loss coefficient b is linear in the electron density.

(c)   The F2 peak occurs at the level where the photochemical processes and transport processes (especially plasma diffusion) are of equal importance. Up to this peak, the daytime electron density approximates to the photochemical ratio q/β. Well above the peak, the electron distribution is controlled by diffusion.

(d)   The height hmF2 is altered if a horizontal wind drives the plasma along geomagnetic field lines (upward to a level of smaller β by an equatorward wind, as generally by night; downward to a level of greater β by a poleward wind, as generally by day).

(e)   Plasma exchange between the F2-layer and protonosphere may influence the daily variations of the F2-layer.

The above processes are probably the main ones that control the mid-latitude F2 layer, where electromagnetic drifts due to electric fields seem to be of minor importance. At high latitudes, principally in the auroral oval, particle ionization may be important, while the F2-layer in the polar cap is controlled largely by plasma convection driven by the magnetospheric electric field.

Detailed modelling based on (a-d) accounts for many features of the quiet-day variations of NmF2 and hmF2 at mid-latitudes. From (c) it may be deduced that the F2 peak tends to form at a constant pressure level, i.e. a constant "reduced height" z as defined in [A.3] [14,33]. Changes of thermospheric temperature with season, solar cycle or magnetic disturbance affect the real height of the peak (hmF2), but not its "reduced height" (zmF2), nor do they have much effect on NmF2. However, thermospheric winds, acting as in (d), cause marked local-time variations of zmF2, and hence affect NmF2 also.

A.2 COMPOSITION AND CIRCULATION EFFECTS IN THE F2-LAYER

The principles (a-c) of [A.1] imply that NmF2 depends strongly on chemical composition. Assuming for simplicity that N_2 makes the major contribution to the loss coefficient β (though O_2 may also be quite important):

$$NmF2 \sim q/\beta \sim [O]/[N_2] \qquad (A1)$$

(where the brackets [ ] represent gas concentration, and all quantities are evaluated at the F2 peak). Hence NmF2 is strongly controlled by the atomic/molecular ratio of the neutral gas. (A more complete analysis, taking proper account of how hmF2 is controlled by the plasma diffusion coefficient which mainly depends on [O], leads instead to the relation $NmF2 \sim [O]^2/[N_2]$, but this does not affect the subsequent arguments in substance). Considering for simplicity a two-gas thermosphere, (A1) gives a relation between changes of NmF2 and of mean molar mass M. If M is in the range 18-22, as usually applies at the F2 peak, then:

$$[O]/[N_2] = (28 - M)/(M - 16) \qquad (A2)$$

$$\Delta(\ln NmF2) = - 0.35 \, \Delta M \qquad (A3)$$

The seasonal variations of NmF2 are linked to the seasonal composition changes, which depend on the circulation patterns shown in Fig.2. Up-welling in the summer hemisphere, driven by the strong and geographically widespread input of solar radiation, decreases the $[O]/[N_2]$ ratio; the corresponding down-welling in the winter hemisphere tends to increase it. The actual pattern of seasonal, annual and semiannual variations in NmF2 is more complicated than could be produced by the idealized picture of Fig. 2, and must be influenced by other factors, such as the dependence of chemical rate coefficients on neutral or electron temperature [34].

A.3 BAROMETRIC AND DIFFUSIVE EQUILIBRIUM.

The atmosphere as a whole is almost precisely in "hydrostatic" or "barometric" equilibrium, i.e. gravity balances the vertical pressure gradient, so the variation of air pressure p with height h is given by:

$$- p \, dh/dp = RT/Mg = H \qquad (A4)$$

(where R = gas constant, T = temperature, M = mean molar mass, g = acceleration due to gravity). H is known as the "atmospheric scale height". Eq. (A4) holds accurately despite the effects of chemical processes, the Earth's rotation, and winds. Any individual gas that conforms to a similar equation, with its proper "scale height", is in "diffusive equilibrium" which, however, can be perturbed by chemical processes or vertical winds. In the F-region, only the major neutral gases are usually in diffusive equilibrium.

In discussing the physics, a "pressure coordinate" is often more useful than ordinary height h; the appropriate parameter is "reduced height" z, related to h by:

$$z(h) = \int (dh/H) = -\ln p \tag{A5}$$

in which the integration extends from the base of the thermosphere (taken as 80 km) up to the height in question. Fortuitously, the constant of integration in (A5) is zero because the pressure at 80 km is almost exactly 1 Pa. Thus at any real height h, z(h) is "the number of scale heights above 80 km", bearing in mind that H varies with height. As the distribution of temperature and composition (and thus of scale height) varies with latitude, longitude and local time, as well as solar and geomagnetic activity, the height h(z) of a given pressure level can vary considerably.

A.4 BAROMETRIC AND DIVERGENCE VELOCITIES

There are two physically different components of the vertical wind velocity: the "barometric" velocity W_B and the "divergence velocity" W_D [35,36]. W_B represents the rise or fall of fixed pressure-levels, in response to thermal expansion or contraction as the temperature changes. Then if h(z) is the height of any fixed pressure level z:

$$W_B = dh(z)/dt \tag{A6}$$

Thermal expansion or contraction does not change the chemical composition at fixed z (provided the composition is unchanged at the lower boundary), because all neutral constituents have the same W_B. The "divergence velocity" W_D represents motion of air relative to the pressure levels. For any given cell of air:

$$W_D = -(dp/dt)/\rho g \tag{A7}$$

where ρ = density. Through the continuity equation for the air, W_D is related to divergence or convergence of the horizontal winds, and thus forms part of the three-dimensional circulation. Heating produces "up-welling" (upward W_D) which absorbs energy that is carried away by the horizontal winds and released as compressional heating in regions of "down-welling" (downward W_D).

The individual gases may have different divergence velocities, and their scale heights are slightly modified by vertical motion. An upward W_D increases the proportion of heavier gases (such as N_2) in the upper thermosphere, and decreases the proportion of lighter gases (such as O). A downward W_D has opposite effects. This is the basis of the relation between vertical motion and composition changes.

The individual gases may have different divergence velocities, and their scale heights are slightly modified by vertical motion. An upward W_D increases the proportion of heavier gases (such as N_2) in the upper thermosphere, and decreases the proportion of lighter gases (such as O). A downward W_D has opposite effects. In theoretical investigations of this process [8], it has proved useful to define a composition parameter P in terms of the partial pressures of O and N_2:

$$P(z) = 16 \ln p(N_2) - 28 \ln p(O) \tag{A8}$$

The P-parameter is related to the mean molar mass M, though not in a very straightforward way. It is easy to see that $P(z)$ is independent of height if the thermosphere is in diffusive equilibrium; it is actually the same everywhere in a simple "static" model with fixed lower boundary conditions, and it is completely unaffected by thermal expansion.

If P does vary with height, there must exist a departure from diffusive equilibrium. Molecular diffusion will act, in opposition to the upwelling or downwelling that causes the perturbation, to restore diffusive equilibrium. It appears (though it has yet to be proved) that changes of P represent the cumulative effect of upwelling or downwelling, according to a relationship of the type:

$$\Delta P \sim \int W_D \, dt \tag{A9}$$

(A9) expresses the idea that composition changes are closely related to vertical motions, and thus to the input and release of energy. From the point of view of storm effects, the important conclusion is that only vertical motions - not horizontal winds - are effective in changing the composition of the neutral air, and thus the F2-layer electron density.

THERMOSPHERIC DYNAMICS IN THE POLAR E- AND F REGION: RESULTS OF A NONLINEAR, SPECTRAL MODEL

I.S. MIKKELSEN
Division of Geophysics
Danish Meteorological Institute
Lyngbyvej 100
DK-2100 Copenhagen Ø Denmark

ABSTRACT. A thermosphere general circulation model based on the spectral technique, and including realistic plasma convection patterns, electron densities, and solar heating is presented. The dominance of the dusk wind-vortex in the high-latitude F region is due to a secondary circulation established by the Joule heating. The divergent winds advect planetary vorticity equatorward from the pole, leading to a strengthening of the dusk vortex. The opposite Coriolis forces at dusk and dawn affect the shapes, but not the magnitudes, of the vortices. The secondary circulation, driven by the Joule heating, closes in the lower E region. The convergent winds advect planetary vorticity poleward, resulting in a strengthening of the dawn wind-vortex. This asymmetry is also enhanced by the night-day electron density gradient which changes the otherwise divergent Hall force to a counterclockwise rotational force.

1. INTRODUCTION

The thermospheric dynamics of the high-latitude E region, especially the polar cap, is still open to debate. Several observations exist using the chemical-trail technique (Heppner and Miller (1982), Mikkelsen et al. (1987)), but a systematic picture, like of the F-region winds, has not emerged.

In the auroral zone the incoherent scatter radar observations have revealed that the upward propagating, semidiurnal tide dominates during quiet geomagnetic conditions (Brekke et al. (1973, 1974), Johnson et al. (1987), Kirkwood (1986)). According to Johnson et al. (1987) enhanced magnetic activity is accompanied by mainly eastward winds at 107 km altitude at all local times. At 115 km altitude the disturbed winds are also eastward between dawn and noon, but between dusk and midnight have a small westward component that may be interpreted as due to ion drag from the sunward convecting plasma.

The main purpose of this paper is to discuss how the electrodynamic forcing of magnetospheric origin affects the region between 100 and 130 km of altitude using a new thermosphere general circulation model based on the spectral technique. Although better understood, the F-region winds will also be discussed. The separation of the wind field in a stream function and a velocity potential, to be used here, sheds new light on the nonlinear mechanisms present with the large wind speeds above 120 km altitude. The model comprises the same set of differential equations like in the other thermosphere general circulation models including auroral dynamics (Fuller-Rowell and Rees (1980), Roble et al. (1982)). The spectral model will be described in chapter 2.

407

P. E. Sandholt and A. Egeland (eds.), Electromagnetic Coupling in the Polar Clefts and Caps, 407–419.
© *1989 by Kluwer Academic Publishers.*

The theory of geostrophic adjustment deals with the response of a fluid, like the thermosphere, to forces and heat sources of varying horizontal and vertical wavelengths. Linearized versions of the theory have been applied to the high-latitude thermospheric dynamics in a series of papers (Larsen and Mikkelsen (1983, 1987), Mikkelsen and Larsen (1983), Walterscheid and Boucher (1984), and Clark et al. (1988)). The results of these papers, summarized in chapter 3, will be taken as a starting point in the interpretation of the nonlinear effects, present in the solutions of the spectral model. In chapter 4.1 these solutions will be discussed, obtained with a constant global electron density and without solar heating. Finally the general case with a variable electron density and solar heating will be discussed in chapter 4.2 followed by a short discussion and summary.

2. THE SPECTRAL MODEL

All fields are expanded in series in terms of the spherical harmonics in the horizontal direction. The spectral coefficients thus define the fields. This is combined with a descrete vertical representation, in the calculations to be presented here, at 27 isobars spaced by one half scaleheight and extending from the mesopause to the top of the F region. There is a series expansion for each isobar, or equivalently the spectral coefficients may be considered descrete functions of the vertical pressure coordinate. Normally the wind velocity components in spherical coordinates, discontinuous at the poles, are replaced by the divergence and rotation of the wind field. Together with the temperature these are the prognostic fields of the model, and as usual the pressure tendency and the geopotential are the diagnostic fields. For a general discussion of the spectral method applied to atmospheric models see Machenhauer (1979). The prediction equations become a set of ordinary differential equations for the spectral coefficients in the horizontal direction. The vertical differentiations are still represented by a finite difference scheme. The nonlinear terms, like the Joule heating, are computed using the transform method, and the gravity wave, heat conduction, and viscous terms are treated implicitly in the time scheme to improve the performance of the model.

Like the thermospheric fields the magnetospheric electrostatic potential and the three-dimensional electron density and solar heating must be expanded in series in terms of the spherical harmonics. The shortest horizontal wavelength, as defined by the truncation of these series, may be compared to the shortest wavelength present in a grid-point model which is at least two grid spacings and probably more because of the strong numerical damping of the shortest wavelengths.

To see the significance of this a Heppner-Maynard Model A plasma convection pattern (Heppner and Maynard (1987)) together with its truncated spectral version is shown in figure 1. The r21 truncation will be used in all solutions to be presented in this paper. Heuristically this corresponds to a longitude-latitude grid with zonal and meridional grid spacings of $8.6°$ and $4.4°$ respectively. The amplitude and location of the extrema of the electrostatic potential are well reproduced in the spectral version, but the steep gradients corresponding to plasma speeds of 1900 m/s in the dusk return flow and the throat are halfed in the spectral version because of the truncation. Although a single plasma velocity vector may be assigned a very large magnitude, this problem is also present in grid-point models, but is not so evident, because it first manifests itself as a numerical damping of the solutions. The effects of such plasma jet-streams, beyond the resolution of the models, have been examined in a two-dimensional, nested-grid model (Fuller-Rowell (1984,1985)); however, their effects in a three-dimensional thermosphere model are still to be examined.

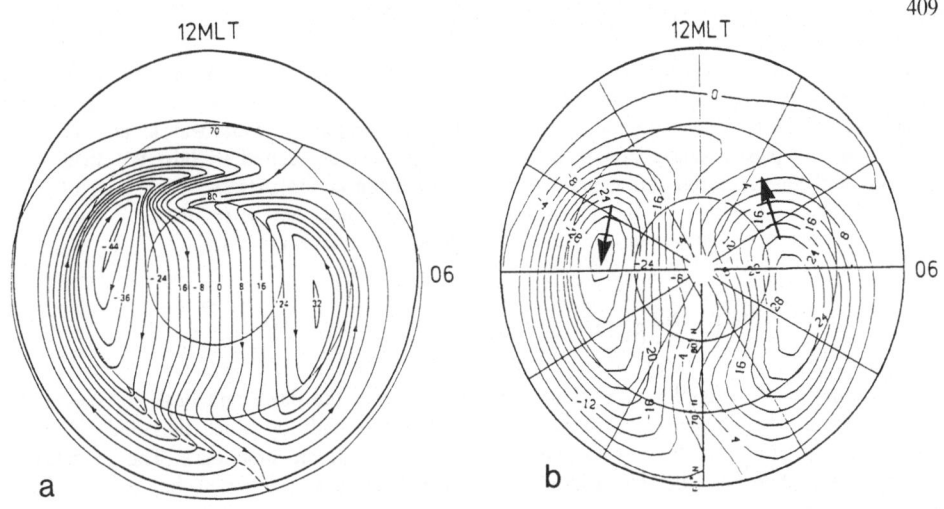

Figure 1. Polar plots of Heppner-Maynard Model A magnetospheric electrostatic potential (a), and spectral version (b) using a rhomboidal truncation with maximum zonal and meridional wavenumbers of 21 and 41 respectively. The outer latitude circle is at 60°, and the contour interval 4 kV in both plots. The heavy arrows in (b) are the Hall forces left, when the nightside electron density is removed.

3. SUMMARY OF THE GEOSTROPHIC ADJUSTMENT THEORY

The high-latitude Lorentz forces due to the Pedersen- and Hall-currents, here called Pedersen- and Hall-drag (or -force) respectively, have longitudinal extents of several thousand kilometers and transverse, normal to the auroral oval, extents of 100-1000 km. The size of the socalled Rossby radius of deformation with respect to these scales determines how the thermosphere adjusts itself to the forces. The Rossby radius, a function of the thermospheric structure and the vertical wavelength of the perturbation, varies from a few thousand kilometer in the F region to 500-1000 km in the E region (Larsen and Mikkelsen (1983)). For a constant, vertical, perturbation wavelength the Rossby radius would reach a maximum at 100-150 km altitude, but the Hall drag and the Joule heating as well as the Pedersen drag enhanced by auroral precipitation all create perturbations of shorter vertical wavelengths in the E region leading to a net decrease of the Rossby radius.

The Pedersen drag, in the direction of the plasma flow, accelerates rotational winds leading to the formation of the twin-vortex, F-region neutral wind pattern similar to the plasma convection pattern. The E-region Hall drag, normal to the plasma flow, is partly balanced by a back-pressure, and partly drives rotational and irrotational winds of similar magnitudes. This leads to the formation of a twin-vortex that may be rotated 180° with respect to the plasma convection pattern. These effects were estimated by a one-layer, shallow-water model, (Mikkelsen and Larsen (1983)) and later by a multi-layer model (Larsen and Mikkelsen (1987)). Common to these models is, that the equations are linearized, and that all terms coupling different horizontal wavelengths are disregarded. As a consequence the Joule heating must be disregarded, and the electron density must be considered constant in the horizontal direction, although vertical variations may be specified. For simplicity the analysis will be limited to a constant vertical electron density profile of $5.10^{10}/m^3$.

Expanding the fields in series in terms of the spherical harmonics the partial differential equations may be separated into sets of ordinary differential equations for each horizontal wavenumber of the spherical harmonics. Since the diurnal component, zonal wavenumber $m = 1$, is dominant in the plasma convection pattern, the analysis will be limited to this. We shall study the wind- and temperature-fields generated by a plasma convection pattern which is constant in the solar frame. At time $t = 0$ the wind- and temperature-fields are zero. In the F region the spectral component of the stream function ψ in the solar frame is given by,

$$\psi = [1 - e^{-(\alpha + i\Omega)t}] \frac{\alpha}{\alpha + i\Omega} \tau \tag{1}$$

The magnitude of the complex number ψ is the amplitude of the dawn and dusk vortices of the stream function. The, in general negative, phase of ψ is the angle by which the wind pattern is rotated towards later local times with respect to the plasma stream function τ. The geometrical meaning of the phase is indicated in figure 2, showing a nonlinear solution to be discussed later. The bracket contains the temporal development of ψ. α is the time constant of the Pedersen drag, and Ω is the angular rotation rate of the earth. i is the imaginary unit. As the time approaches infinity, the bracket approaches unity. The next factor is the complex number by which τ is multiplied to obtain the steady state solution.

Because of the large Rossby radius in the F region the divergent wind component is negligible and the rotational wind field, as expressed by (1), is in geostrophic balance. Another effect of the large Rossby radius is, that the spectral amplitude ψ is independent of the wavenumber of the spherical harmonic in consideration. It means that the neutral wind pattern and the plasma convection pattern, although of different magnitudes, will have the same shape.

(1) is applicable above 125 km altitude. Below this altitude α decreases sharply and β, the time constant of the Hall drag, becomes dominant. As a result the phase of ψ decreases gradually to $-180°$ at the bottom of the E region, and the divergent wind, represented by the velocity potential χ, becomes significant during the period of geostrophic adjustment which lasts for the first few hours. A phase of $-180°$ means that the counterclockwise wind vortex associated with the dawn vortex of the plasma convection pattern is placed at the dusk side, and the clockwise wind vortex associated with the dusk plasma convection vortex at the dawnside of the polar cap. The amplitude of the E-region wind also exhibits the diurnal variation contained in the bracket of (1). Because of the smaller value of α these diurnal oscillations last longer than in the F region.

4. NONLINEAR SOLUTIONS

4.1 Asymmetries of the dawn and dusk wind vortices with a symmetric plasma convection pattern and a constant electron density.

In the linear theory, outlined in chapter 3, all terms, coupling different wavelengths of the solutions, are disregarded. As a result the solutions have the same symmetry properties as the plasma convection pattern. To study the effects generated by the nonlinear terms, like the Joule heating and the centrifugal force, a Volland plasma convection pattern (Volland (1978)) is applied in the nonlinear spectral model. This has a diurnal, zonal wavenumber $m = 1$, component only. The dawn and dusk plasma convection vortices are therefore completely symmetric. The dawn-dusk electrostatic potential difference is 59 kV, corresponding to a plasma speed of 900 m/s at the pole. Combined with a constant electron density of $5.10^{10} el/m^3$ the resulting Pedersen- and Hall-drag will be symmetric as well.

The thermospheric parameters used to compute the various terms are obtained from a global mean, vertical profile of the MSIS-86 model (Hedin (1987)) corresponding to solar minimum conditions. The prognostic fields, the divergence D and the vorticity Z, contain a lot of short-wavelength information, and plots of these are therefore rather confusing. Instead the velocity potential χ and the streamfunction ψ are plotted. These contain the same information like in D and Z, but in a much more perceptible form. Since, $D = \Delta\chi$, and $Z = \Delta\psi$, and since the spherical harmonics are eigenfunctions of the Laplace operator Δ, χ and ψ are readily obtained from the series expansions of D and Z. Figure 2 is a plot of χ and ψ at 350 km altitude.

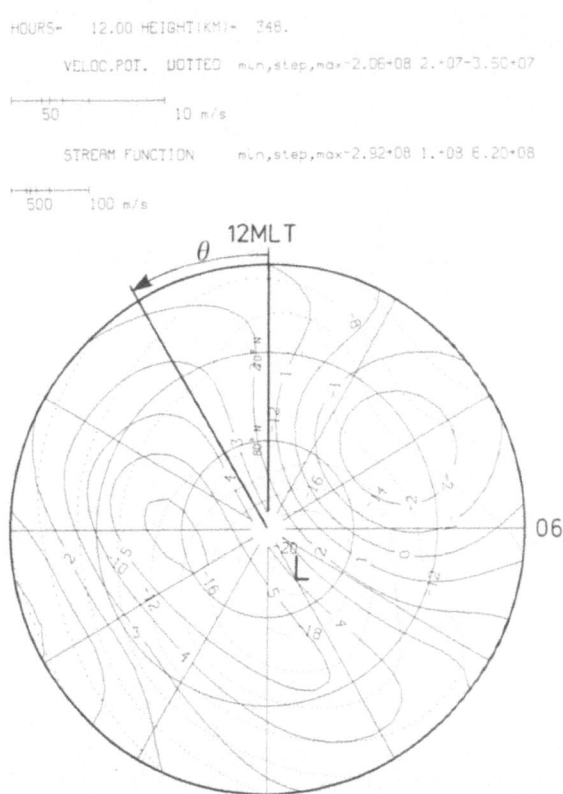

Figure 2. Polar plot, in a format like figure 1, of contours of the stream function ψ (solid) and the velocity potential χ (dotted) at 350 km altitude and after 12 hours of integration. 59 kV Volland potential. $n_e = 5.10^{10}$el/m^2. θ is the negative phase of the diurnal component of ψ.

The divergent wind is normal to the dotted contours of the velocity potential χ, and the rotational wind is along the solid contours of the stream function ψ. Different contour intervals are used to plot the two fields. The wind speeds are inversely proportional to the transverse spacing of the contours and may be read from the two scales on top of the plot. The divergent winds blow radially outward from L, the deep minimum of χ near the pole. These winds, reaching speeds of 50 m/s, are due to the Joule heating in the thermosphere below.

The rotational winds form a circular, counterclockwise vortex between dawn and noon, the circular shape having the overall largest radius of curvature. The dusk vortex intermittently becomes very elongated, as is the case at 12 hours, apparently because the more stable clockwise vorticity may be advected over larger distances before breaking down into smaller vortices. This larger stability is attributed to the partial balance of the centrifugal- and -Coriolis forces in the dusk vortex. For a rotational flow having a radius of curvature of 1000 km, typical of the dawn vortex, the centrifugal force will dominate over the Coriolis force for wind speeds exceeding 150 m/s. Figure 2 shows that significantly larger wind speeds exist only in regions, like the central polar cap and the dusk return flow, in which the radius of curvature is much larger than 1000 km. The elongated dusk vortex later breaks into two minor vortices and a new one starts to form. This instability is repeated with a period close to 12 hours and must be related to the centrifugal forces and the Joule heating which both have a 12-hour period because of the special Volland potential and constant electron density used. To separate the effects of the centrifugal forces and the Joule heating the run has been repeated without the Joule heating term in the thermodynamic equation. The two runs are compared in figure 3.

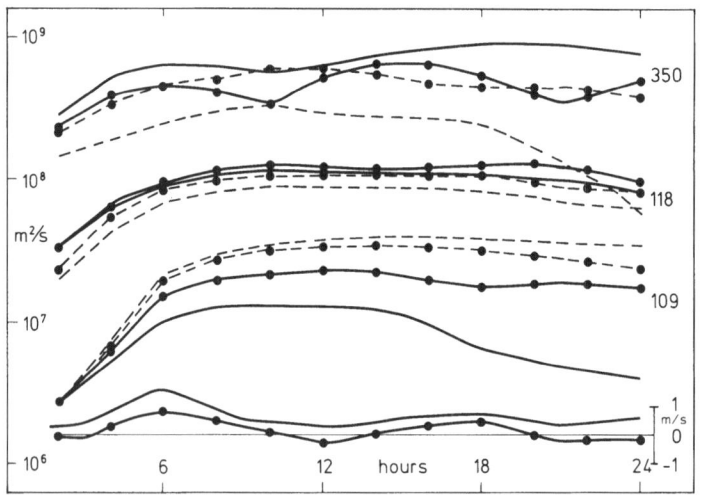

Figure 3 Numerical values of the extrema of the stream function (m²/s) at dusk (solid) and dawn (dashed) versus integration time for 350, 118, and 109 km altitude. At the bottom of the figure: The average, poleward, meridional wind (m/s) through the 60° parallel at 109 km. Results without Joule heating are marked with filled circles.

Because the Joule heating term contains the square of the diurnal plasma velocity it has a very large zonally invariant, m = 0, component. This creates the upwelling and, because of mass-continuity, divergent wind shown in figure 2. At the top of the F region an average equatorward meridional wind is 15-20 m/s. Over an extended time interval this moves a significant amount of planetary vorticity equatorward and creates the dominant, negative, dusk vortex. According to equation 1 the amplitudes of the vortices oscillate with a diurnal period after the onset of the plasma convection at time, t = 0. In the second half of the first oscillation (12-24 hours) the difference between the plasma- and neutral velocity becomes very large again, like it is initially, causing a large Joule heating. The extra secondary circulation, generated by this, probably is the reason for the sharp drop of the F-region dawn vortex after 18 hours. Note that the strength becomes lower than that of the vortices at 118 km altitude.

With the Joule heating included the average meridional wind through the 60° parallel is equatorward at all altitudes above 118 km and poleward below. The secondary circulation thus must close by a sinking of the air at lower latitudes. Below 118 km altitude the dawn vortex dominates over the dusk vortex, opposite to what is seen in the F region (figure 3). This is due to the secondary circulation. The average poleward velocity at 109 km altitude, although only 1 m/s of magnitude, advects a significant amount of planetary vorticity poleward, creating the dominant, positive, dawn vortex.

Without the Joule heating the dusk vortex in the F region oscillates in a very pronounced 12-hour instability cycle and in the average does not exceed the dawn vortex, but in the lower E region the dawn vortex is still dominant. The average, meridional wind at 109 km altitude is positive for the first few hours (figure 3). This may be due to a transient secondary circulation created by the centrifugal forces acting on the F-region winds, but this question needs further examination. Later the average meridional wind becomes oscillatory.

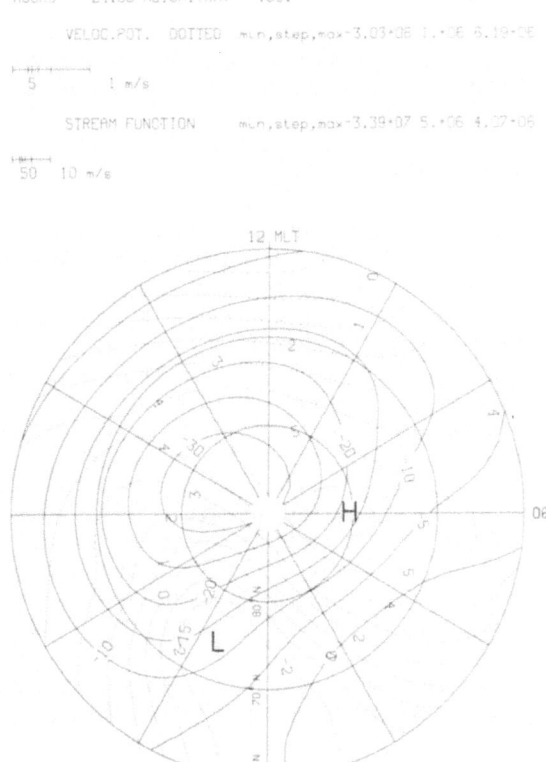

Figure 4. Same as figure 2, but after 24 hours of integration and at 109 km altitude.

After 24 hours of integration the dawn vortex in the lower E region, figure 4, is dominant. The phase angle of the, now weak, diurnal component of the stream function has reached a value of-120° in qualitative agreement with the linear theory. The divergent winds, blowing from L to H, show a strong asymmetry, like the rotational winds. The exact vertical route of the secondary circulation through the E region is complicated and will not be discussed here.

Even after 24 hours of integration the thermosphere is far from a steady state. This is because of the unrealistic, large electron density in the lower E region. Because of this, the calculated winds will probably become unrealistic for longer integration times and will not be analyzed. However, the secondary circulation due to the Joule heating, and the characteristic circular and elongated shapes of the F-region vortices are also present in a more realistic model to be presented in the next chapter.

4.2 Solutions with solar heating and variable electron density

The general solution including solar heating, a complex plasma convection pattern, and a three-dimensional electron density distribution has been extensively discussed in the literature. Roble et al. (1983) analyzed the nonlinear interaction between the wind- and temperature fields generated by the solar heating and the magnetospheric forcing. Let SM be the solution with both energy sources present, S with solar heating alone, and M with magnetospheric forcing alone. They looked at the difference fields,

SM - (S + M)

In the F region these difference fields are small compared to the base solution SM, but in the upper E region at 130 km altitude the difference field is characterized by a large flow encircling the pole. The vorticity of this flow has the same sign, in both hemispheres, as the planetary vorticity.

When the SM solution is compared to the M solution the role of the solar-driven wind is to decrease the numerical value of the phase angle θ, (see figure 2 for the meaning of θ). Reducing this angle also reduces the vector difference beteen the plasma- and neutral velocity. This in turn diminishes the Joule heating and therefore the secondary circulation, discussed in chapter 4.1. The SM solution, at all altitudes where the secondary circulation is equatorward, must therefore contain more planetary vorticity than the (S + M) solution, and most in the lower F- and upper E region, where the numerical value of the angle θ is largest in the M solution in agreement with the finding of Roble et al. (1983).

To illustrate these linearity problems, and other questions, the spectral model has been integrated using the Heppner-Maynard Model-A plasma convection pattern shown in figure 1b and a Chiu empirical model of the ionospheric density (Chiu (1975)). The MSIS-86 thermospheric parameters, the solar heating, and the electron density have been computed for solar minimum- and equinox conditions. It is tempting in the SM solution, figure 5, to relate the rotational wind field to the magnetospheric forcing, and the divergent wind, blowing from post-noon to-midnight, to the solar heating, but it is not quite correct. However, the circular dawn- and elongated dusk vortex are seen again, like in the cases without solar heating. The solar-driven wind prevents the F-region dusk vortex from developing extreme shapes.

Figure 5. Polar plot, in a format similar to figure 2, of velocity potential and stream function after 44 hours of integration at 350 km altitude. Heppner-Maynard Model-A plasma convection pattern and solar heating. Coincident geographic and magnetic poles. Chiu-model. Solar minimum. Equinox conditions. (Figure on next page).

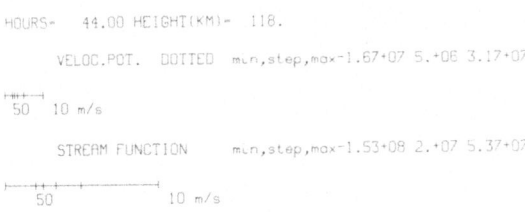

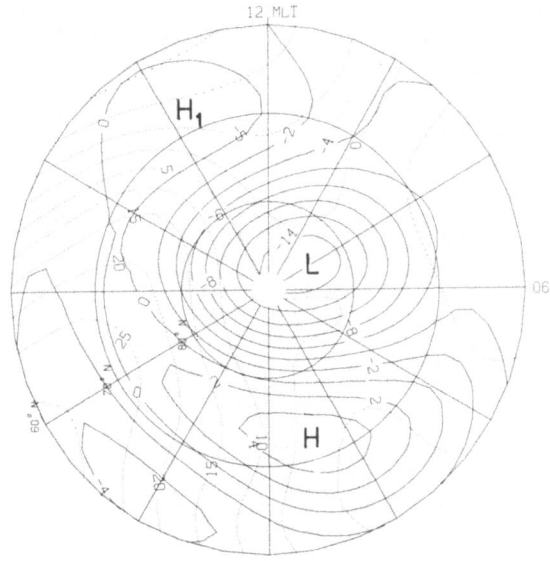

The model contains a rich spectrum of nonlinear instabilities. When the E region is loaded with energy, like after 44 hours, the dusk vortex develops the very elongated shape, seen in the F region without solar heating.

In figure 6 a minor vortex, labeled H_1, has divided from the main part, labeled H. With wind speeds exceeding 100 m/s and with radii of curvature of 500-1000 km, like in the counterclockwise vortex, labeled L, the nonlinear terms, like the centrifugal force, are important and make the vortices unstable. The instabilities may be seen in the rate of change of the temperature, probably due to variations in the Joule heating. Figure 7 shows the growth of the global mean, perturbation temperature, T, and dT/dt at the upper isobar of the model. The thermosphere is seen to oscillate with inertial periods close to 12 hours. As the steady state is approached the amplitudes of these oscillations diminish.

416

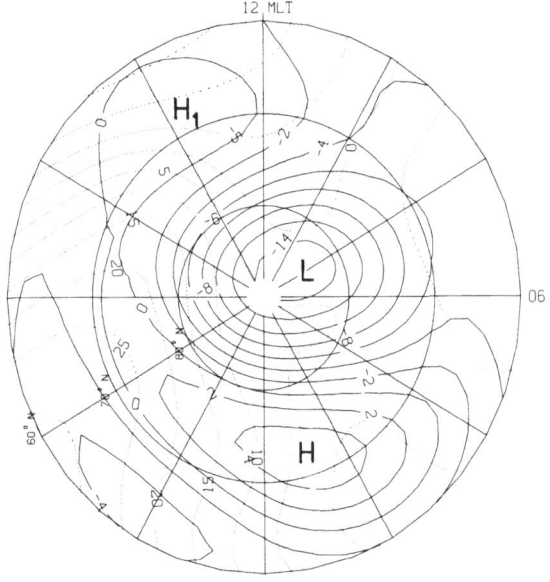

Figure 6. Same as figure 5, but at 118 km altitude.

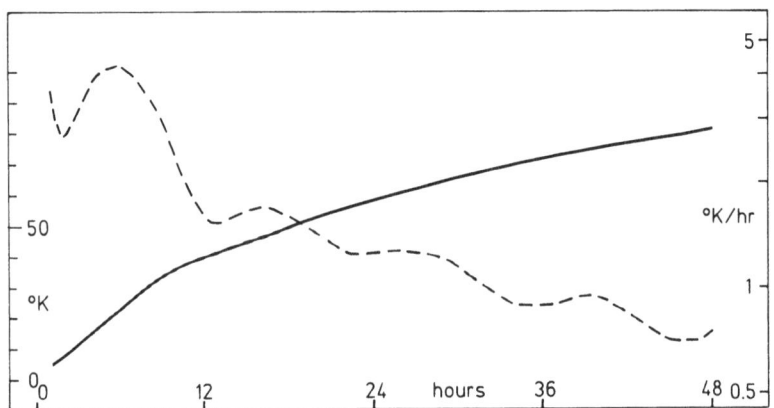

Figure 7. Temporal development of global mean temperature T (solid), and dT/dt (dashed) at 350 km altitude from same run presented in figures 5 and 6.

To see the amount of secondary circulation, driven by the Joule heating, the average, poleward, meridional wind through the 60° parallel has been computed. This net wind is negative above, and positive below, 135 km altitude at 44 hours. As discussed in chapter 4.1 the secondary circulation increases the counterclockwise, positive vorticity in the lower E region by poleward advection of planetary vorticity. Because of the night-day gradient in the E-region electron density in the Chiu model, the Hall force is changed from a divergent force, as it would be with a homogenous density, to a partly rotational force. This is illustrated in figure 1b. When the night side of the electron density is removed, the remaining Hall forces in the day side form a rotational force field. This is more effective in accelerating winds, important in the lower E region, where the Hall force dominates over the Pedersen force. The total effect is illustrated in figure 8, a vertical cross section of the dusk-dawn meridian. The component of the rotational wind, normal to the paper, is inversely proportional to the horizontal spacing of the solid contours of the stream function. The wind speeds may be read from the velocity scale, like in the polar plots.

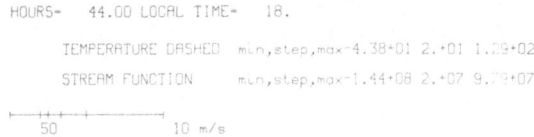

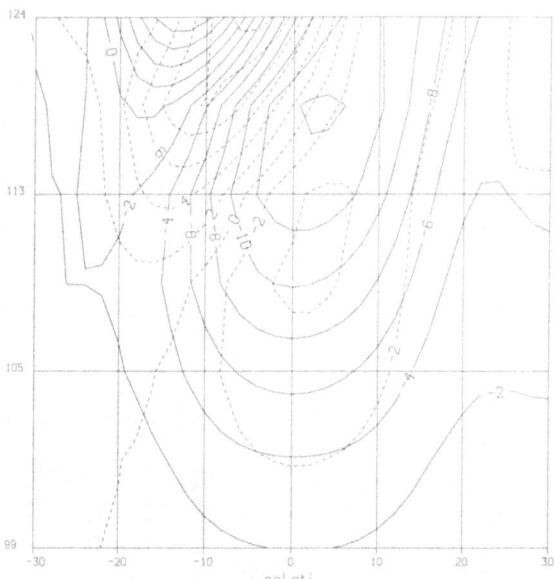

Figure 8. Vertical contouring of the stream function (solid) and temperature field (dashed). +-30° colatitude, dusk at left, and dawn at right. The horizontal lines at 99, 105, 113, and 124 km altitude are spaced by one scale height. From same run presented in figures 5-7.

Wind is coming out of the paper at the pole at 124 km altitude. This wind separates the clockwise dusk-vortex to the left from the weaker counterclockwise dawn-vortex to the right. Below 124 km altitude the dawn vortex becomes dominant and centered at the pole. The temperature contours illustrate that the wind field in the lower E region is in geostrophic balance.

The relative importance of the secondary circulation and the rotational Hall force in creating the counterclockwise wind vortex in the lower E region has not been determined. However, since the vortex is centered at the pole, and since it is similar to what is obtained with a constant electron density, the secondary circulation is probably the most important contributor.

5. DISCUSSION AND SUMMARY

The opposite directions of the Coriolis force in the high-latitude neutral-wind vortices at dusk and dawn affect the shapes, but probably not the magnitudes, of the two vortices. These results, present only in a two-dimensional horizontal analysis, are not in contradiction with the findings of a simpler one-dimensional analysis of the centrifugal- and Coriolis forces (Gundlach et al. (1988)).

The main reason for the dominant dusk wind-vortex in the upper E- and F region is a secondary circulation generated by the Joule heating that advects planetary vorticity equatorward. The solar-driven wind across the pole forces the neutral wind pattern towards an orientation along the noon-midnight meridian. This reduces the Joule heating and the associated secondary circulation.

In the lower E region the poleward, secondary circulation generates a counterclockwise vortex by advecting planetary vorticity poleward. The counterclockwise vortex is seen in results from other TGCM's (Fuller-Rowell et al. (1987)). Roble and Ridley (1987) showed that an addition of auroral ionization to the Chiu model, although it changes the Hall force, has a moderate effect upon the vortex in the lower E region. This supports the hypothesis that the secondary circulation due to the Joule heating is the most important contributor.

Examples of internal thermospheric instabilities have been shown. These will add to the variability generated by variations in the magnetospheric forcing.

Acknowledgment. The solar heating code was obtained from Drs. Roble and Emery at NCAR.

References

Brekke, A., Doupnik, J. R., Banks, P. M., (1973) 'A preliminary study of the neutral wind in the auroral E-region', J. Geophys. Res., 78,8235-8250.

Brekke, A., Doupnik, J. R., Banks, P. M., (1974) 'Observations of neutral winds in the auroral E-region during the magnetospheric storm of August 3-9, 1972', J. Geophys. Res., 79, 2448-2456.

Chiu, Y. T. (1975) 'An improved phenomenological model of ionospheric density', J. Atmos. Terr. Phys., 37, 1563-1570.

Clark, M. A., Larsen, M. F., and Mikkelsen, I. S. (1988) 'Analysis of the response of the thermospheric normal modes to temporally-varying convection at high latitudes', J. Geophys. Res., 93, 12893-12900.

Fuller-Rowell, T. J. (1984) 'A two-dimensional, high-resolution, nested-grid model of the thermosphere, 1, Neutral response to an electric field "spike"', J. Geophys. Res., 89, 2971-2990.

Fuller-Rowell, T. J. (1985) 'A two-dimensional, high-resolution, nested-grid model of the thermosphere, 2, Response of the thermosphere to narrow and broad electrodynamic features, J. Geophys. Res., 90, 6567-6586.

Fuller-Rowell, T. J., and Rees D. (1980) 'A three-dimensional, time-dependent, global model of the thermosphere', J. Atmos. Sci., 37, 2545-2567.

Fuller-Rowell, T. J., Rees, D., Quegan, S., Moffett, R. J., Bailey, G. J. (1987) 'Interactions between neutral thermospheric composition and the polar ionosphere using a coupled ionosphere-thermosphere model', J. Geophys. Res., 92, 7744-7748.

Gundlach, J. P., Larsen, M. F., and Mikkelsen, I. S. (1988) 'A simple model describing the nonlinear dynamics of the dusk/dawn asymmetry in the high-latitude thermospheric flow', Geophys. Res. Lett., 15, 307-310.

Hedin, A. E. (1987) 'MSIS-86 thermospheric model', J. Geophys. Res., 92, 4649-4662.

Heppner, J. P., Maynard, N. C. (1987) 'Empirical high-latitude electric field models', J. Geophys. Res., 92, 4467-4489.

Heppner, J. P., Miller, M. L. (1982) 'Thermospheric winds at high latitudes from chemical release observations', J. Geophys. Res., 87, 1633-1647.

Johnson, R. M., Wickwar, V. B., Roble, R. G., Luhmann, J. G. (1987) 'Lower-thermosphere winds at high latitude: Chatanika radar observations', Annales Geophysicae, 5A, 383-404.

Kirkwood, S., (1986) 'Seasonal and tidal variations of neutral temperatures and densities in the high latitude lower thermosphere measured by EISCAT', J. Atmos. Terr. Phys., 48, 817-826.

Larsen, M. F. and Mikkelsen, I. S. (1983) 'The dynamic response of the high-latitude thermosphere and geostrophic adjustment', J. Geophys. Res., 88, 3158-3168.

Larsen, M. F. and Mikkelsen, I. S. (1987) 'The normal modes of the thermosphere', J. Geophys. Res., 92, 6023-6043.

Machenhauer, B. (1979) 'The spectral method', in Numerical Methods used in Atmospheric Models. Volume II. GARP Publ. Series, No. 17. ICSU/WMO, Geneva, 121-275.

Mikkelsen, I. S. and Larsen, M. F. (1983) 'An analytic solution for the response of the neutral atmosphere to the high-latitude convection pattern', J. Geophys. Res., 88, 8073-8080.

Mikkelsen, I. S., Larsen, M. F., Kelley, M. C., Vickrey, J., Friis-Christensen, E., Meriwether, J., Shih, P. (1987) 'Simultaneous measurements of the thermospheric wind profile at three separate positions in the dusk auroral oval', J. Geophys. Res., 92, 4639-4648.

Roble, R. G., Ridley, E. C. (1987) 'An auroral model for the NCAR thermospheric general circulation model (TGCM)', Annales Geophysicae, 5A, 369-382.

Roble, R. G., Dickinson, R. E., and Ridley, E. C. (1982) 'Global circulation and temperature structure of thermosphere with high- latitude plasma convection', J. Geophys. Res., 87, 1599-1614.

Roble, R. G., Dickinson, R. E., Ridley, E. C., Emery, B. A., Hays, P. B., Killeen, T. L., Spencer, N. W. (1983) The high latitude circulation and temperature structure of the thermosphere near solstice', Planet. Space Sci., 31, 1479-1499.

Volland, H. (1978) 'A model of the magnetospheric electric convection field', J. Geophys. Res., 82, 2695-2699.

Walterscheid, R. L., Boucher, Jr., D. S., (1984) 'A simple model of the transient response of the thermosphere to impulsive forcing', J. Atmos. Sci., 41, 1062-1072.

GRAVITY WAVE STUDIES AT POLAR LATITUDES

M.J. TAYLOR
Physics Department,
Southampton University,
Southampton,
SO9 5 NH, U.K.

and

K. HENRIKSEN
The Auroral Observatory,
University of Tromsö,
N-9001 Tromsö,
Norway.

ABSTRACT. Images of structure in the OH nightglow emission have been used to establish the presence of short period gravity waves in the winter polar mesosphere. The observations, which were made from the auroral station at Longyearbyen, Svalbard (78°N), are currently being used to investigate the generation of small scale gravity waves by auroral processes. Results so far show that the airglow wave patterns appear similar in both morphology and dynamics to those commonly seen at mid-latitudes indicating tropospheric sources for the waves. However, on one occasion there is good evidence for the generation of gravity waves by an auroral electric field that exhibited an intrinsic periodicity close to that observed in the airglow wave pattern. Further studies are in hand.

1. Introduction

Much of the data presented at this workshop has been concerned with the coupling between the magnetosphere and the ionosphere within the polar clefts and caps. This theme will be extended by considering some aspects of wave induced coupling between the disturbed ionosphere and the lower regions of the neutral atmosphere at polar latitudes. The primary aim is to investigate any connection between auroral disturbances and short period (<1 hour) mesospheric gravity waves.

The existence of gravity waves in the mid-latitude upper atmosphere is well established (see review by Francis, 1975). Several methods of studying their propagation through the mesosphere and lower thermosphere have been used in the past and with varying degrees of success. Imaging of the naturally occurring airglow emissions in the altitude range 80-100 km has proven to be a particularly useful tool. The data give a unique 2-dimensional "snap shot" of the gravity wave field which can be used to study

P. E. Sandholt and A. Egeland (eds.), Electromagnetic Coupling in the Polar Clefts and Caps, 421–434.

the wave propagation in both time and space. For example, Hapgood and Taylor (1982) and Taylor et al. (1987) have used images of airglow structure in the near infra-red hydroxyl (OH) nightglow emission (which has a peak altitude at ~87 km and a halfwidth of ~8 km; Baker and Stair, 1987), to study the morphology and dynamics of mesospheric gravity waves in detail. Many attempts have been made to correlate the occurrence of such wave structure with various upper atmospheric and tropospheric disturbances. Recently Taylor and Hapgood (1988) have shown that thunderstorms are an important source of the short period gravity waves seen at mid-latitudes. However, the question "whether such gravity waves exist at auroral latitudes and if so what are their main sources" has yet to be properly answered.

It is now recognised that most of the large scale and some of the medium scale disturbances that are seen travelling equatorward in the mid-latitude F-region ionosphere (altitude ~350 km) are associated with the occurrence of strong magnetic storms (e.g. Davies and da Rosa, 1969). Myrabö et al. (1983) has shown that the OH emission at polar latitudes can also exhibit large amplitude quasi periodic variations (period few hours), similar to those associated with mid-latitude travelling ionospheric disturbances (TID's). Three mechanisms have been suggested for the generation of gravity waves by auroral processes. The first is due to the direct effect of Joule heating of the neutral atmosphere which arises when the magnetospheric electric field drives a large scale current (the auroral electrojet) through the ionosphere. The second mechanism, termed "Lorentz force" arises from the interaction between the electrojet and the earth's magnetic field (Chimonas and Hines, 1974). Finally the impulsive influx of charged particles (of energy >1keV) can cause significant frictional heating in the vicinity of optical arcs. Although the relative importance of each of these "sources" is dependent upon the type of auroral disturbance (i.e. hard or soft precipitation; electric field strength) it can be assumed than in most instances the primary source region will reside at an altitude of 100-120 km. As this is less than 40 km above the OH emission layer, mesospheric observations of short period gravity waves generated by auroral processes must necessarily be made in close proximity to the source region.

In 1980 and again in 1984 we successfully imaged OH wave structure at auroral latitudes from a site near Tromsö, Norway (69°N). However the frequent recurrence of auroral precipitation at these latitudes often masked the much fainter airglow emissions making it difficult to study individual sources of the wave structures. In 1986/7 we repeated these measurements from the polar latitude station at Longyearbyen, Svalbard (78°N). Since this site is usually well within the polar cap around magnetic midnight we have been able to make a more detailed study of isolated auroral activity and airglow wave structure.

2. Summary of Mid-Latitude Observations

For various reasons the majority of the airglow observations contained in the literature have been made from mid-latitude sites. A wealth of wave-like fluctuations of varying periods, durations and scale sizes have been reported. Remarkably, images of structure in the nightglow emissions often show the wave patterns to be extensive and highly coherent. A study of these data has revealed many similarities in the patterns produced by the gravity wave perturbations. These have been summarised into three main categories (Taylor 1986a):

(a) Broad bands   Horizontal wavelength >100 km
                                Horizontal velocity ?
                                Lateral extent >1000 km
                                Duration: several hours
                                No. of wave crests: typically <5

(b) Narrow Stripes   Horizontal wavelength: mainly 20-70 km
                                Horizontal velocity: from 10-50 m/s
                                Lateral extent > 500 km
                                Duration: 0.5-2 hours
                                No. of wave crests: typically 2-15

(c) Ripples   Horizontal wavelength: 5-15 km
                                Horizontal velocity: 70-90 m/s
                                Lateral extent <50 km
                                Duration: typically <45 minutes
                                No. of wave crests: 3-20+   (after Peterson, 1979)

The wave forms usually appear uniform and elongated but may on occasions exhibit pronounced curvature. This latter property makes the task of identifying the source much easier (see Taylor and Hapgood, 1988). From the data currently available there appears to be no seasonal preference in either the abundance or the type of structure present on any one night. More than one wave pattern can be present in the sky at any instant.

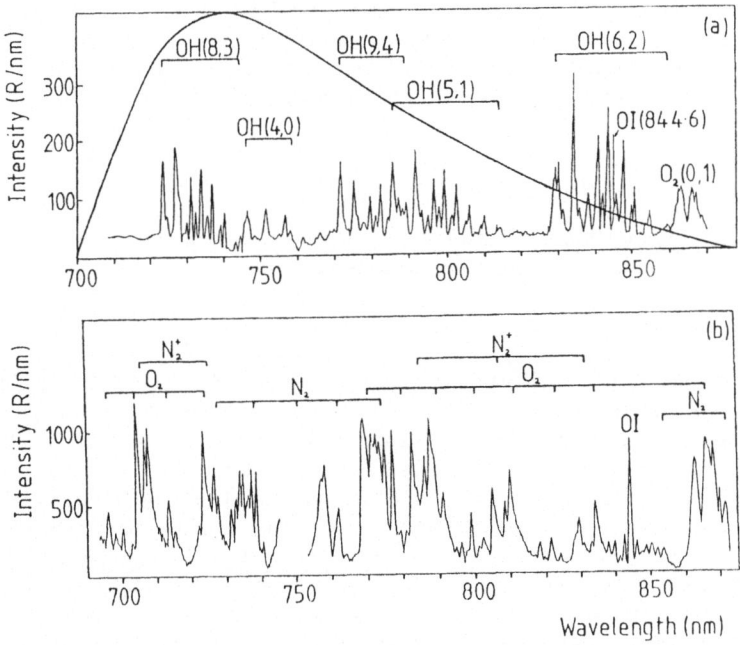

Figure 1. Spectral response of the filtered TV system showing (a) the night sky airglow emissions and (b) the principal auroral emissions for Longyearbyen, Svalbard.

3. Polar Latitude Airglow Measurements

The observations were made from a site near Longyearbyen, Svalbard (78.4°N, 15.8°E) during January and February 1986 and 1987. A low light television system capable of integrating the video signal for intervals of up to 1 second was used to search for structure in the OH emission (Taylor, 1986b). The camera had a 30° horizontal by 20° vertical field of view and was filtered to limit the bandwidth of the measurements to 715-830 nm (half maximum). Several airglow and auroral features exist within this spectral range. Figure 1a shows the night sky spectrum from 700 to 870 nm for Svalbard during a quiet magnetic period. The smooth curve overlaying the spectrum shows the relative response of the TV system. The main contributions to the airglow emission arise from the (8,3), (9,4) and (5,1) OH Meinel bands. The contributions to the image from the OI auroral line at 844.6 nm and the O_2 Atmospheric band centred at ~ 865 nm are minimal. Figure 1b shows the night-time spectrum for Longyearbyen during an IBC type II aurora. Several auroral bands, originating mainly from neutral and ionised molecular nitrogen and neutral oxygen now dominate this spectral region. In this example the contribution to the spectrum from the OH emission is less than 15% of the total auroral luminosity. The camera was able to observe the sky at all azimuths. Measurements were usually made at an elevation of 15° (field centre) to take advantage of line of sight enhancement in both the contrast and brightness of the airglow structures.

Figure 2 shows four examples of the airglow image data obtained from Svalbard. In each image several well defined, coherent wave forms can be seen against the star background. The orientation, shape and location of these patterns is different in each example yet the patterns all look similar. In figure 2c some near vertical, faint auroral rays are also present. During times of intense precipitation it was necessary to either reduce the camera gain or point to another azimuth. Because the auroral arcs observed from Svalbard were usually discrete and rarely occupied the whole sky it was often possible to image airglow structure even during magnetically active periods.

4. Results

To investigate any relationship between the occurrence of mesospheric wave structure and auroral activity three analyses have been performed:

4.1 Σ Kp COMPARISON

Figure 3 is a plot of the total magnetic activity index, Σ Kp, versus the occurrence of airglow wave structure for the Svalbard data set. (Note average Kp = Σ Kp/8.) The black regions indicate days containing strong auroral emissions and the hatched areas show the days when airglow structure was observed. Aurora of some form was present on every night that airglow structure was observed. However, on three occasions there was significant auroral precipitation and no apparent airglow structure. The 1986 observations were dominated by the exceptionally strong magnetic storm of 8/9 February (Σ Kp~61, Ap >200). Although it was not possible to image airglow structure during this near continuous auroral display, wave structure was detected on the following night when Σ Kp was still very high ~47 (see Figure 2a). Clairemidi et al. (1985) deduced that "OH structures were clearly distinguishable (at auroral latitudes) only when the average Kp was higher than 1.80". Applying this criterion we might

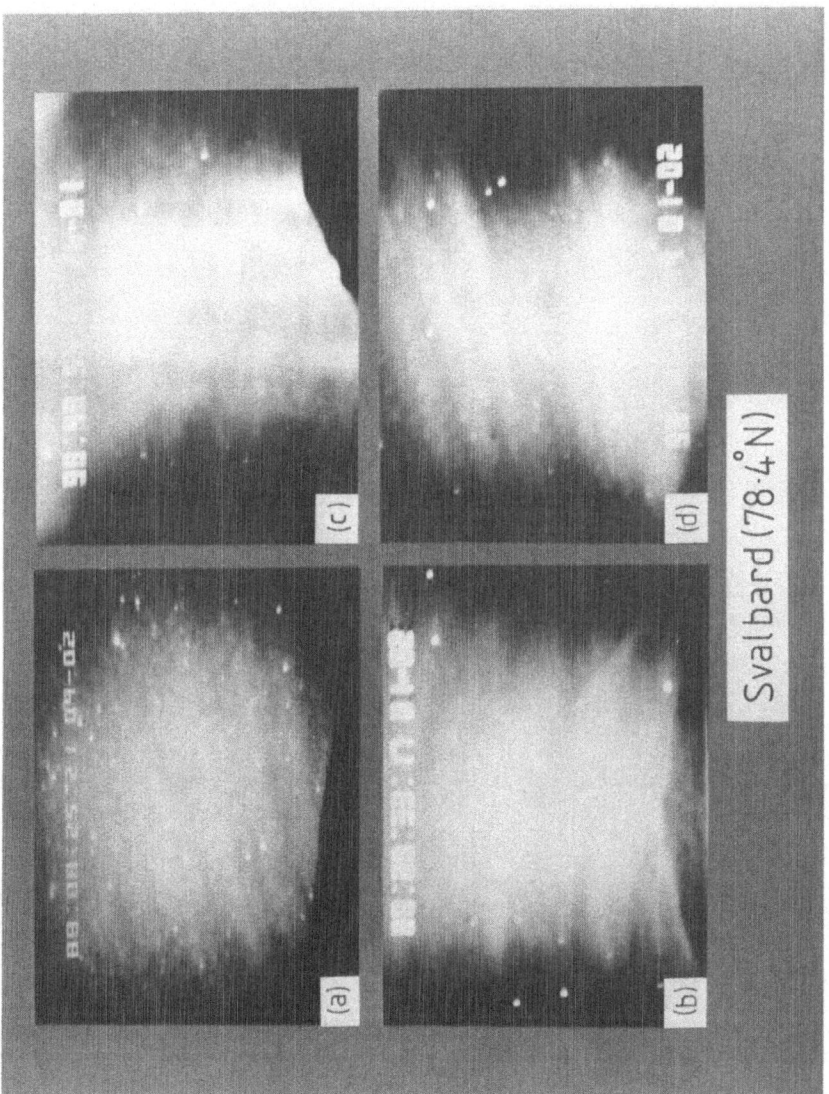

Figure 2. Photographs taken from the TV monitor showing four examples of OH nightglow structure imaged during the Svalbard campaigns.

expect the wave pattern to be clearly visible (the average Kp was 5.9) yet the structures present in Figure 2a are relatively faint and exhibit low contrast <10%.

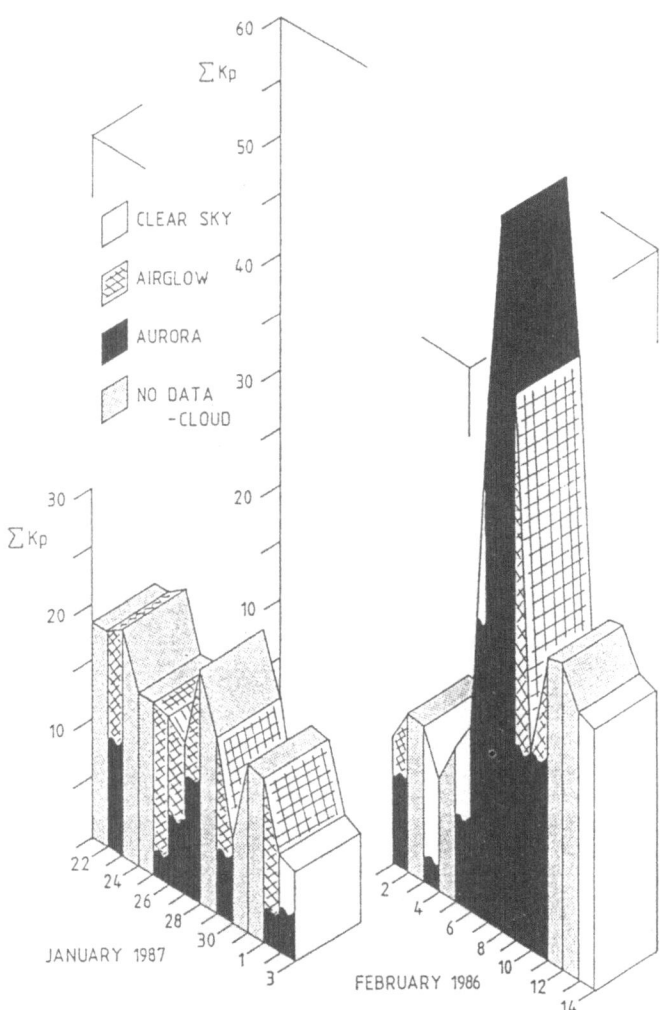

Figure 3. Plot of the magnetic activity index Σ Kp versus the occurrence of airglow wave structure. (Note the hatched areas only indicates the presence of airglow structure and not the amount or clarity.)

During the 1987 observations Σ Kp ranged from 8 to 20. Airglow structure was imaged on 6 occasions during the 13 day period but it was usually of low contrast and short duration. However, the night of 1/2 February was exceptionally, good. Several bright wave patterns were imaged during the 13.5 hour observing time. Figure 2b and 2d show examples of the structures observed. The wave pattern in (d) lasted for about an hour and exhibited a horizontal wavelength of 19 ± 1 km and a velocity of 31 ± 3 ms^{-1} (wave period 10 ± 1 min, assuming no background wind). The structures were moving away

427

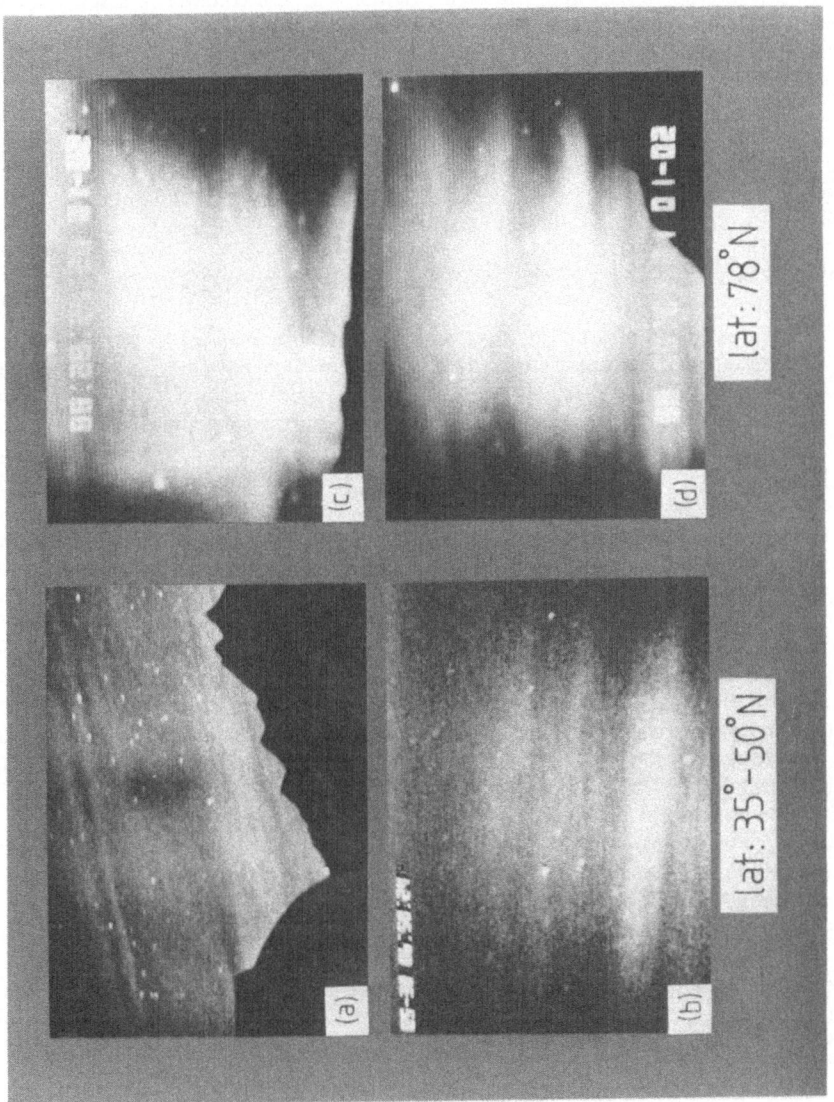

Figure 4. Examples of "broad band" structure imaged at mid and polar latitudes.

from the observing site in a southerly direction. In contrast the wave pattern shown in (b) was seen to the north east and although brighter was less coherent. Σ Kp was about 14 (average Kp~1.8) for this night and yet it only just falls within the "clearly distinguishable" region as proposed by Clairemidi et al.

These observations firmly establish the existence of short period airglow wave structure at polar latitudes. Although we do not have examples of structure during very quiet auroral conditions (i.e. Σ Kp <10) there appears to be no obvious relationship between the magnitude of Kp and either the occurrence or clarity of the wave structure. This result is in contrast to the observations of Clairemidi et al. and implies little or no association between the general auroral activity and the mesospheric wave abundance.

4.2 COMPARISON OF POLAR AND MID-LATITUDE AIRGLOW DATA

Figure 4 shows two examples of the airglow wave structure that we have termed "broad bands" as seen at mid-latitudes and their equivalent imaged at polar latitudes (see section 2). The mid-latitude data (pictures a and b) were obtained from high altitude sites and are generally of better clarity than the images obtained from Svalbard which is at sea level. Nevertheless the similarity in morphology of the two data sets is evident. The large dark band in (a) is over 200 km wide and has some finer "narrow stripe" structure associated with it whilst in (b) three well defined elongated bands are present. For comparison the bands seen in (c) and (d) have similar scale sizes (horizontal wavelengths ~100 km), shape, orientation and duration (few hours). (Note the faint auroral rays to the left in Figure 4c.)

Figure 5 contains examples of mid and polar latitude "narrow stripe"-type airglow wave structure. In (a) the growth of a set of stripes in a region of sky that once contained broad bands is shown. In (b) two isolated stripes are present (the structure in the top right of the image is the Milky Way). The polar latitude data shown in (c) and (d) not only manifest similar periodic structure but also illustrate that stripes can exist in pairs at high latitudes as well. Figure 6 gives three examples of airglow "ripples". Due to the short duration of these events we only have a limited amount of polar data. The mid-latitude examples (a) and (b) show their unique spatial features well. In (c) the polar latitude data are partly obscured by diffuse auroral light but two isolated sets of ripples similar in morphology to those present in (a) are distinguishable. Within 30 minutes the display had faded completely.

These data show that the wave structures imaged at polar latitudes exhibit many similarities to those commonly observed at lower latitudes. As the mid-latitude wave structure is almost certainly not auroral in origin (due to the extreme range of the source from the observing volume) most of the short period wave structure imaged over Svalbard was probably caused by gravity waves that originated from sources located in the lower atmosphere. This does not rule out the aurora as a source of short period gravity waves. However, it does indicate that any waves generated by auroral processes should have similar properties.

4.3 CASE STUDY OF THE AIRGLOW DISPLAY OF 9/10 FEBRUARY 1986

To investigate whether the aurora is a source of short period gravity waves we have made a case study of mesospheric wave structure following isolated precipitation events. On the 9/10 February 1986, after strong precipitation in the early evening, airglow wave structure was observed to grow in a region of sky to the west of Svalbard that had previously been clear. The wave pattern as seen at 21:25 UT is shown in Figure 2a. The structures were first detected at 21:00 UT and lasted for 40 minutes before they

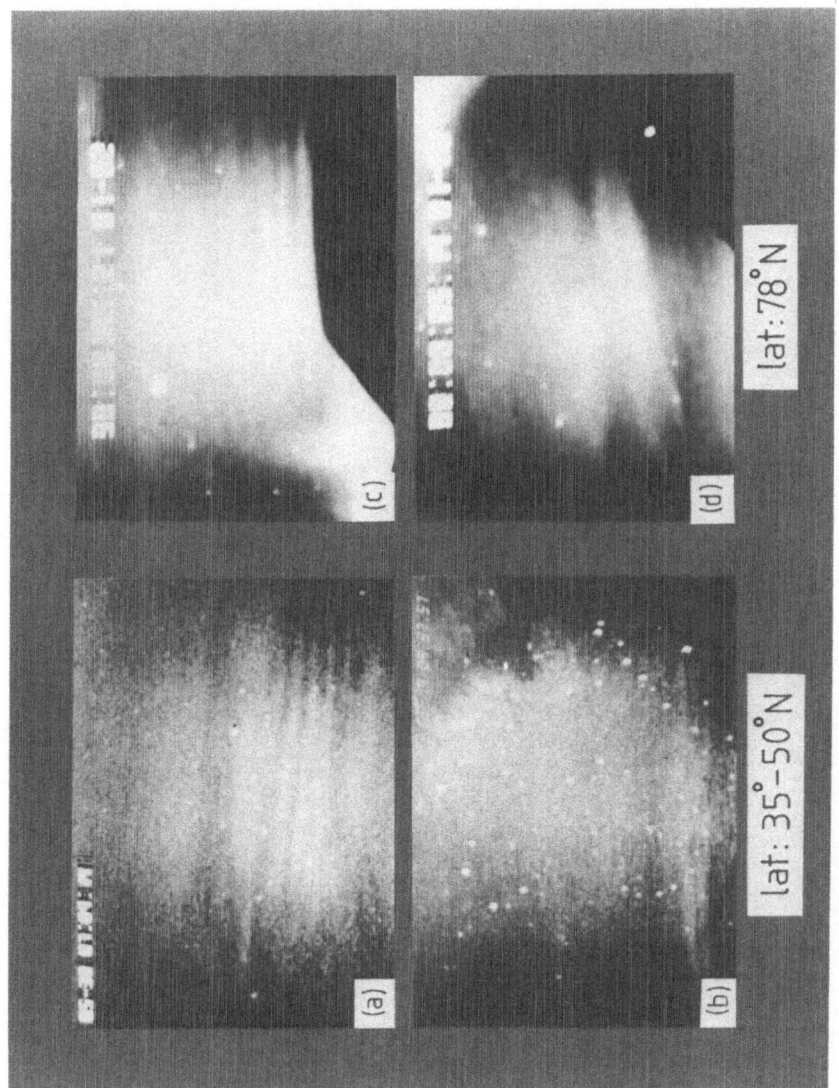

Figure 5. Comparison of "narrow stripe" structure at mid and polar latitudes.

430

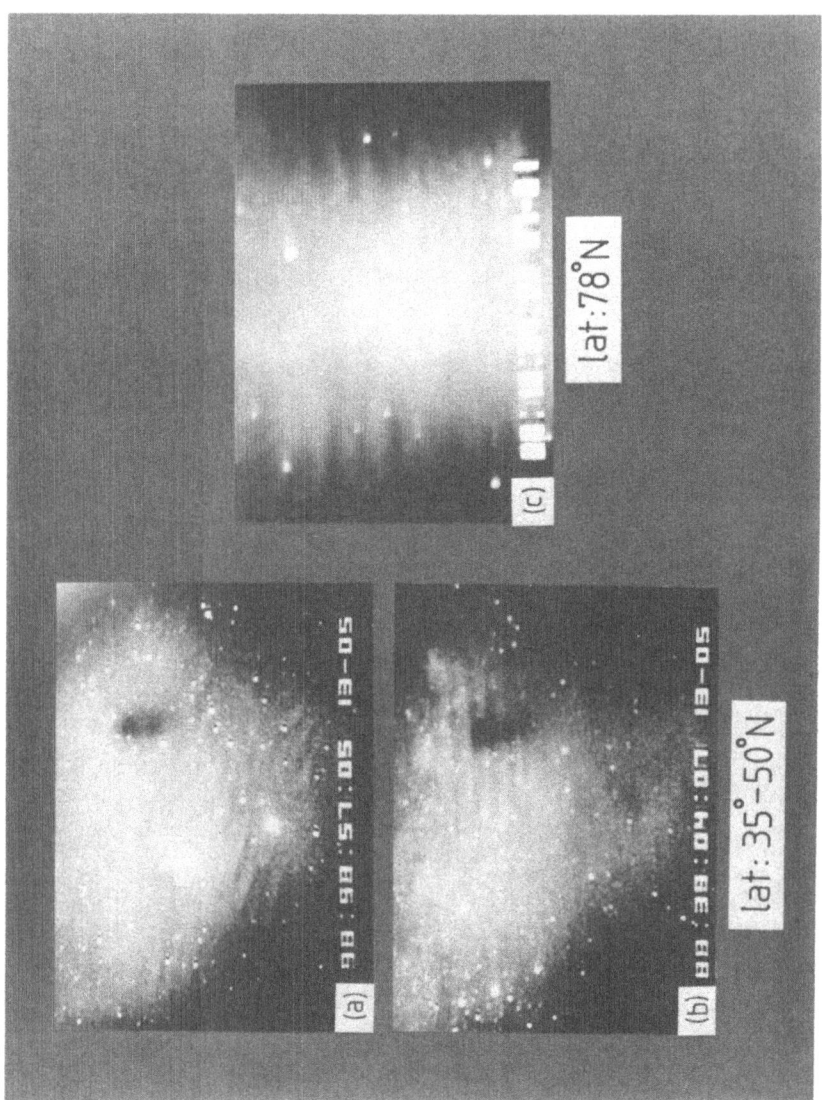

Figure 6. Airglow "ripples" as observed at mid and polar latitudes.

were masked by strong aurora. Figure 7 is a map showing the geographical location, shape and orientation of the four bright wave crests assuming an emission altitude of 87 km. The structures are near North-South aligned and exhibit slight curvature indicating a source region for the gravity waves to their west and at about the same geographic latitude. The pattern exhibited a horizontal wavelength of 34 ± 1 km and a velocity of 40 ± 2 ms^{-1} (giving a wave period of 14 ± 1 minutes, assuming zero background wind).

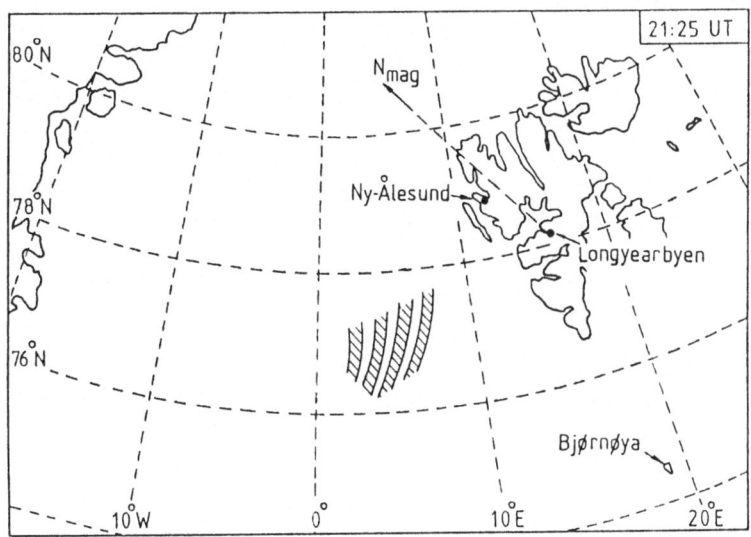

Figure 7. Ground projection of the airglow wave display of 9/10 February, 1986 (as shown in Figure 2a). The hatched areas indicate the location, shape and orientation of the four bright structures imaged at 21:25 UT.

If the structures were generated by the passage of a linear gravity wave train then the energy would have propagated at a maximum angle of about 20° to the horizontal (assuming the vertical wavelength was not substantially greater than the local atmospheric scale height) and at a group velocity of ~35 ms^{-1} (Taylor and Hapgood, 1988). For an auroral source at 120 km the time for the waves to propagate down to the OH layer would therefore have been ~ 1 hour. All-sky camera data for Ny-Ålesund (a research station approximately 115 km due magnetic north of Longyearbyen) show that prior to this display there was one major precipitation event from 18:00 to 18:30 UT and two weaker events centred at 15:50 and 16:50 UT. The main auroral display consisted of several bright arcs well to the south of Svalbard, in the vicinity of Bear Island (Björnöya). The arcs were near geographic East-West aligned and exhibited some poleward motion. As these arcs bore no resemblance in either location, orientation or shape to the airglow wave structure it does not seem likely that heating due to particle precipitation associated with the optical arcs was the source of the gravity waves. The arcs observed earlier were closer to the source region but were generally weaker and occurred far too soon to have been the source.

As mentioned earlier an alternative auroral source could have been Joule heating or Lorentz forcing associated with the electrojet electric field. Recently Crowley and

432

Williams (1987) have related periodicity in the auroral electric field with gravity wave structure observed in the F-region ionosphere to the south of the auroral zone. Thus if the 14 min period airglow structures were caused by an oscillating electric field the magnetometer date should exhibit a similar wave-like variation. (As the induced changes in the ionospheric Hall current should be closely related to the Pedersen currents responsible for the Joule heating.) Figure 8 shows the horizontal (H) component of the magnetometer data for Ny-Ålesund and Bear Island for this night. The times when auroral precipitation and airglow structure were observed from Svalbard are indicated in graphical form.

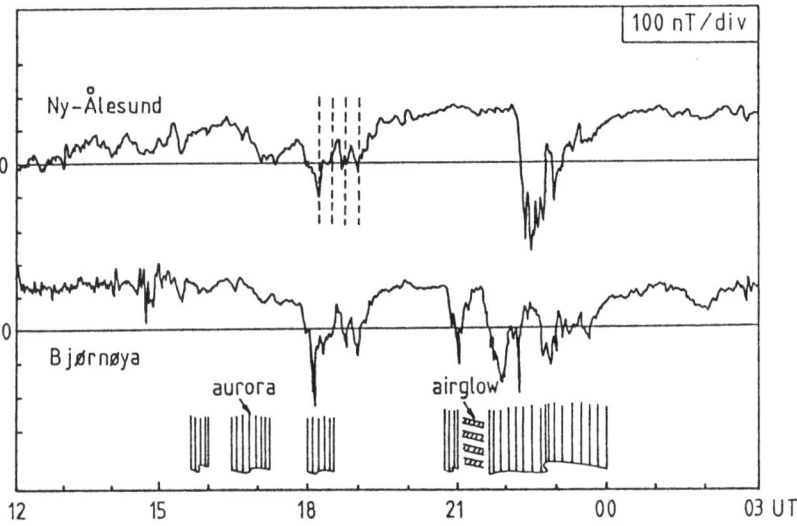

Figure 8. H component magnetometer data for Ny-Ålesund and Björnöya on the night of 9/10 February, 1986. The four dashed lines on the Ny-Ålesund plot illustrate the region that contained oscillations of similar period to that seen in the airglow data.

The structures were seen in a short gap before the onset of strong northward expansion of the oval (Σ Kp ~ 47). For nearly 2 hours prior to the airglow display both magnetometer traces were positive and almost flat indicating very little activity. In contrast to this the auroral disturbance at 18:00 UT contained oscillations of similar period to that determined for the mesospheric wave structure. This is indicated by the four dashed vertical lines on the Ny-Ålesund plot separated by 14 minute intervals.

If indeed this substorm was the source then the gravity waves took over 2.5 hours to propagate, indicating that either the source occurred at a higher altitude than ~120 km or that it was at a greater range than we estimated. The latter may well have been the case as the theory used only gives the minimum range to the source (i.e. the steepest angle of energy propagation). Thus it is possible that the source of the OH structure observed on this occasion was located to the west of Svalbard and was due to a quasi-periodic variation in the magnetospheric electric field associated with the substorm at 18:00 UT. Whether the waves were generated by Joule heating or Lorentz forces remains unknown.

5. Discussion and Summary

This data set is unique. Although several attempts have been made in the past to measure gravity wave "structure" at polar latitudes using spectrometers (that can resolve the airglow signal from the auroral contamination) the signal to noise ratio of the data is usually so low that the spectra have to be summed for several minutes. The image data presented here were obtained in less than 1 second and clearly show the presence of short period wave structure.

The fact that similar type structure is present at such high latitudes as well as at mid and low latitudes indicates that it is a global phenomenon. Structure was observed during times of high and low auroral activity and showed no obvious association with the global magnetic activity index Kp. This fact together with the result that structure seen at polar latitudes bears great similarity to that present at lower latitudes leads to the conclusion that the aurora is not a prominent source of short period gravity waves. However the case study of 9/10 February provides good evidence for the generation of gravity waves by some auroral processes.

Whether all substorms (exhibiting intrinsic periodicity) are strong enough to generate gravity waves that can penetrate down to the airglow layer (against the atmospheric density gradient) is an open question. Waves generated within the troposphere are amplified as they propagate upwards (assuming no dissipation) and thus can constitute a sizeable presence at airglow levels (Hines, 1960). Thus, much of the structure seen at polar latitudes may have been generated by gravity waves originating in the lower atmosphere. We are currently investigating this possibility using satellite and weather data. However, some of the waves appear to have an auroral origin. Wave coupling between the ionosphere and the lower neutral atmosphere at polar latitudes therefore looks likely but is not marked. Further case studies are in hand.

ACKNOWLEDGEMENTS: We are grateful to Mr. M. Maurdrel and Wing Commander Cooper (RAE, Farnborough) for arranging shipment of the video equipment to Norway for the 1987 measurements. Funding for this study was provided in part by the U.K. Science & Engineering Research Council, grant numbers SGD 10320 and GR/E/30959. The data analysis was supported by the U.S. Air Force Office of Scientific Research as part of the MAPSTAR programme.

References

Baker, D.J. and Stair, Jr., A.T. (1987) 'Rocket measurements of the altitude distributions of the hydroxyl airglow', Physica Scripta TID-83, Stockholm, Sweden.

Chimonas, G. and Hines, C.O. (1974) 'Atmospheric gravity waves launched by auroral currents', in C.O. Hines and Colleagues, The Upper Atmosphere in Motion, A.G.U. Publication, Washington, D.C, pp. 672-673.

Clairemidi, J., Hersé, M. and Moreels, G. (1985) 'Bi-dimensional observations of waves near the mesopause at auroral latitudes', Planet. Space Sci., 33, 1013-1022.

Crowley, G. and Williams, P.J.S. (1987) 'Observations of the source and propagation of atmospheric gravity waves', Nature, 328, 231-233.

Davies, M.J. and da Rosa, A.V. (1969) 'Travelling ionospheric disturbances originating in the auroral oval during polar substorms', J. geophys. Res., 74, 5721.

Francis, S.H. (1975) 'Global propagation of atmospheric gravity waves: a review', J. atmos. terr. Phys., 37, 1011-1054.

Hapgood, M.A. and Taylor, M.J. (1982) 'Analysis of airglow image data', Ann. Geophys., 38, 805-813.

Hines, C.O. (1960) 'Internal atmospheric gravity waves', Can. J. Phys., 38, 1441-1481.

Myrabö, H.K., Deehr, C.S. and Sivjee, G.G. (1983) 'Large amplitude nightglow OH(8-3) band intensity and rotational temperature variations during a 24-hour period at 78°N', J. geophys. Res., 88, 9255-9259.

Peterson, A.W. (1979) 'Airglow events visible to the naked eye', Applied Optics, 18, 3390-3393.

Taylor, M.J. (1986a) 'TV observations of mesospheric wave structure', in Collection of Works of the International Workshop of Noctilucent Clouds, Tallinn, Estonian SSR, U.S.S.R., August, 1984, Tallinn "Valgus", pp 153-172.

Taylor, M.J. (1986b) 'Observation and analysis of wave-like structures in the lower thermospheric nightglow emissions', Ph.D. Thesis, Department of Physics, Southampton University, U.K., pp 44-60.

Taylor, M.J., Hapgood, M.A. and Rothwell, P. (1987) 'Observations of gravity wave propagation in the OI(557.7 nm), Na(589.2 nm) and the near infra-red OH nightglow emissions', Planet. Space Sci., 35, 413-427.

Taylor, M.J. and Hapgood, M.A. (1988) 'Identification of a thunderstorm as a source of short period gravity waves in the upper atmospheric nightglow emissions', Planet. Space Sci., 36, 975-985.

50 MHz BACKSCATTER OBSERVATIONS IN THE POLAR CAP IONOSPHERIC E REGION

C. HALDOUPIS[1], M. J. McKIBBEN, J. A. KOEHLER, G. J. SOFKO
Institute of Space and Atmospheric Studies
Physics Department, University of Saskatchewan
Saskatoon, S7N 0W0 Canada

1 : Physics Department, University of Crete
Iraklion, Crete, 714 09 Greece

ABSTRACT. Results of VHF Doppler measurements made at an invariant geomagnetic latitude near 77° are presented. The experiment was set up so that high resolution spectral measurements could be made from a common E region volume along two bistatic CW radio links at magnetic aspect angles of 9.5° and 13.5° from perpendicularity to the earth's magnetic field. For the events considered in this study the evidence, based on K_p index and meridional magnetograms, shows the backscatter region to be located mostly poleward of the auroral electrojet boundary inside the polar cap. The most striking feature of the data is the temporal morphology of the polar cap echo occurrence which differs from that in the auroral zone. In particular there is a complete absence of echoes at midnight and early morning hours, unlike the auroral zone where strong westward electrojet echoes are often present. The strongest and most frequently occurring echoes were observed for a few hours prior to midnight and were always associated with southward Hall drift motions. Comparison with ionosonde data shows a close connection between echo occurrence and the presence of a strong sporadic E_s layer. Magnetometer and riometer data suggest that, at times, other factors, than electrojet currents and absorption, control the echo occurrence at these high latitudes. On the other hand, the Doppler spectral signatures of the echoes suggest the same instability mechanisms to be responsible for the generation of short scale irregularities in the polar cap as in the auroral zone. Finally, the results show the magnetic aspect control is not strong at 50 MHz, thus VHF coherent radars can be useful diagnostic tools in probing the lower ionosphere at polar cap regions.

1. Introduction

Over the last three decades, a large number of ground based VHF radar experiments have been carried out in both hemispheres, in order to investigate the physical properties of meter scale irregularities in the E-region of the high latitude ionosphere (e.g. see recent reviews by Fejer and Providakes, 1987; Haldoupis, 1989). These high latitude backscatter phenomena, which are reffered to as 'Radio Aurora', are closely associated to the electrojet current systems in the auroral zone. This and the property that echoes are aspect sensitive, i.e. the received power intensifies when the transmitted wavevector is perpendicular to the

435

P. E. Sandholt and A. Egeland (eds.), Electromagnetic Coupling in the Polar Clefts and Caps, 435–447.
© *1989 by Kluwer Academic Publishers.*

earth's magnetic field, has caused the vast majority of the VHF radars to be located at sites equatorward of the auroral zone because this would ensure reception of strong echoes and maximum echo occurrence. As a result, latitudinal coverage of the phenomena has been limited in the 55° to 70° magnetic latitude range, i.e. to L shell values between $\sim$ 3 to 10.

To our knowledge, the only measurements of meter scale irregularities at magnetic latitudes higher than 70° were made by A. G. McNamara in Canada during the 1957 - 59 IGY period by using a network of identical 48 MHz non-Doppler radars. McNamara, (1972) reported that frequent E region echoes were seen inside the polar cap and even at latitudes close the magnetic dip pole. There are several other measurements which indicate that meter scale electrostatic turbulence exists in the E region poleward of the auroral zone; HF radars, ionosondes, and rocket probes (e.g. Primdhal et al., 1974; Tsunoda et al., 1976; Bahnsen et al., 1978). All these measurements have suggested that the prime agent for the generation of these polar cap short scale irregularities is the modified two stream (or Farley-Buneman) instability, i.e. the same mechanism which is responsible for short wavelength plasma waves in both the auroral and equatorial electrojet E regions (e.g. see Fejer and Kelley, 1980). This, of cource, is not surprising, because the polar cap E region would be expected to have strong Hall currents carried by the electrons under the action of significant dawn to dusk electric fields which could lead to electron drift velocities relative to the ions larger than the ion sound velocity threshold needed for the plasma to become unstable.

Given the existence of E-region turbulence, the fundamental question to be asked here, however, is if these strongly magnetic field aligned irregularities could be detected in the polar cap by VHF radars, because aspect angles at E region heights are fairly large, possibly, larger than 15°. As it will be shown in the present report, the large aspect angles do not seem to be a prohibitive factor for echo occurrence at these geomagnetic latitudes; in sharp contrast to the existing theories.

Obviously, VHF radio measurements in regions poleward of the auroral oval are highly desirable. These regions, which constitute the polar cap, have unique geomagnetic properties which lead to distinctive forms of phenomena to take place. The special interest to the polar cap regions emanates from the notion of an open magnetospheric model. The polar cap ionosphere, in contrast to the earth's ionosphere anywhere else, it is possible to be exposed to direct interaction processes between the earth's magnetic field, the solar wind and the interplanetary magnetic field (IMF). Thus, VHF radio measurements in this region, if realizable, would have an important diagnostic role to play not only for the level of ionospheric turbulence but also for the dynamic state of the ionosphere and its direct response to solar wind and IMF changes.

In the present paper we present measurements of 3-meter irregularities, made with a CW bistatic Doppler system operating at 50 MHz and observing at an invariant geomagnetic latitude near 76.5°. The scatter region, which extends in the 76° to 77° magnetic latitude range, is located near the polar cusp in the dayside sector (e.g. Sandholt et al. 1985) and well into the polar cap in the nightside when the auroral oval expands equatorwards. This paper deals only with general characteristics of the echoes. A study on the Doppler spectrum properties of the data can be found in a paper by Sofko et al. (1987).

2. The Experiment

In July 1982, a 50 MHz CW radio Doppler system had been operated by the University of Saskatchewan Radar Auroral Group in the Canadian Arctic for two weeks in order to investigate the E region backscatter. Detailed accounts of the basic principles of the experiment and the equipment in general can be found in previous publications (e.g. see Koehler et al., (1985). Here we provide only those experimental details which are pertinent in the present study.

The basic experimental setup included two transmitters and a common receiver, all units being located at different sites which were widely separated.The radio links between transmitters and receiver were established using narrow beamwidth (5.5°) interferometric arrays of Yagi antennas all having their maximum radiation directed to the same scattering volume at E region heights. Figure 1 summarizes the experimental geometry relative to both, geographic and geomagnetic coordinates and provides several other experimental details. The experiment was designed so that the scattering region was over Sachs Harbour, North West Territories (geographic coordinates, 72.0° N; 125.2° W; Invariant geomagnetic latitude, 76.5° and L value, ~ 18.5°).

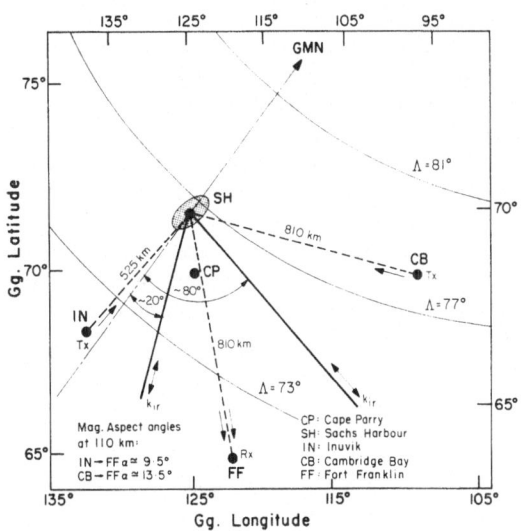

Figure 1. Experimental geometry and various other experimental details

The observing directions were along the bisectors of the angles between transmitter-scattering region-receiver, and formed angles of 20° and 80° east of magnetic south for the Inuvik-Fort Franklin (INF) and the Cambridge Bay-Fort Franklin (CBF) links, respectively. For the geometry of the experiment shown in Figure 1, the INF and CBF links are sensitive to irregularity wavelengths equal to 3.5 m and 3.7 m, respectively. The high magnetic latitude location of the experiment produced rather unfavorable magnetic aspect angles (i.e. the angles between the observing directions and the normal to the magnetic field). Here, the mean aspect angle within the scattering region at about 110 km altitude is 9.5° for the INF link and 13.5° for the CBF link. The aspect angle calculations were based on the 1980 MAGSAT data and they are accurate within a degree, but contain no correction for possible refraction. Refraction will certainly lower the actual aspect angles but Moorcroft (1985) has shown that this correction would be expected to be small, usually lower than a degree or so.

3. Location of the Scattering Region

Next we will comment on the following question: Is the scattering region, which is confined in the 76° to 77° invariant magnetic latitude, located inside the polar cap? The answer to this question would depend on the definition of the polar cap as well as the level of the geomagnetic activity. The most common meaning attached to the term 'polar cap' refers to the region enclosed by the auroral oval, the latter defined as the belt around the geomagnetic pole of the true location of aurora at any time (e.g. Acasofu and Chapman, 1972). The auroral oval however, which is assymmetric about the pole, it is also a highly dynamic feature, thus its location depends on the time sector and the level of magnetic activity.

In order to provide some kind of an estimate about the polar cap status of our scattering region, we follow the procedure of Fraser-Smith (1982) and make use of the auroral oval statistics for different K_p indexes of Chubb and Hicks (1970). According to these authors, for $K_p = 3$ the mean geomagnetic latitudinal limits of the oval is 65° to 69° on the nightside and 74° to 78° on the dayside. In this case (i.e. for $K_p = 3$) the scatter region above Sachs Harbour is located inside the dayside oval and well into the polar cap in the magnetic night sector, whereas in the early afternoon and late morning sectors possibly it is in the vicinity of the dynamic interface between the oval and polar cap. For more disturbed conditions, the inner and outer oval limits move equatorward and according to the results of Chubbs and Hicks (1970), Sachs Harbour is most likely inside the polar cap for $K_p \geq 3^+$. Another way to check on the location of the scattering region is to examine meridian chain magnetograms which are useful in identifying where the region is relative to the auroral electrojets and the E-field convection reversal boundary. For nearly all the events of strong backscatter during the entire 2 week long campaign, there was found an opposite sense of polarity in the H magnetogram in the high latitude stations around Sachs Harbour as compared to lower latitude (auroral) stations. This suggested Sachs harbour to be located poleward of the auroral zone, most likely inside the cap.

4. Observations

The experiment had been in operation on a continuous basis for two weeks, from July 5 (day 186) to July 19 (day 200) of 1982. During this summer period the atmosphere above Sachs Harbour is always sunlit with the sun's altitude varying, on the average, from a low of 4° to a high of 40°. The three hourly K_p index underwent drastic variations during the campaign period, from extremely quite ($\Sigma K_p = 7^+$ during July 5) to extremely disturbed conditions ($\Sigma K_p = 56$ in July 13). At Sachs Harbour, local midnight is 08 30 UT whereas magnetic local midnight was near 11 00 UT.

The procedures of data recording and processing have been described elsewhere in detail (e.g. Koehler et al., 1985). In the present work, in order to study the echo occurrence we have examined Esterline Angus chart records and temporal variations of the receiver gain settings during the whole campaign period. To study the type of irregularities and their mean motion we used one minute spectral estimates obtained by averaging a large number (290) of unit spectra each obtained by Fourier transforming a 2048 point time series. In the digitizing process a fixed sampling interval of 0.1 ms was used resulting in 4.9 Hz frequency resolution. In addition to the radio data we also used magnetometer and riometer recordings made available by the World data center (WDC), as well as ionogram data obtained with a digital ionosonde operating at Sachs Harbour during the campaign.

4.1 DIURNAL FREQUENCY OCCURRENCE OF THE ECHOES

A characteristic pattern has been emerged in the daily variation of the echo occurrence. The great majority of echoes and the strongest backscatter took place in the time sector a few hours prior to local magnetic midnight, whereas an unexpected null in reception followed in the postmidnight early morning hours. Also, some subsidiary echo activity was observed soon after dawn followed by another minimum in the near noon early afternoon hours. This diurnal pattern persisted in a regular fashion for all the observational period (i.e. 2 weeks). The mean daily backscatter occurrence, based on half hour intervals, is illustrated in histogram form in Figure 2. Notice that a strong maximum occurs about 2 to 3 hours after magnetic dusk (i.e. after 18 h LMT) while a secondary maximum is seen a couple of hours after magnetic dawn. This behaviour is different from that observed in the auroral zone (i.e. $L \simeq 4$ to 10) where a strong maximum is seen near magnetic midnight with a great deal of echoes received during the early morning hours (e.g. Unwin, 1972; McNamara, 1972; Haldoupis and Sofko, 1976, among several others).

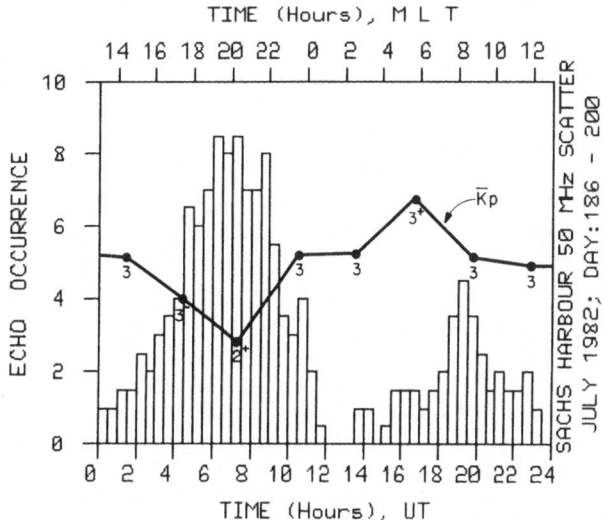

Figure 2. Mean diurnal distribution of backscatter occurrence. Also shown is the mean variation in K_p during the period of the experiment.

In Figure 2, the diurnal variation of the 3 hourly K_p index averaged over the whole campaign period is also plotted. Although it appears that the maximum in scatter occurrence seemed to correspond to weaker rather than stronger magnetic activity, we believe, our data base is not large enough for a definitive relationship to be established between the frequency of occurrence of echoes and K_p. In general, however, we can state that the dependence of backscatter on K_p index is rather poor at this large geomagnetic latitude. This, again, differs from what is known for radio aurora where the probability of echo occurrence increases with increasing K_p index (e.g Unwin,1972).

4.2 COMPARISON WITH IONOSONDE DATA

A digital ionosonde was in operation at Sachs Harbour (beneath the scattering region) for several days in July 1982 and provided coverage for about 70 percent of the campaign period. The sounder was sweeping its 1.6 to 16 MHz frequency range in 20 s, and instant ionograms were plotted out in real time, every 10 min. Unfortunately, the ionogram quality was poor because of low sensitivity, making the data inappropriate for a quantitative analysis.

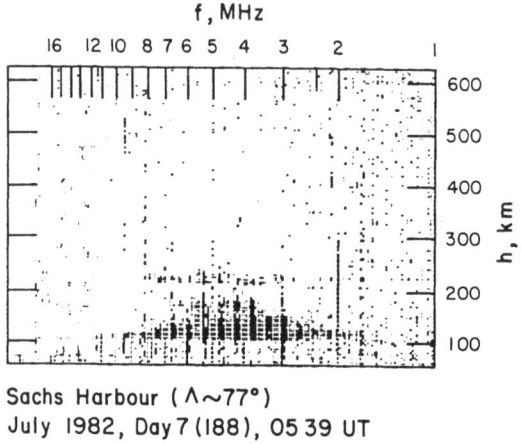

Sachs Harbour (Λ∼77°)
July 1982, Day 7 (188), 05 39 UT

Figure 3. Typical sporadic E_s ionogram signature observed nearly always when strong 50 MHz echoes were received

Inspection of the ionosonde data, in conjunction with the 50 MHz data, revealed a close correlation between the occurrence of VHF echoes and the appearence of a pronounced sporadic E_s type trace in the ionograms. In general, the comparison indicates a nearly one to one correspondence, especially in the premidnight sector, suggesting a close interelationship between these two phenomena. A typical example of the ionosonde signature observed during periods of strong 50 MHz backscatter is shown in Figure 3. This is a well known 'spread' E_s type of sporadic E layer seen often during disturbed periods in high latitudes (e.g. Piggot and Rawer, 1975). Usually there is complete blanketing (i.e. no reflections appear from higher layers at all), whereas often there are equally spaced multiple traces due to ground reflections, which become progressivly weaker with height. These observations are supportive of a strong E_s layer and of relatively weak non deviative absorption in the D region.

Figure 4 shows the daily occurrence distribution of the sporadic E_s layers and blackout events (i.e. absence of any trace), based on one hour intervals during the whole period the ionosonde was in operation (notice,this period extends a week after the end of the radio campaign). Comparison of Figures 2 and 4, shows a great deal of similarity between the probability of occurrence of sporadic E_s and 50 MHz echoes while on the other hand, the largest frequency of blackouts occurs at times when the backscatter is greatly reduced. The latter could again suggests that D region absorption may have a serious attenuating effect on the 50 MHz echoes, because an ionosonde blackout is attributed to strong non deviative absorption. This option will be pursued further in the following section.

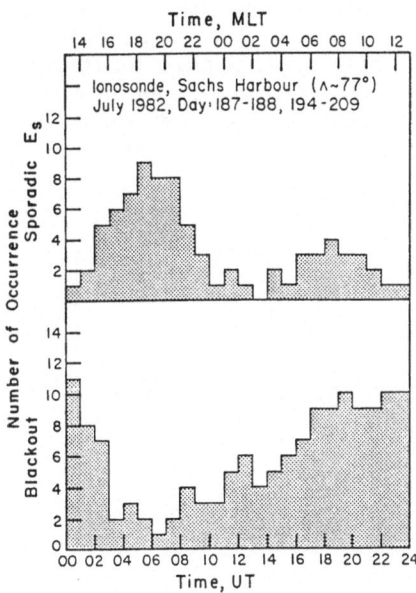

Figure 4. Occurrence distributions of sporadic E_s (upper panel) and ionosonde blackout events (lower panel) observed underneath the scattering region at Sachs Harbour

4.3 COMPARISON WITH MAGNETOMETER AND RIOMETER DATA

To gain some understanding on the unusual echo occurrence distribution, we have examined magnetometer and riometer data from the Alaska meridian chain, supplied by the WDC. Unfortunately, the instruments at Sachs Harbour were not in operation during July 1982. As a result, we had to use available data from neighboring stations to draw some conclusions. The nearest station to Sachs Harbour was Cape Parry located 180 km south of the scattering region center. In the following statistics we use magnetometer and riometer records from Cape Parry to compare with the radio data. We base the validity of this comparison on the good deal of similarity that exists between the magnetic variations seen in the northern stations of Inuvic, Cape Parry and the northest station at Mould Bay, therefore one would expect the hourly averaged variations at Sachs Harbour to be similar, although often weaker, to those at the nearest station of Cape Parry.

Next we show that the presence of strong equivalent current activity is often not accompanied with 50 MHz backscatter. Figure 5 summarizes riometer magnetometer, and echo amplitude backscatter data for the whole campaign period. In the upper two panels of Figure 5 we compare the temporal variations in echo strength and equivalent horizontal current density $(H^2 + D^2)^{1/2}$, based on hourly mean values. Also in the upper panel the 3 hourly K_p index variations are also shown. A close inspection of this figure shows poor correlation between the echo occurrence and strength, and the intensity of magnetic perturbations observed on the ground as well as the level of geomagnetic activity represented by K_p. Often, when the magnetometer shows the presence of strong currents (e.g. day 194-195) there is no echo activity, whereas at times fairly strong scatter is observed during periods of relatively weak magnetic variations. Also, at days the equivalent current is quite strong in the early morning hours when echoes were almost never observed. These findings

442

are in contrast to what has been observed for auroral VHF echoes received at small magnetic aspect angles. Namely, that the received power is proportional to the horizontal component of the equivalent current system measured by ground magnetometers (e.g Greenwald et al., 1975; Haldoupis et al., 1982, among others). Obviously, there are other factors, in addition to the electrojet currents, which at times dominate and control entirely the strength and probability of echo occurrence at these large magnetic latitudes.

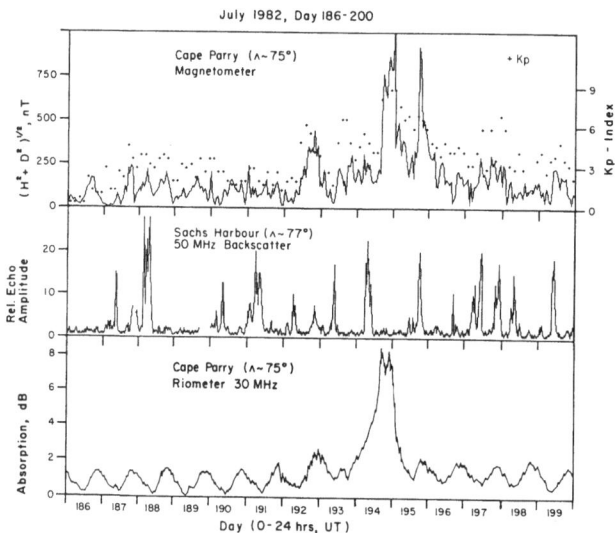

Figure 5. Time series summary of magnetometer, riometer and 50 MHz backscatter amplitude variations during the two week campaign period. See section 4.3 for details.

Next we examine the importance of non-deviative ionospheric absorption as a potential source of attenuation of the 50 MHz backscatter. This option may indeed be significant especially in this experiment, because all sites are located at large geomagnetic latitudes. In these latitudes the ionospheric absorption is subject to large increases, relative to its background levels, during disturbed periods caused by auroral energetic particles (i.e. auroral type absorption), and/or during SID type events when relativistic protons can cause dramatic enhancements in D region electron densities and therefore widespread strong polar cap absorption (PCA) events (e.g. see Hultqvist, 1966). Since this abnormal non-deviative absorption takes place mainly at D region altitudes one would expect both the transmitted and received signals to suffer a great deal of attenuation because the radiowave spends a large part of its propagation time in penetrating obliquely the D layer.

To estimate the effect of absorption on the echo strength we considered 30 MHz broadband riometer data from Cape Parry which is located on the way between the scattering region at Sachs Harbour and the receiver at Fort Franklin The lower panel in Figure 5 shows the mean variation of 30 MHz cosmic noise absorption at Cape Parry based on hourly averaged values for the whole campaign. Other than a small time lag, a similar variation is also observed further south at Inuvik. For the location of Inuvik and Cape Parry relative to the experiment sites, see Figure 1. Inspection of Figure 5 shows: 1) A pronounced diurnal variation in absorption with a maximum near local noon and a minimum near midnight,

2) The absorption level increases somewhat during the time period from day 192 to 196 in line with increases in the level of geomagnetic activity, and 3) A dramatic absorption enhancement occurs during day 194 and 195, apparently caused by a strong PCA event as suggested by the WDC neutron monitor data. In the same period there was a continuous ionosonde blackout and virtually no 50 MHz backscatter despite the very strong current system that must have existed in the E region (as suggested by the large magnetogram deflections).

In order to obtain a rough estimate of radio wave attenuation due to absorption at 50 MHz, we use the Appleton's formula of non deviative absorption for a vertically propagating wave, i.e.:

$$A(dB) = 4.5 \times 10^{-8} \int_h \frac{N_e + \nu_e}{\nu_e^2 + \omega^2} dh, \tag{1}$$

where N_e and ν_e are the height dependent values of electron density and electron collision frequency respectively, and ω is the angular wave frequency. Because $\nu_e^2 \ll \omega^2$, we can obtain an approximate estimate for the total oblique absorption at 50 MHz from the observed vertical absorption at 30 MHz through the expression

$$A_{50}(dB) = 2A_{30}(dB)(\frac{\omega_{30}}{\omega_{50}})^2 sec(90° - \phi), \tag{2}$$

where ϕ is a mean elevation angle of the propagation path and the factor 2, at the beginning of the right hand side expression is to account for both transmitted and received paths. By taking a representative angle $\phi = 10°$, we estimate the total attenuation in echo strength at 50 MHz to be about 4 times that observed by the 30 MHz riometer (i.e. $A_{50}(dB) \sim 4A_{30}(dB)$). Provided that on the average the signal level is of the order of about 10 dB, we see that the attenuation of 50 MHz echoes due to increased ionospheric absorption may be important. As an example, during day 194 the absorption causes an attenuation in the 50 MHz scatter higher than 20 dB, so this alone could possibly explain the absence of echoes during this period of large magnetic disturbance.

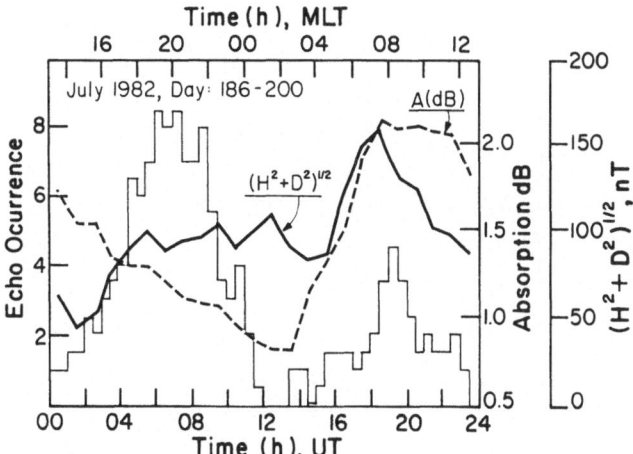

Figure 6. Mean diurnal variation of magnetometer and riometer data superimposed on the echo occurrence distribution of Figure 2 for a direct comparison

444

Although absorption may have a considerable attenuating effect on the backscatter strength, it cannot account, as in the case of the magnetometer variations, for the observed diurnal pattern in the observation of backscatter. This is emphasized in Figure 6 where the mean diurnal variations of riometer and manetometer data of Figure 5 are superimposed on the observed diurnal distribution of echo occurrence for comparison purposes. Certainly, one cannot reconcile the fact that the scatter null after local magnetic midnight coinsides with the minimum of absorption. Also, it is difficult to explain the absence, or reduction, of echoes during times when there is strong horizontal current activity. Obviously, the presented evidence cannot account for an adequate explanation as to what causes the characteristic diurnal backscatter pattern.

4.4 SPECTRAL SIGNATURES AND MEAN IRREGULARITY MOTIONS

To complete the observations of the main data characteristics we briefly discuss the Doppler spectral signatures of the echoes. The Doppler spectra are useful in identifying the instability mechanisms which generate the irregularities. A more detailed study on the Doppler spectra of these data, is given by Sofko et al. (1987) who found that, in spite of the large magnetic aspect angles, the Doppler spectra show all the usual types (i.e. Type 1, 2, and 3) as in the auroral zone. Typical examples of normalized power specta indicating the existence of three distinct irregularity types are shown in Figure 7. The uppermost left corner panel shows a broad type 2 spectrum for the CBF link and a mixture of what were likely ion acoustic and ion cyclotron type irregularities for the INF link. A similar situation but with the ion acoustic peak being dominant, is given in the lower left pannel. The spectra at the right panels show that type 3 peaks can have mean peak phase velocities near either 100 m/s or 200 m/s. The lower value corresponds to the gyrofrequency of the molecular ions (either NO^+ or O_2^+), while the higher value corresponds to their second harmonics and/or the gyrofrequency of the O^+ ion.

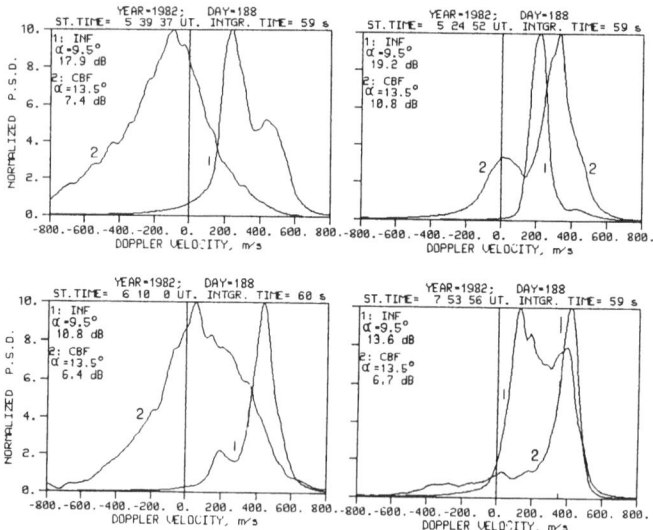

Figure 7. Typical examples of the observed Doppler spectral types

Finally, Figure 8 represents in a summarized form the general behaviour of the mean irregularity drift motions that prevail in the premidnight period of strong echo reception. The irregularity drift velocity vector were computed from simultaneously observed spectra at both radio links by combining the mean Doppler shifts. Figure 8 shows predominantly southward motions for all nine different events of backscatter. Under the assumption that irregularity drifts follow closely the $\vec{E} \times \vec{B}$ drift of the electrons, we conclude that the convection electric field is quite strong and points mainly westward during the times of strong echo occurrence.

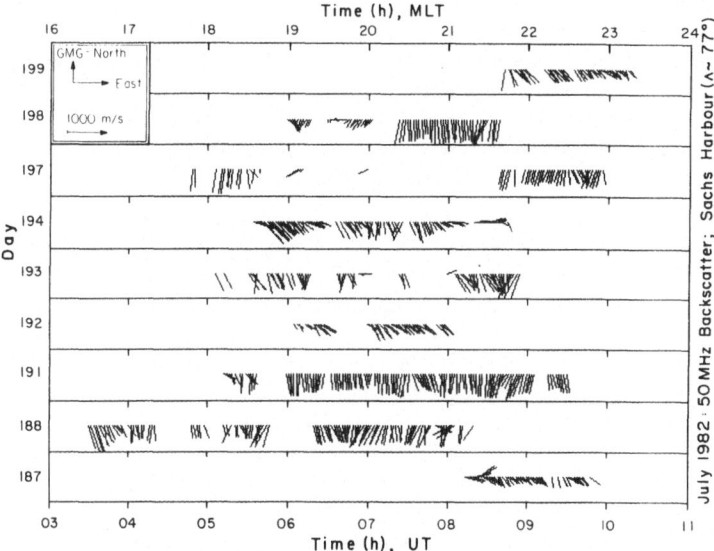

Figure 8. Summary of the mean irregularity drift motions observed in the premagnetic midnight sector during periods of strong backscatter received in both radio links.

At this point, it is interesting to mention that the above premidnight period of strong backscatter coinsides with the time sector of the Harang discontinuity (HD) occurrence. The HD is a dynamic auroral region that often intrudes into the polar cap which is characterized, among other things, by a westward turning of the electric field and intense upward field aligned current activity (e.g.see Sofko et al. (1985) and papers cited therein).

5. Summary and Concluding Remarks

In this paper we presented results obtained in the polar cap E region with a Doppler experiment using two bistatic radio links operating at 50 MHz. The results are summarized as follows:

1) VHF radio auroral zone phenomena, associated with E region electrostatic plasma wave turbulence, extend also into polar cap regions.

2) In the geomagnetic altitude near 77° deg, the diurnal echo occurrence at 50 MHz follows a pattern that differs from that observed in auroral zone latitudes. The most characteristic difference is the complete absence of echoes at midnight and early morning hours

regardless of geomagnetic activity.

3) There is a close interdependence between echo occurrence and the presence of a strong sporadic E_s layer.

4) The correlation between 50 MHz backscatter and geomagnetic activity, inferred from both, K_p index and magnetogram equivalent current densities, is ruther poor contrary to the auroral zone.

5) Although, non-deviative ionospheric absorption can have important effects on the 50 MHz echo strength and occurrence, absorption alone cannot account for the mean diurnal variation in backscatter.

6) The strongest echoes are observed for a few hours in the premagnetic midnight sector and are always associated with southward Hall drift motions.

7) Magnetic aspect angle control is not strong at 50 MHz; plasma waves with $k_\parallel/k_\perp$ ratios as large as 0.25 can exist in the polar cap E region.

8) At polar cap latitudes the same instability mechanisms seem to be responsible for the short scale irregularities as in lower, auroral zone, latitudes. The existing instabilty theories, however, cannot explain these large magnetic aspect angle observations.

From this work and the work of Sofko et al. (1987), it has become clear that the polar cap and auroral zone spectral properties of 50 MHz E region echoes are similar but that the temporal morphology of echo occurrence differs. The fact that polar cap echoes maximize during few hours in the premidnight sector at the same times as the Harang discontinuity, which occurs in the auroral zone and intrudes north into the cap, may be of importance in understanding the mechanism responsible for polar cap echoes. One possibility to consider is the presence of strong upward field parallel currents, which are known to be present in the Harang discontinuity latitudinal sector, as an alternative destabilizing source down in the E region plasma. This argument is favored by the present ground magnetogram data which suggests that $\vec{E} \times \vec{B}$ drifts may be necessary but not sufficient for the appearance of the echoes. Another important observation that needs to be exploited further is the one to one relationship betwenn 50 MHz echoes and E_s. The present data cannot tell if both phenomena have the same common cause or if one leads to the other and which. We know, however that a strong E_s is related to a narrow layer of enhanced electron density and consequently to a sharp vertical density gradient. On the other hand, both N_e and gradN_e play an important role in the generation of radio aurora although this role is far from being fully studied and understood. An additional parameter that needs also to be taken into consideration in a viable interpretation, is the magnetic aspect conditions at these high latitudes.

The present results pose a number of questions to which our data alone cannot provide many answers. Certainly, more of VHF radar experiments are needed at these large geomagnetic latitudes. In addition, supporting evidence from groung based instrumentation and satelite data (e.g. IMF data) are necessary. Finally, an important point brought up with this experiment has to do with the usefulness of VHF radars in studying the polar cap ionospheric E region. As it has been shown, the large magnetic aspect angles are not as prohibitive as they thought to be in the detection of echoes. Consequently, VHF coherent radar systems can probe the lower ionosphere in the polar cap and provide valuable information of geophysical importance in studying the coupling of the system: Solar Wind-IMF-Magnetosphere-Ionosphere.

Aknowledgements.This work was supported by National Science and Engineering Research Council of Canada (NSERC) grants to J. A. Koehler and G. J. Sofko.

6. List of References

Acasofu, S.-I. and Chapman, S. (1972) Solar-Terrestrial Physics, University Press, Oxford.

Bahnsen, A., Ungstrup, C., Falthammar, C.-G., Fahleson, U., Olesen, J. K., Primdahl, F., Spasnglev, F. and Pedersen, A. (1978) 'Electrostatic waves observed in an unstable polar cap ionosphere', J. Geophys. Res., 83, 5191-5197.

Chubb, T. A. and Hicks, G. T. (1970) 'Observations of the aurora in the far ultraviolet from OGO 4', J. Geophys. Res., 75,1290-1311.

Fejer, B. G. and Kelley, M. C. (1980) 'Ionospheric irregularities', Rev. Geophys., 18, 401-454.

Fejer, B. G. and Providakes, J. F. (1987) 'High latitude E region irregularities: new results', Phys. Scripta, Vol T 18, 167-178.

Fraser-Smith, A. C. (1982) 'ULF/lower-ELF electromagnetic field measurements in the polar caps', Rev. Geophys. Space Phys., 20, 497-512.

Greenwald, R. A., Ecklund, W. L. and Balsley, B. B. (1975) 'Radar observations of auroral electrojet currents', 80, 3635-3641.

Haldoupis, C. (1989) 'A review on radio studies of auroral E region ionospheric irregularities', Ann. Geophysicae, 7.3.

Haldoupis, C., Nielsen, E. and Goertz, C. K. (1982) 'Experimental evidence on the dependence of 140 MHz auroral backscatter characteristics on ionospheric conductivity', J. Geophys. Res., 87, 7666-7670.

Haldoupis, C. and Sofko, G. J. (1976) 'Doppler spectrum of 42 MHz CW auroral backscatter', Can. J. Phys., 54, 1571-1584.

Hultqvist, B.(1966) 'Ionospheric Absorption of cosmic radio noise', Space Sci. Rev.,5. 771-817.

Koehler, J. A., Sofko, G. J., Mehta, V., McNamara, A. G. and McDiarmid, D. R. (1985) 'Observations of magnetic aspect effects in auroral radar backscatter', Can. J. Phys., 63, 402-408.

McNamara, A. G.(1972) 'The occurence of radio aurora at high latitudes: The IGY period, 1957-1959', Geophys. Publ., 29,135-149.

Moorcroft, D. R. (1985) 'An examination of radio auroral aspect sensitivity', Can. J. Phys., 63, 1005-1012.

Piggot, W. R. and Rawer, K. (1975) URSI handbook of ionogram interpretation and reduction, Elsvier Publ., Amsterdam.

Primdahl, F., Olesen, J. K. and Spangslev, F. (1974) 'Backscatter from a postulated plasma instability in the polar cap ionosphere and the direct measurement of a horizontal E region current', J. Geophys. Res., 79, 4262-4268.

Sandholt, P. E., Egeland, A., Holtet, J. A., Lybekk, B., Svenes, K. and Asheim, S. (1985) 'Large- and small- scale dynamics in the polar cusp', 90, 4407-4414.

Sofko, G. J., Koehler, J. A., Haldoupis, C., McKibben, M. J. and McNamara, A. G. (1987) 'Doppler radio observations of 3-meter irregularities in the polar cap E region', J. Geophys. Res., 92, 1271-1276.

Sofko, G. J., Koehler, J. A., Prikryl, P. and McDiarmid, D. R. (1985) '50-MHz auroral Doppler spectra dynamics during the Harang discontinuity', Radio Sci., 20, 696-708.

Tsunoda, R. T., Perreault, R. D. and Hodges, J. C. (1976) 'Azimuthal distribution of HF slant E region echoes and its relationship to the polar cap electric field', J. Geophys. Res., 81, 3834-3844.

Unwin, R. S. (1972) 'The radio aurora', Ann. Geophys., 28(1), 111-127.

THE MIDDLE AND HIGH LATITUDE IONOSPHERE AT ~ 550 km ALTITUDE

Y. K. TULUNAY
İ.T.Ü. Uçak ve Uzay Bilimleri Fakültesi
Maslak, İstanbul, Turkey
and
J. M. GREBOWSKY
NASA/Goddard Space Flight Center, Greenbelt, MD 20771, USA

SUMMARY

Ambient electron density measurements made by the radio frequency
capacitance probe on the Ariel 4 satellite have been analysed in
order to study the structures present at the satellite altitudes,
i.e., between ~ 470 and ~ 600 km, at latitudes poleward of ± 50° Λ
both during winter and equinox. Data obtained on 94 days centered on
the solstices, and March 1972 have been used. The general morphology
of the extreme densities, the mid-latitude electron density trough,
the polar cap depletions and electron density enhancements associated
with the cusp auroral zone were determined statistically. Analysis
of the winter solstice and the equinox data acquired during quiet
magnetic conditions ($K_p \leq 2+$) show that the Northern and Southern
Hemisphere ionospheres were significantly different. The global
characteristics of the winter solstice and the equinox electron
densities were similar although they differ from each other in fine
details.

1. INTRODUCTION

Tulunay and Grebowsky [1] and Tulunay [2] after analysing electron
density data obtained by means of the Ariel 4 satellite during two
magnetically quiet ($K_p \leq 2+$) periods centered on the winter solstice
and the March 1972 equinox at high latitudes in both the Northern and
Southern Hemispheres, reported that the Northern and the Southern
Hemisphere winter and equinox ionospheres were significantly
different. The analysis technique used by Brinton et al. [3] was
adopted in both of these studies. In this paper the results of the
above mentioned papers are reported in order to facilitate a
simultaneous comparative examination of the earth's ionosphere at
about 550 km.

P. E. Sandholt and A. Egeland (eds.), Electromagnetic Coupling in the Polar Clefts and Caps, 449–454.
© *1989 by Kluwer Academic Publishers.*

2. THE ARIEL 4 SATELLITE AND THE DATA ANALYSIS TECHNIQUE

Ariel 4 was launched in December 1971 into a near-circular orbit
having an inclination of 83°; the orbital period was 95 min, with
perigee and apogee occurring at approximately 470 and 600 km,
respectively. The electron density data were obtained by means of a
radio-frequency capacitance probe |4| and all the results used were
obtained from tape-recorded data. Measurements of the electron
density were made at latitude intervals of 1.9° ± 0.5° at low and
middle latitudes |5|. The winter periods investigated were from
16 December 1971 to 19 March 1972 for the Northern Hemisphere and
from 26 April to 29 July 1972 for the Southern Hemisphere. The
equinox period investigated was from 4 February 1972 to 8 May 1972
for both the Northern and Southern Hemispheres. All the data were
restricted to magnetically quiet periods having the 3-h planetary
magnetic activity index $K_p \le 2+$.

The maximum and minimum electron densities at any given location were
obtained using a technique similar to that described by |3|. The
electron densities observed by Ariel 4 during the winter solstice
ranged between about 9×10^2 and 2×10^5 cm^{-3} and the actual values were
divided into the following seven ranges: (1) 10^2-4×10^3, (2) 4×10^3-
8×10^3, (3) 8×10^3-1.6×10^4, (4) 1.6×10^4-3.2×10^4, (5) 3.2×10^4-6.4×10^4,
(6) 6.4×10^4-1.3×10^5 and (7) greater than 1.3×10^5 cm^{-3}. The electron
densities observed by Ariel 4 during the equinox ranged between about
1×10^3 and 3×10^5 cm^{-3} and the actual values were divided into the
following six ranges: (1) 10^3-8×10^3, (2) 8×10^3-1.6×10^4, (3) 1.6×10^4-
3.2×10^4, (4) 3.2×10^4-6.4×10^4, (5) 6.4×10^4-1.3×10^5, (6) greater than
1.3×10^5 cm^{-3}. The spatial locations of the measurements made within
each of the electron density ranges were plotted on "maps" based on
magnetic local time and invariant magnetic latitude, i.e., in M.L.T.
-Λ space. By overlaying the plots corresponding to all six electron
density ranges in order of decreasing number density, it was possible
to obtain the "extreme electron density" topographical maps shown in
Fig. 1a or Fig. 2a for both hemispheres; these maps show, for any
M.L.T. -Λ location, the maximum electron density observed during the
94-day period. Similarly, by overlaying the six plots in order of
increasing electron density, it is possible to obtain topographical
maps (Fig. 1b or Fig. 2b) which indicate the minimum electron density
observed at each M.L.T. -Λ location.

3. RESULTS AND CONCLUSIONS

The main purpose of this paper is to present a comprehensive
topographical picture of the extreme ambient electron densities
observed over the high magnetic latitudes at about 550 km altitude,
using a similar approach to that adopted by Brinton et al. |3| for
the ion composition near 300 km. Figures 1a and 2a show the
topographical maps of the Northern Hemisphere and Southern Hemisphere
electron densities indicating the maximum electron densities measured

when $K_p \leq 2+$ at all M.L.T. $-\Lambda$ locations during the relevant 94-day winter solstice and March equinox periods respectively. Figures 1b and 2b are the topographical maps of the Northern Hemisphere and the Southern Hemisphere electron densities indicating the minimum electron densities during quiet magnetic conditions ($K_p \leq 2+$) at all M.L.T. $-\Lambda$ locations during the relevant 94-day winter solstice and March equinox periods. The following results and conclusions were reached concerning magnetically quiet conditions ($K_p \leq 2+$) during the 1972 winter solstice periods |1|.

(i) The highest electron densities were observed in a region that is symmetrical with respect to the 02-14 M.L.T. meridian in the Northern Hemisphere but symmetrical with respect to the 00-12 M.L.T. meridian in the Southern Hemisphere.
(ii) Electron densities were generally smaller in the Southern Hemisphere than in the Northern Hemisphere.
(iii) The mid-latitude trough is the main structure at night in both hemispheres. In the Northern Hemisphere the low densities did not persist into the dayside, but in the Southern Hemisphere the region of reduced electron density extended almost throughout the day and towards higher latitudes.
(iv) A polar cavity is observed in both hemispheres.
(v) In the Southern Hemisphere the highest winter electron densities observed ($>1.3 \times 10^5$ cm^{-3}) occurred only in the dayside cusp region whereas in the Northern Hemisphere the maximum density region extended to low latitudes in the daytime.

Data for the 1972 March equinox periods revealed the following results and conclusions during quiet magnetic conditions ($K_p \leq 2+$) |2|.

(i) The highest electron densities were observed in a region that is symmetrical with respect to the 02-15 M.L.T. meridian in the Northern Hemisphere but symmetrical with respect to the 00-12 M.L.T. meridian in the Southern Hemisphere.
(ii) Electron densities were generally smaller in the Southern Hemisphere than in the Northern Hemisphere. The extreme values of the electron densities were also observed in the Southern Hemisphere.
(iii) The mid-latitude trough is the main structure at night in both hemispheres.
(iv) A polar cavity is observed in both hemispheres.
(v) In the Southern Hemisphere the highest winter electron densities observed ($>1.3 \times 10^5$ cm^{-3}) occurred only in the dayside cusp region whereas in the Northern Hemisphere the maximum density region extended to low latitudes between 06 and 18 M.L.T.

A comparison of the winter solstice results with the March equinox results revealed that electron densities were smaller in the winter data in both hemispheres in general. However, both sets of data exhibited a similar pattern statistically on the M.L.T. $-\Lambda$ space. The results showed that during quiet magnetic conditions ($K_p \leq 2+$) the Northern and the Southern Hemisphere winter and equinox ionospheres were significantly different; in particular, the Southern Hemisphere densities were lower than those in the Northern Hemisphere during both 94-day periods.

4. REFERENCES

|1| Tulunay, Y. T., Grebowsky, J. M., Hemispheric Differences in the Morphology of the High Latitude Ionosphere Measured at ~ 500 km, Planet. Space Sci., Vol. 35, No. 6, pp. 821-826, 1987.

|2| Tulunay, Y. T., Ambient Electron Density Distribution at About 500 km Altitude at the Earth's Ionosphere, Reidel Pub. Co. (in press) 1988.

|3| Brinton, H. C., Grebowsky, J. M., and Brace, L. H., The High Latitude Winter F Region at 300 km: Thermal Plasma Observations from AE-C, J. Geophys. Res., Vol. 83, p. 4767, 1978.

|4| Sayers, J., Proc. R. Soc. Lond., A281, p. 450, 1964.

|5| Goodall, C. V., Wall, J., and Hopkins, H. D., The Electron Density Experiment on Board the Ariel 4 Satellite, J. Brit. Interplanet. Soc., Vol. 26, p. 135, 1973.

FIGURE CAPTIONS

Fig. 1a Topographical "maps" of the Northern Hemisphere (left-hand diagram) and the Southern Hemisphere electron densities indicating the maximum electron densities measured when $K_p \leq 2+$ at all M.L.T. -Λ locations during the relevant 94-day solstice periods.

Fig. 1b Topographical "maps" of the Northern Hemisphere (left-hand diagram) and the Southern Hemisphere electron densities indicating the minimum electron densities measured when $K_p \leq 2+$ at all M.L.T. -Λ locations during the relevant solstice periods.

Fig. 2a Topographical "maps" of the Northern Hemisphere (left-hand diagram) and the Southern Hemisphere electron densities indicating the maximum electron densities measured when $K_p \leq 2+$ at all M.L.T. -Λ locations during the relevant 94-day equinox periods.

Fig. 2b Topographical "maps" of the Northern Hemisphere (left-hand diagram) and the Southern Hemisphere electron densities indicating the minimum electron densities measured when $K_p \leq 2+$ at all M.L.T. -Λ locations during the relevant equinox periods.

453

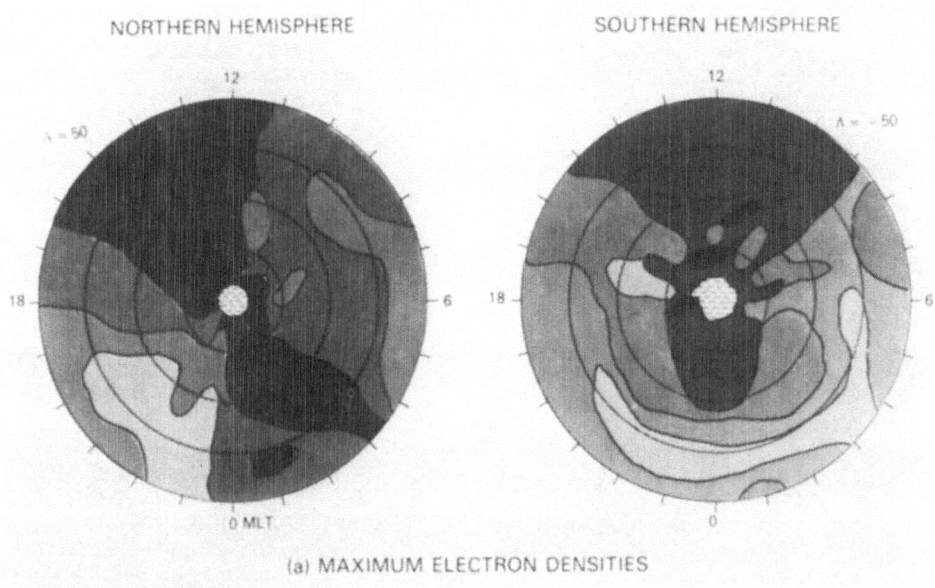

(a) MAXIMUM ELECTRON DENSITIES

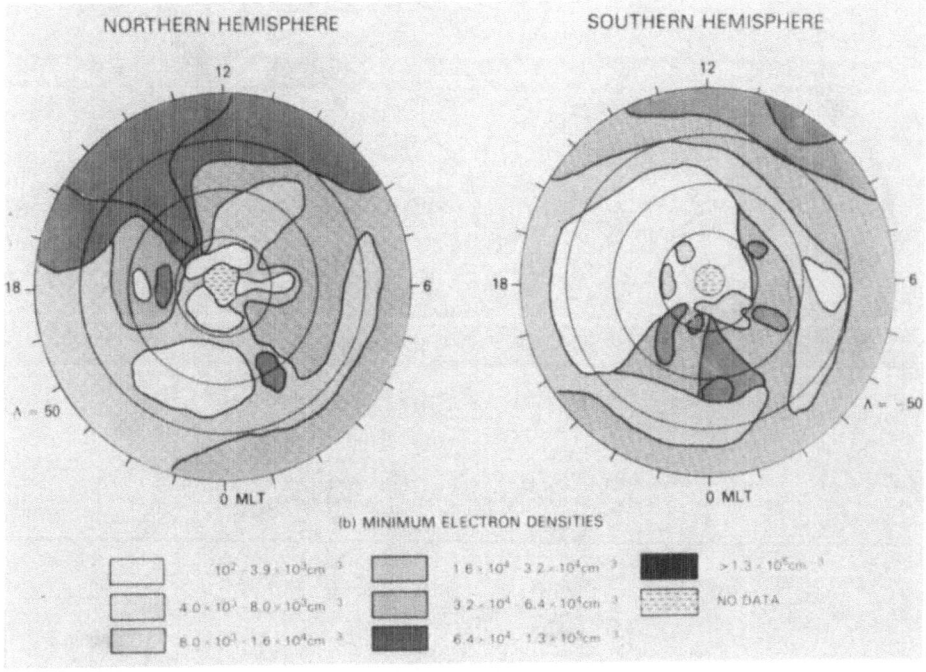

(b) MINIMUM ELECTRON DENSITIES

Figure 1

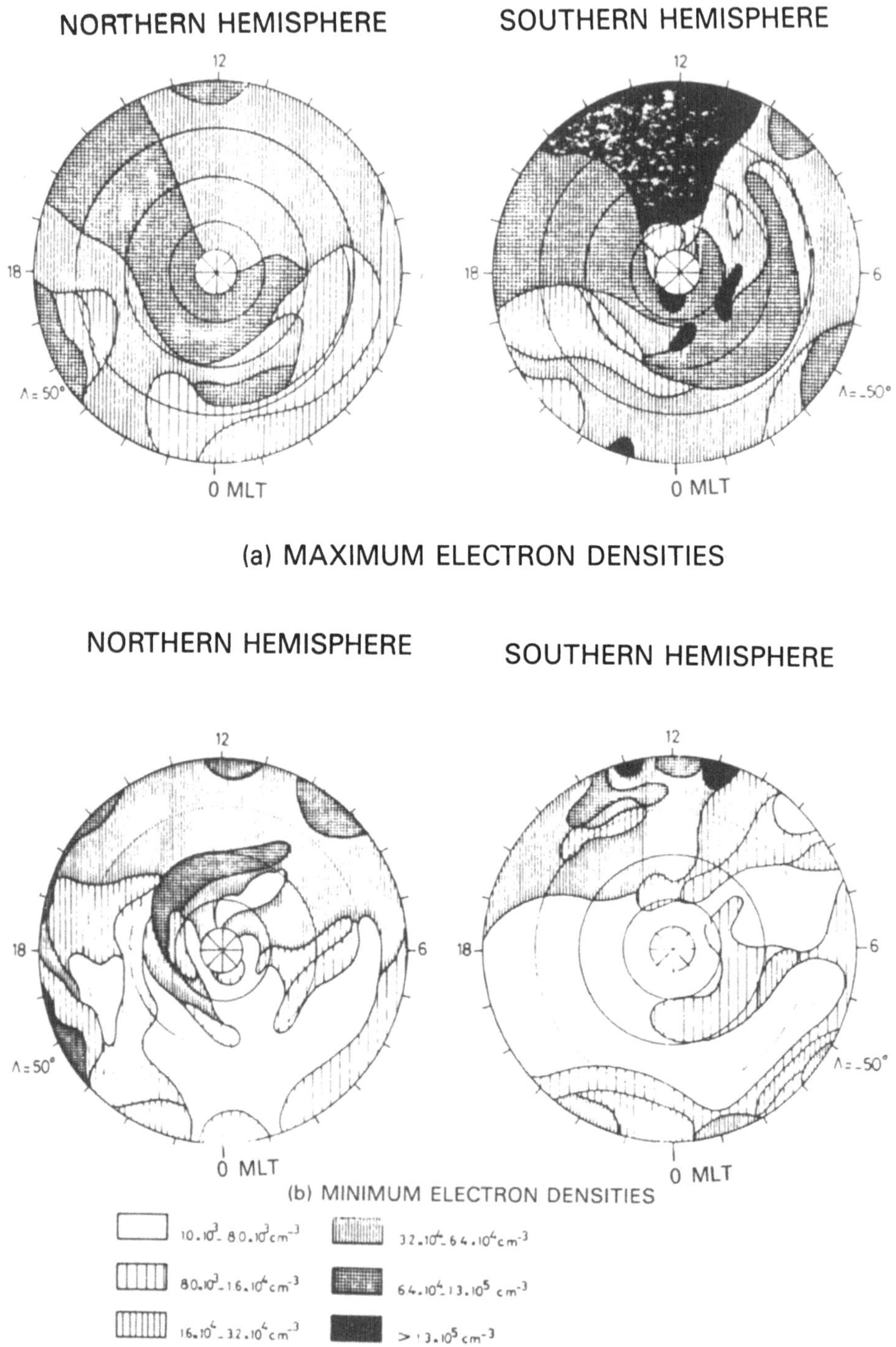

NORTHERN HEMISPHERE
SOUTHERN HEMISPHERE
12
18
6
Λ = 50°
Λ = -50°
0 MLT
(a) MAXIMUM ELECTRON DENSITIES
NORTHERN HEMISPHERE
SOUTHERN HEMISPHERE
12
18
6
Λ = 50°
Λ = -50°
0 MLT
(b) MINIMUM ELECTRON DENSITIES
1.0.10^3 - 8.0.10^3 cm^-3
8.0.10^3 - 1.6.10^4 cm^-3
1.6.10^4 - 3.2.10^4 cm^-3
3.2.10^4 - 6.4.10^4 cm^-3
6.4.10^4 - 1.3.10^5 cm^-3
> 1.3.10^5 cm^-3

Figure 2

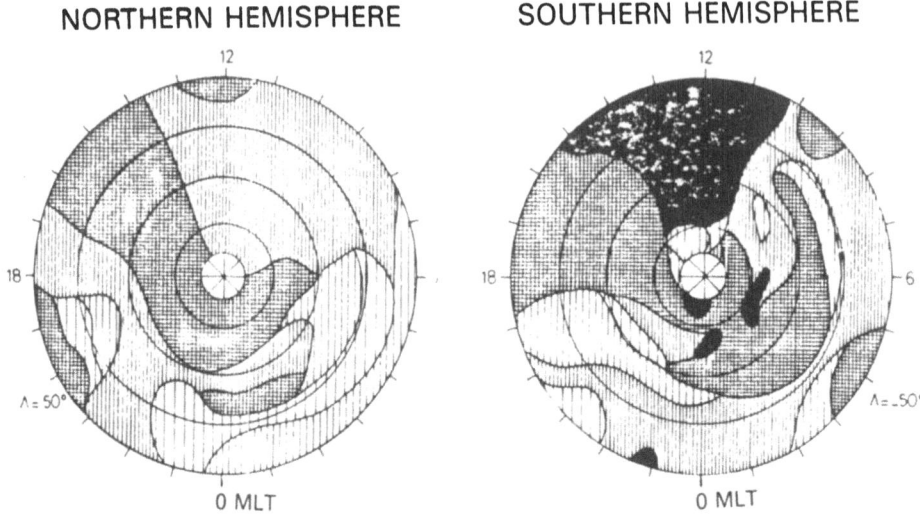

SUMMARY

R.W. Smith
Geophysical Institute
University of Alaska
Fairbanks, Alaska

INTRODUCTION

Papers and discussion at this workshop have shown clearly
that one of the most exciting and yet illusive areas of
research in solar-terrestrial coupling is the transfer of
momentum and energy from the solar wind into the magnetospheric cavity. The key to an understanding of the mechanisms lies in the magnetospheric boundary layer sunward of
the dawn-dusk plane. Although B-field and plasma measurements from ISEE and AMPTE satellites have been interpreted
in terms of reconnection processes during which solar wind
plasma may enter the magnetosphere, events before and after
are not well known. Indeed, recent work in the literature
challenges the idea that the B-field perturbations were
caused by reconnection, rather favouring an interpretation
in terms of density fluctuations in the solar wind. Further
progress in the study of these events has been reported at
this meeting following the identification by Sandholt et
al. (1989); Lockwood and Cowley (1989), and Lockwood et al.
(1989) of optical and ion drift signatures of these boundary events (cf. also Rairden and Mende, 1989). Pulsations
in the geomagnetic field and the detailed structure and
motions of discrete auroral forms seen at the dayside polar
cap boundary have been searched for behaviour matching the
theoretically proposed signatures. Despite recent success
with ionospheric signatures there remains an information
void about the connection between magnetospheric boundary
layer events and their ionospheric signatures. Discussion
in the final session centered on experimental and theoretical work which is already planned or could be proposed
which will provide the missing information.

The sequence of topics in this report of the discussion begins with a new international proposal GAMBLE which
is a part of the Global Environmental Modelling (GEM) pro

P. E. Sandholt and A. Egeland (eds.), Electromagnetic Coupling in the Polar Clefts and Caps, 455–460.
© 1989 by Kluwer Academic Publishers.

gram, followed by discussions of a possible polar cap incoherent scatter radar on Svalbard, new spacecraft and
rocket initiations and experiments, the importance of
routine monitoring of the geomagnetic field and ionosphere,
and the need for predictive theories to stimulate further
experiments.

1.   THE GAMBLE (Groundbased and Airborne Magnetospheric
      Boundary Layer Experiment) PROGRAM

Based on the idea that processes at the boundary layer of
the magnetosphere may have ionospheric counterparts in the
polar cap, a program of observations in The Eastern Arctic
has been planned which is centered at the Longyearbyen and
Ny Alesund stations on Svalbard. Dayside aurora can be
readily observed in the Eastern Arctic from Thule to Heiss
Island using low light level TV or CCD cameras. A region of
adjacent or overlapping all-sky camera views of the dayside
aurora may be obtained during clear-sky conditions from
Thule, Nord (Greenland), Ny Alesund, Longyearbyen, Hornsund

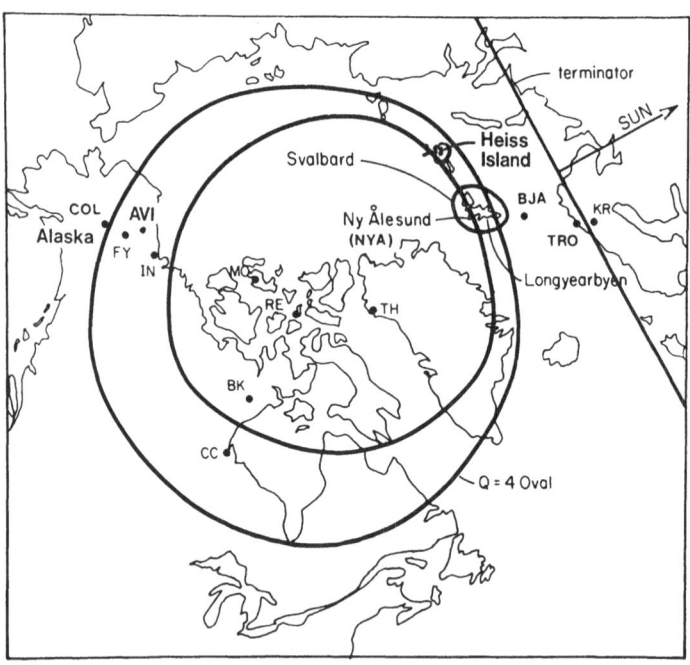

Fig. 1. Relationship of Svalbard, Heiss Island (USSR), and
Greenland to the Q = 4 auroral oval (Feldstein and Starkov,
1967) and the sunlit Earth in December at 09 UT. North
American observing chains are also indicated.

(Svalbard, Norway), and Heiss Island, Franz Josefs Land (USSR) (cf. Figure 1). With these stations operating simultaneously, the dimensions and dynamics of coherent forms may be found, and interesting events selected for detailed study. Meridian scanning, monochromatic photometers are also proposed for these stations (some of them are already equipped) which will enable the relative intensities of auroral emissions to be found and hence the general nature of the precipitation spectrum inferred. The location of the cusp/cleft can be monitored at high time-resolution.

The stations would also have flux-gate and pulsation magnetometers for the study of disturbance events in the ionosphere due to the dayside plasma-entry phenomena. Some stations may also have riometers. A proper description of ionospheric convection is needed and cannot be obtained without some radio or radar measurements. Lockwood and Cowley have shown the importance of a knowledge of the ion drift or electric field signature. A polar cap convection radar is an essential part of the GAMBLE program, but is not yet in place.

2.   THE POLAR CAP RADAR

Dr. Oguti stated that for proper studies of the electrodynamics of the magnetosphere, knowledge is required of the convection pattern with simultaneous data on the magnetic field and optical emissions. A new Institute in Japan is presently preparing plans for an incoherent scatter radar on Svalbard and will seek National funds for their implementation.

Dr. Lockwood explained a different plan put forward within the UK which would combine forward scatter from EISCAT with a new radar on Spitzbergen giving five possible lines of sight and five possible slices through the ion velocity function. Such an installation may also measure field-aligned flows. A proposal for such a complex, extending the EISCAT radar range and grafting on a new polar cap facility extending to 85° invariant latitude has been received very favourably by the UK national funding agency. The plan is to instal the forward scatter receiver first followed by the incoherent scatter radar transmitter for the polar cap as the second stage.

Dr. Greenwald commented that an efficient means to sample forward scatter from EISCAT would be to use a phased array antenna with 16 beams to give multiple directions of view.

3. NEW SPACE EXPERIMENTS

Dr. Winningham reported on a recent meeting which took a forward look into new experiments in Space relevant to the Earth's magnetosphere. Imaging was a major topic. Techniques to produce an image of the structure by remote sensing include the neutral particle method and EUV imaging of solar resonance from magnetospheric constituents such as He^+ or O^+. This is an essential complement to in situ sampling and measurement which identifies the context of the observation.

A new solar physics mission called JANUS, a satellite in orbit above the Lagrangian point, was highlighted as an opportunity for the imaging technique. The satellite would detect the events in the solar wind as it passed and remotely sense their impact on the dayside magnetosphere by imaging. Good theory and modelling would be needed to support this.

Dr. Potemra commented that the emphasis in studying planetary magnetospheres may shift away from the Earth to Mercury after ISTP. Dr. Mende added that active (release) experiments will have a role to play, best done from multiple, free flying, spacecraft. Dr. Smith commented that recent work at The Geophysical Institute in Alaska has indicated the value of magnetospheric imaging using resonantly scattered sunlight from O^+. This ion traces regions of mass flow between the ionosphere and magnetosphere. A camera operating at the 834 A line of O^+ has been proposed. Simulations of the experiment show details such as the radiation belts and the plasma sheet with 15 minute exposures from about 60 R_E. Dr. Vasyliunas remarked that imaging of the ring current and plasma sheet is less useful than high temporal and spatial resolution images of the aurora. Through a proper interpretation it may be used to learn what happens at the magnetospheric boundary layer. On the other hand, magnetospheric images will not show boundary layers well. Dr. Mende cautioned that magnetospheric images could be very ill-defined with poor time resolution because of low intensities. Large apertures will be required to obtain adequate signal.

Dr. Winningham continued by describing the CENTAUR project. This consists of a pair of Black Brant X rockets to be flow from Andøya, Norway in December 1989, having a comprehensive payload of plasma and field instruments well suited to the investigation of ionospheric signatures of plasma-entry events. The launches will take the payload through the cusp at altitudes above the F-region peak up to 800 km. This project will use ground-based optical instruments from a meridian-scanning monochromatic photometer to determine the cusp position. Other ground-based measurements of the configuration and dynamics of the cusp and cleft auroras will be needed to complete the data set gathered by the rockets. Collaborations are planned with the EISCAT radar and optical stations downrange of Andøya.

Dr. Smith outlined an experimental proposal called "Barflite" by Drs. Westcott and Nielsen of the Geophysical Institute in Alaska. A high-altitude Scout rocket would be used to place a pair of oppositely directed Barium shaped charges up at about 1 R_E to be detonated with one jet going up and the other going down the field line. Field aligned electric fields and plasma instabilities will be investigated by optical backing of the Barium streaks.

Dr. Reiff added that CLUSTER should make a critical contribution to this area of research. This would be particularly effective if the CLUSTER group of satellites intersects the boundary layer just tailward of the dayside. It is this area which is the principle region of momentum transfer (where $J \cdot E$ has the right sense). She also foresaw that CLUSTER observations of tail-site reconnection under B_z positive conditions may be expected to be very interesting. In all these cases, ground-based supporting measurements will be needed.

4. ROUTINE MONITORING

New and exciting developments in magnetospheric and ionospheric research may occur because of a new instrument or program, but seldom would such work have come to tuition without data which is gathered by routine monitoring. Dr. Rishbeth developed this theme, stating that by a recent study, about 1/2 of all papers in issues of major refereed journals in the area of solar-terrestrial physics used data or indices which result from routine monitoring. This is data which is hardly noticed until it is needed.

Despite the utility of the results, in the UK and other countries funding for routine measurements is continually threatened because of possible irreversible short term measures to save money. A long term approach is needed to safeguard our routine monitoring activities. Constant vigilance is essential. Dr. Basu informed us that IMF data from IMP 8, at periods when it is in the solar wind, is not collected for the full period available because some ground-tracking stations operate only during business hours. An effort is needed to gain more coverage of the satellite in the Pacific sector, possibly by adding a receiving station by collaboration with Japan.

5. PREDICTIVE THEORIES

Dr. Primdahl commented that he could recall only one test-
able prediction which had been made during the meeting.
Such predictions are the stimulants of new thought and
activity which are continually needed. Dr. Vasyliunas
responded that many of the easier predictions, requiring
little work to make, have been made. Today, it is more
difficult to come up with a worthwhile prediction without
considerable effort. Dr. Lockwood reminded us that so many
programs have within them the comprehensive instrumentation
needed for new ideas to be postulated and tested without
being noticed so much in the literature and meetings. Ideas
are still flowing but get proposed and checked out in small
group meetings. Dr. Lemaire predicted that new information
would come by extending our attention to other solar wind
quantities besides B_X, B_Y, and B_Z. He emphasised the possible importance of $\Delta B/B$, solar wind velocity and density
considered at high time resolution (certainly higher than
hourly values).

REFERENCES

Lockwood, M., and S.W.H. Cowley, Observations at the magne-
    topause and in the auroral ionosphere of momentum
    transfer from the solar wind, Adv. in Space Res., in
    press, 1989.
Lockwood, M., P.E. Sandholt, and S.W.H. Cowley, Dayside
    auroral activity and magnetic flux transfer from the
    solar wind, Geophys. Res. Lett., 16, 33-36, 1989.
Rairden, R.L., and S.B. Mende, 6300 Å auroral emission at
    South Pole: Dayside poleward motion and sun-aligned
    arcs, this volume.
Sandholt, P.E., B. Lybekk, A. Egeland, R. Nakamura, and T.
    Oguti, Midday auroral breakup, J. Geomagn. Geoelectr.,
    in press, 1989.

SUBJECT INDEX

(Where a subject is treated over more than one page in the same article only the first page is given.)

Acceleration mechanisms 312
AC fields - electric 103,299
         - magnetic 103,167,187,299
Airglow 127,421
Air pressure 404
Alfvén waves 4,176,191,300
Aurora   - cusp 253,299,311,459
          - cleft 274,311,319,459
          - polar cap 47,61,269,314,319
          - sun-aligned arcs 61,277,327
Auroral   - dynamics (cusp/cleft) 253,269,314,323,455
          - electrojet 155,245
          - emission rates 195,319,423
          - intensity ratios 313,457
          - morphology 68,248,312,319
          - oval 1,61,134,259,272,312,319,356,432
          - pulsations 195
          - spectrum,423,458
Averaging technique 239

Ballon observations 137
Barometric equation 404
Birkeland currents 1,133,156,168,198,237,240,270,286,299,
           311,321,355
           - classification 4,246
Boundary layers (dayside magnetosphere): See also low
          latitude boundary layer, entry layer, exterior
          cusp, 31,44,87,163,188,199,253,312,455

Circulation patters/models: See convection
Cleft: see Polar Cleft
Conductivity, ionospheric 28,198,240,263,279,299,372
Convection - cells 7,137,155,244,255,309,323,343,408
           - electric field 46,115,134,345,355
           - flow 254,295
           - pattern 115,137,242,321,344,407,457
           - reversal 123,142,244,255,345
           - velocity 127,304,344
           - vortices 159,260,364,407
Coupling, magnetosphere-ionosphere 1,151,167,356,372,455
           - solar wind - magnetosphere 1,151,239,356,455

Hydromagnetic waves: See waves, MHD

Impulsive penetration 27
Incoherent scattering 117,127,269,285,343,456
Interplanetary magnetic field (IMF), influence:
      See also solar wind-magnetosphere interaction
      3,11,27,43,61,87,115,137,151,167,203,239,285,304
      311,319,343,355

Ion   - composition: See plasma composition
          - conics 104
          - density 127,288,300,355,370
          - drag forces 5,356,410
          - drift 127,138,240,253,289,304,355,455
          - heating 103
          - precipitation 118,311
          - temperature 103,118,285,368
Ionization blobs 131
Ionospheric currents: See Currents
Irregularities - ionospheric 286,436

Joule heating 28,355,372,409,422

Kelvin-Helmholtz instability 167,191
Keogram 320

Laboratory plasma 27
Low-latitude boundary layer 121,151,168,242,256,311,455

Magnetic - activity: See also Mag. perturbations 253,302,
            355
          - flux ropes 11
          - flux tubes 11
          - impulses 152,253
          - indices 61,117,137,204,244,258,338,343,424,
            435,449
          - perturbations/disturbances: See also waves,
            ULF 151,239,353,432,438,455
          - substorms 173,422
          - pulsations 159,167,187,229,455
          - topology 6,242
Magnetopause 4,11,30 43,87,115,163,187,203,239,256
Magnetosheath 43,87,103,115,167,191,312
Magnetosphere-ionosphere coupling: See coupling
Magnetospheric convection 2,151,258
          - substorm 259
Magnetotail 1,43,183,253
Merging, magnetic field line 6,115,121,155,259,320
Mesosphere 299,421
MHD generator 181